SYMMETRY: A BASIS
FOR SYNTHESIS DESIGN

SYMMETRY: A BASIS FOR SYNTHESIS DESIGN

TSE-LOK HO
National Chiao Tung University
Hsinchu, Taiwan
Republic of China

A Wiley-Interscience Publication

JOHN WILEY & SONS, INC.

New York ● Chichester ● Brisbane ● Toronto ● Singapore

CHEM

Library of Congress Cataloging in Publication Data:

Ho, Tse-Lok.
    Symmetry: a basis for synthesis design / Tse-Lok Ho.
        p.   cm.
    "A Wiley–Interscience Publication."
    ISBN 0-471-57376-0 (acid-free)
    1. Organic compounds—Synthesis.   2. Symmetry (Physics)
    I. Title.
    QD262.H617 1995
    547'.2—dc20
                                                        95–5007

Printed in the United States of America

10 9 8 7 6 5 4 3 2 1

To Honor
Symmetrizer of my life
for twenty-five years

CONTENTS

PREFACE

Symmetry is a vast subject, significant in art and nature.

—Hermann Weyl

Symmetry, usually associated with beauty, is a pervasive feature of many objects, natural and artificial. Sometimes obvious, sometimes more subtle, the symmetry of molecular structures transcends phenomenology because it inherently affects molecular properties. An asymmetrical molecule may embody some hidden symmetrical subunits, but in the supramolecular realm, asymmetrical molecules may align themselves to form symmetrical aggregates. Profound consequences are witnessed in such circumstances.

Synthesis must conform to the fundamental rules of symmetry as specified by target compounds. Exploitation of symmetry rules is no mere simplification of a synthetic adventure; adherence is often an absolute imperative. In recent times the extravagant growth of asymmetric synthesis attests to the many conceptual breakthroughs and to the practical significance of organic synthesis in general, but asymmetry is only one face of the coin. Although there is no argument to say the universe is chiral, and that chiral molecules including those making up life are dominant, the synthetic chemist will reap enormous rewards by paying attention to symmetry. Symmetry principles hold true even in dealing with chiral compounds, as will be repeatedly demonstrated in the present monograph. Numerous examples illustrate the benefits conferred by symmetrical precursors in the elaboration of asymmetric target molecules by desymmetrization processes.

This book includes synthetic works directed toward symmetrical molecules and the usage of symmetrical starting materials and/or intermediates irrespective of the symmetry characteristics of the target compounds. It is impossible to survey all the cases in the chemical literature, particularly when the criteria for symmetrical starting materials can be set so arbitrarily; therefore, syntheses starting from rather simple compounds are selected to serve as illustratives. On this premise, criticism by some readers on the scope of this book is expected, but it remains my cherished hope to inspire readers with the magnificence and relevance of symmetry to chemical synthesis.

Taiwan, ROC

Tse-Lok Ho

ABBREVIATIONS

Ac acetyl

acac acetylacetonate

AIBN 2,2′-azobisisobutyronitrile

Ar aryl

9-BBN 9-borabicyclo[3.3.1]nonyl

9-BBN-H 9-borabicyclo[3.3.1]nonane

BINAP 2,2′-bis(diphenylphosphino)-1,1′-binaphthyl

bipy 2,2′-bipyridyl

Bn benzyl

t-BOC *t*-butoxycarbonyl

BOM benzyloxymethyl

BSA *N,O*-bis(trimethylsilyl)acetamide

BTAF benzyltrimethylammonium fluoride

Bz benzoyl

CAN ceric ammonium nitrate

COD 1,5-cyclooctadiene

COT cyclooctatetraene

Cp cyclopentadienyl

18-crown-6 1,4,7,10,13,16-hexaoxacyclooctadecane

CSA camphorsulfonic acid

CSI chlorosulfonyl isocyanate

DABCO 1,4-diazabicyclo[2.2.2]octane

dba dibenzylideneacetone
DBN 1,5-diazabicyclo[4.3.0]non-5-ene
DBU 1,8-diazabicyclo[5.4.0]undec-7-ene
DCC dicyclohexylcarbodiimide
DDQ 2,3-dichloro-5,6-dicyano-1,4-benzoquinone
DEAC diethylaluminum chloride
DEAD diethyl azodicarboxylate
DET diethyl tartrate (+ or −)
DHP dihydropyran
DIBAL-H diisobutylaluminum hydride
diglyme diethylene glycol dimethyl ether
dimsyl Na sodium methylsulfinylmethide
DIPT diisopropyl tartrate (+ or −)
DMAP 4-dimethylaminopyridine
DME dimethoxyethane
DMF dimethylformamide
DMSO dimethyl sulfoxide
dppb 1,4-bis(diphenylphosphino)butane
dppe 1,2-bis(diphenylphosphino)ethane
dppp 1,3-bis(diphenylphosphino)propane
EE 1-ethoxyethyl
HMPA hexamethylphosphoric triamide
HMPT hexamethylphosphorous triamide
IpcBH$_2$ isopinocampheylborane
Ipc$_2$BH diisopinocampheylborane
KAPA potassium 3-aminopropylamide
K-selectride potassium tri-*s*-butylborohydride
LAH lithium aluminum hydride
LDA lithium diisopropylamide
LICA lithium isopropylcyclohexylamide
LITMP lithium tetramethylpiperidide
L-selectride lithium tri-*s*-butylborohydride
LTA lead tetraacetate
MCPBA *m*-chloroperbenzoic acid
MEM methoxyethoxymethyl
MPM *p*-methoxyphenylmethyl
Ms methanesulfonyl
MsCl methanesulfonyl chloride
MTM methylthiomethyl

MVK   methyl vinyl ketone
NBS   *N*-bromosuccinimide
NCS   *N*-chlorosuccinimide
NIS   *N*-iodosuccinimide
NMO   *N*-methylmorpholine *N*-oxide
NMP   *N*-methyl-2-pyrrolidone
Nu   nucleophile
PPA   polyphosphoric acid
PCC   pyridinium chlorochromate
PDC   pyridinium dichromate
PMBOM   *p*-methoxybenzyloxymethyl
Pht   phthaloyl
PPA   polyphosphic acid
PPTS   pyridinium *p*-toluenesulfonate
Red-Al   sodium bis(methoxyethoxy)aluminum dihydride
SEM   *β*-trimethylsilylethoxymethyl
Sia$_2$BH   disiamylborane
TBAF   tetra-*n*-butylammonium fluoride
TBDMS or TBS   *t*-butyldimethylsilyl
TBDPS   *t*-butyldiphenylsilyl
TBHP   *t*-butyl hydroperoxide
TCE   2,2,2-trichloroethanol
TES   triethylsilyl
Tf   triflyl (trifluoromethanesulfonyl)
TFA   trifluoroacetic acid
TFAA   trifluoroacetic anhydride
THF   tetrahydrofuran
THP   tetrahydropyranyl
TIPS   triisopropylsilyl
TMEDA   tetramethylethylenediamine [1,2-bis(dimethylamino)ethane]
TMS   trimethylsilyl
Tol   tolyl
Tr   trityl (triphenylmethyl)
Trisyl   2,4,6-triisopropylbenzenesulfonyl
Ts   tosyl (*p*-toluenesulfonyl)
TTFA   thallium trifluoroacetate
TTN   thallium(III) nitrate
Vn   vinyl

SYMMETRY: A BASIS
FOR SYNTHESIS DESIGN

1

INTRODUCTION

The world is composed of symmetrical and unsymmetrical objects, therefore symmetry is essential to our comprehension of the universe, ranging from subnuclear processes to celestial phenomena. Perhaps few people are aware that symmetry lies at the foundations of science—reproducibility and predictability—but it is widely recognized that great expansion of our knowledge in physics and chemistry often attends discoveries of symmetry rules.

The downfall of the parity tenet in weak nuclear interactions shook the foundation of modern physics. This notion of broken symmetry in physical laws was so important that its conceptualizers, Chen Ning Yang and Tsung-Dao Lee, were awarded a Nobel prize in 1957. Since then, a theory (Vester–Ulbricht Hypothesis) on the causal effect of parity violation in nuclear events on the observed asymmetry of molecules in the biosphere has been advanced.

The community of organic chemistry had its share of excitements over matters of symmetry only a few years later. First, the existence of a symmetrical nonclassical structure for the norbornyl cation became a lively polemic which lasted for at least two decades. Then R.B. Woodward and Roald Hoffmann announced in 1965 the fundamental rules of orbital symmetry conservation for pericyclic reactions. Previously, the stereochemical courses for a great number of these reactions were shrouded in mystery, so much so the reactions were called "no-mechanism reactions". The significance of the symmetry rules to the development of chemistry was immediately recognized by all as truly deserving a Nobel prize for chemistry, which eventually went to Hoffmann (and to K. Fukui). Only his untimely death prevented Woodward from winning a second time.

Another epochal event that took place more than a century ago was Louis Pasteur's manual separation of racemic acid (Latin *racemus* "grape") crystals and detection of the ability of their solutions to rotate a plane of polarized light to the same degree but in opposite directions. This discovery of two optically active forms of tartaric acid was revolutionary and it foreshadowed the theorization of the tetrahedral carbon atom by J.A. Le Bel and J.H. van't Hoff.

The second great discovery of Pasteur pertains to the consumption of one isomer of racemic acid by a certain plant mold, leaving behind an optically active solution. Here, a living organism accomplished the tedious task of crystal separation, although by selective depletion of one molecular form.

There is an unmistakable interplay between symmetry and asymmetry in the material world. Symmetry is required for stability whereas asymmetry is the condition for storage and transfer of unambiguous information. In this context it is amusing to contemplate the double helix structure of the DNA molecules in which the base pairings provide local symmetry and hence stability, however, during replication or transcription into RNAs the symmetry is temporarily broken by dissociation of the base pairings. Another example is heme, the iron–porphyrin complex responsible for the red color of arterial blood. Its core is symmetrical as far as the immediate surrounding of the metal ion is concerned, although the porphyrin periphery is adorned with three different types of substituents. Of course the local symmetry is completely overwhelmed by the large peptide chain when the subunit of hemoglobin is taken into consideration as the fundamental entity.

Enhancement of stability due to symmetry is witnessed in simple molecules also. For example, isovalent conjugation vs. sacrificial conjugation is reflected in the stability of the conjugated acids and bases of azoles, and the high acidity and basicity of a carboxylic acid and guanidine, respectively. The inevitable, ultimate transformation of various tricyclic $C_{10}H_{16}$ hydrocarbons to adamantane under the influence of Lewis acids also attests to the dominant effect of symmetry.

The reaction between molecules is dependent of molecular symmetry. Thus the activation enthalpies for diastereoisomeric transition states of two chiral molecules are different, due to varying degree of steric accessibility. Such stereodifferentiating effects become very prominent when more complex structures are involved.

The enormous implications of symmetry to the development of organic chemistry are such that it is virtually impossible to discuss any facet of the subject without reference to stereochemistry. In the context of synthesis the molecular symmetry of the target, or the lack thereof, is of prime concern. Although it is likely that symmetrical substances are more readily constructed from intermediates of high symmetry, the synthesis of unsymmetrical molecules may also benefit from the use of symmetrical precursors, and vice versa. Symmetrization and desymmetrization are chemical operations invaluable to the simplification of synthetic sequences, especially when dealing with synthetic targets that possess local or pseudosymmetry. It is amusing that very effective desymmetrization methods are those performed enzymatically, in a manner reminiscent of Pasteur's observation of microbial effect on this racemic acid, i.e., selective action on or production of one enantiomer per chirality center.

It is also fitting that asymmetric synthesis of organic compounds nowadays relies heavily on tartaric acid, by using one of its enantiomers as building block or, in the form of an ester, as catalyst in the Sharpless epoxidation that confers both functionality and chirality to allylic and homoallylic alcohols. Furthermore, the intertwinement of synthesis and symmetry can hardly be more aptly illustrated by the fact that the probe leading to the Woodward–Hoffmann Rules originated from synthetic studies of vitamin B_{12}.

It has been recognized that symmetry in a synthesis plan is a powerful simplifying principle [Corey, 1989a], although certain conditions must be heeded [Bertz, 1984], and it was claimed that symmetry may complicate synthetic operations when approaching less symmetrical targets [Bertz, 1983]. Perhaps a good example would be the formation of chiral 3-substituted 3,5-cyclohexadiene-1,2-diols by microbial oxidation of benzene derivatives [S.M. Brown, 1993] for which benzene itself is not suitable because the product is meso. However, in the following chapters various syntheses and the strategies that pertain to symmetry will be discussed. While it is true that all syntheses can be traced to a symmetrical precursor (ultimately a carbon atom), obviously only the more intriguing and enlightening examples need be addressed.

1.1. SYMMETRY ELEMENTS AND SYMMETRY OPERATIONS

A symmetrical object contains some identical parts. It is possible for certain operations on the object, which do not involve breaking the whole, such as reflection, rotation, and inversion, to return the object to exactly the same arrangement of the various parts. These operations are called symmetry operations, and the geometrical element about which a symmetry operation occurs is a symmetry element.

A symmetry axis is a line through an object about which rotation through an angle of $2\pi/n$ radians results in an identical array. The rotation element is C_n (n-fold axis of symmetry). For example, water has a twofold axis of symmetry, whereas ammonia, methane, boron trifluoride, and carbonate ion have threefold axes of symmetry. Many molecules have more than one axis of symmetry; for example, there is one sixfold axis of symmetry and six twofold axes in benzene. The axis having the largest value of n is the principal axis. Rotation of the axis of symmetry is the symmetry operation. Of course all molecules have an infinite number of C_1 axes, denoted E. If a molecule has only E as the symmetry element it cannot be considered as a symmetrical molecule.

The plane of symmetry (σ) is another symmetry element which is associated with reflection as a symmetry operation. When this plane passes through a molecule possessing such a symmetry element, two indistinguishable mirror images are observed. There are two planes of symmetry for the water molecule, one passing through all the atoms and one orthogonal to the first plane and passing through the oxygen atom. The plane containing the principal axis of symmetry is σ_v, that perpendicular to this axis is σ_h, and the plane bisecting the angle between the two perpendicular n-fold axes is σ_d. Ammonia has three reflection planes which intersect at 120° at the C_3 axis whereas benzene has six such planes intersecting at 30° at the C_6 axis besides the single σ_h.

A center of symmetry (i) is the point in a molecule which is equidistant to a pair of equivalent atoms in a straight line. Remarkably, methane does not possess a center of symmetry, while ethylene, benzene, and cyclohexane do. The corresponding operation is inversion through the center of symmetry.

An alternating axis of symmetry (S_n) which operates by rotation of 360°/n and subsequent reflection through a plane perpendicular to the rotational axis. The results are that S_1 is equivalent to a mirror plane, S_2 to a center of inversion, S_3 to $C_3 + \sigma_h$, S_5 to $C_5 + \sigma_h$. This composite operation for C_n^1 and σ works on methane in the following way. Take the intersecting line of two perpendicular planes, each containing CH_2, as the rotational axis, rotate 90° and then reflect through a plane that passes through the carbon atom and orthogonal to the original planes.

Point groups with one principal axis C_n include C_n, C_{nv}, C_{nh}, D_n, D_{nh}, D_{nd}, and S_n ($C_s = S_1$ and $C_i = S_2$); those with multiple axes of high order represent the five Platonic solids. The cubic groups include the cube and the octahedron (O_h, $6C_4$, $8C_3$) and the tetrahedron (T_d, $8C_3$). The rotational groups are O and T. Adding a center of symmetry to O gives O_h, and $T \times C_i = T_h$. Included in the icosahedral groups are the regular icosahedron and dodecahedron (I_h, $24C_5$, $20C_3$) and the rotational subgroup I.

There are isomorphisms such as $C_{nv} \sim D_n$ and $D_{2n} \sim D_{nd}$ (for all n), $C_n \sim S_n$ (for even n), $D_{nh} \sim D_{nd}$ and $C_{nh} \sim C_{2n}$ (for odd n), and special cases $C_2 \sim C_s \sim C_i$; $D_2 \sim C_{2h}$; $O \sim T_d$.

Figure 1 can be used to help assign molecules to point groups. The groups

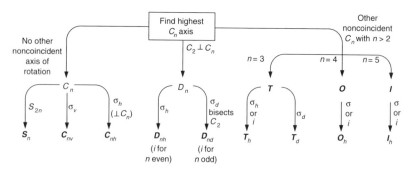

Figure 1 Assignment of molecules to point groups. Special nomenclature: $S_1 \equiv C_{1v} \equiv C_{1h} \equiv C_s$, $S_n \equiv C_{nh}$ for odd n, $S_2 \equiv C_i$. (From B. Douglas, D.H. McDaniel, and J.J. Alexander, *Concepts and Models of Inorganic Chemistry*, 2nd ed., Wiley, New York, 1983, p. 101. © 1983 John Wiley & Sons, Inc.; reprinted by permission.)

C_n, C_{nv}, C_{nh}, and S have only one axis of rotation, and among them the C_n group possesses the C_n axis as the only symmetry element.

1.2. MAINTENANCE OF SYMMETRY AND SYMMETRIZATION

Benefits for maintaining symmetry in synthetic intermediates are many when target compounds or subtargets are symmetrical or exhibit only slight deviations. Efficiency increases should be emphasized. A simple example is the preparation of a lactone precursor for the spiroacetal of the milbemycins [Van Bac, 1988]. The C_2-symmetric bicyclo[2.2.1]heptane-2,5-dione was oxidized to a dilactone which was methanolyzed. Relactonization then served to desymmetrize the molecule.

Papuamine is C_2-symmetric, therefore the most expedient synthetic approaches must be those adhering to this structural characteristic. Such a process, for producing an enantiomer of the natural product [Barrett, 1994], required only temporary desymmetrization before the intramolecular coupling.

ent- papuamine

Another synthesis was also based on the symmetry elements of papuamine [Borzilleri, 1994]. A key feature of this approach is an intramolecular imino ene cyclization which delivered the indan derivative possessing all the stereocenters.

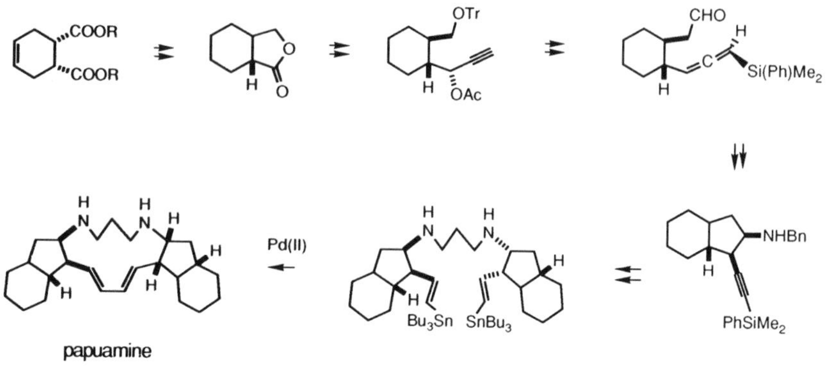

papuamine

In connection with synthesis of teurilene [Hoye, 1987] a unique transformation of a (+/−)-diepoxide into a *meso*-tetraol by means of Payne rearrangement was observed. The *meso*-selectivity was due to the exclusive pathway of intramolecular attack by a primary alcohol on the neighboring epoxide to initiate the reaction.

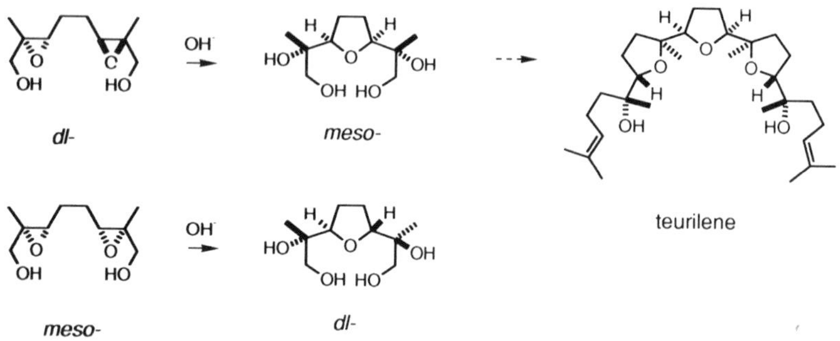

teurilene

The triepoxides obtained by asymmetric epoxidation of (*E,Z,E*)-dodeca-2,6,10-triene-1,12-diol underwent cascade reactions to give 2,5-linked bistetrahydrofurans [Hoye, 1985]. In methanol the processes were terminated by opening of the terminal epoxides with methoxide ion, thus preventing the optical purity loss of the precursor (as compared to "racemization" in aqueous medium in which the epoxides were attacked by hydroxide ion). In a complementary pathway, the tetrahydrofuran formation started from an internal oxide ion.

Concurrent and interdependent equilibration and cyclization of 4,6-dimethyl-5-oxononanedioic acid provided a spirodilactone [Hoye, 1981] which underwent bismethylation. Only one isomer was obtained (at least as major product), apparently all the intermediates in this series of compounds showed a preference for C_2-symmetry. The spirolactone can be hydrolyzed on exposure to trifluoroacetic acid without epimerization, and the tetramethylated keto diacid has a configuration similar to that of the C-15 to C-23 segment of venturicidin macrolide.

The reductive conversion of papaverine to give the symmetrical pavine [Battersby, 1955] involved a biomimetic cyclization.

Symmetrization of a chiral intermediate during a synthesis of patchouli alcohol [Büchi, 1964] must be mentioned. The undesirable result was due to Wagner–Meerwein rearrangement. Naturally the discovery of this phenomenon enforced a change of reaction sequence.

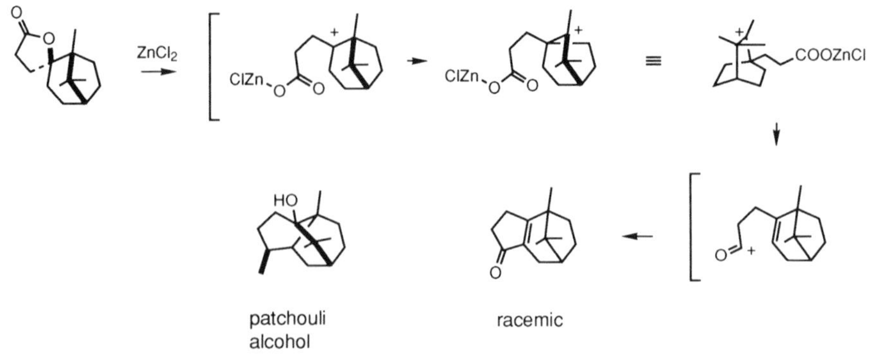

patchouli
alcohol

racemic

Symmetrization tactics may simplify the synthesis of certain compounds via the cationic polyene cyclization protocol. Thus setting up the cationic initiator as a symmetrical allyl cation was most logical in the routes to fichtelite [W.S. Johnson, 1966], taxodione [W.S. Johnson, 1982], and a precursor of progesterone [W.S. Johnson, 1971].

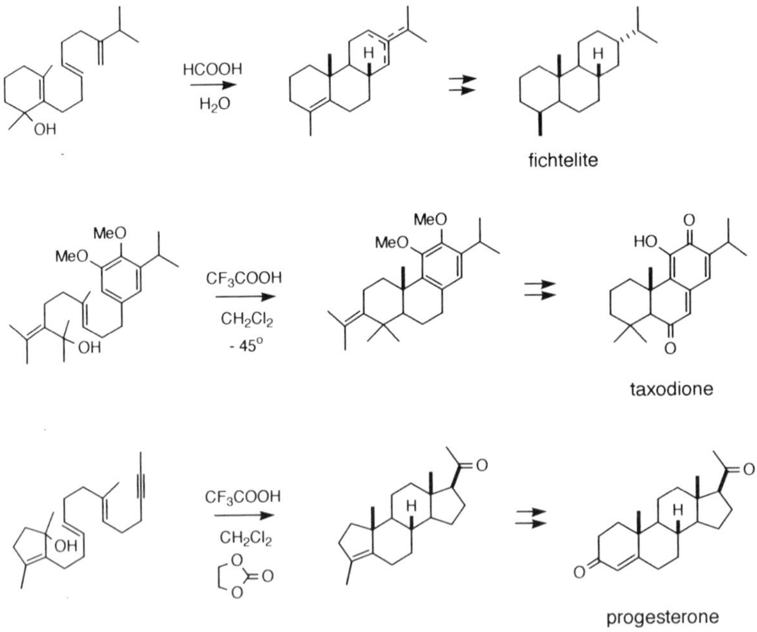

fichtelite

taxodione

progesterone

1.3. DESYMMETRIZATION

Asymmetric synthesis has almost become an imperative for any significant construction of target compounds which are devoid of symmetry. One way of inducing asymmetry is by incorporation of chiral auxiliaries in the reactants.

It is important to note that enantiofacial differentiation during the reaction does not always require an auxiliary lacking mirror or inversion symmetry, instead dissymmetric (e.g., C_2-symmetric) auxiliaries can play the essential role [Whitesell, 1989; Noyori, 1994]. However, this enormous subject will not be discussed.

Off-mirror plane reaction of a σ-symmetric compound always leads to one or more unsymmetrical products. This consequence can often be taken into retrosynthetic consideration such that a shorter route may be disclosed. It is thus more convenient to prepare a symmetrical intermediate and perform desymmetrization to further the synthetic progression. To illustrate this point reference can be made to a synthesis of eburnamonine [M.F. Bartlett, 1960]. Thus the cleavage of a 4,4-disubstituted cyclohexanone provided a diacid differing in chain length from the branch-point, each was to combine with a nitrogen atom to complete the pentacyclic system.

eburnamonine

An analogous case is the synthesis of α-trans-bergamotene [Corey, 1971]. A symmetrical bridged ketone underwent α-alkoxymethylenation on being converted to the trisubstituted olefin. These last few steps were quite straightforward due to self-termination of the desymmetrizing process.

α-trans-bergamotene

The desymmetrization principle was also exploited in the access of a tricyclic intermediate of gibberellic acid [Stork, Unpublished] from cis-8a-propargyl-2,7-decalindione. Reductive cyclization involving the triple bond and one of the two equivalent ketone groups was followed by contraction of the ring with the residual carbonyl group.

Elimination or extrusion reactions can convert a conveniently accessible symmetrical intermediate into a synthetic target or precursor. For example, 2-substituted 2-chloro-1,3-cyclohexanediones undergo elimination of hydrogen chloride and carbon monoxide on treatment with sodium carbonate in a high boiling aromatic solvent. This method formed the basis of a synthesis of cis-jasmone [Büchi, 1971a] and prostaglandin $F_{2\alpha}$ [Kienzle, 1973].

cis-jasmone

A related reaction is the rearrangement of certain 2,3-epoxy-1,4-cyclohexanediones [Herz, 1975] which has found many synthetic applications, for example, trichodermol [Still, 1980c].

The use of an unsymmetrical reagent in the functionalization of a symmetrical substance destroys the molecular symmetry. Thus it is remarkable that protonation of 4-hydroxyheptanedioic acid with 10-camphorsulfonic acid led eventually to a chiral lactone in 94% enantiomeric excess (ee) [Fuji, 1985].

With the aid of chiral amide bases enantioselective deprotonation of prochiral ketones can be achieved [Cox, 1991; Izawa, 1989]. For example, the allylation of cis-2,6-dimethyl-cyclohexanone [Simpkins, 1986] led to different enantiomers by varying the base of lithium N-benzyl-α-phenylethylamide in the (R)- or (S)-form. The synthesis of (+)-brasilenol [Greene, 1987] from 4-isopropylcyclohexanone via (R)(−)-cryptone, the elaboration of (+)-monomorine-I [Momose, 1990] from an N-protected tropinone, that of (+)-α-cuparenone [T. Honda, 1993] from 4-methyl-4-p-tolylcyclohexanone, and the transformation of a monoacetal of bicyclo[3.3.0]octane-3,7-dione into a synthon of the Corynanthe alkaloids [Leonard, 1990] involved the corresponding chiral enolates.

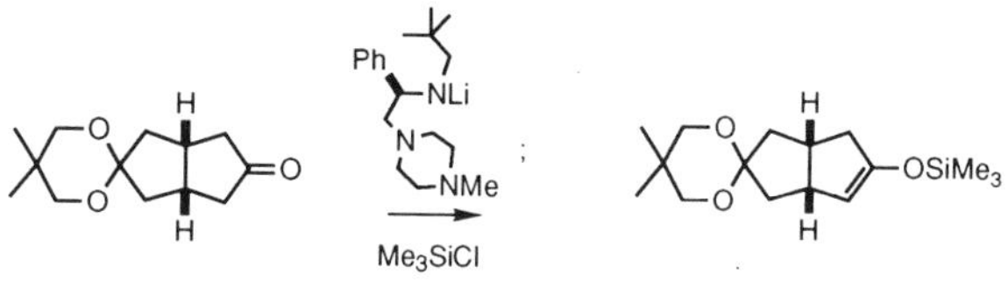

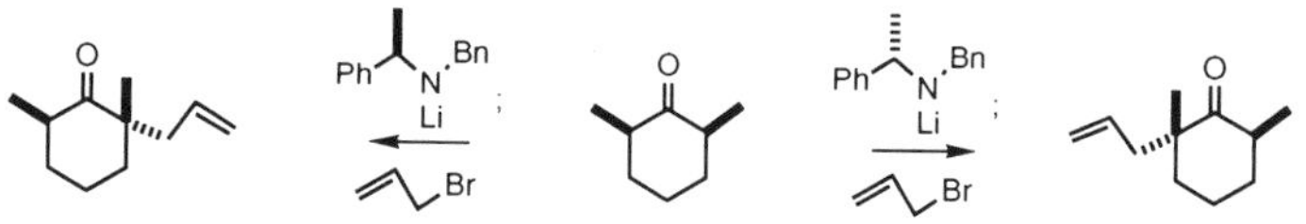

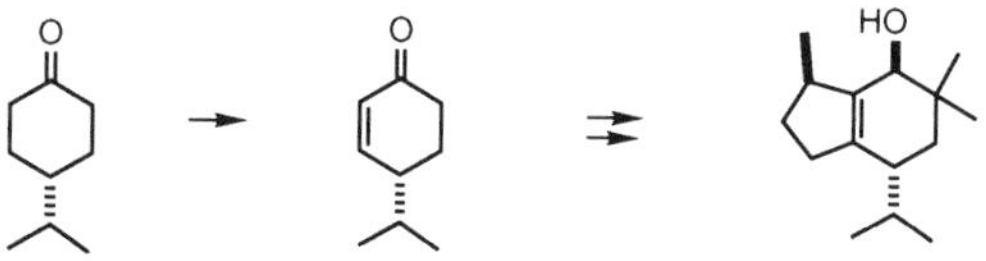

(+)-brasilenol

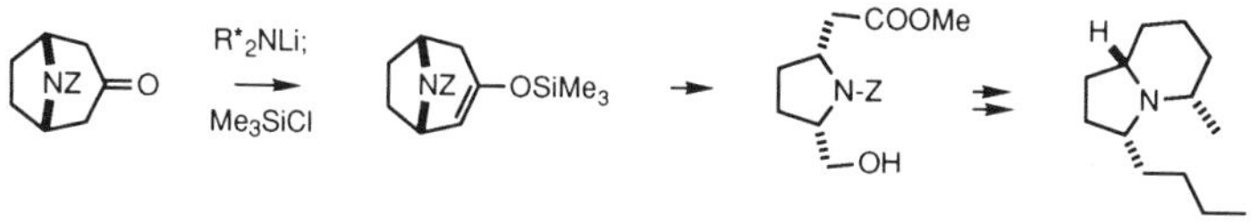

(+)-monomorine-I

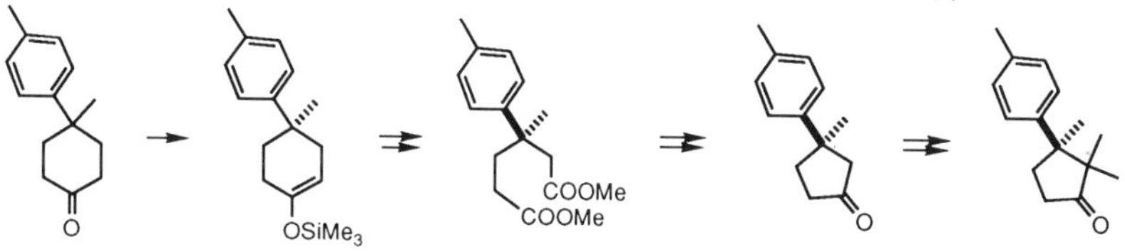

(+)-α-cuparenone

Acetalization of σ-symmetric ketones with diols having a C_2 axis of symmetry and using such acetals in asymmetric organic synthesis [Alexakis, 1990] constitutes an area of fruitful research. Interestingly, enantiodifferentiation of meso-1,3,5-pentanetriols has been accomplished [T. Harada, 1991] in acetal formation with (−)-menthone. Among many interesting transformations is the diastereoselective cleavage of acetals on treatment with triisobutylaluminum to give chiral enol ethers [Naruse, 1988]. Selective monoacetalization of β-diketones with chiral diols has been accomplished [Duthaler, 1982, 1984].

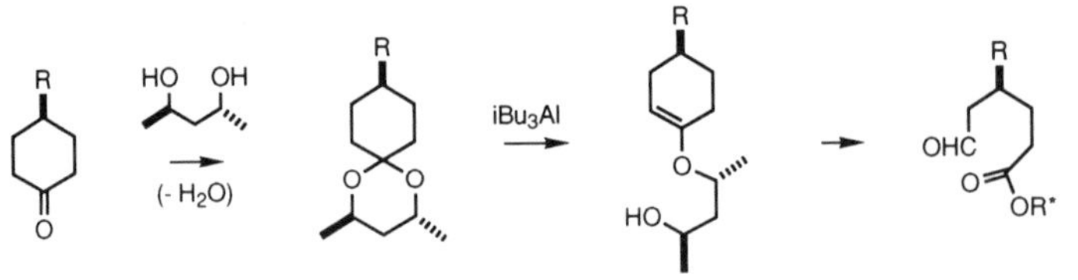

Axially dissymmetric compounds such as the cyclohexylideneacetic acid shown below have been acquired from the racemic mixture through a hydrohalogenation–dehydrohalogenation sequence; the dehydrohalogenation was achieved with a chiral amide [Duhamel, 1987].

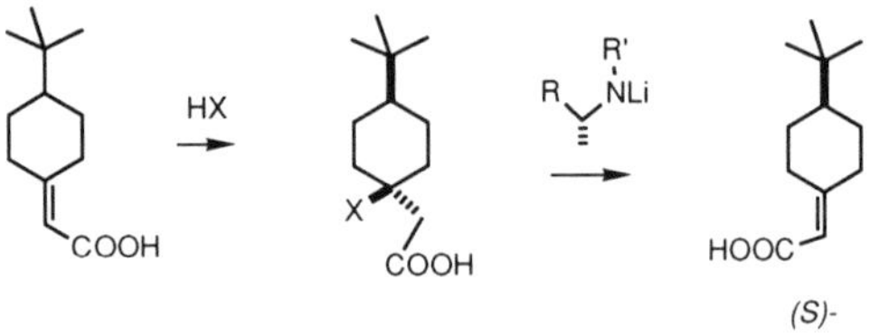

An excellent enantioselective β-elimination of a bicyclic triflate has also been observed [Kashihara, 1987]. Opening of a *meso*-epoxide with a chiral amide base which was accompanied by high enantiomeric excess [Hodgson, 1994] has been ascribed to an effect of a complexed alkoxide ion.

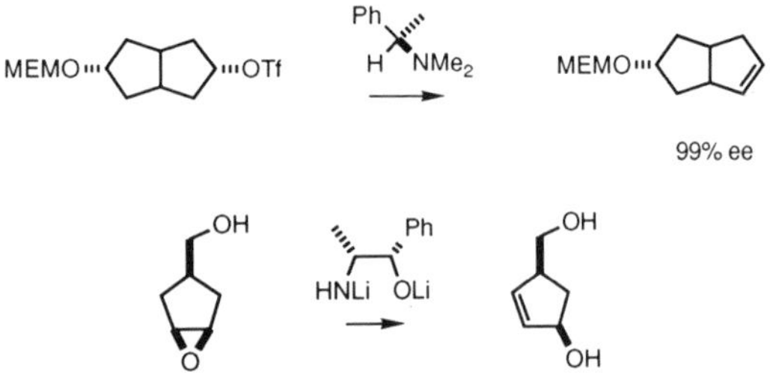

Addition to a symmetrical multiple bond at which center the symmetry element lies may result in a symmetrical or unsymmetrical product. The nature of the addend is the determining factor. Accordingly, hydration of 1,4-diacetoxy-2-butyne gives 1-acetoxy-3-buten-2-one, and 1,4-bisallyloxy-2-butyne to the unsymmetrical ketone; they were used in a simple synthesis of the flavorant 3-methyl-2-cyclopenten-2-ol-1-one [T.L. Ho, 1983] and an approach to valeranone [Ponaras, 1990], respectively.

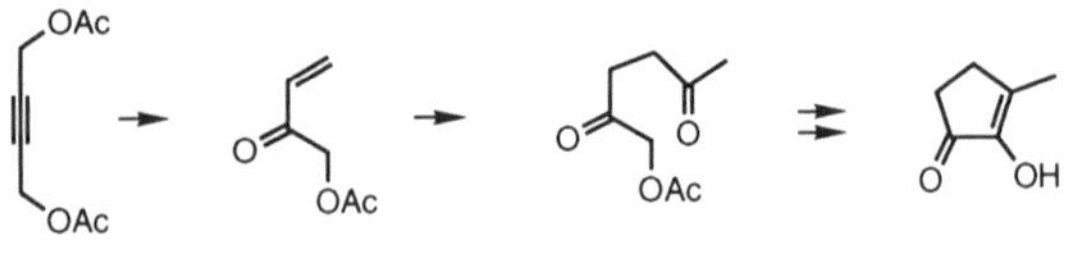

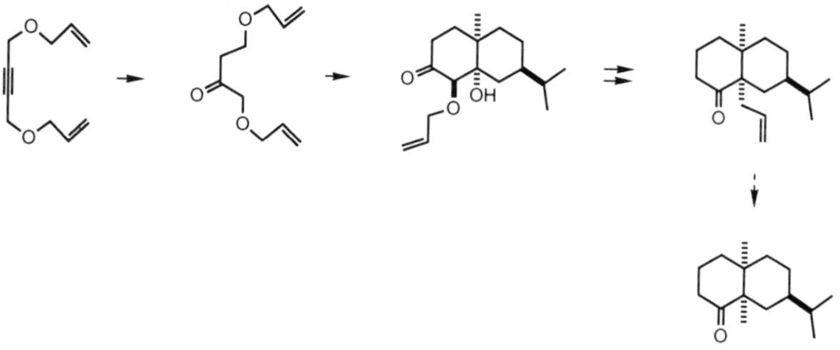

valeranone

Actually the desymmetrization by an addition reaction is not restricted to a typical multiple bond to gain advantage in synthesis. A formal synthesis of *O*-methyljoubertiamine [Schulte, 1988] was completed on the basis of reductive allylation of a bis(tricarbonylchromium) complex of 4,4′-dimethoxybiphenyl with which the two halves of the molecule were differentiated.

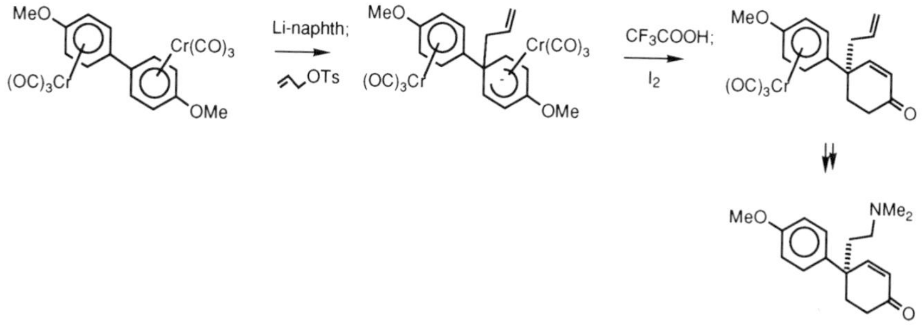

O- methyljoubertiamine

The ozonolytic cleavage of cycloalkenes can give rise to terminally differentiated products [Schreiber, 1982]. It is interesting to note that the initial ozonides maintain the σ-symmetry, and only the subsequent transformations effect desymmetrization.

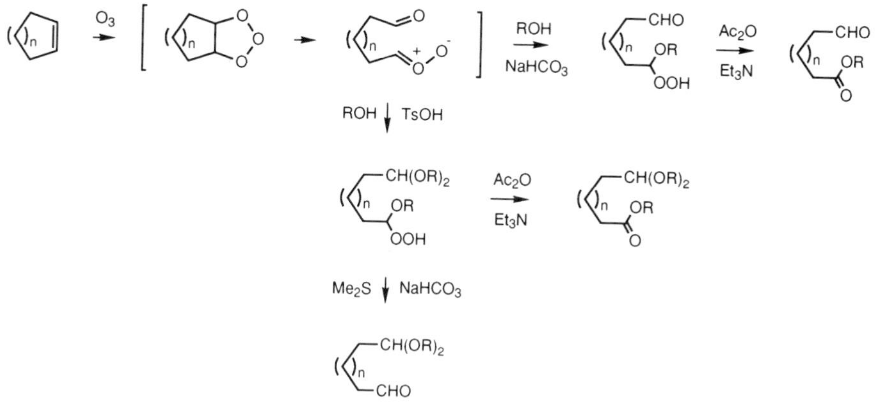

A synthesis of mevalonolactone [Fetizon, 1975] via ozonolysis of 4-methyl-1,6-heptadien-4-ol and a redox sequence involved desymmetrization at the final step. The heterogenous silver carbonate on Celite reagent has a low probability of effecting oxidation on both primary alcohol groups of the same molecule so that the hydroxy aldehyde cyclizes immediately. Thus irrespective of the rates of the second-stage oxidation the predominant product was the lactone.

Differentiation of symmetrical bifunctional compounds is often achieved by protection of one of the functional groups. A better option, although not always feasible, is to conduct a controlled reaction at one site by limiting the quantity of the reagent and/or fine-tuning the reaction conditions. Thus it has been shown that certain diketones can be converted by the Baeyer–Villiger reaction into the ketolactones [Garlaschelli, 1992; Feng, 1992] which were used in the synthesis of loganin and the CD-ring component of hemibrevitoxin-B.

loganin

hemibrevetoxin-B
CD-synthon

In the case that the two functional groups of a symmetrical compound are separated by a convenient distance to permit formation of cyclic derivatives, differentiation may be possible by methods involving attack of a reagent on only one of the functional units while the ring structure suffers cleavage. Thus the derivatization of one hydroxyl group of a diol via dioxastannacycloalkane [Shanzer, 1980; Reginato, 1990] and that of diamines from the silacyclic intermediates [Schwartz, 1982] have helped solve many synthetic problems.

Reductive or alkylative cleavage of cyclic acetals derived from C_2-symmetric diols [A. Mori, 1987] constitutes a useful method for synthesis of chiral alcohols from carbonyl compounds.

The Grignard reaction of symmetrical cyclic anhydrides is interesting in the context of desymmetrization. The ketone group of the initial products reacts further while the cogenerated carboxylate ion is immune to attack by the reagent. Thus the indiscriminate first reaction creates a situation that is automatically conducive to desymmetrization (in the sense of reactivity).

The synthesis of chinensin/retrochinensin and taiwanin-C/justicidin-E—isomeric lactone pairs—from their anhydride precursors [Ishii, 1986] involved selective manipulation at one of the two carbonyl sites.

The Horner–Wadsworth–Emmons reaction at one end of a *meso*-dialdehyde has been effected [Kann, 1993]. This example is one of many methods critically required in differentiation of chain termini in connection with the powerful concept concerning chain elongation in both directions [Poss, 1994]. Enantiotopic differentiation of a bridged α-diketone with a chiral phosphonate ester has also been reported [K. Tanaka, 1993].

The problem concerning terminal differentiation for synthesis of a 2,5-linked bistetrahydrofuran derivative by alcoholysis of terminal esters of a symmetrical diepoxide was solved by using the bispivalate ester as substrate [Hoye, 1992]. In other words, desymmetrization was rendered by selecting a protecting group whose cleavage was slower than cyclization to the furan ring, so formation of the unsymmetrical product predominated.

R = H
R = Piv

| | 60 | 40 |
| | 98 | 2 |

Still of synthetic value for differentiation of diesters is partial saponification. Thus when an investigation of the electronic properties of *anti*-1,6:7,12-bismethano[14]annulene was launched, its synthesis was initiated by conversion of dimethyl cycloheptatriene-1,6-dicarboxylate into the monoester for decarboxylative bromination and coupling [Vogel, 1986].

However, the most popular and widely applicable method for desymmetrization is that involving esters and alcohols with σ-symmetry. Enzyme-catalyzed hydrolysis or transacylation usually gives chiral monoesters. Asymmetric hydrolysis of such diesters (most frequently the methyl esters) can often be achieved with the aid of pig liver esterase (PLE) [Ohno, 1989], consequently many chiral building blocks have become available. It should be noted that by converting the free carboxyl group of the chiral monoester thus obtained into the *t*-butyl ester and saponifying the methyl ester under mild conditions, the antipodal *t*-butyl ester is produced. Various saturated and unsaturated aliphatic and alicyclic (including bridged systems) and heterocyclic diesters have been desymmetrized.

The high optical purity of monoesters may be explained as follows. Suppose the prochiral substrate M undergoes hydrolysis with a faster rate k_{R1} leading to monoester R, and a slower rate k_{S1} leading to monoester S, but the succeeding hydrolysis of these monoesters proceeds with a slower k_{S2} for R, and a faster rate k_{R2} for S. The net result is that the slowly formed minor monoester S is rapidly hydrolyzed to give the diacid while the monoester R which is produced (faster therefore) in larger amount accumulates as the reactions continue [Y.-F. Wang, 1984b]. The relative rates are expected because the enantiotopic specificity of an enzyme must be consistent. In other words, the consecutive kinetic resolution phenomenon makes this kind of hydrolysis practical even using enzymes of low to moderate specificity.

Lipases effect hydrolysis of diesters of lower carboxylic acids by accommodation of one specific ester group in the cleavage site. An alternative method is by selective transacetylation of the symmetrical diols with vinyl acetate to supply the acetyl group, as shown in a route to the spined citrus bug pheromone [K. Mori, 1994a] which has the (3R,4S)-configuration. Fairly complex substrates, such as a decalindiol [Toyooka, 1993], a useful precursor for drimane-type sesquiterpenes, have been subjected to this desymmetrization.

Sometimes it is possible to find two different lipases which exhibit opposite selectivity. Thus *meso*-hydrobenzoin gave enantiomeric monoacetates using lipases from *Candida cylindracea* or *Rhizopus javanicus* [Nicolosi, 1994].

Perhaps a synthesis of (+)-meroquinene [Danieli, 1991], the well-known precursor of quinine alkaloids, deserves special attention. Two enzymatic processes were involved this route, the first one to hydrolyze the diacetate with the *pro(S)* porcine pancreatic lipase to allow the preparation of a chiral octahydroisoquinoline derivative which was cleaved to the *cis*-piperidine-3,4-diacetic acid. The dimethyl ester of the latter compound then underwent PLE-catalyzed hydrolysis to differentiate the two functional groups. It is interesting that the presence of a nitrogen atom in the cyclic moiety had no effect on the enantiotopic selectivity of the enzymatic reaction. It must be emphasized that stereoisomerism is the major determining factor in these reactions, so product structure is generally well predicted. This is an important aspect, quite unlike courses of ordinary reactions in which electronic perturbations may be of overriding importance.

(+)-meroquinene

Oxidoreductases discriminate the functionality of dioxygenated *meso*-compounds such that chiral products are produced. The most intensively investigated enzyme in connection with organic synthesis is horse liver alcohol dehydrogenase (HLADH) and the very inexpensive fermenting yeast which effects ketone reduction to afford chiral alcohols, 1,3-diketones to chiral aldols [K. Mori, 1993a].

The access of chiral substances by a combination of chemoenzymatic methods often renders a synthetic transformation effortless. For example, (2R,5S)-2,5-diacetoxy-(3E)-hexene undergoes hydrolysis with acetylcholinesterase to afford the (2S)-alcohol, and the remaining acetoxy group is readily replaced by a nucleophile with retention of configuration; on the other hand, the secondary alcohol acquires a higher electrofugal activity on conversion into a carbonate during an analogous palladium(0)-catalyzed substitution reaction [Schink, 1992].

An intramolecular reaction at one of two identical sites of a symmetrical molecule is always desymmetrizing. Many syntheses have taken advantage of this principle, often facilitating functionalization and obviating protection with external reagents. Take for example a route to boschnialactone [Callant, 1983] from the cyclopentadiene–maleic anhydride adduct, the heterocycle exchange step enabled deoxygenation of the remaining primary alcohol.

boschnialactone

Biomimetic cyclization of poly-(β-diketones) leading to aromatic natural products has been quite successful [Harris, 1977]. While the cyclization step converts a symmetrical substrate into an unsymmetrical product, it is interesting to note that chemical interventions such as functional group protection may further diverge an intermediate into different skeletons, as illustrated in the elaboration of barakol and 6-hydroxymusizin [Harris, 1975].

6-hydroxymusizin

barakol

Biogenetic consideration of serratenediol has led to a proposal of its synthesis [Tsuda, 1964] from α-onocerindione. Thus desymmetrization also occurred during cyclization.

serratenediol

The Baeyer–Villiger oxidation of cyclic ketones leads to unsymmetrical lactones. It is also known that when the α-positions of such ketones carry different numbers of substituents the products are predictable. A route to palustrine [Momose, 1979] was designed to go through a symmetrical 9-azabicyclo[3.3.1]nonane-3,7-dione, which on C-ethylation and regioselective Baeyer–Villiger reaction gave the proper lactone intermediate. The advantage of this scheme is that the monoethylation afforded only one product and the site of the Baeyer–Villiger reaction was found to be exclusively at the most highly substituted α-position of the molecule (one out of four). Completion of the synthesis requires a configuration inversion of the carbinol center after lactone opening, introduction of the double bond, and formation of the macrocyclic heterocycle.

Meso-cycloalkanones are subject to asymmetric Baeyer–Villiger oxidation by enzymatic methods [Taschner, 1988].

A synthesis of (+)-yohimbine [Aubé, 1994] via (−)-yohimbone starting from a C_2-symmetric hydrindenone is meritorious. The scheme called for insertion of the tryptamine moiety into either side of the ketone, which led to the same chiral lactam. But as the process developed, it was noted that the regioselective functionalization of the double bond could take advantage of conformational effects in the oxaziridine formation and the photoinduced ring expansion to give mainly the C-secoyohimbane derivative in which C-17 and C-18 were β-acetoxylated. Saponification of acetoxy groups followed by pivalylation enabled differentiation of the two functionalities because the equatorial hydroxyl group has a preference for pivalylation. Separation of the 17-pivaloxy-18-ol from the diastereomer paved the way to reach Δ^{18}-yohimbenone via dehydration, Bischler–Napieralski reaction, and redox maneuver. Carbomethoxylation and two-stage reduction completed the yohimbine synthesis.

(+)-yohimbine

Sometimes subtle effects such as ground-state conformation may lead to preferred formation of one product over the other, as shown in the lactonization of a monohydroxy dicarboxylic acid [Hoye, 1984]. Kinetic iodolactonization studies of three heptadienoic acids [Kurth, 1987] have shown substantial group and face selectivities. Conformational bias provided a powerful control element in the desymmetrizing reaction.

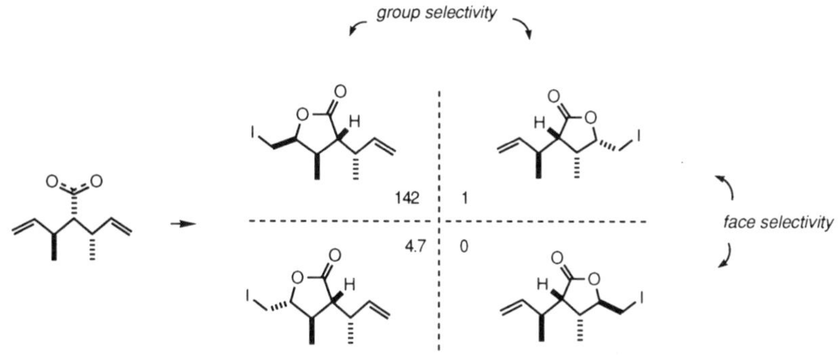

group selectivity

	142	1
	4.7	0

face selectivity

If a chiral auxiliary is present in the vicinity of the reactive moiety, the product will be chiral. Iodolactonization reactions [Fuji, 1990b; Yokomatsu, 1992], hydroformylation [X.-M. Wu, 1993], and an organoyttrium-catalyzed cyclization of dienes [Molander, 1992] are shown below. In the adopted transition states of the iodolactonization steric interactions are minimized. Related to the iodolactonization is the treatment of 1,10-undecadien-6-one with N-iodosuccinimide [Gourcy, 1991] which forms a 2,8-disubstituted 1,7-dioxaspiro[5.5]undecane. The two side chains can be differentiated, after replacement of the iodine atoms with butyroxy groups, by selective hydrolysis with PPL.

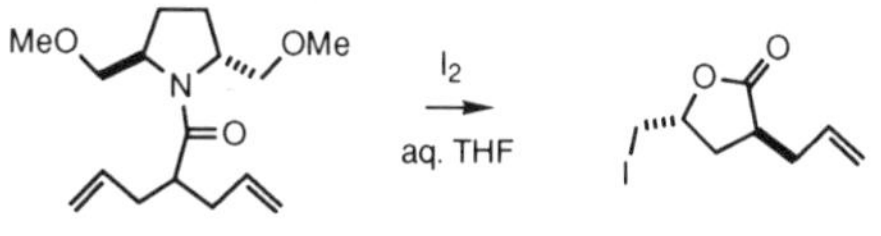

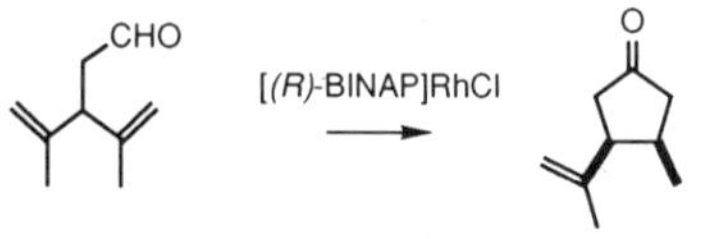

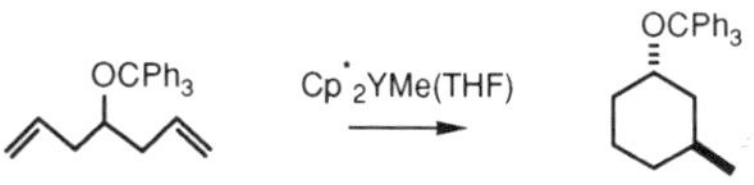

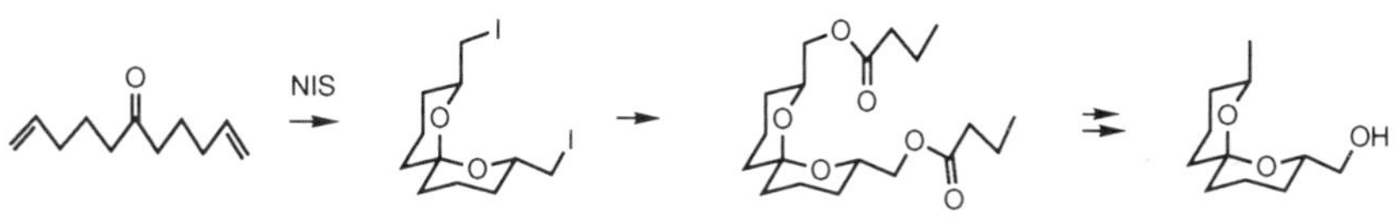

Enantiotopic group discrimination such as represented by a lactone formation [T. Suzuki, 1991] which employed catalytic methods is a promising area for research.

57% ee

Chain homologation concomitant with spiroacetal formation from the same dienone has also been effected [Yadav, 1990]. But here the product is still symmetrical.

Asymmetric desymmetrization in carbonyl-ene reaction has been demonstrated [Mikami, 1992] using a chiral titanium complex as catalyst. There were high levels of remote asymmetric induction.

> 99% *syn*
(> 99% ee)

(R)-

The biomimetic assemblage of the skeleton characteristic of the aromatic *Erythrina* alkaloids by oxidative phenolic coupling of a symmetrical bisarylethylamine [Gervay, 1966; Mondon, 1966; Barton, 1967] initially gave rise to a dibenzoazonine. But on further oxidation to a biquinonoid structure an intramolecular Michael addition involving the secondary amino group desymmetrized the molecule. The dibenzoazonine has been prepared in better yields using an Ullmann-type reaction to demonstrate the subsequent transformation [Hewgill, 1985].

erysodienone

While six-membered rings are very readily accessible by the Diels–Alder reaction, the corresponding chemistry leading to cyclopentenes is much less highly developed. An interesting method [Corey, 1972b] which constitutes a formal analogy to the Diels–Alder reaction is that using Δ^2-1,3-dithianium ion as the dienophile, effecting rearrangement of the adducts with *n*-butyllithium and by thermolysis. As far as molecular structures are concerned, at least in the transformation involving butadiene, there is a transition from a symmetrical state to an unsymmetrical state and finally back to a symmetrical state.

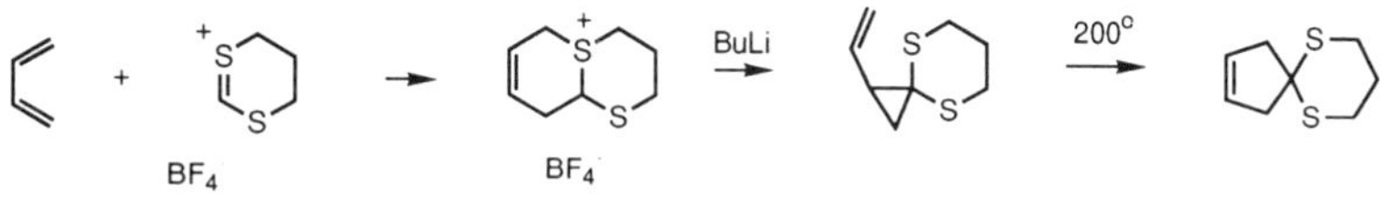

The change between states of molecular symmetry is a interesting phenomenon. An example is the conversion of cyclododecanone into muscone [Gray, 1977] via an unsymmetrical bicyclic enone. Through a sequence of reactions the latter compound was transformed into the symmetrical 3-methyl-1,5-cyclopentadecanedione, and eventually to the target molecule.

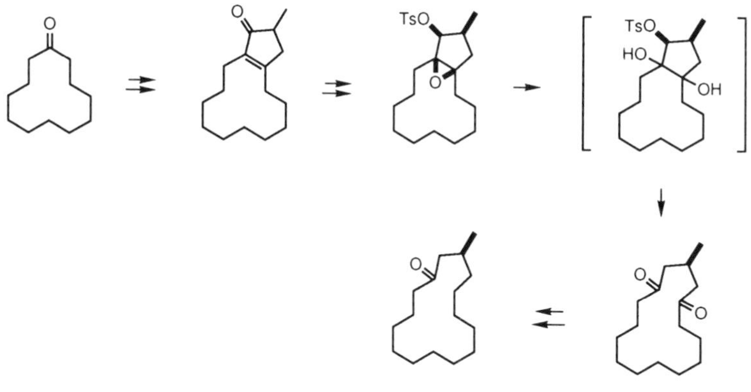

muscone

The elegant method for desymmetrization of glycerol [Boons, 1993] is shown below. Formation of the tricyclic adduct is controlled by anomeric effects—the preference of the methyl groups and the hydroxymethyl substituent to occupy equatorial positions.

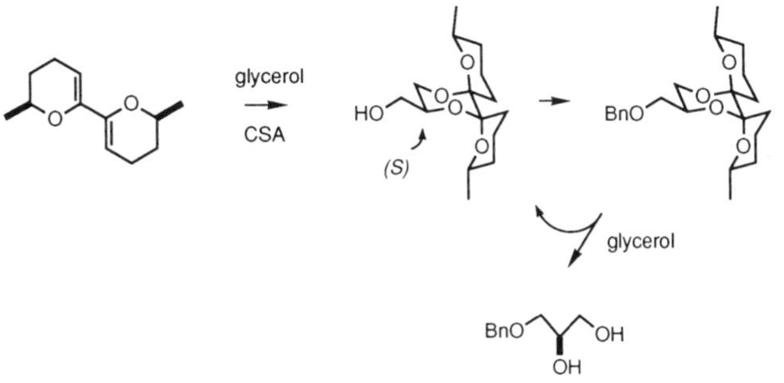

A special desymmetrization process involves halving a C_2-symmetric molecule in which each half is chiral. D-Mannitol has been extensively exploited to acquire chiral products via the halving process [Takano, 1982e; T. Suzuki, 1986].

D-mannitol

Deannulation of σ-symmetric compounds by cleavage of two bonds which do not intersect the mirror plane is a desymmetrizing process. The thermolysis of 1,1-dimethyl-1-silacyclobutane to give ethene and 1,1-dimethylsilaethene [Gutowsky, 1991] is representative.

Certain symmetrical molecules made up of unsymmetrical subunits may be deoligomerized. For example, the cyclic trimer of 1-pyrroline has been converted into N-trimethylsilylmethylpyrrolinium triflate and used in a synthesis of trachelanthamidine [Terao, 1982].

trachelanthamidine

1.4. ENANTIODIVERGENCE

Enantioselective synthesis can suffer from inflexibility when only one form of the chiral building block, catalyst, or auxiliary is available. Sometimes this problem can be solved by functional group manipulations in a different order or via an entirely new reaction sequence to afford a target or subtarget molecule having all centers of chirality in the opposite sense. In other words, by paying due attention to the symmetry properties of a molecule, it may be possible to design synthesis of both enantiomers from a common chiral intermediate. Usually, inversion of a single center of chirality is involved, but if the target molecule has a multitude of such centers and asymmetric induction of the others by one center is feasible, enantiodivergent synthesis may be realized.

Whenever applicable, functional group differentiation in the absolute stereochemical sense is the most effective technique. Although currently limited to a few classes of substrates, the combination of enzymatic desymmetrization of *meso*-compounds and subsequent chemical operations is a powerful methodology. Thus the synthesis of mevalonolactone [F.-C. Huang, 1975] and β-cuparenone [Canet, 1992] may be cited as representatives. Although a route to only (−)-carvone has been demonstrated [Takano, 1993a] the access of the enantiomeric pair of the tricyclic enone intermediate left no doubt about the enantiodivergent feasibility. See also the approach to (−)-eutypoxide-B and its enantiomer [Takano, 1993b].

A chiral *O*-benzylglycidol has been successfully converted into (*R*)- and (*S*)-sulcatol [Takano, 1982c, 1983a] through different processing of a diol intermediate. A C_{10} aldehyde in either optical series is available from a chiral arylglycidol, the latter via asymmetric epoxidation of the arylpropenol and reaction of the epoxide with a methylcuprate reagent or trimethylaluminum to establish the benzylic center. Further reactions on the two aldehydes led to (+)- and (−)-α-curcumene, respectively [Takano, 1988h].

It must be emphasized that for the most authentic enantiodivergent synthesis the same kinds of chemical operations are involved. The pathways towards the two enantiomeric targets consist of the same steps in different orders. A clear demonstration of this principle is found in the synthesis of pinitol [Hudlicky, 1990] which employed a halocyclohexadienediol at the point of divergence. The halogen atom was a reference site for asymmetric dihydroxylation of the aromatic compound by the microorganism, and it also affected the relative reactivity and hence regiochemistry of reactions on the two double bonds.

Stereocontrol in the reaction at a prochiral center due to chelation effect is a rather common phenomenon. The access to the precursors of (*R*)- and (*S*)-4-hydroxy-2-cyclohexenones [Carreño, 1990] is readily accomplished by reduction of the β-keto sulfoxide with diisobutylaluminum hydride with or without addition of zinc chloride.

Another method for optical resolution of a compound is via selective reaction with a chiral reagent that involves separation of the unreacted enantiomer and antipode regeneration by decomposition of the reaction product. Particularly effective processes are those which only require a chiral catalyst. For example, the access to (S)(+)-5-trimethylsilyl-2-cyclohexenone by a Michael addition using half equivalent of 4-toluenethiol as addend in the presence of cinchonidine. Under these conditions the (R)(−)-isomer forms the adduct preferentially, the enone is recovered on treatment of the adduct with 1,8-diazabicyclo-[5.4.0]undec-7-ene [Asaoka, 1990b]. These chiral building blocks have proved their value in serving as starting material for (−)-enterolactone [Asaoka, 1988a], (+)-curcumene [Asaoka, 1988c], (+)-α-cuparenone [Asaoka, 1988d], (−)-β-vetivone [Asaoka, 1988f], (+)-ramulosin [Asaoka, 1989a], (+)-magydardienediol [Asaoka, 1991], (+)-4-butyl-2,6-cycloheptadienone [Asaoka, 1988e] from (R)(−)-trimethylsilyl-2-cyclohexenone; and (+)-quebrachamine [Asaoka, 1989b], (+)-ptilocaulin [Asaoka, 1990a], (−)-O-methyljoubertiamine [Asaoka, 1988b] from the (S)(+)-enone. Stereocontrol in all these syntheses was effectively exerted by the bulky trimethylsilyl group during reaction on the enone side of the molecule. Furthermore, the silyl substituent also has subtle electronic effects to direct regiochemistry such as in Baeyer–Villiger reaction.

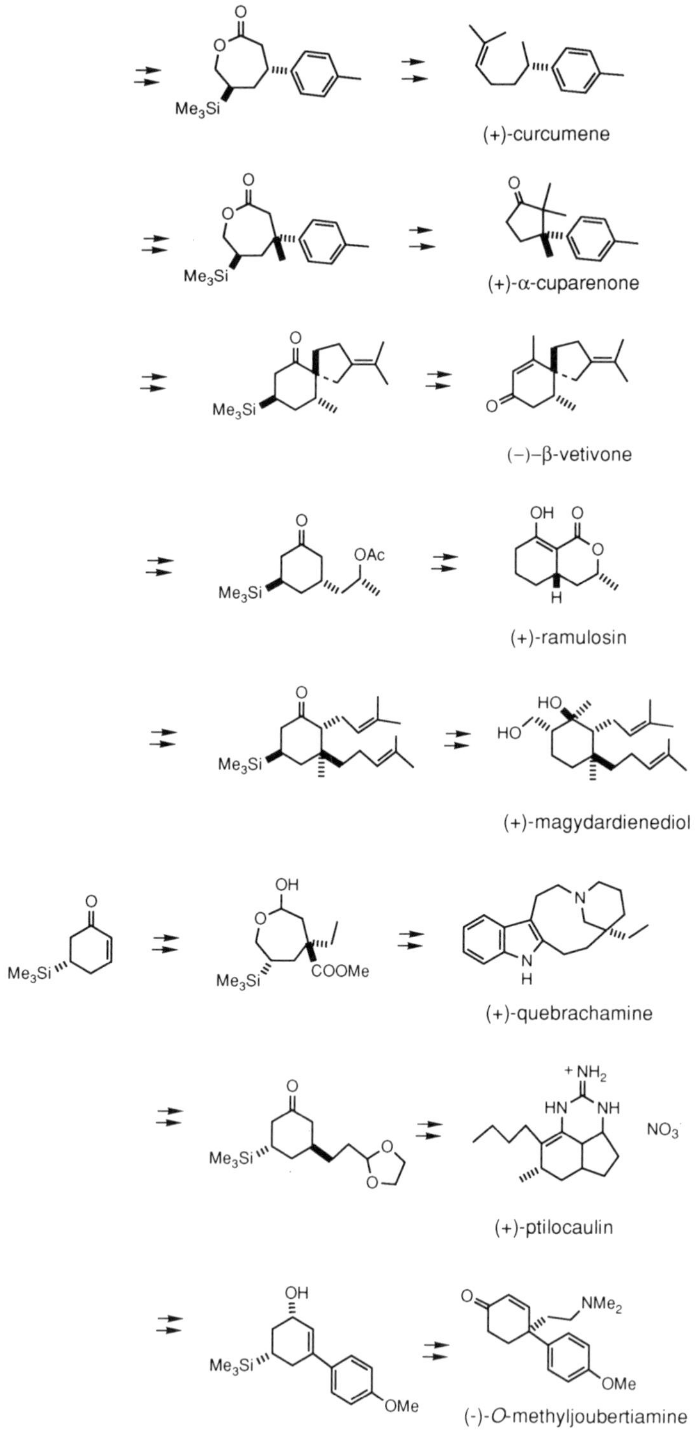

(+)-curcumene

(+)-α-cuparenone

(−)-β-vetivone

(+)-ramulosin

(+)-magydardienediol

(+)-quebrachamine

(+)-ptilocaulin

(−)-O-methyljoubertiamine

Both enantiomers of 1,2:5,6-diepoxyhexane are now available from the same (*S,S*)-tetraol [Machinaga, 1990] through selective sulfonylation of the primary or secondary hydroxyl groups so that epoxide ring closure involves retention or inversion of configuration of the existing stereogenic centers.

A single chiral sulfoxide is the source for 1,7-dioxaspiro[5.5]undecane in the (*R*)- or (*S*)-form [Iwata, 1985]. Spiroacetalization and reductive removal of the phenylsufinyl group from the product gives one isomer, whereas equilibration of the initial product before the reduction step leads to the enantiomer.

Reduction of the ketone groups of a C_2-symmetric bis(β-ketosulfoxide) with diisobutylaluminum hydride in the presence or absence of zinc chloride led to two diastereomers which retained symmetry [Solladié, 1994]. Spiroacetalization and desulfurization gave enantiomeric products.

A valuable chiral synthon is (S)-4-trityloxymethyl-4-butanolide which is readily prepared from either (S)-glutamic acid or D-mannitol. Excellent practice in enantiodivergent synthesis by utilizing this building block has been demonstrated in the cases of quebrachamine [Takano, 1980a, 1981c] and velbanamine [Takano, 1980b, 1982b].

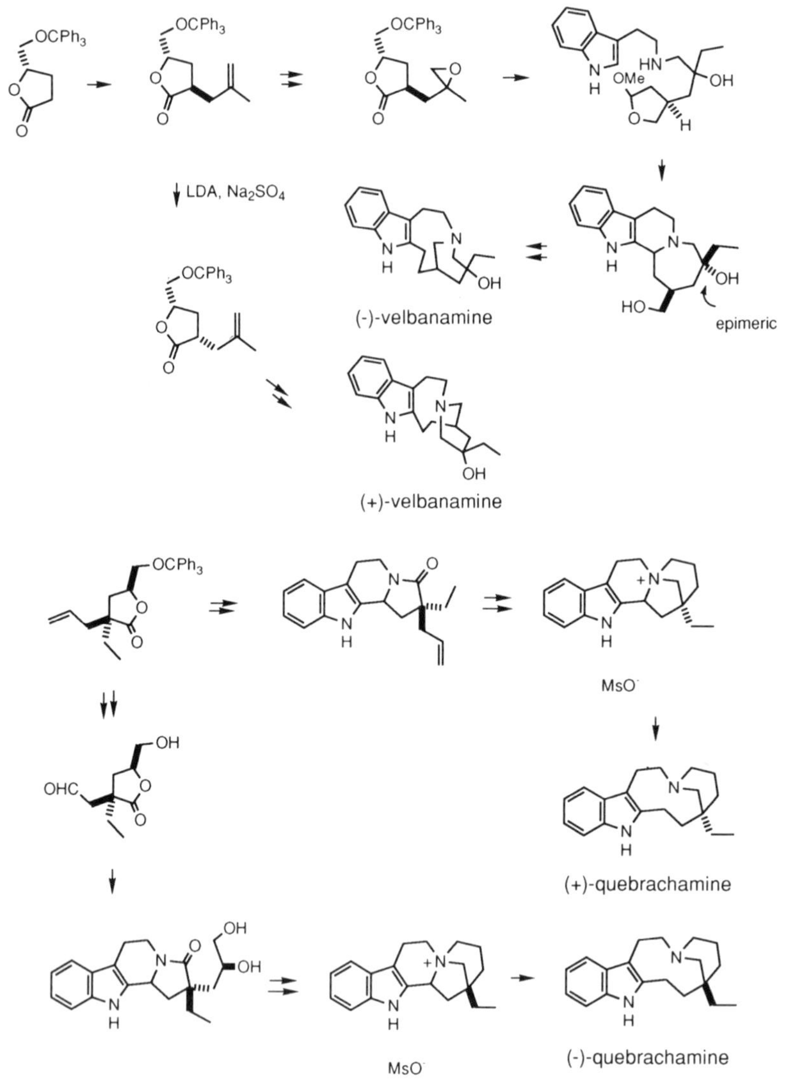

Besides epimerization to invert a stereogenic center and application of different reaction sequences to elaborate carbon chains emanating from a quaternary asymmetric carbon atom to achieve enantiodivergent syntheses as shown above, it is interesting to note that the proper establishment of one stereogenic center was all it needed to reach (+)- and (−)-pinidines [Yamazaki, 1989]. The critical step was the hydride reduction as the resulting secondary alcohol underwent Mitsunobu displacement stereospecifically, and this center exerted asymmetric induction after the heterocycle was formed.

Both enantiomers of β-santalene and epi-β-santalene have been acquired [Takano, 1987b] from the Diels–Alder adduct of cyclopentadiene and (S)-5-hydroxymethyl-2-butenolide. The latter compound is available from D-mannitol.

Manipulation of two different functionalized chains emanating from a center of chirality to create diastereoisomeric intermediates, which eventually leads to enantiomeric targets, is also involved in a synthesis of eburnamonine [Hakam, 1987].

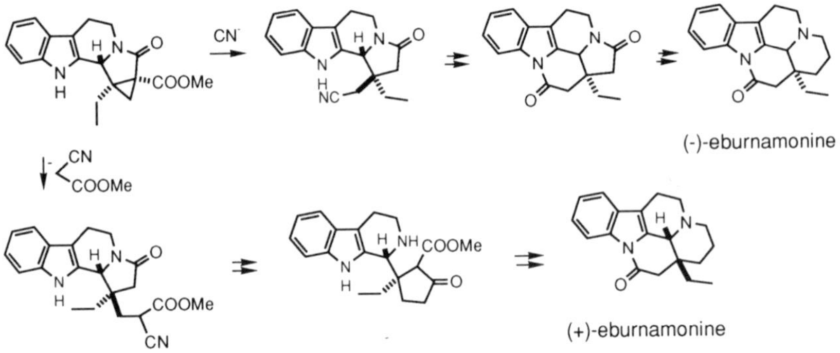

(-)-eburnamonine

(+)-eburnamonine

Some chiral templates (auxiliaries) are flexible in the sense that the reversed order of reactions leads to diastereomeric products which yield enantiomers when the auxiliary component is detached. For example, *C*-acylation of a 1,3-oxathiane derived from pulegone followed by Grignard reaction and cleavage of the heterocycle gave a chiral tertiary alcohol. With proper follow-up, the synthesis of both optical isomers of linalool can be accomplished [Ohwa, 1986]. A bicyclic imidazolidine derived from proline has been used in the synthesis of (*R*)- and (*S*)-frontalin [Mukaiyama, 1979; Sakito, 1979] by changing the order of two Grignard reactions. On the other hand, the enantiodivergent elaboration of 3-substituted *cis*-4,5-dihydroxy-2-cyclopentenone [Bestmann, 1986] depended on chemoselective reagents.

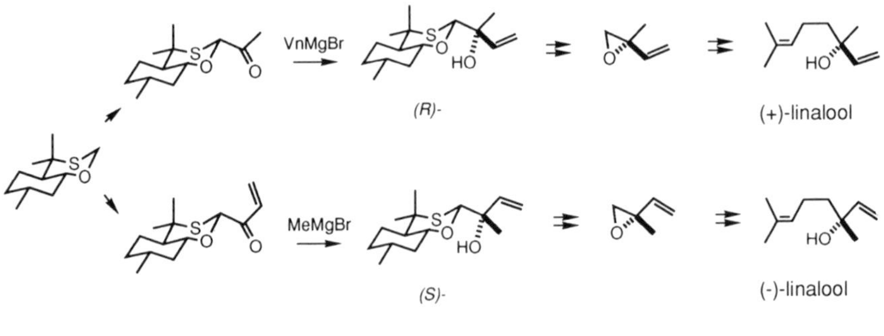

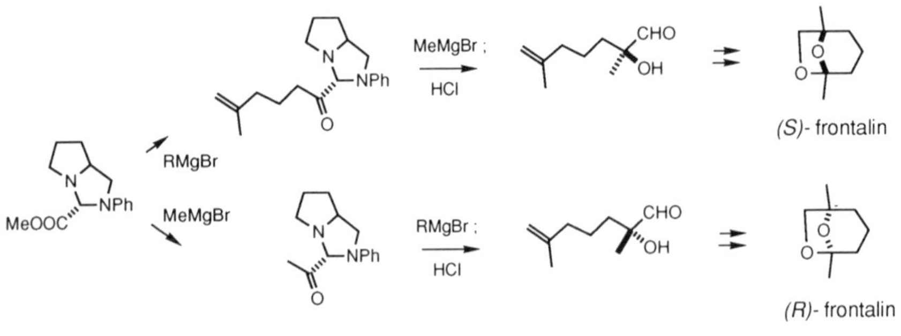

(*R*)-

(+)-linalool

(*S*)-

(-)-linalool

(*S*)- frontalin

(*R*)- frontalin

A chiral adduct derived from glutaraldehyde and (R)-2-aminophenylethanol has been transformed into both enantiomers of isonitramine [Quiron, 1988]. In the adduct, one of the two secondary hydroxyl groups was internally protected, so their selective manipulation was straightforward.

(+)-isonitramine   (-)-isonitramine

The acquisition of (+)- and (−)-Geissman–Waiss lactone [Thanning, 1990] starting from allylation of a pyrrolidine derivative involved separation of two diastereoisomers. The configuration of the oxygen functionality initially attached to C-3 was either retained or inverted in the process.

Geissman-Waiss
lactone

The subtle appreciation accorded to the structural features of the indolizidinediol shown below underlay the access to both enantiomers [Heitz, 1989], starting from D-isoascorbic acid via a γ-lactone. The direction of piperidine ring closure determined the outcome, while the stereochemistry at the ring juncture was affixed at the final stage. The *exo* preference of the two hydroxyl groups was responsible for inducing a *cis* relationship with the α-methine hydrogen to the nitrogen atom.

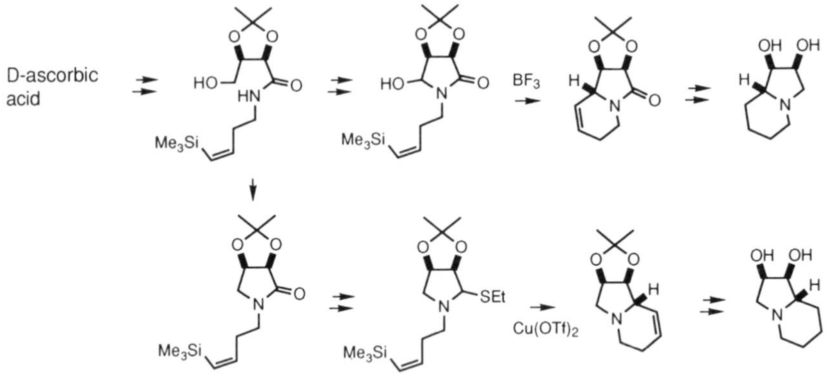

A very intriguing route diverging from a single *syn*-1,3-dihydroxy keto ester and leading to (+)- and (−)-nonactic acid synthons [P.A. Bartlett, 1984] relied on the roles of the ketone group and the proximal hydroxyl function as acceptor or donor site, respectively, in the formation of a 2-alkylidenetetrahydrofuran. The next two asymmetric centers were established in a subsequent hydrogenation of the double bond which was favored to give a 2,5-*cis* substitution pattern. In one series the configuration of the secondary alcohol needed inversion.

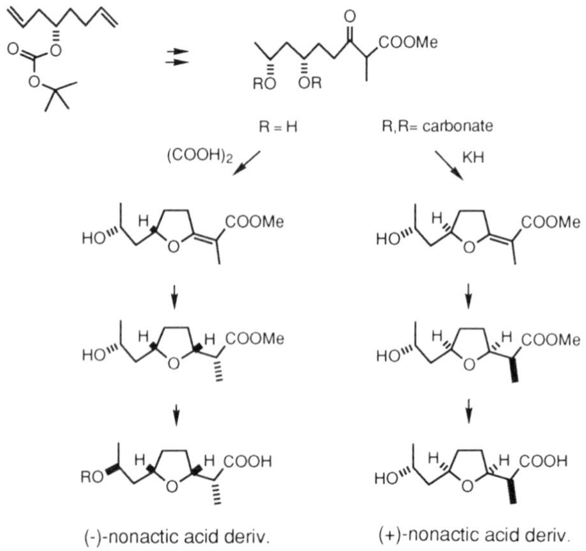

(-)-nonactic acid deriv.   (+)-nonactic acid deriv.

1.5. ENANTIOCONVERGENCE

In recent years convergency has virtually become a vogue as well as an imperative when dealing with synthesis of complex molecules. Enantioconvergency is a step further, in view of the importance of obtaining only one isomer of a chiral target molecule. It has even higher economic and aesthetic implications than ordinary convergency.

The classical method of deracemization when achievable still deserved consideration. Thus it was a pleasant surprise to obtain an optically active tricyclic ketone precursor of (−)-emetine [Openshaw, 1963] by refluxing the racemate with (−)-10-camphorsulfonic acid in ethyl acetate. A reversible Mannich reaction was set up under the equilibration conditions. Similarly, equilibration of an aldehyde was the key to making full use of material in a synthesis of vincamine [Oppolzer, 1977].

A very unusual role for dihydropyranyl ethers was assigned for a synthesis of (−)-monic acid-C [Kuroda, 1986] in the which diastereomers were formed and converged to a single chiral lactone.

(-)-monic acid-C

cis-5-Hydroxy-3-cyclohexenecarboxylic acid belongs to a class of compounds in which the two enantiomers are interconvertible on suprafacial transposition of the allylic substituent. A route to a potential chiral precursor of the prostaglandins has been developed [Trost, 1978]. The diastereoisomers prepared with a chiral isocyanate were separated and isomerization of the "unwanted" urethane by mercuric trifluoroacetate catalysis led to the useful enantiomer.

The lipase-catalyzed hydrolysis of one of the *endo*-dicyclopentadienyl acetates permitted separation and processing into the enantiomeric enones. Both are useful for synthesizing (+)-α-cuparenone [Takano, 1989a], using different reaction sequences. Note the related but different method for gaining access to the enantiomeric enones [Z.Y. Liu, 1993].

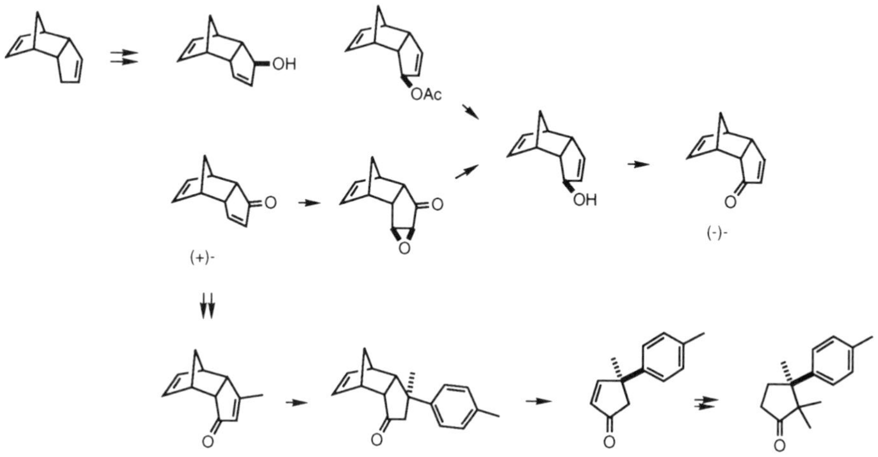

(+)-α-cuparenone

The so-called *meso*-trick for separation of racemates has become very popular because of its effectiveness. During synthesis of enantiopure (+)-biotin [Gerecke, 1970] and prostanoids [Terashima, 1977; Fischli, 1975], acquisition of the desired intermediates with or without addition of a resolving agent has been shown.

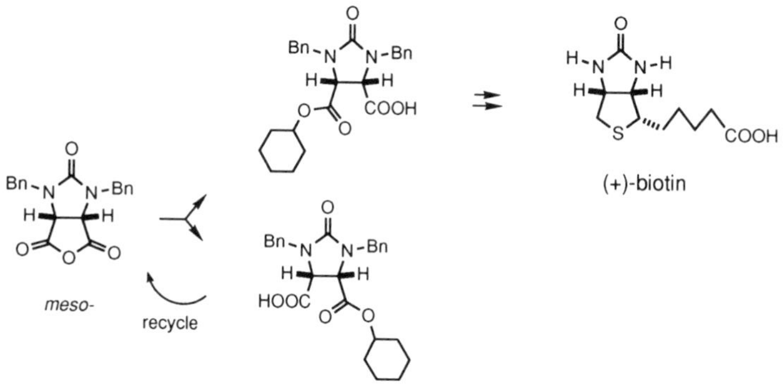

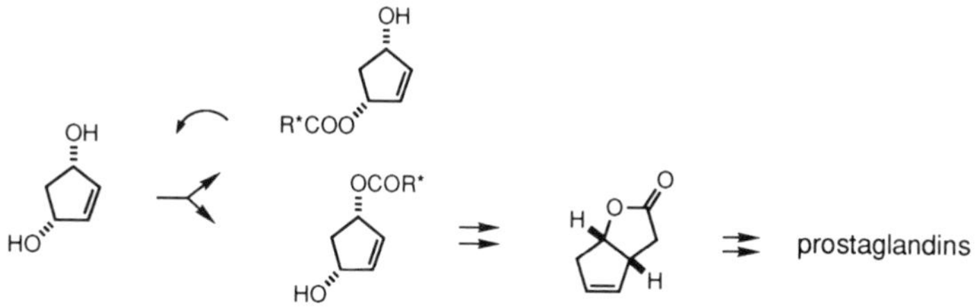

(+)-biotin

prostaglandins

prostaglandin-F$_2$

Of interest is the demonstration of a double *meso*-trick [Breuilles, 1993] on the enzymatic acetylation of a *meso*-tetraol which is also C_2-symmetric.

Carvone possesses only one chirality center and the two enantiomers are related by allylomerism of the enone system. The preparation of (+)-*trans*-chrysanthemic acid [Torii, 1983] from both (R)- and (S)-carvone is an instructive exercise in enantioconvergent synthesis.

trans-chrysanthemic acid

Perhaps the most intriguing possibility of enantioconvergent synthesis presented to date is an approach to protoemetinol (hence emetine) [Takano, 1979]. The (+)- and (−)-isomers of norcamphor can be manipulated individually to the same target molecule. The major difference is the choice of the chain for the Pictet–Spengler cyclization. While this selection led to a product with opposite configuration at one' asymmetric carbon atom, important factors contributing to the successful inversion of the two other stereogenic centers are the presence of the lactam carbonyl, which enabled equilibration of the ethyl group to a *trans*-relationship with the vicinal side chain, and the phenomenon of reversible ring cleavage/closure of the quinolizidine system which was initiated by a Lewis acid. The lactam and the conjoint methoxy group in the aromatic ring were critical requirements.

protoemetine

1.6. PSEUDOSYMMETRY AND LOCAL SYMMETRY

Pseudosymmetry refers to the presence of slight structural perturbations that cause an otherwise symmetrical molecule to be asymmetrical. In contemplating the synthesis of a pseudosymmetric molecule it is often advisable to begin with the synthesis of a similar symmetric molecule, perhaps with a slight modification of the route or the addition of a few steps. Local symmetry pertains to molecular segments but not the whole molecule.

The almost σ-symmetric aminocyclohexane portion of hygromycin-A challenges chemists to design syntheses. In an interesting solution a 5-enopyranoside derivative of D-glucose fulfilled many requirements [Chida, 1991].

hygromycin-A

Albene is a tricyclic sesquiterpene in which the double bond position determines enantiomerism. Thus it is expedient to synthesize the molecule via symmetrical intermediates [Trost, 1982b]. On the other hand, it would be difficult to modify such synthetic routes to gain access to the chiral product.

albene

Regarding synthesis of pseudosymmetrical compounds, those possessing almost C_2-symmetry and differing only in substituents have the best solution. For example, the fire ant venom component known as pyrrolidine-197B can be elaborated from (S,S)-1,2:5,6-diepoxyhexane [Machinaga, 1991a] by chain extension on both sides to the length required for the shorter of the two side chains in the target molecule, while keeping the other end functionalized. Deoxygenation of one end after proper protection of the other groups was straightforward thanks to the C_2-symmetry, which permitted processing of the other end and the formation of the heterocycle. Note that the route passes through two azido alcohols but both converged into the desired product.

(+)-pyrrolidine-197B

Great benefit in operational simplification accrues when it is possible to devise synthetic intermediates which contain more than one identical functional group so that each reaction proceeding simultaneously at all the exposed sites increases molecular complexity twofold or more. A case in point is a synthesis of cularimine [Kametani, 1964] in which a dialdehyde was formed and homologated to a diacid. Structural limitation ensured the subsequent formation of a tricyclic ketone, the designated precursor of the alkaloid.

cularimine

In talaromycin-B one of the spirocyclic component rings contains an extra hydroxyl group, and a hydroxymethyl substituent corresponding to the ethyl group in the other ring. Thus it is possible to use a nor-compound in which the ethyl group is replaced with another hydroxymethyl chain. The structural feature of this potential intermediate is favorable for the ultimate elaboration of the metabolite [Schreiber, 1983], since the 1,3-diol system in one ring is readily protected as an acetal, leaving the other primary alcohol free to undergo reductive methylation. A further advantage of the synthetic route formulated according to the concept is the possibility of twice using the same allylic chloride as a building block.

talaromycin-B

The synthesis of fredericamycin-A is perhaps a less formidable task than that expected of a molecule possessing so many functional groups, thanks to the local symmetry about the spirocyclic center. This reduces the number of isomers even if the process means the chemist has to be content with mixtures in the intermediate stages. In such an achievement [T.R. Kelly, 1988] the importance of symmetry was further demonstrated in the spirocyclization step, which initially failed. The aldol reaction saved the effort, although it produced a mixture and lengthened the route.

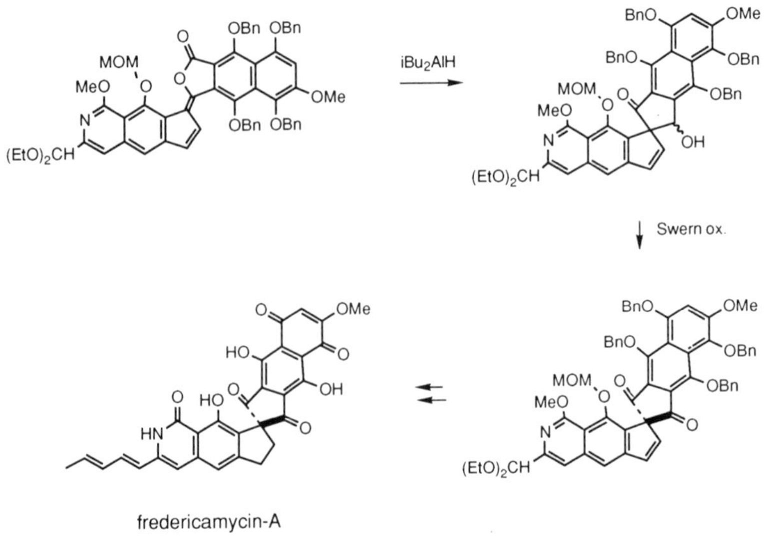

fredericamycin-A

The existence of pseudosymmetry in some molecules is more subtle than others. Recognition and exploitation of this aspect usually afford an intelluctually satisfying experience to synthetic chemists and their audience. Thus the elaboration of the tricarbocyclic portion of ikarugamycin [Whitesell, 1987] from the two enantiomers of a bicyclo[2.2.0]octa-2,6-diene-2-carboxylic ester is enlightening.

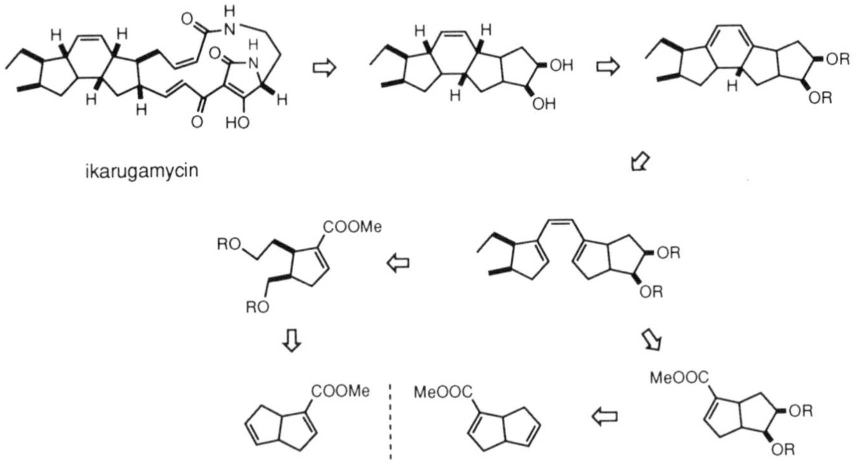

ikarugamycin

The simplication of synthetic routes by consideration of local symmetry is illustrated in a synthesis of cis-trikentrin-A [MacLeod, 1988]. A cis-1,3-dimethylindan is a suitable intermediate because C-5 and C-6 are equivalent.

Once the ethyl group was introduced, formylation at the less hindered *ortho* position was preordained. The pyrrole moiety of the indole system was constructed without any difficulty. Thus in the present case the local symmetry removed any regiochemical problems and steric effects imposed the subsequent reaction site.

cis- trikentrin-A

Also of interest is an approach to *cis*-trikentrin-A [Boger, 1991] which similarly took advantage of a symmetrical bissulfone so that the pressure-induced displacement of one sulfonyl group with an amine could lead to only one product. Subsequent intramolecular Diels–Alder reaction established the indole nucleus.

cis- trikentrin-A

Dynemicin-A exhibit local symmetry in the far end of the anthraquinone chromophore. Regiochemical problems were not encountered or foreseen in a synthesis of this important antitumor agent [Myers, 1995] based on a Diels–Alder reaction. The structural feature was reassuring to success.

R= SiMe₃

dynemicin-A

Interestingly a synthetic approach to hirsutic acid-C [Schuda, 1986] recognized a symmetrical diquinane ketone as intermediate. Although unsymmetrical starting material was used and desymmetrization occurred upon alkylation of the diquinane ketone in its annulation process, any regiochemical issue associated with the alkylation was removed because of the σ-symmetry.

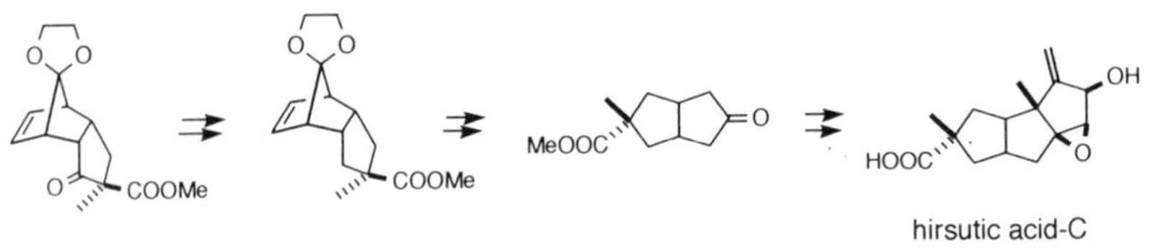

hirsutic acid-C

Although molecular symmetry of arcyriaflavin-B is destroyed by the presence of a hydroxyl substituent, its synthesis can still benefit from symmetry considerations, e.g., via a Diels–Alder route [Hughes, 1983] (cf. an approach to rebeccamycin [Kaneko, 1985]), and in the case of arcyriarubin-B reactions of N-methyldibromomaleimide with indolylmagnesium bromide and later with an alkoxyindolylmagnesium bromide [Brenner, 1988].

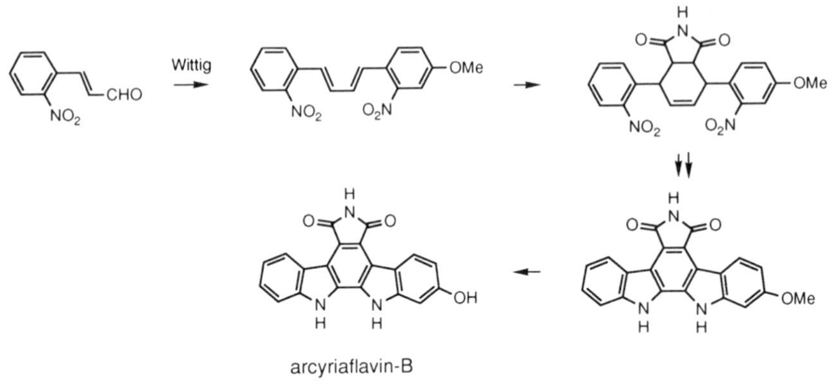

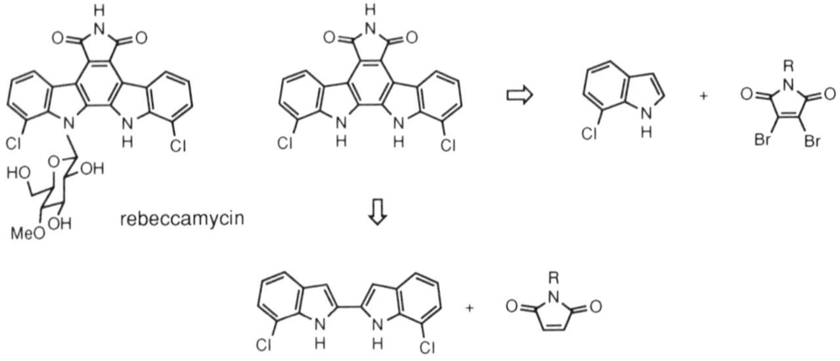

arcyriaflavin-B

rebeccamycin

In an effort to synthesize sesquiterpenes of both cuparane and herbertane series by a divergent tactic, a locally symmetrical dihydroisobenzofuran was considered as the common intermediate [T.-L. Ho, 1994]. This compound was prepared by an intramolecular Diels–Alder reaction.

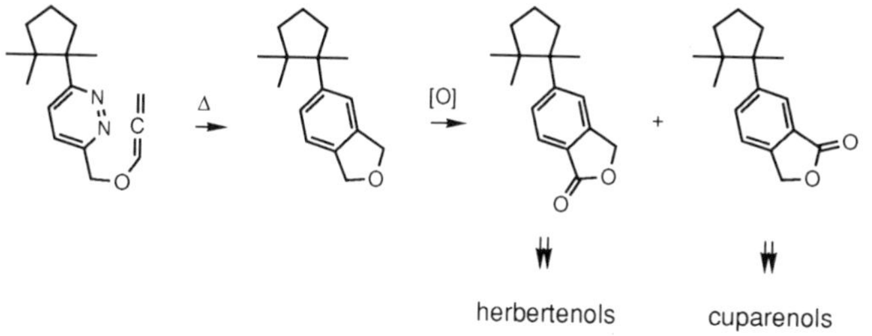

herbertenols   cuparenols

The access to occidol is facilitated by its pseudosymmetry, as evidenced by Diels–Alder approaches [T.-L. Ho, 1972, 1973] and Wittig cyclization [Dauben, 1973].

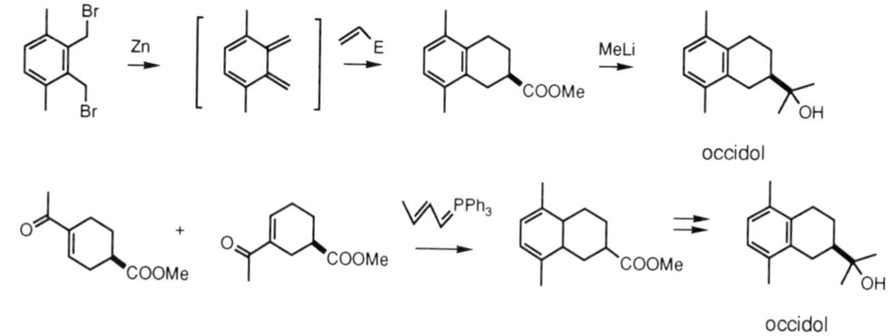

occidol

occidol

The elaboration of both enantiomers of calystegine-B$_2$ [Boyer, 1994] from a common intermediate was based on the observation of pseudosymmetry arising from the positioning of a bridgehead hydroxyl group. Thus it required stereocontrolled deoxyamination of the 4-hydroxycycloheptenone derivative by directing toward either the carbonyl function (via reduction) or the hydroxyl group.

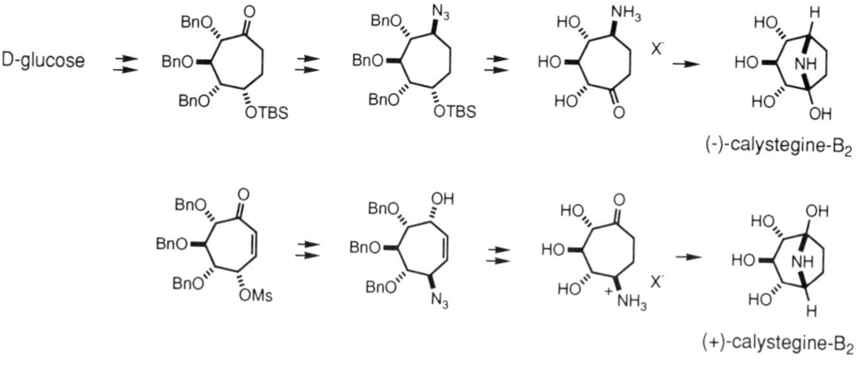

(-)-calystegine-B$_2$

(+)-calystegine-B$_2$

It is important to note pseudosymmetry which has a dynamic feature. Allylic and more highly conjugated compounds in which the α,ω-isomerism is degenerate yield symmetrical ionic or free radical species by group detachment, therefore regioselectivity of reactions involving these compounds does not arise. However, the reaction products are permanently desymmetrized. There are subtle advantages of the pertinent synthetic operations such as the availability of the reagents. An example is the use of [(3-hydroxymethyl)-2,4-pentadienyl]trimethylstannane in a synthesis of nitraraine [R. Yamaguchi, 1991].

nitraraine

A key step in a synthesis of pentalenolactone-E methyl ester [Taber, 1985] involved intramolecular carbene insertion into a C—H bond to form the diquinane skeleton. The precursor of the spirocyclic tetrahydropyran is easy to prepare by a double alkylation. Cyclization to give product of the desired stereochemistry was ensured by proximity effects.

pentalenolactone-E
methyl ester

Exploitation of local symmetry in synthesis must conform to strict criteria such as reaction types. For example, in a synthesis of matrine [Mandell, 1963] the quinolizidine system was created by reductive condensation of a six-membered ketone. Biscyanoethylation via enamine established two side chains at the α- and α'-positions, and since two newly formed stereogenic centers are subjected to equilibration the ultimate elaboration of matrine was assured.

matrine

A triumph of biogenetic speculation is the successful assemblage of protodaphniphylline [Piettre, 1990] from an (18*E*)- or (18*Z*)-10,11-dihydrosqualene-27,28-dialdehyde by reaction with ammonia, ammonium acetate, and triethylamine hydrochloride. The slightly desymmetrized dihydro compound was required to accomplish the intial intramolecular Michael reaction.

X = CHO   protodaphniphylline

A synthesis of gephyrotoxin [Hart, 1983] was based on a penetrating analysis which recognized a potential early intermediate with local symmetry. It was considered possible to create vicinal two-carbon pendants on the cyclohexane unit, i.e., a segment of the unsaturated side chain and ring members of the piperidine moiety, from a cyclohexane and by virtue of proximity effect to desymmetrize the two branches. In other words, the subsequent cyclization to complete the decahydroquinoline framework would involve only the proximal branches leaving the distal chain to be fashioned into the enyne pendant.

gephyrotoxin

The hidden symmetry in a molecular segment of emetine was exploited in the construction of a symmetrical 1,3-bis(tetrahydroisoquinolinyl)acetone intermediate which was reacted with methyl vinyl ketone to form a 1:2 adduct [Clark, 1962a, b]. Since the original ketone group was available for intramolecular aldol condensation only once, there were no regiochemical problems. The unreacted *N*-oxobutyl group was subsequently removed.

(series B)

emetine

The presence of a *cis*-1,3-cyclopentanediol system in F prostaglandins invoked synthesis designs that took advantage of bicyclo[2.2.1]heptene intermediates which were to undergo cleavage and degradation. The route culminated in prostaglandin-$F_{2\alpha}$ [Fischli, 1975] mentioned above, and that concerned with the *meso*-trick is representative.

One of the most convenient synthetic routes to pyridoxine (vitamin-B_6) is via a Diels–Alder reaction between 4-methyl-5-alkoxyoxazole and maleic anhydride [Firestone, 1967]. The local symmetry about the two hydroxymethyl substituents on the pyridine ring is responsible for the succinctness of this approach.

pyridoxine

Byssochlamic acid contains two maleic anhydride units. It is of interest that a synthesis [Stork, 1972b] was devised on the premise that each of these functional groups would be derived from a 1,4-dimethoxybenzene, therefore using the same oxidation steps. Note also the symmetrical bisalkylation agent used in the formation of the bridged intermediate.

byssochlamic acid

Benzocyclobutene formation by the cobalt-mediated $[2+2+2]$-cycloaddition of a 1,5-diyne and another alkyne was the basis of an estrone synthesis [Funk, 1979]. The cotrimerization using methoxytrimethylsilylethyne was seriously faulty due to catalyst depletion and poor regioselectivity, however, bistrimethylsilylethyne could be used instead. The regioselectivity problem vanished and it was possible to remove the trimethylsilyl group at C-2 of the estrone analog by protonolysis, leaving the group at C-3 for oxidative replacement.

estrone

Consideration of local symmetry also touches on desymmetrization. Thus the disturbance of local symmetry may be crucial to synthetic progress. Enolizable 1,3-diketones are frequently encountered as synthetic intermediates, and the direction of the enolization must be recognized in order to apply the appropriate reactions. Relevant to this aspect are operations involved in the elaboration of drimenin and confertifolin [Wenkert, 1964], cedrol [Stork, 1961], and $(-)$-acoradiene [Solas, 1983] from the corresponding 1,3-diketone precursors.

84%   13%

drimenin   confertifolin

cedrol

(-)-acoradiene

There was really no regiochemical problem for a synthesis of alnusone [Semmelhack, 1981b] employing a 1,3-dithiane alkylation to assemble the enone and the eventual intramolecular coupling. The two aromatic units are locally symmetrical (1,4-disubstitution) so that iodination *ortho* to the oxygen functionality yielded one product. Note also the same arylacetaldehyde furnished the dithiane and its alkylating agent.

alnusone

Although not directly concerned with synthetic problems, the stereochemical correlation of (−)-kaurene with (+)-phyllocladene [Cross, 1963], thereby completing elucidation of the stereochemistry of the newer diterpene, warrants a note here. The tricyclic keto ester derived from (−)-kaurene is a diastereoisomer of the degradation product of (+)-phyllocladene. Interestingly, treatment of the former with sodium methoxide led to the enantiomer of the latter. Apparently a tetracyclic diketone with the bicyclo[2.2.2]octane skeleton was formed and it broke down subsequently to give the more stable isomer. Despite the local symmetry of the β-diketone unit the reversible reaction conditions ensured generation of the tricyclic keto ester with a *trans* BC-ring juncture. The thermodynamic factor determined the course of desymmetrization.

(-)-*ent*-kaurene

(+)-phyllocladene

enantiomeric

Unsymmetrical and enolizable α-ketols usually exist in two dynamic isomers. Desymmetrization of such molecules would be detrimental as two products are expected. However, a sacrificial ketone synthesis [Wakamatsu, 1980] is based on C-alkylation at the carbinol site of symmetrical acyloins, with reduction and glycol cleavage to follow. The ester serving as precursor of the acyloin must be inexpensive to meet the economy criterion.

A very effective strategy in synthesis is the breaking of local symmetry by proximity effects, that is to engage one of two identical functional units into reaction with a neighboring group. The application is exemplified in an approach to daunomycinone [Sih, 1979].

E = COOMe

daunomycinone

A more complicated case is a synthesis of chanoclavine-I [Plieninger, 1976a, b]. The key intermediate, a benzobicyclo[2.2.2]octadiene which contains a nitro substituent in the aromatic ring and a ketone on the ethano branch of the bridged ring portion, is unsymmetrical beyond the four carbon atoms comprising the bridgeheads and the alkene. The unsaturated bridge was constructed from a symmetrical addend and that portion of the molecule maintained a local symmetry upon oxidative cleavage. However, the intramolecular trapping of the aldehyde group by a *peri* carbamate to form a heterocycle solved several synthetic problems.

chanoclavine-I

1.7. EFFICIENCY VERSUS SYMMETRY

Convergence [Velluz, 1967; Hendrickson, 1977] has been preached as a cornerstone of synthetic efficiency. However, since the degree of convergence can vary a great deal the truly guaranteed efficient syntheses are those involving assemblage of two identical units in the final step, even if perfect convergence cannot be achieved in forming these units. Compared with an ordinary convergent synthesis, in which the ultimate synthons are different, it calls for approximately half the work. But this *reflexive* course [Bertz, 1984] operates only when the synthetic target is symmetrical.

A magnificent case is usnic acid. This seemingly unsymmetrical molecule is actually made up of two parts, more readily identified on examination of the aromatic moiety. When disconnected from the aromatic unit the cyclohexadienone is revealed as a tautomer of the former compound, and the whole molecule is an *o,p*-coupling product of one and the same phenol [Barton, 1956]. The forbidden aromatization of the ketonic portion caused an automatic dehydration and the molecular symmetry was hidden more deeply.

usnic acid

Another intriguing and efficient synthesis is that of carpanone [D.L. Chapman, 1971] which consisted of a Pd(II) oxidation of a substituted styrene. The dimerization was succeeded by an intramolecular Diels–Alder reaction.

The dynamic desymmetrization has also been exploited in a synthesis of yuehchukene [Cheng, 1985] by treatment of β-(3-methyl-1,3-butadienyl)indole with acid. Coupling was initiated by protonation of one molecule, which resulted in a formal Diels–Alder adduct, and cascaded down another cyclization pathway.

carpanone

yuehchukene

The nonoxidative [6+4]-dimerization of a substituted maleic anhydride to afford the skeleton of the nonadrides [Sutherland, 1968] is remarkable. The only imperfection was related to the stereochemical aspect.

The dimeric nature of (+)-xestospongin has invoked a combinative approach [Hoye, 1994]. The most interesting aspects are concerned with formation of the oxaquinolizidine system from a precursor having only one asymmetric carbon atom; the other centers were established through equilibration. The use of a thiophene ring facilitated the macrocyclization by reducing the freedom of four carbon units in each half of the condensing chains, so it simplified the chain-building process.

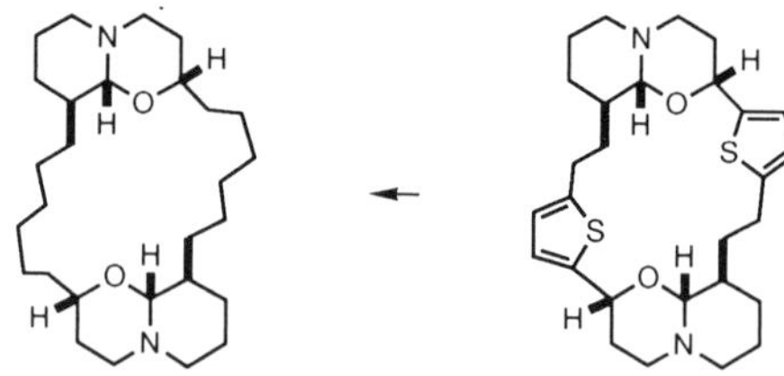

(+)-xestospongin-A

It is transparent that reflexive schemes for the synthesis of molecules such as squalene are the simplest. Two such syntheses [Biellmann, 1969; Blackburn, 1969] utilized sulfide/disulfide-mediated reactions to unite two farnesyl units.

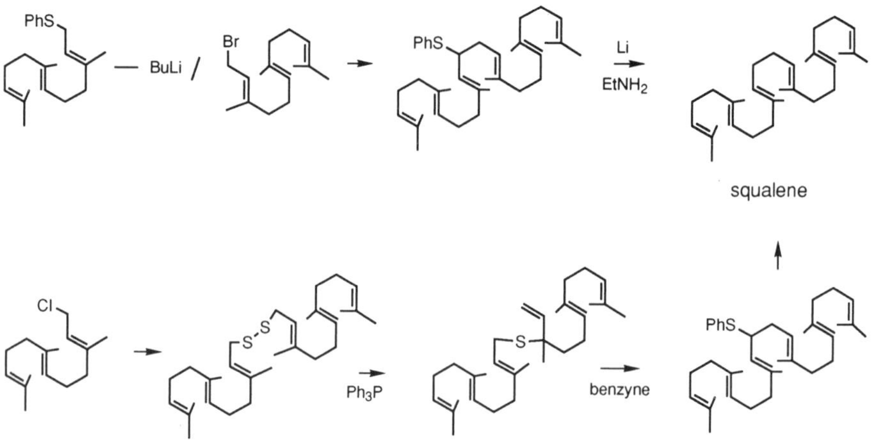

Finally, it must be mentioned that the most rational approach to C_2-symmetric molecules including diolides would involve elaboration of the "half-molecule" and assemblage of two of them. A particularly interesting tactic which was employed in a synthesis of vermiculine [Y. Fukuyama, 1977] is to place the two carbonyl groups in latency in the form of a 1,2-cyclohexanediol.

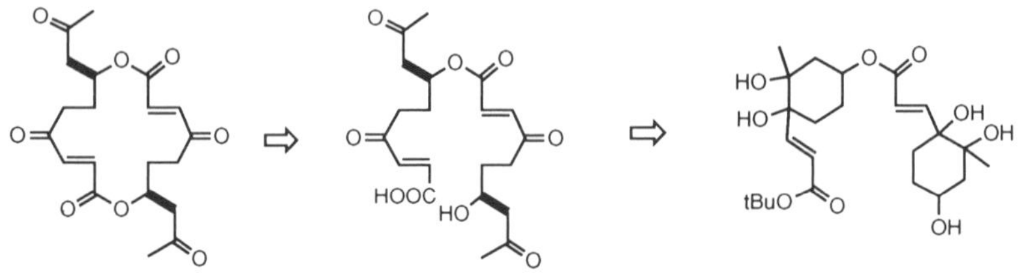

vermiculine

2

SYNTHESIS OF SYMMETRICAL MOLECULES

Highly symmetrical molecules are very attractive synthetic targets. They are also challenging in the sense that viable plans demand maintenance or increase of symmetry while execution is in progress. Unavoidable deviations must be corrected in a few steps, otherwise the synthesis will become unmanageable. The fascinating possibility of combining isometric segments (segments having the same arrangement of atoms) must be given priority in retrosynthetic analysis of such molecules, and symmetry provides a guide to planning. The bisection of an achiral molecule into identical enantiopure segments is referred to as "la coupe du roi" [Anet, 1983] and synthesis following this analysis may be most rewarding [S. Alvarez, 1992].

2.1. CONCOCTIVE SPECIES

2.1.1. Dodecahedrane and Fragments

Dodecahedrane is a hydrocarbon analog of the most complex of the Platonic solids. The I_h symmetry of this $C_{20}H_{20}$ compound is reflected in the display of a single line absorption in its proton and carbon nmr spectra.

This supreme symmetry of dodecahedrane necessarily demands utmost attention in synthetic design, including choice of starting material and execution of individual steps. Intermediates which have been desymmetrized temporarily must be readily remedied in the next few steps. But with a meticulous plan, rapid ascent to the structural summit can be achieved by elaboration of a polyfunctional precursor. A corollary to this principle is that a synthesis will go hopelessly astray if the symmetry feature is not strictly observed.

Several rational approaches* to dodecahedrane have been delineated. While those involving dimerization of triquinacene [Woodward, 1964] and peristylane capping [Eaton, 1977, 1979] have yet to be realized, triumph of a scaffold synthesis [Ternansky, 1982; Paquette, 1984] indeed was founded on symmetry considerations in all phases.

A serious problem pertaining to synthesis of dodecahedrane is the maintenance of molecular convexity from start to finish. Any step that allows protonation of bowl-shaped intermediates can subvert an approach by undesirable proton transfer from a solvent molecule trapped inside the cavity. Note that such a process is favored in advanced intermediates of dodecahedrane in view of the fact that convex molecule construction is always accompanied by increasing nonbonding interactions. Placing the C—C bond in an *exo* orientation removes steric congestion from such a molecule.

The Paquette synthesis started from a domino Diels–Alder reaction of 9,10-dihydrofulvalene and dimethyl acetylenedicarboxylate. One of the adducts has a C_{2v}-symmetry and contains four interlocking five-membered rings of the target molecule. Moreover, the high symmetry of the diene diester and the pairwise proximity of the functionalities render its conversion to a symmetrical diketone very easy. Annulation of the two cyclopentanone moieties of the resulting compound can then be undertaken to give an intermediate encompassing all the skeletal carbon atoms of dodecahedrane.

It should be noted that the annulation via a bis(spirocyclobutanone), a dilactone, and a biscyclopentenone was followed by catalytic hydrogenation, with the extra C—C bond between the α-carbon atoms of the two ester groups left intact during all the manipulations. This C—C bond served as a locking device for the polyquinane system and as an effective shield against invasion of external reagents into the core of the various intermediates. Thus the hydrogenation from the *exo* face delivered a useful precursor of dodecahedrane.

Upon conversion of the diketone to a dichloro compound a reductive cleavage of the extra C—C bond also caused ring closure in the desire manner. Addition of an alkylating agent after the reduction forced the unreacted ester group inside the molecular surface as defined by the incomplete sphere, enabling involvement of this ester group in subsequent skeletal C—C bond formation formation. Of course the alkylating agent must be equipped with a removable mechanism in order that dodecahedrane itself be unraveled.

From the heptacyclic stage onward, thermodynamic factors became an ally to coordinate closure of the remaining carbocycles. The rigidity of the ring system permitted formation of only one isomer in each of the C—C formation processes. In fact, photochemical cyclization was used thrice, each time to create a new ring. The final step was a palladium-catalyzed transannular dehydrogenation.

---

*Isomerization of pagodane to dodecahedrane has been accomplished catalytically [Fessner, 1987] and in a stepwise manner [Spurr, 1987], albeit in relatively small yields.

dodecahedrane

A bonus from the established route to dodecahedrane is the acquisition of C_{16}-hexaquinacene [Paquette, 1979], another potential precursor of dodecahedrane. The divergence of reaction pathways resulted in the net formation of two C—C bonds from the termini of the bishomologated diacid to one of the double bonds. The actual route consisted of photocycloaddition of an enedione and reductive cleavage of the four-membered ring as well as scission of the C—C bond lying inside the molecule. The change in the degree of symmetry as a result of the reaction sequence should be noted.

C_{16}-hexaquinacene

[5]Peristylane (C_{15}-hexaquinane) lacks only a cyclopentane cap to complete dodecahedrane. This molecule has been acquired via a series of symmetrical intermediates [Eaton, 1977].

[5]peristylane

A hexacyclic diketo diester which is a potential precursor of dodecahedrane has been elaborated [Mehta, 1985, 1988b]. The (C_{2v})-C_{12}-tetraquinanedione intermediate was obtained by pyrolytic decomposition of a cyclobutane unit.

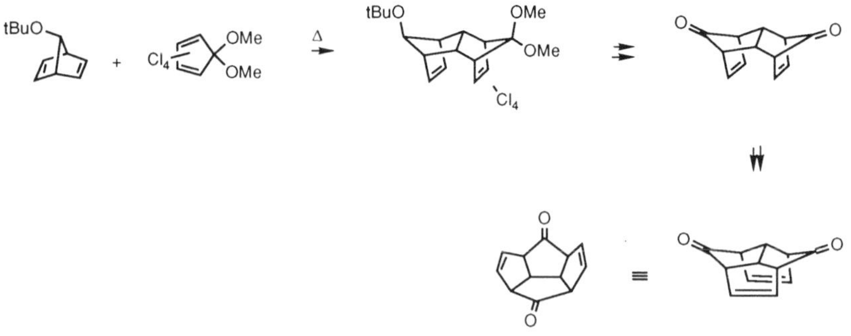

[5]Peristylane also belongs to another series of compounds which feature an *n*-membered ring joining by alternate carbon atoms to a 2*n*-membered ring. The simplest homolog of this hydrocarbon series is [3]peristylane, or triaxane [Nickon, 1968; Garratt, 1977]. The three approaches summarized below are characterized by an increasing symmetrization of precursors having essentially a plane of symmetry.

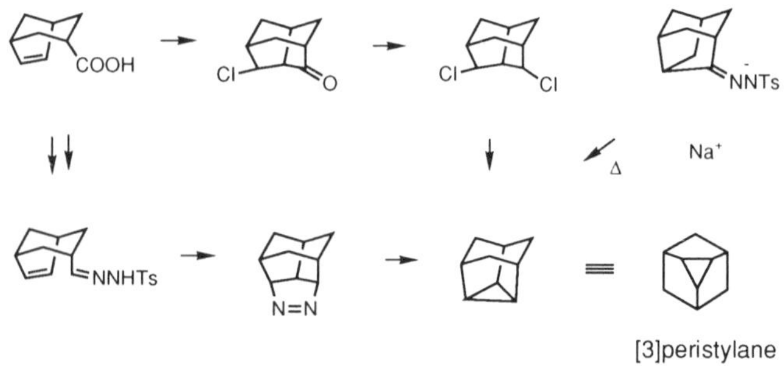

[3]peristylane

The presence of a cyclobutane platform in [4]peristylane logically invokes a photocycloaddition approach [Paquette, 1983]. The route that was cleared for the synthesis traversed several unsymmetrical intermediates only because an activated acetylene was needed as a dienophile.

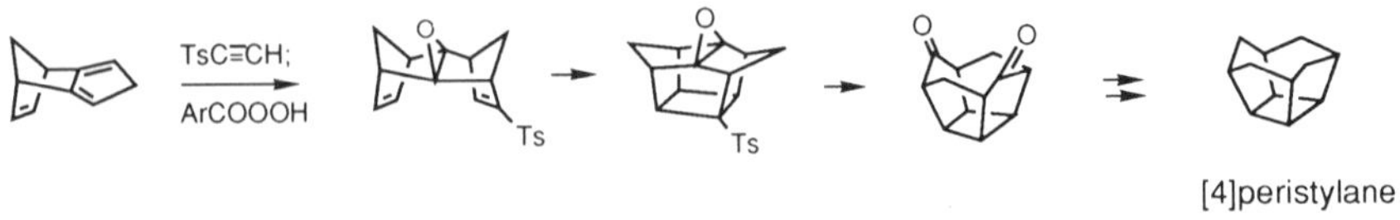

[4]peristylane

Another successful synthesis of dodechadrane proceeded via pagodane. The rationale of the approach is shown in the following scheme (path X) which also summarizes other possibilities.

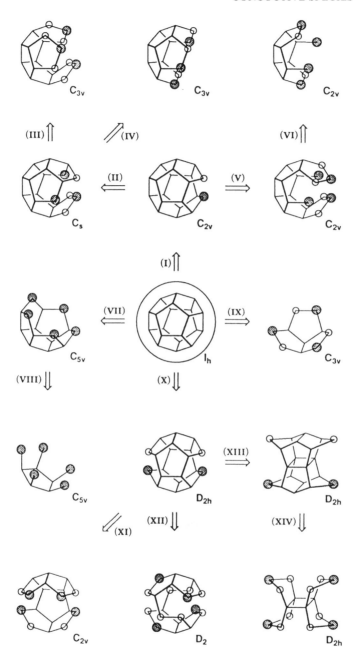

For the elaboration of pagodane [Fessner, 1983] reactions that tend to give symmetrical products or products amenable to rapid resymmetrization were employed. These include the Diels–Alder reaction and photocycloaddition.

pagodane

2.1.2. Cubane, Prismanes, and Related Cage Molecules

The hydrocarbon, cubane, which corresponds to the more familiar and simpler Platonic solid, yielded to synthesis at an earlier date [Eaton, 1964]. It must be noted that, for achieving control of reactivity of suitable starting material and intermediates, a compromise was made with regard to strict maintenance of symmetry.

cubane

An improved version [Barborak, 1966] involved trapping cyclobutadiene with 2,5-dibromo-p-benzoquinone. Photocycloaddition of the adduct gave a symmetrical diketone which was subjected to Favorskii rearrangement and decarboxylation. The return to a C_2-symmetric molecule upon photocycloaddition enabled twofold ring contraction in one operation.

cubane

(Note the transformation of basketene (see below) into cubane [Chin, 1966] via oxidative cleavage of the double bond, Dieckmann condensation of the derived diester, and conversion of the ketoester into a substrate for the Favorskii rearrangement.)

The feasibility of creating cyclobutane units by Favorskii ring contraction has had profound influence in designing syntheses of many cage molecules. In a route to garudane (1,4-bishomohexaprismane) [Mehta, 1987, 1991] such a method was exploited in conjunction with intramolecular photocycloaddition.

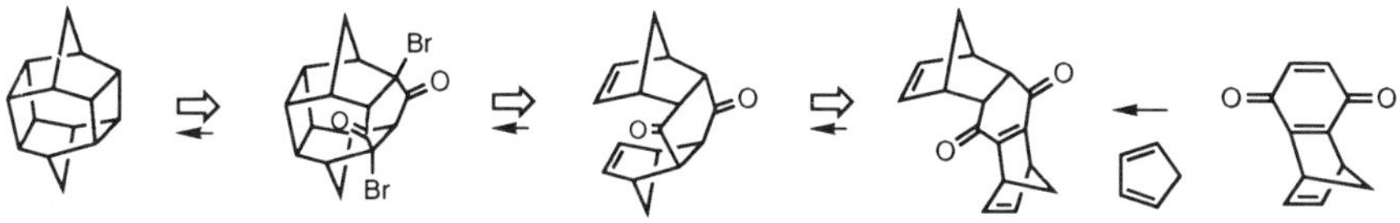

garudane

Disecopentaprismadiene, known by the trivial name hypostrophene [McKennis, 1971], is a fluxional molecule. The narcissistic reaction is a Cope rearrangement.

The preparation of hypostrophene starting from a Diels–Alder reaction of cyclobutadiene and *p*-benzoquinone, was via dioxo-, dihydroxy-, and dibromo-secopentaprismanes. It should be emphasized that this route traversed very symmetrical intermediates so it is shorter than another synthesis [Paquette, 1974] which was developed later.

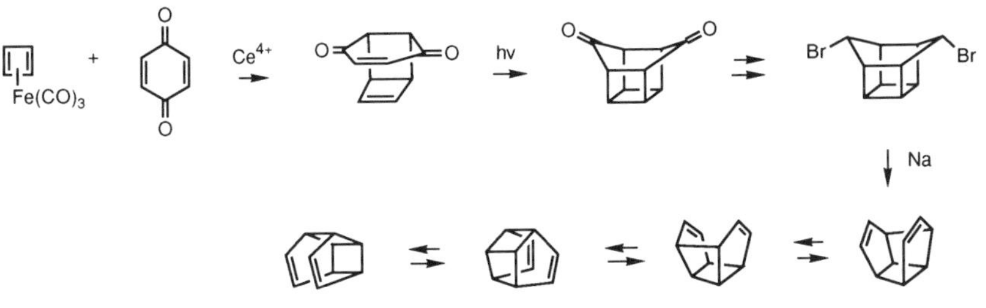

hypostrophene

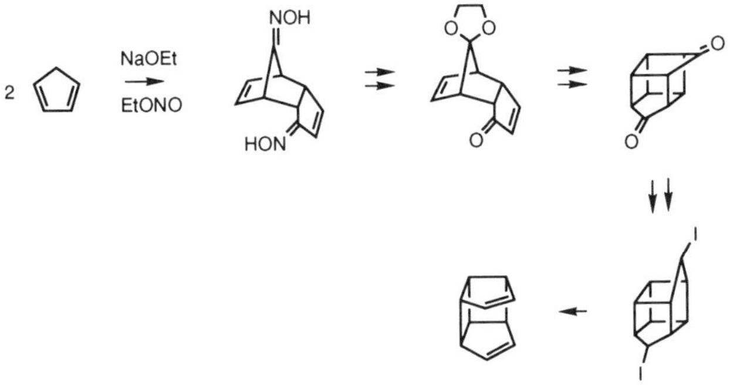

hypostrophene

The saturated isomer of hypostrophene is pentaprismane. Due to the proximity and perfect alignment of the two double bonds of hypostrophene it is logical to expect that an intramolecular photocycloaddition of the diene would give pentaprismane, but all attempts in this conversion met with failure. A synthesis of pentaprismane [Eaton, 1981] was achieved instead by following a route analogous to the approach for cubane, i.e., involving intramolecular photocycloaddition of tri- and tetracyclic precursors, and Favorskii ring contraction. All the intermediates are symmetrical.

pentaprismane

Note that homopentaprismanone had been previously obtained [Ward, 1971]. Another pathway to pentaprismane [Dauben, 1983] also employed cyclobutadiene as a building block.

pentaprismane

Structurally speaking, several polycyclic hydrocarbons constitute the class of n-prismanes. Each member contains $2n$ identical methine units arranged at the corners of a regular prism with D_{nh} symmetry, and defined by two parallel n-membered rings cojoined by n four-membered rings. Accordingly, cubane is tetraprismane, but the hydrocarbon corresponding to the most familiar prismatic objects is triprismane.

The structure of triprismane has a long history. It was proposed by Ladenburg to represent benzene. Interestingly, an access to this highly strained and symmetrical compound was gained through a new benzene isomer, benzvalene [Katz, 1973]. Benzvalene acted as a 1,3-dipole in a stepwise cycloaddition in which a 1,2-shift of a cyclopropane C—C bond served to establish four of the five edges of triprismane. Photochemical extrusion of dinitrogen from a diaza compound culminated in the product, albeit in a miniscule yield.

benzvalene   triprismane

Basketene has the shape of a basket in which a double bond resembles the handle. A very convenient access to this hydrocarbon started from a Diels–Alder reaction of cyclooctatetraene and maleic anhydride [Masamune, 1966; Dauben, 1966].

Usually cyclooctatetraene undergoes Diels–Alder reactions in its tautomeric bicyclic form. The two double bonds of its adduct with maleic anhydride are well disposed to participate in an intramolecular photocycloaddition. Oxidative removal of the elements of CO and CO_2 from the cage compound rendered it symmetrical.

basketene

Tautomerization of cyclooctatetrene by electrocyclization lowered its symmetry. Further desymmetrization accompanied the Diels–Alder reaction. However, the process was reversed in the subsequent steps.

Many synthetic routes have been developed for D_3-trishomocubane [Underwood, 1970; Eaton, 1974; Marchand, 1976; E.C. Smith, 1976]. The isomerization method [Kent, 1977] is convenient, and it appears that D_3-trishomocubane is the most stable pentacycloundecane. The compound lacks a three- or four-membered ring and other strain features.

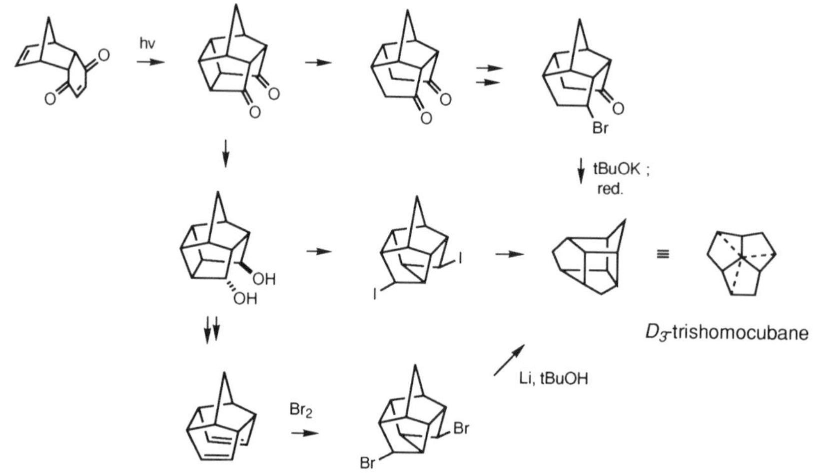

D_3-trishomocubane

D_3-trishomocubane

Triblattanes are bridge-expanded basketanes. Thus synthesis of [2.2.2]tri-battatriene has been achieved by ring enlargement of the ketonic derivatives [Müller, 1988]. A crucial precursor is D_3-trishomocubanetrione [Fessner, 1986]. Note that [2.2.2]triblattane is also known as [8]tritwistane [Hirao, 1980].

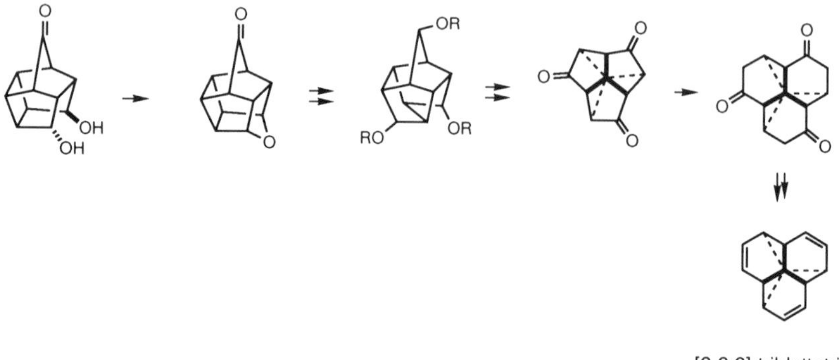

[2.2.2]-triblattatrier

The heterocyclic analog of basketane in which the handle is replaced by nitrogen atoms is readily prepared by using an azodicarboxylic ester instead of maleic anhydride in the reaction with cyclooctatetraene. Interestingly, the compound could be transformed into semibullvalene [Askani, 1970; Paquette, 1970]. Semibullvalene is a fluxional compound that undergoes rapid autogenous rearrangement.

R = COOEt
R + R = (CO)$_2$NPh

semibullvalene

The new hydrocarbon from a silver ion-catalyzed skeletal isomerization (formally a dyotropic rearrangement) of basketane derivative is a precursor of snoutene [Paquette, 1971], and that of cubane itself is called cuneane [Cassar, 1970] because of its wedge shape.

snoutene

cuneane

Snoutene is photochemically labile; on irradiation it is transformed into a crown-like hydrocarbon, diademane [deMeijere, 1971]. And as expected, the presence of three cyclopropane rings in diademane contributes to its conversion to triquinacene at its melting point [Bosse, 1974].

snoutene   diademane   triquinacene

2.1.3. Tetrahedrane

Returning to the realm of "Platonic hydrocarbons" the ultimate challenge of synthesis is tetrahedrane. This is due to the extreme strain inherent in the molecule, its calculated enthalpy of formation amounting to 126.3 kcal/mol. While the parent compound remains elusive, tetra-*t*-butyltetrahedrane has been obtained by photolysis of tetra-*t*-butylcyclopentadienone [Maier, 1978].

Tetra-t-butyltetrahedrane forms air-stable, colorless, volatile crystals (mp 135°C). Its structure is corroborated by spectroscopic studies including x-ray diffraction. The nmr spectra show only three ^{13}C signals and one proton absorption. The compound owes its stability to the shielding effects of the t-butyl groups against many reagents. Placement of the bulky substituents on the apices of a tetrahedron minimizes intramolecular repulsion; breaking of a tetrahedrane bond is bound to increase the steric strain. It seems that only protonation and oxidation would initiate the destruction of the tetrahedrane.

A consequence of such a "corset effect" is that unsymmetrical tetrahedranes would be unstable. Interestingly, tetrakis(trimethylsilyl)tetrahedrane cannot be prepared in the same manner, due to the longer C—Si bond which weakens the corset effect. Photolysis of the corresponding cyclopentadienone led to the butatriene instead [Maier, 1981].

A more readily accessible isomer of tetrahedrane is cyclobutadiene. Based on its calculated enthalpy of formation of 94.4 kcal/mol, it is not surprising that melting tetra-t-butyltetrahedrane gives rise to the cyclobutadiene. The parent cyclobutadiene is a singlet diene, its generation and detection by matrix isolation techniques have been well developed [Maier, 1988]. In terms of preparative value the protocol of oxidative liberation from its tricarbonyliron complex [Emerson, 1965, Watts, 1965, 1966] is without peer. Virtually all other routes involve photolytic decomposition of a less symmetrical precursor. The synthetic applications of nascent cyclobutadiene are extensive, some of them having already been mentioned.

2.1.4. Cyclohexane-based Polycyclic Compounds

Adamantane and its "homologs" are strain-free hydrocarbons whose skeletons resemble the diamond lattice. One of the most efficient rational syntheses of the adamantane skeleton involves α,α'-dialkylation and Dieckmann cyclization [Stetter, 1968]. All the intermediates maintain a plane of symmetry.

An intriguing perturbation of the symmetry of adamantane and derivatives is the selective placement of one carbonyl group of the *meso*-2,4-adamantanedione in the excited state [Meijer, 1988]. A preparation of the chiral molecule was achieved by thermal decomposition of an optically active 1,2-dioxetane.

Access to the adamantanoid hydrocarbons based on thermodynamic properties is relatively simple. Polycyclic hydrocarbons which are isomeric to these compounds undergo rearrangements to the latter species under extreme conditions. Thus, tetrahydro-dicyclopentadiene was converted into adamantane [Schleyer, 1960] and norbornene dimer into diamantane (congressane) [C. Cupas, 1965]. After addition of two CH_2 groups a dimer of cyclooctatetraene underwent isomerization to afford triamantane [V.Z. Williams, 1966]. Symmetry is the ultimate arbiter of these transformations.

adamantane

congressane

Methods for the construction of the noradamantane system have been developed. One approach starts from a perhydropentalene-3,7-dione [Hofmann, 1986].

A formal bisnordiamantane is p-$[3^2.5^6]$octahedrane which is also analogous to dodecahedrane in having a central n-membered ring whose carbon atoms are alternately connected to the corners of two $n/2$-membered rings lying above and below it. The starting material for a synthesis of p-$[3^2.5^6]$octahedrane [Lee, 1993] is the tetracyclic diester which served previously in the dodecahedrane synthesis. Entry into the skeleton was achieved by 1,3-dehydrobromination of the C_2-symmetric dibromide.

E = COOMe

(1 : 2)

III

p-$[3^2.5^6]$octahedrane

Not only tricyclic hydrocarbons seek the stability saddle in diamonoid structures, the formation of hexamethylenetetramine from a mixture of ammonia and formaldehyde attests to a common structural basis. It is also interesting to note that Tröger's bases are readily prepared from anilines and formaldehyde. However, the synthesis of unsymmetrical analogs requires multistep maneuvers [Webb, 1990].

Tröger's bases

Iceane is a hydrocaron composed of two parallel chair-shaped cyclohexane rings linked at alternate carbon atoms by axial bonds. The three six-membered rings so created are in a boat form. Such a molecule was named iceane because it resembles the shape of ice (actually a segment of H_2O dodecamer of ice crystals). A synthesis of iceane [C.L. Cupas, 1974] exploited the driving force from strain reduction and symmetrization by rearrangement of a molecule containing seven- and five-membered rings.

iceane

Twistane is an isomer of adamantane in which the cyclohexane rings are permanently forced into the twist conformation. A short route to the twistane skeleton [Belanger, 1968] is by an intramolecular aldol condensation of a cis-decalindione which was rendered irreversible by trapping the bridgehead hydroxyl group in situ as the acetate. The symmetrical decalindione was critical to the formation of only one product.

It is interesting that hydrogenation of basketane gave twistane [Osawa, 1974].

The more complex [8]tritwistane is actually quite readily available [Hirao, 1980]. The half-cage diketone, obtained from 1,3-cyclohexadiene and p-benzoquinone via a thermal and photochemical cycloaddition reaction sequence, was transformed into a less strained isomer. Both cyclopentanone units underwent ring expansion to effectuate the carbon framework.

Bicyclo[2.2.2]octane is a well-known compound consisting of two cyclohexane rings in the boat form, the bridged ring system has also been used as a connector of carbon chain termini to form a series of [n.2.2.2]paddlanes [Eaton, 1983].

[n.2.2.2]paddlanes

The fusion of two bicyclo[2.2.2]octanes leads to [2.2.2]^{2}geminane [H. Park, 1979] and the tetrabenzo analog, lepidopterene [G. Felix, 1975; Becker, 1979].

syn-[3.2.1]^{2}geminane

[2.2.2]^{2}geminane

lepidopterene

The core of dibenzoequinene can be considered as embodying two chair or two boat cyclohexane rings which share four central carbon units. The compound was readily obtained by a photochemical reaction of [2.2]paracyclonaphthane [Wasserman, 1967].

dibenzoequinene

A symmetrical hexacyclic hydrocarbon $C_{16}H_{18}$ has been synthesized [T.-C. Chou, 1994]. This compound is made up of two chair cyclohexane units connected by five six-membered rings in the boat conformation. All synthetic intermediates possessed at least one plane of symmetry.

In the asterane family of polycyclic hydrocarbons the cyclohexane rings are essentially in the boat form. Triasterane is the first member to be synthesized using the carbene insertion method [Musso, 1967, 1970], but a convenient route to triasteranedione is via bromination and debromination of bicyclo[3.3.1]nonane-3,7-dione [McDonald, 1973]. A key to the facility of this synthesis is the high symmetry of both substrate and product.

triasterane

triasterane-3,7-dione

The most prominent reaction for an approach to tetraasterane [Hutmacher, 1975] is photocycloaddition.

tetraasterane

Another way of locking a cyclohexane ring in the boat shape is by 1,4-dehydro-oligomerization. Two compounds composed of such units are tricyclo[4.2.2.2^{2,5}]dodeca-1,5-diene [Wiberg, 1984] and tetracyclo-[8.2.2.2^{2,5}.2^{6,9}]octadeca-1,5,9-triene [McMurry, 1986].

Barrelene commands theoretical interests due to the three unconjugated double bonds which are fixed in a bicyclo[2.2.2]octane nucleus. The proximity of these double bonds suggests inevitable through-space interactions. However, the three double bonds cannot overlap simultaneously without causing destabilization of the molecule.

The presence of symmetry axes and planes in barrelene indicates that synthetic approaches based on construction of symmetrical intermediates are favored, particularly when intermediates already containing the bicyclic skeleton can be prepared in a single step. Accordingly, a route was developed from the reaction of α-pyrone with methyl acrylate [Zimmerman, 1969]. The 1:1 cycloadduct lost carbon dioxide to regenerate a conjugated diene unit which can undergo the Diels–Alder reaction for the second time. The final product was a diester from which the introduction of the two double bonds via proper degradation methods was readily conceived.

barrelene

Also of theoretical interest is [2.2.2]hericene which has been acquired [Pilet, 1983] by extension of the synthetic method applied to barrelene. Note that a threefold Diels–Alder reaction of [2.2.2]hericene with 1-bromo-2-chlorocyclopropene followed by complete dehydrohalogenation led to the symmetrical triscyclopropanotriptycene [Billups, 1994].

[2.2.2]hericene

2.1.5. Radialenes, Rotanes, Coronanes, Propellanes, Fenestranes, and Others

There is a partial correlation in structural characteristics of the hericenes and the [*n*]radialenes. The most convenient methods for the synthesis of the latter class of highly symmetrical compounds are those involving elimination: [3]radialene [Waitkus, 1966; Mandel'shtam, 1973], [4]radialene [Griffin, 1962, 1963; Trabert, 1980], and [6]radialene [Harruff, 1978; Schiess, 1978]. Less general pathways are based on cycloaddition [Köbrich, 1965; Uhler, 1960] and rearrangement [Dower, 1982].

[3]-radialene

X = OAc; Me₃N⁺ OH⁻

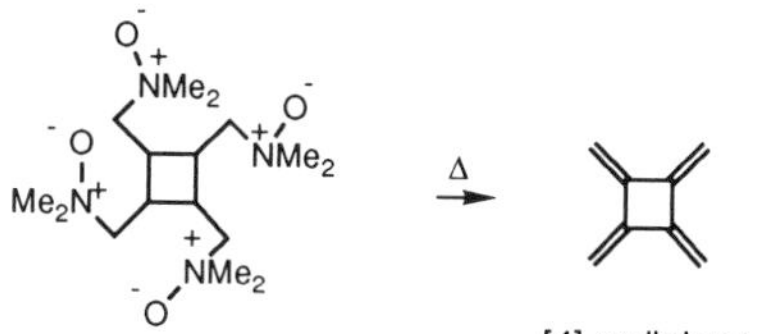

[4]-radialene

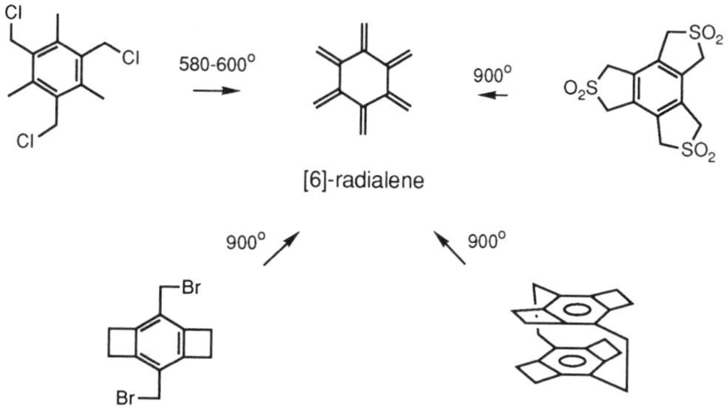

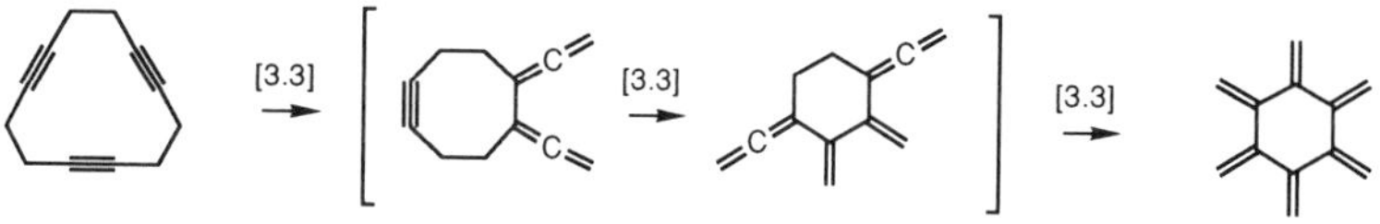

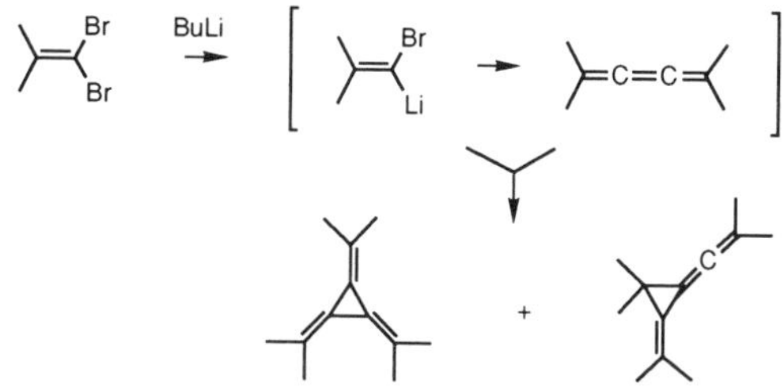

[6]-radialene

hexamethyl[3]radialene

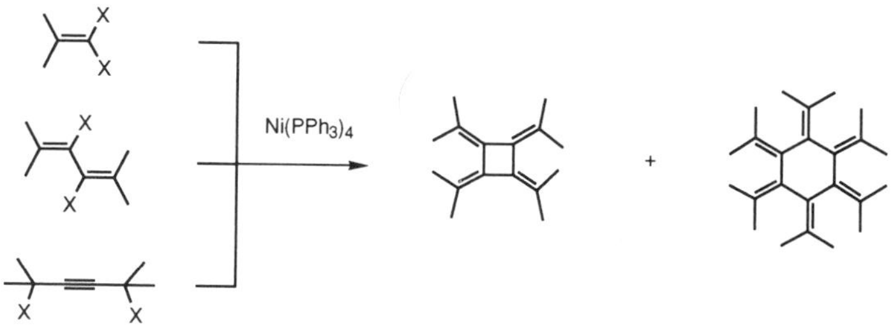

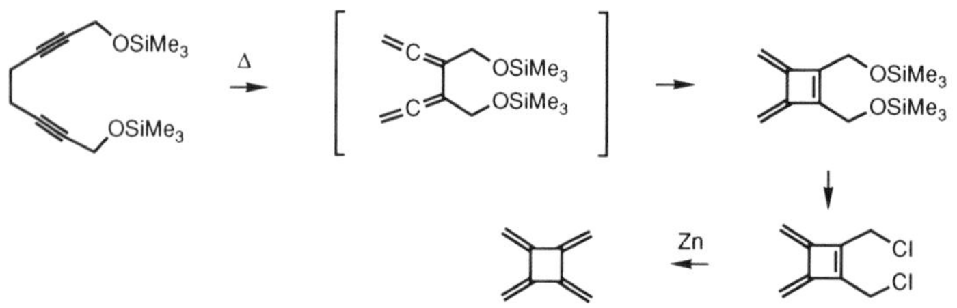

When all the exocyclic methylene groups of an [n]radialene are replaced by cyclopropane units the resulting compound is an [n]rotane: $n=3$ [Fitjer, 1973; Erden, 1986], $n=4$ [Le Perchec, 1970], $n=5$ [Ripoli, 1971; Fitjer, 1982], and $n=6$ [Proksch, 1976; Fitjer, 1982].

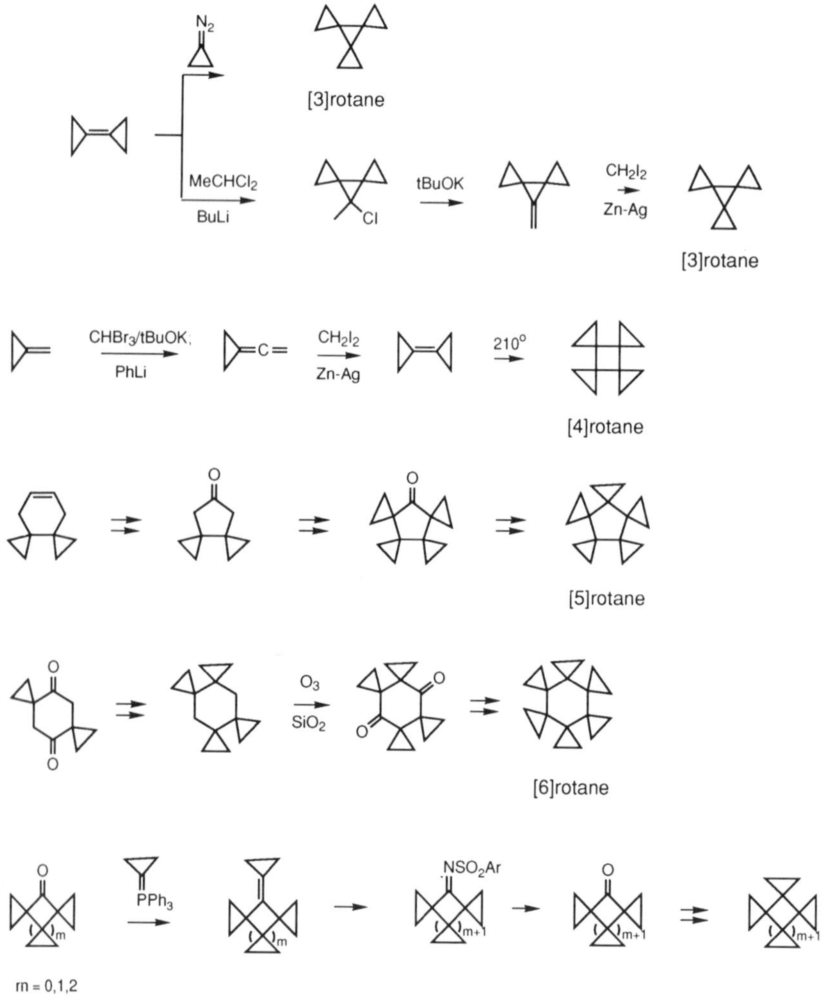

[3]rotane

[3]rotane

[4]rotane

[5]rotane

[6]rotane

rn = 0,1,2

Extension of [3]rotane by spirocyclopropanation leads to another fascinating class of hydrocarbons of branched triangulanes [Zefirov, 1992; Kozhushkov, 1993]. High symmetry is responsible for the rapid construction of rotanes and triangulanes.

The [m.n]coronanes are saturated hydrocarbons possessing a central m-membered ring (m=4, 6, 8 ...) coupled to m peripheral n-membered rings (n=3, 4, 5 ...) such that central ring share edges with each of the peripheral rings. The difficulty in synthesizing [6.5]coronane via the cascade rearrangement of a polyspirane lies in the closure of the final step [Wehle, 1987].

[6.5]coronane

Propellanes are condensed tricyclic compounds with a single skeletal bond shared by all three rings. [1.1.1]Propellane is the most highly strained symmetrical molecule in which the two quaternary carbon atoms have an "inverted" configuration. Its remarkable existence was demonstrated in a synthesis [Wiberg, 1982] starting from 3-oxocyclobutanecarboxylic acid via dichlorocarbene insertion into the central C—C bond of a bicyclo[1.1.0]butane intermediate. A phenyl group was used as a latent carboxyl in the early part of the synthesis and its degradation by ruthenium(VIII) oxidation and esterification returned the molecule to a symmetrical state. Symmetry was further improved on dechlorination. The bridgehead carboxyl groups were transformed into the bromo substituents by a modified Hunsdiecker reaction and cyclization was achieved on treatment of the dibromide with t-butyllithium.

[1.1.1]propellane

1,4-Trimethylene(Dewar benzene) has been obtained from 1,1'-trimethylene-bicyclopropenyl [Landheer, 1975] by rearrangement of another symmetrical molecule.

Many methods have been developed for the synthesis of [3.3.3]propellane. A particularly distinguished route, because of various highly symmetrical intermediates involved, is that via the 3,7,10-trione [Altman, 1969].

1,2-Bis(trimethylsiloxy)-5-methylenecycloheptene has been prepared, ring expanded and eventually converted into [3.3.1]propellane-2,8-dione [Reingold, 1989]. Every intermediate of the synthesis possessed a plane of symmetry.

An intriguing propellane [Kaiser, 1985] is a potential precursor of the fused bis(triasterane). However, by far the most interesting compound is propalla[3_4]prismane [Gleiter, 1988] which has cubane as the core.

propella[3_4]prismane

The efficient transmutation of polyspiro compounds into propellanes [Fitjer, 1988, 1989] by a manifold cascade rearrangement demonstrated the inherent relationship of structural (hence symmetry) nature and reactivity.

Considerable angular strain is expected for the fenestranes. The central atom of the four rings is forced to planarity from a tetrahedral configuration. One of the most easily accessible symmetrical fenestranes is [5.5.5.5]fenestrane, which has four cyclopentane rings sharing a common vertex, emerged from a route featuring carbene insertion [Pfenninger, 1985]. The corresponding tetraene (staura-2,5,8,11-tetraene) [Venkatachalam, 1986] has been prepared using carbonyl condensation reactions as methods of skeletal construction.

The tetrabenzo[5.5.5.5]fenestrane is readily constructed from 1,3-indandione [Kuck, 1986]. The synthetic advantage due to molecular symmetry of centrohexaindane is also evident. This centropolyquinane with six indan units mutually fused around a common carbon atom has been acquired in a few simple steps [Kuck, 1988, 1994].

fenestrindan

The first synthesis of triquinacene [Woodward, 1964], on which the dimerization approach to dodecahedrane was based, is a paradigm of symmetry exploitation in terms of starting material selection and operation. From the pentacyclic diketone onward all the intermediates possess a plane of symmetry, consequently the route was shortened. Degradation occurred at both enantiotopic positions.

The domino Diels–Alder reaction of dihydrofulvalene with diethyl azodicarboxylate via an "exo" approach gave an adduct useful for elaboration of triquinacene [Paquette, 1978]. Thus it required saponification, oxidation, and photolysis to complete the transformation. The Diels–Alder adduct is symmetrical and none of the reactions disturbed this feature.

A step in the direction of "dimerizing" triquinacene is the preparation of a dithiatriquinacenophane [W.P. Roberts, 1981].

Also of significance is a synthesis of the triketotriquinacane [Carcellar, 1985] which did not depend on symmetrical features to abbreviate the reaction sequence. A Pauson–Khand reaction served to form two rings.

2.1.6. Other Interesting Molecules

Semibullvalene referred to above, like hypostrophene, is a fluxional molecule. Its derivation was from the remarkable bullvalene [Doering, 1963]. While the first synthesis of bullvalene [Schröder, 1963] from a cyclooctatetraene dimer indeed furnished strong evidence about the rapid interconversion among the 10!/3 (i.e., 1,239,600) isomers, a rational approach [Serratosa, 1972] via intramolecular cyclopropanation from a tris(diazoketone) also demonstrated the ramification of symmetry in synthesis.

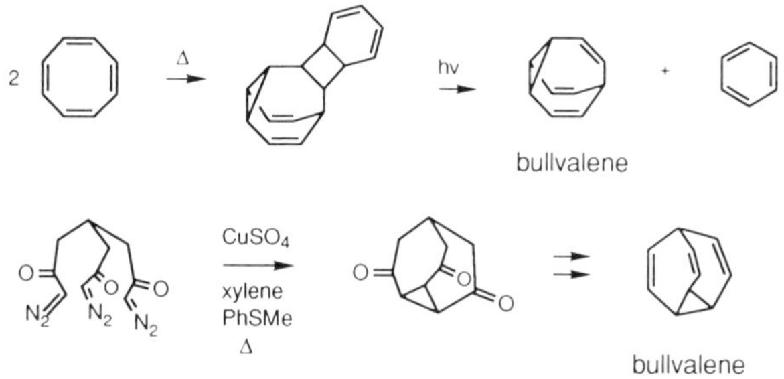

bullvalene

bullvalene

Corannulene is a bowl-shaped molecule. Several recent routes to this compound [Borchardt, 1992; Scott, 1991, 1992] owe their efficiency to adoption of symmetry considerations.

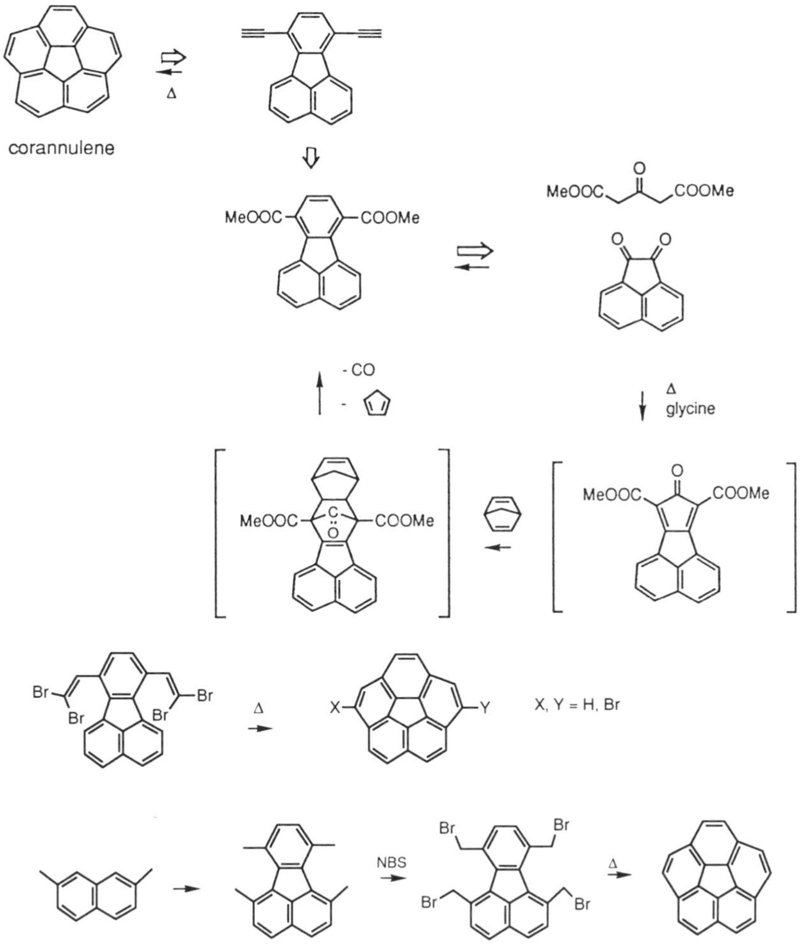

corannulene

Perhaps it should be pointed out that the brexanes have an intriguing symmetry [Nickon, 1965]. A substituent at C-2 is *exo* to one norbornyl unit and *endo* to the other. In fact, 2-*exo*- and 2-*endo*-brexanes are one and the same compound.

A molecular belt in a shape suggested by the side view of its model has been synthesized [Kohnke, 1989; Ashton, 1992] from two building blocks based on two slightly different Diels–Alder schemes. Removal of the oxa bridges and reorganization of the aromatic rings produced [12]collarene, which is made up of six cyclocondensed tetralin units. The collarene is an obvious precursor of the parent, [12]beltene. The high efficiency of the cycloaddition reactions reflects the inherent symmetry of the reactants so that the adduct(s) can be controlled stoichiometrically. It is also important that only the *exo*-side of the dienophile is attacked by the approaching diene.

A molecular Möbius strip with three double bonds to maintain its integrity and help its formation was created by intramolecular alkylation of a molecular ladder [Walba, 1982a, 1985]. Crossover cyclization at high dilutions appears to be as efficient as the collateral reaction which results in the cylindircal topoisomer.

There are two kinds of C_2 conformations for this molecule. In one the C_2 axis passes through the center of a double bond and parallel to the π-orbitals, in the other C_2 conformation one of the double bonds is a the twist and its sp^2-hybridized carbon atoms lie in the axis. The most intriguing aspect of the system is that despite its flexibility and the absence of chiral centers the Möbius strip is chiral.

2.2. NATURAL PRODUCTS

2.2.1. Aliphatic Compounds and Fragments

The carotenoids are pigments of plants (carrot, etc.) and certain animals (e.g., birds and lobster) characterized by the presence of a conjugated polyene chain. The majority of these oligoterpenes are symmetrical, i.e., they have two identical end groups. Consequently, strategies for carotenoid synthesis are well defined; main variations lie in the constitution and assembly of the components or building blocks [Isler, 1963]. In the case of C_{40}-carotenoids the most efficient coupling strategies are those involving symmetrical fragments: $20+20$, $19+2+19$, $18+4+18$, $16+8+16$, $15+10+15$, $14+12+14$, $13+14+13$, $10+20+10$, and $5+30+5$. Except for the $5+30+5$ approach, all of them start from building blocks with preformed trimethylcyclohexene end groups. As expected the Wittig reaction plays a very prominent role in many of the pathways.

A classical route to β-carotene [Isler, 1956] relying on Darzens condensation ($C_{13}+C_1$), aldol condensations via acetals ($C_{14}+C_2$ and $C_{16}+C_3$), and coupling with the di-Grignard reagent of acetylene ($C_{19}+C_2+C_{19}$) may be mentioned. The need for selective semihydrogenation of the symmetrical monodehydro-β-carotene intermediate in this synthesis led directly to the development of the Lindlar catalyst. Finally, isomerization of the (Z)-double bond to the (E)-configuration was achieved simply by heat.

β-carotene

Since the two halves of β-carotene (and many other carotenoids) are joined by a double bond, a most direct approach would be to dimerize two C_{20} building blocks with trigonal terminus, for example, by reductive coupling of the aldehyde [McMurry, 1974].

β-carotene

The constitution of squalene is different in that the C—C bond bisecting the molecule in two identical halves is bisallylic. There are many methods available for accomplishing the union of allylic substrates.

A bidirectional, stereoselective chain extension based on two sets of Grignard reactions and Claisen rearrangements to construct a C_{24} intermediate is a very efficient approach to squalene [W.S. Johnson, 1970]. The $C_{11} + C_8 + C_{11}$ assembly mode that features coupling of homo-geranylmagnesium chloride with 3,6-dichloro-2,7-octanedione and dechlorohydroxylation represents an even shorter route [Cornforth, 1959]. Yet another synthetic route to squalene [Bhalerao, 1971] exploits molecular symmetry by starting from functionalization of the terminal carbon atoms of 2,7-dimethyl-2,6-octadiene and using the dibromide to couple with two molecules of geranyltri-n-butylphosphonium ylide.

squalene

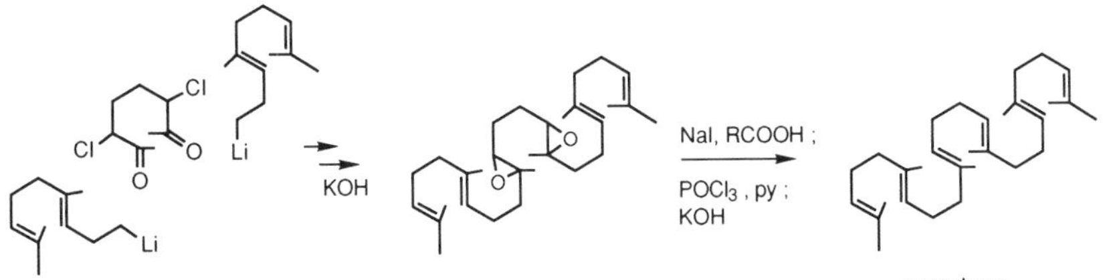

squalene

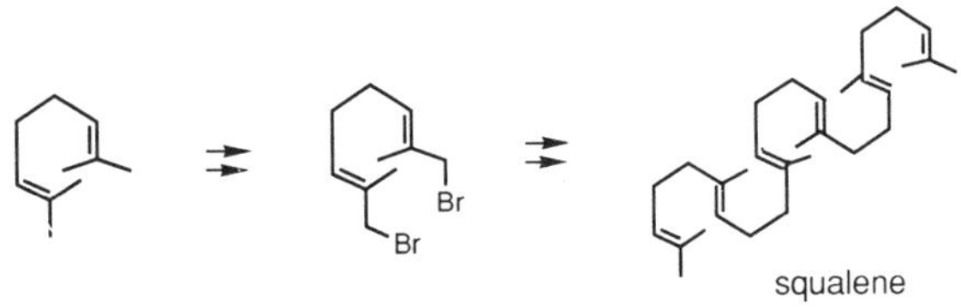

squalene

The C-5/C-15 fragment of streptovaricin embodies nine contiguous asymmetric centers, however, recognition of its C_2-symmetry has greatly simplified its synthetic scheme [Z. Wang, 1990].

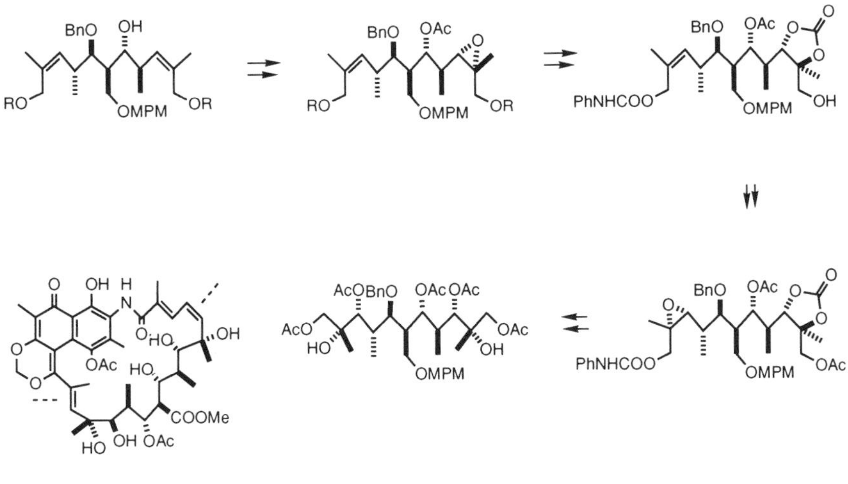

streptovaricin-A

Teurilene is a triterpene with three tetrahydrofuran units and eight asymmetric centers arranged in C_s-symmetry. Its biogenesis, thought to involve a remote Payne rearrangement sequence to establish the tetrahydrofuran array, inspired a synthesis [M. Hashimoto, 1991] involving epoxide opening by a γ-hydroxyl group, although in three stages instead of one.

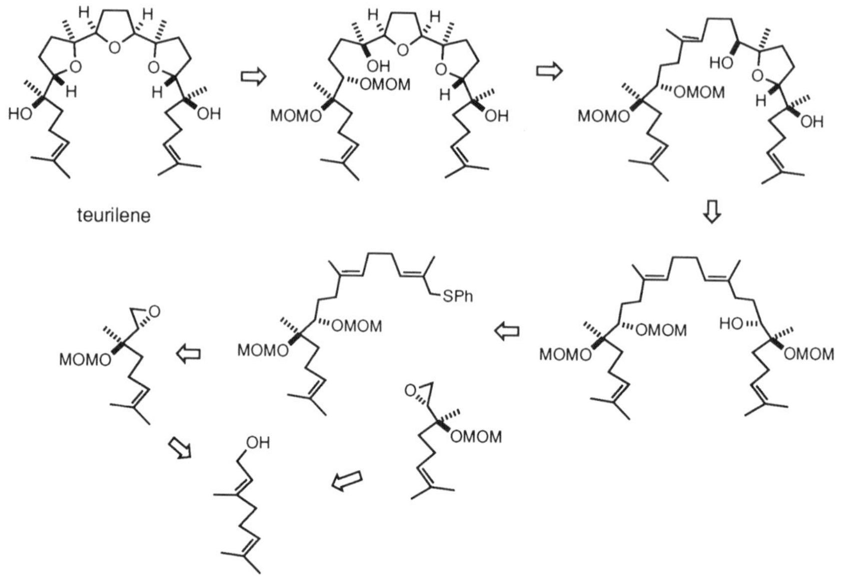

teurilene

With a C_2-symmetry the archaebacterial C_{40}-diol is amenable to convergent synthesis via Kolbe electrolysis [Czeskis, 1993]. Furthermore, the two C_{20} components are readily prepared from a common intermediate.

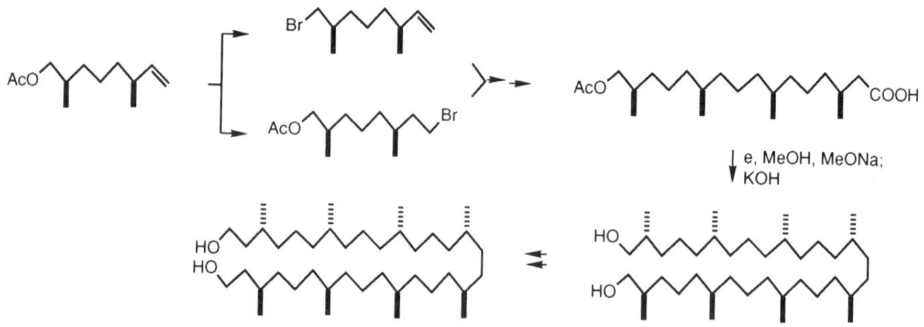

The C_2-symmetric pyrrolidine-197B from a dendrobatid frog has been synthesized from D-mannitol via the (S,S)-diepoxyhexane [Machinaga, 1991a].

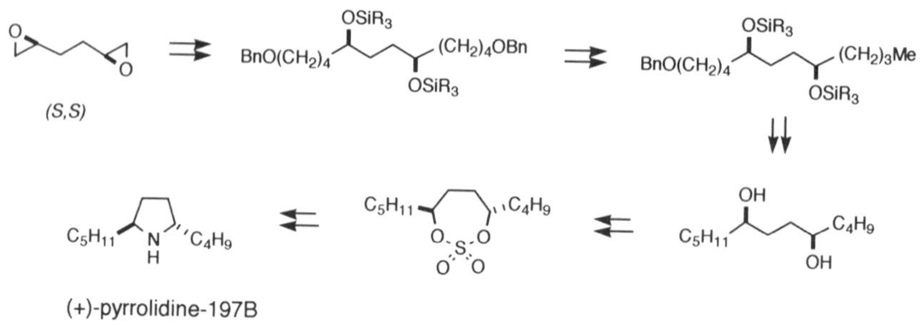

(+)-pyrrolidine-197B

The strategy of elaborating certain aliphatic compounds by cleavage of symmetrical cyclic precursors is also viable, for example, fulvinic acid from a *meso*-cyclobutanone via ring expansion and ring fission [Vedejs, 1984].

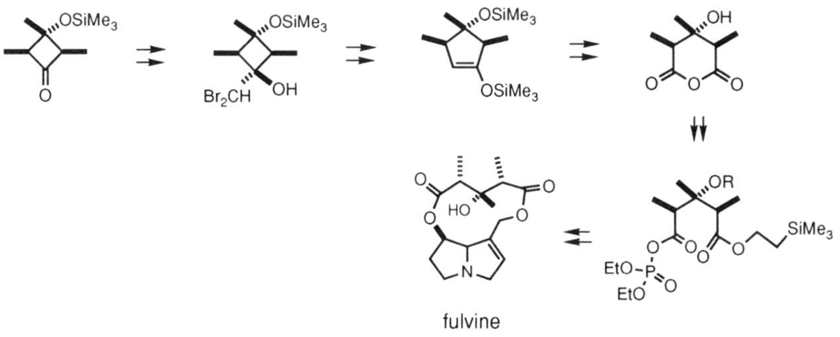

fulvine

2.2.2. Alicyclic Compounds

Civetone is symmetrical about a plane passing through the C=O bond and the center of the (Z)-double bond, therefore α,ω-coupling of a symmetrical linear ketone is the most apparent synthetic approach [Tsuji, 1980a].

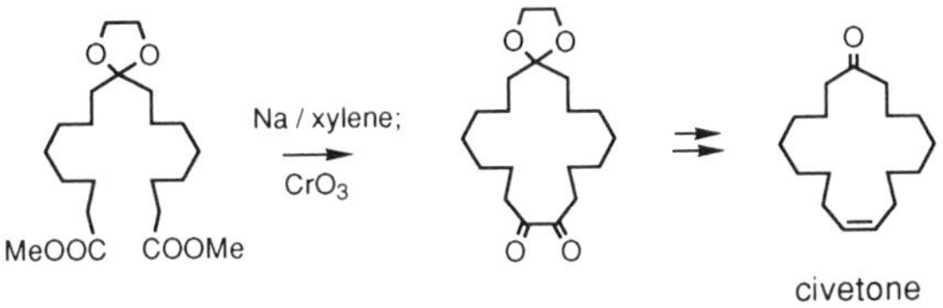

civetone

Other schemes for civetone synthesis include methathesis of di[(Z)-8-heptadecen-1-yl]ketone with a rhenium oxide catalyst [Plugge, 1991] and the one based on Horner–Wadsworth–Emmons reaction of a 1,3-diphosphonylacetone and selective hydrogenation [Büchi, 1979b].

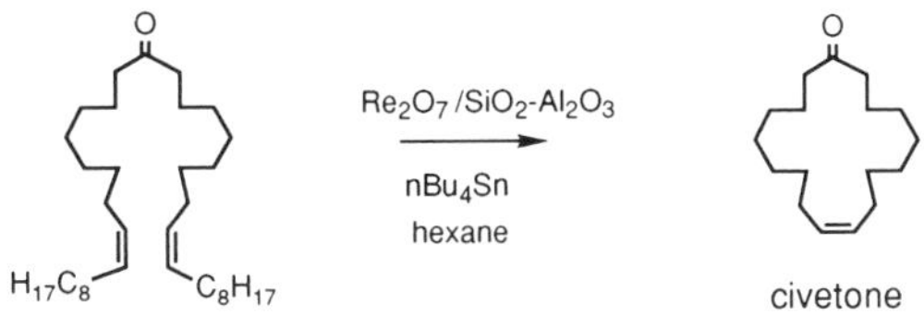

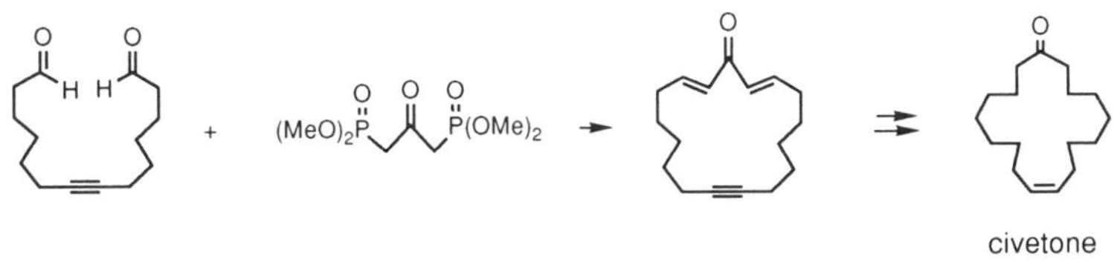

civetone

civetone

Onocerin is a symmetrical tetracyclic triterpene. A total synthesis [Stork, 1963] has made use of Kolbe electrolysis to couple a keto acid to construct the skeleton. Conversion of the resulting diketone to the two methylene units was, however, less straightforward as the Wittig reaction failed.

α-onocerin

Reactivity considerations often determine the use of unsymmetrical molecules to synthesize symmetrical products. For example, it has been demonstrated that the monoterpene tricyclene can be obtained in two steps via bicycloannulation of 3,4,4-trimethyl-2-cyclopentenone with nitroethylene, and Wolff–Kishner reduction [Cory, 1985].

tricyclene

Conversion of camphor into tricyclene [Meerwein, 1920] by oxidation of its hydrazone with yellow mercuric oxide is also a symmetrization reaction.

camphor   tricyclene

A simple synthesis of the C_2-symmetric conduritol-C [Yurev, 1961] involved a Diels–Alder reaction of furan and ethenediol carbonate and hydrolytic cleavage of the heterocycles of the adducts. Streptamine has been obtained [Schwesinger, 1975] in 70–75% yield from 4,5-epoxycyclohexene via the *syn*-benzene triepoxide.

conduritol-C

streptamine

Cantharidin is considered as an alicyclic compound here. This molecule presents synthetic difficulties that defy cursory retrosynthetic analysis. The straightforward synthetic scheme involving Diels–Alder reaction of furan and dimethylmaleic anhydride is not realizable because of mutual unreactivity. Consequently, more circuitous methods were implemented in the early syntheses [Stork, 1953; Schenck, 1953]. However, in keeping with the structural demand of the target molecule, these pathways rarely deviated from symmetrical intermediates.

cantharidin

The initial synthetic proposal has been resuscitated. The development of high pressure techniques for organic synthesis and the modification of the dienophile to the 2,5-dihydrothiophene-3,4-dicarboxylic anhydride enabled its Diels–Alder reaction with furan to occur [Dauben, 1985]. Thus the major cycloadduct was converted to cantharidin by simultaneous hydrogenation and desulfurization. The high pressure condition may be replaced by 5 M lithium perchlorate in ether [Grieco, 1990].

(major)   cantharidin

α-Carophyllene alcohol, an acid-catalyzed rearrangement product of the monocylic sesquiterpene humulene, has a plane of symmetry. Considering that the hydroxyl group can be derived from a carbocation precursor, a synthetic scheme involving cationic rearrangement would exude great appeal. A more highly strained substrate would facilitate the rearrangement and render it unidirectional. The rationale for a 1,2-rearrangement devolves a 5:4:6-fused tertiary alcohol as a proper precursor, migration of a cyclobutane bond relieves large amounts of ring strain.

It is now a matter of devising a route to such an unsymmetrical precursor [Corey, 1964]. One involving photocycloaddition gained focus on account of the limited methodologies for cyclobutane formation. The stereochemistry of the target compound is supportive of the approach in which an *exo* transition state for the cycloaddition is favored. The *endo* transition state suffers severe nonbonding interactions.

Because of reactivity requirements an enone must be used as one of the photocycloaddends. Naturally the other addend is the symmetrical 4,4-dimethylcyclopentene. A striking feature of the synthesis is the molecular symmetrization by rearrangement.

(major)

caryophyllene alcohol

2.2.3. Alkaloids

Eilatin, an unusual marine natural product, has a symmetrical skeleton. A very short synthesis of the compound has been developed [Gellerman, 1993].

eilatin

The reductive cyclization of papaverine to give pavine [Goldschmidt, 1886; Battersby, 1955] is another example of the symmetrization process. Interestingly, treatment of homoveratraldehyde with iodomethane led to the oxa analog of pavine [Jung, 1981].

pavine

A fascinating synthesis of α-isosparteine [Oinuma, 1983] called for reaction of piperideine oxide with 4*H*-pyran to furnish a 2:1 adduct, and hydrogenation.

α-isosparteine

The discovery of an efficient [4+3]-cycloaddition involving conjugated dienes and oxyallyl cations has had important ramifications in the synthesis of cycloheptanones, tropones, and related substances. A measure of its synthetic value should include the entry to 8-oxa- and 8-azabicyclo[3.2.1]oct-6-en-3-ones, an entry which employs furans and pyrrole derivatives as the 4-electron addends [Noyori, 1978]. The alkaloid scopine is thus readily assembled.

scopine

A better-known construction of the tropinone ring system is that based on a biogenetic conjecture. Thus, Mannich reaction of methylamine, succindialdehyde, and acetonedicarboxylic acid indeed afforded tropinone [Robinson, 1917]. A yield of >90% can be reached when the reaction is carried out under *physiological* conditions [Schöpf, 1937].

tropinone

An efficient method for accessing precoccinelline and congeners [Stevens, 1979] is by extension of the Robinson–Schöpf reaction. Thermodynamic factors favor the formation of the symmetrical perhydrophenalenone intermediate with ring juncture stereochemistry corresponding to the first target molecule.

precoccinelline

Tropinone has also been obtained by a Dieckmann cyclization of dimethyl *cis*-pyrrolidine-2,5-diacetate, followed by removal of the ester group [Willstätter, 1921; W. Parker, 1959]. Other routes to tropinone (and cocaine) consisted of reductive cleavage of a 7-azabicyclo[2.2.1]heptane-2,3-dicarboxylic ester [Krapcho, 1985], alkylation of the dianion of 1,4-bisbenzenesulfonylbutane to give a methylenecycloheptane which was subjected to ozonolysis and treatment with methylamine [Lansbury, 1990]. Note that all these syntheses took advantage of the symmetry elements of the target molecule.

The twofold Michael reaction approach has been extended to a synthesis of lobelanine [W. Parker, 1959].

lobelanine

Two different methods for the elaboration of O-protected 3,5-cycloheptadienol into tropane bases such as φ-tropine [Iida, 1984b], scopine and φ-scopine [Schink, 1991] are available. These routes involved desymmetrizing the 1,4-functionalization of the diene unit and in a later step an intramolecular displacement to form the heterocyclic system.

scopine

φ-scopine

A synthesis of arcyriaflavin-A [Bergman, 1989] initiated by a Diels–Alder reaction and completed by a twofold Fischer indolization is well conceived.

R = SiMe$_3$

arcyriaflavin-A

2.2.4. High-Symmetry Compounds Composed of Asymmetric Units

Antibiotic-593A is a 3,6-disubstituted 2,5-dioxopiperazine amenable to synthesis by dimerizing condensation. Using an α-amino-β-lactam building block such a synthesis has been realized [T. Fukuyama, 1980], but because the β-lactam was racemic both cis- and trans-2,5-dioxopiperazines were obtained. This problem did not occur in the elaboration of amauromine [Takase, 1986] starting from L-tryptophan. However, diastereomeric compounds differing in the configuration of the benzylic carbon atoms did arise.

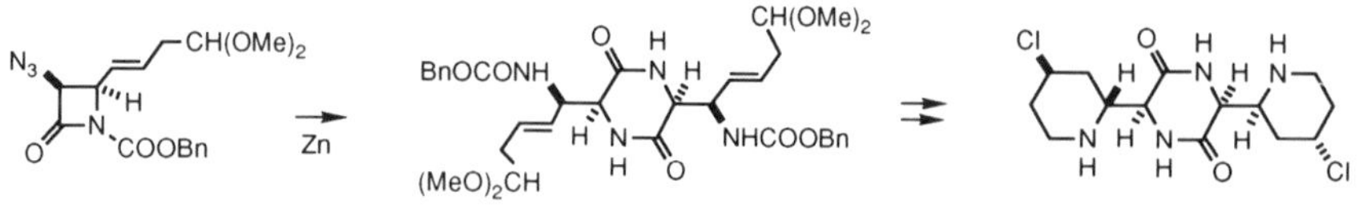

antibiotic 593A

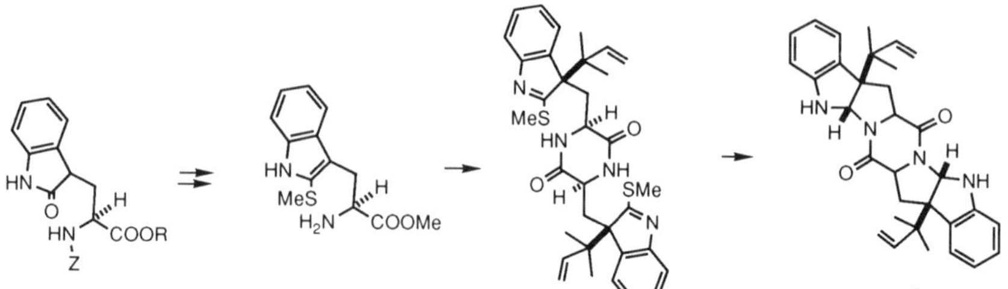

amauromine

A symmetrical 2,5-dioxopiperazine served as a template for the transacylative ring expansion approach to serratamolide [Shemyakin, 1964]. Most of the effort in the synthesis of the condensed 2,5-dioxopiperazine (−)-bipolaramide [Somei, 1989] was actually spent in the hydroxylation step. Note the brevity of such routes is due to molecular symmetry.

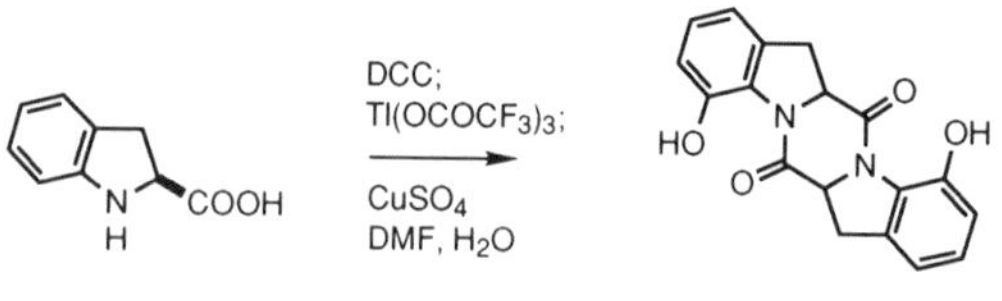

(-)-bipolaramide

A dimerizative approach to (S,S)-palythazine [Jarglis, 1982] is evidently an exploitation of the molecular symmetry.

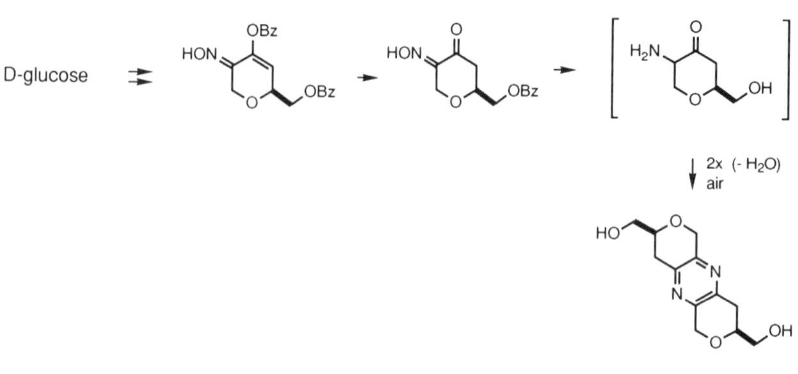

(S,S) -palythazine

The identicalness of two tautomers represented by parazoanthoxanthin-A and the consideration of its biosynthesis led to the cyclodimerization method of its synthesis [Braun, 1978]. Interestingly, the less symmetrical skeleton of pseudozoanthoxanthin was obtained when the primary amino group of at C-2 of the imidazole precursor was protected as a benzamide.

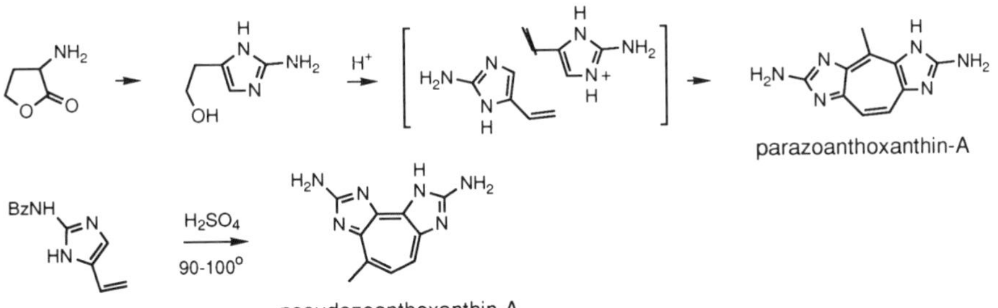

parazoanthoxanthin-A

pseudozoanthoxanthin-A

A straightforward approach to orelline and orellanine [Tiecco, 1986] by coupling of the proper pyridine derivative is evident. Similarly, symmetrical neolignans with the dibenzocyclooctadiene skeleton are readily prepared by dimerization of two C_3 units followed by intramolecular aryl coupling [Schneiders, 1981; Takeya, 1985]. Variants of this scheme include the elaboration of biaryls [Ghera, 1978; J. Xie, 1983] and closure of the eight-membered ring at a later stage.

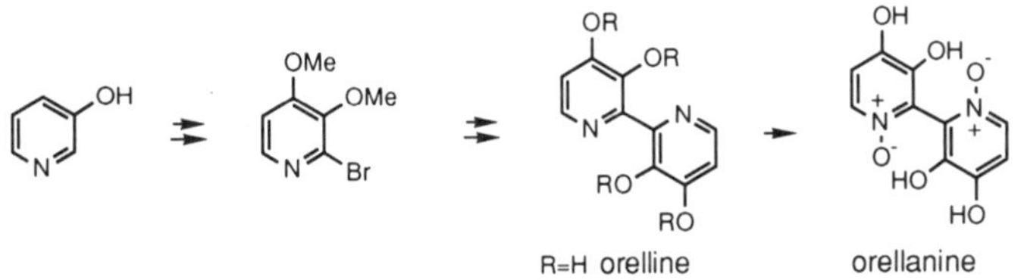

R=H orelline   orellanine

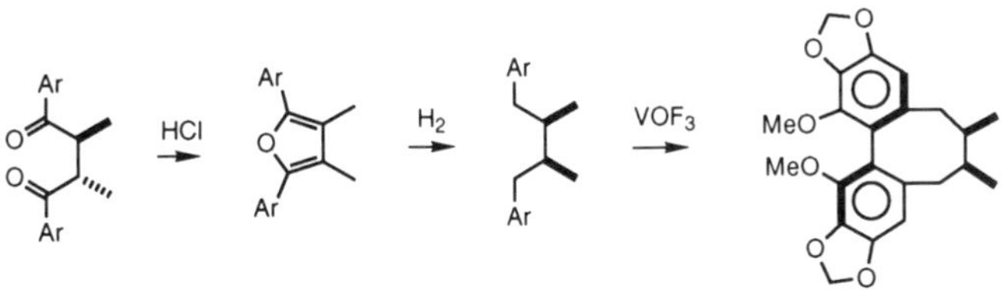

wuweizisu-C

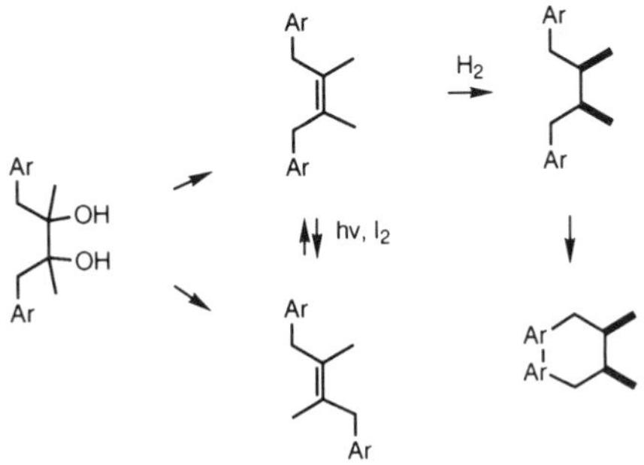

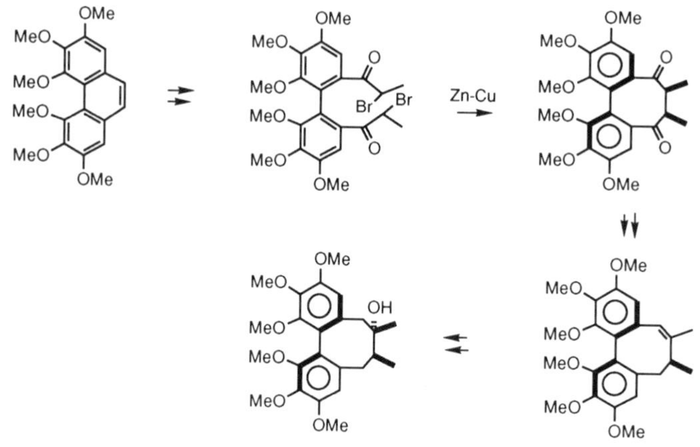

schizandrin

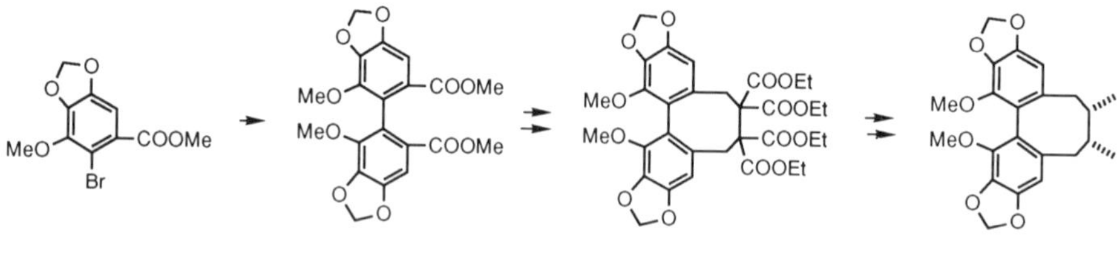

schizandrin-C

The dimeric Calycanthaceous alkaloids have not yet succumbed to conventional assembly. On the other hand, biomimetic oxidative dimerization of tryptamine derivatives proved effective in accessing chimonanthine and calycanthine [Hendrickson, 1964; Hall, 1967].

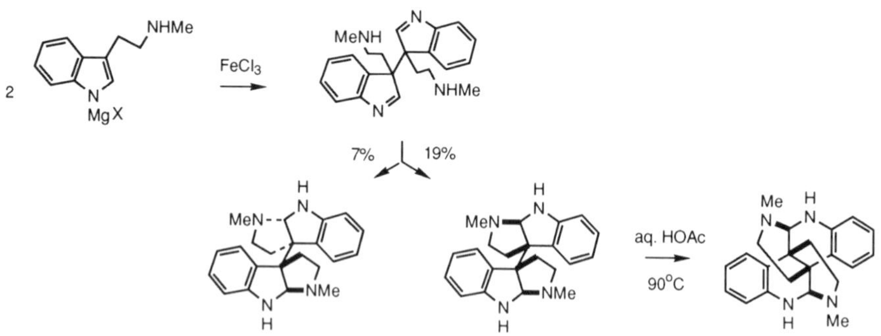

meso-chimonanthine   *rac*-chimonanthine   calycanthine

The problem of indigo synthesis [Baeyer, 1882; Flimm, 1890; Salmony, 1905] is frequently the same as that of indoxyl. In fact indigo is not a natural product; it is an artifact from enzymatic splitting of the glucoside of indoxyl and autoxidative dimerization of the latter compound. Related biomimetic dimerization routes to trichotomine [Iwadare, 1978] and caulerpin [Maiti, 1978] are readily conceived.

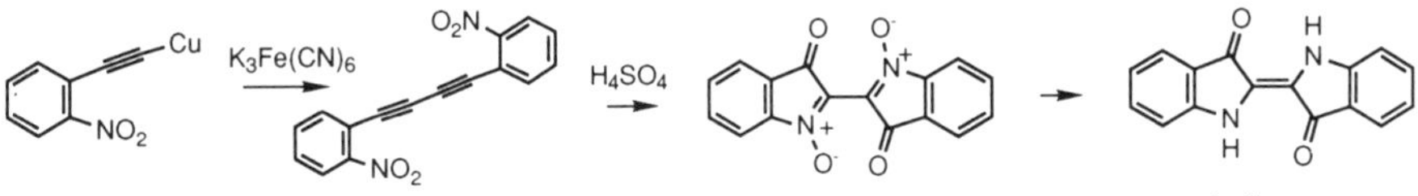

indigo

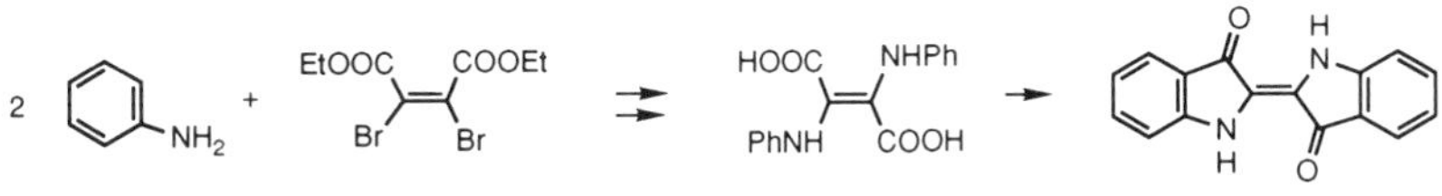

indigo

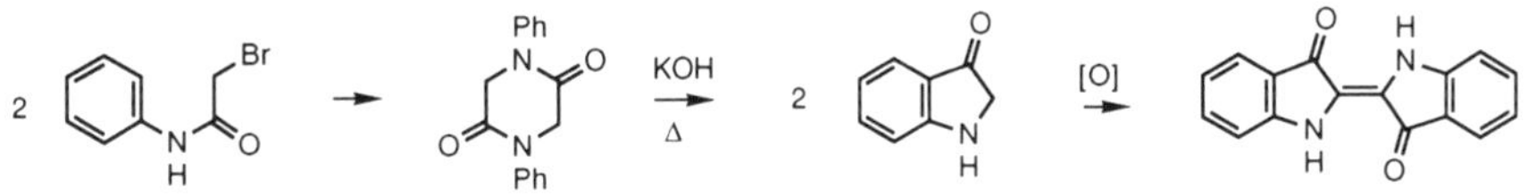

indigo

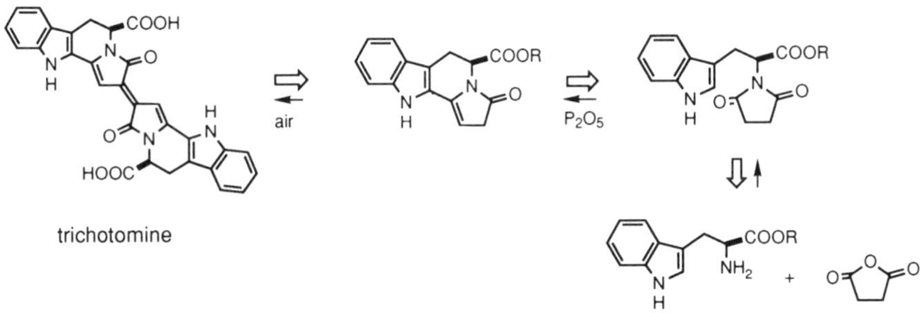

trichotomine

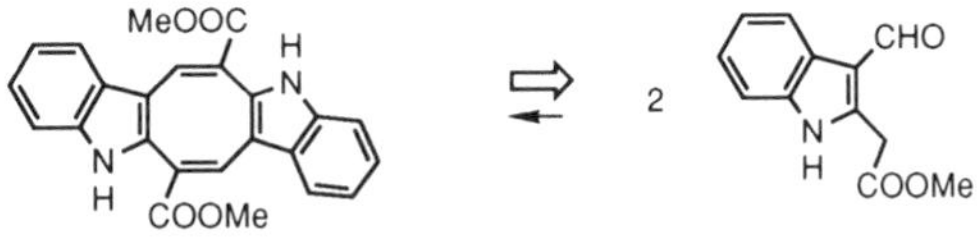

caulerpin

An approach to sesamin [Orito, 1991] took advantage of the symmetry element so that modifications in both rings of the bicyclo[3.3.0]octane-3,7-dione were carried out simultaneously. This two-site promotion is also suitable in the synthesis of certain 3,3′-biflavones [Zhu, 1988].

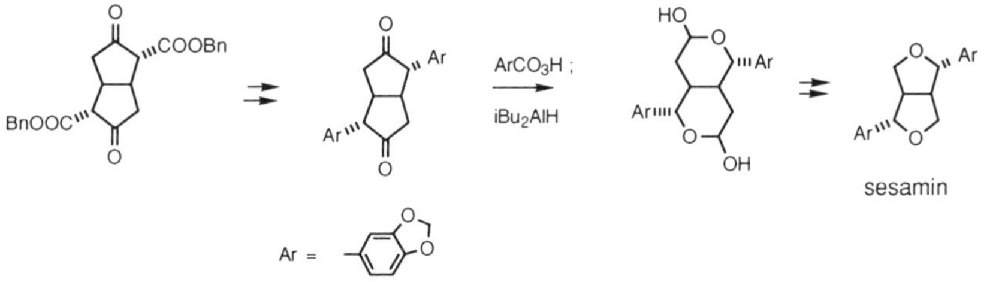

sesamin

Ar =

3,3'-biflavones

Interestingly, the fused dibutenolide-type fungal pitments are accessible from the symmetrical 2,5-diaryl-3,6-dihydroxy-1,4-benzoquinones [Lohrisch, 1986]. The latter compounds can be prepared by treatment of grevillin analogs with methoxide ion.

A number of fungal metabolites are symmetrical or nearly symmetrical perylenequinone derivatives. In synthetic terms these compounds should be most expediently prepared by coupling of two naphthalene subunits [Diwu, 1992] or biphenyl precursors with two identical halves [Chao, 1988]. The total synthesis of phleichrome and calphostin-A [Coleman, 1994] has been accomplished on the basis of such a strategy.

Both *meso*- and DL-limatulone have been obtained via alkylation of a sulfone with a bromide of the same skeleton [K. Mori, 1993b]. The sulfone was prepared from the bromide, therefore the symmetry of these target molecules greatly simplified the synthetic operation.

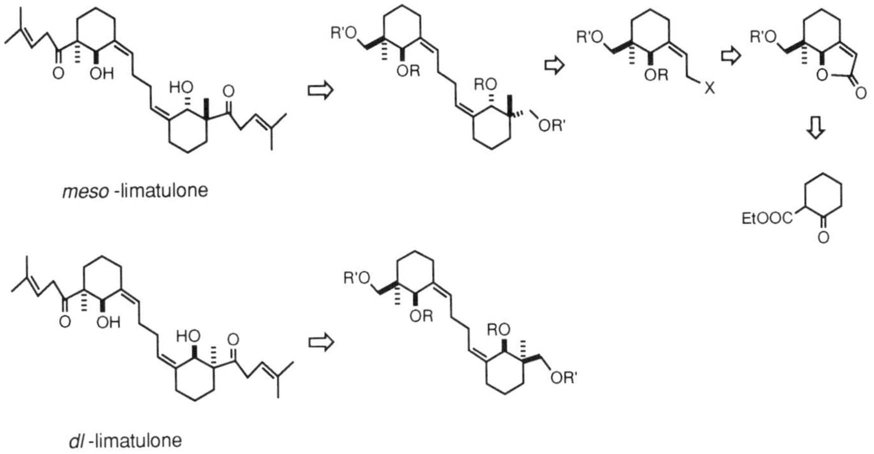

meso-limatulone

dl-limatulone

The interconvertibility of the DL-hazimycin factor-5 and the *meso*-hazimycin factor-6 tends to simplify their synthesis. A most direct method [Wright, 1982] involved oxidative coupling of N-formyltyrosine methyl ester with horseradish peroxidase in the presence of hydrogen peroxide. Equilibration of the two factors occurred during ammonolysis.

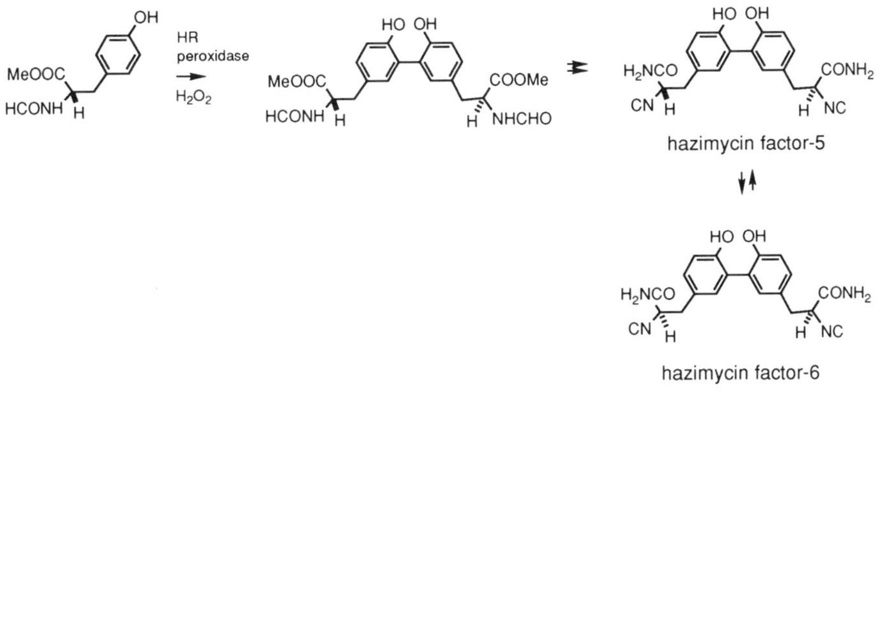

hazimycin factor-5

hazimycin factor-6

3

TWO-CARBON AND THREE-CARBON BUILDING BLOCKS

3.1. TWO-CARBON BUILDING BLOCKS

Symmetrical 1,2-disubstituted epoxides have been frequently employed in synthesis for the purpose of chain elongation. The symmetry dictates formation of one product by attack of a nucleophile, as shown by the role of an (E)-stilbene oxide in a synthesis of reticuline [Hirsenkorn, 1990].

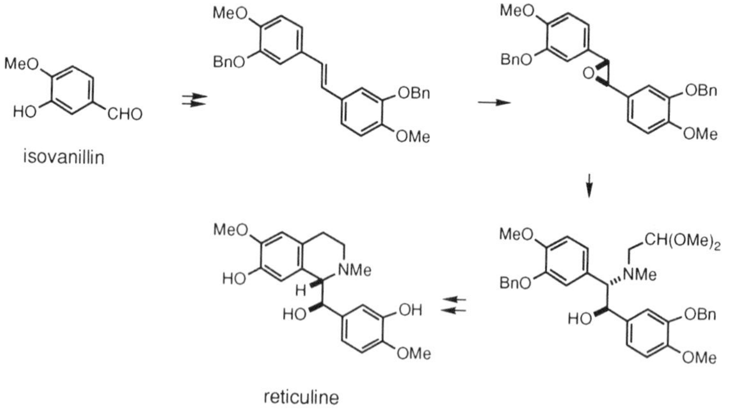

isovanillin

reticuline

The very useful building unit, C_2-symmetric 2,3-epoxybutane, is also mentioned here. For example, the (S,S)-isomer has been integrated into the dimethylpyran ring of the spiroacetal systems of (+)-milbemycin-β_3 [Street, 1985] and lacrimin-A [Takle, 1990].

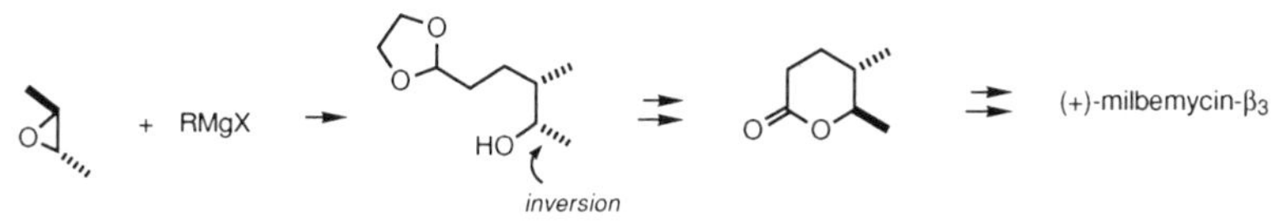

Oxalic esters are excellent two-carbon building blocks with numerous uses. The few examples in the following serve only to illustrate their synthetic potential. An early application is the formation of a cyclopentane precursor of camphor [Komppa, 1903] by a double Claisen condensation. Similar annulation was also witnessed in a synthesis of the E prostaglandins [Katsube, 1971; Sih, 1975].

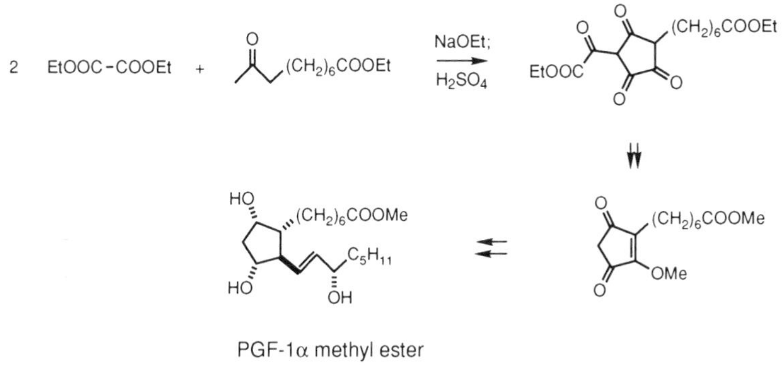

PGF-1α methyl ester

In a synthesis of 5-methoxy-canthin-6-one [Matus, 1985] the condensation of ethyl oxalate with harman completed the cyclic system.

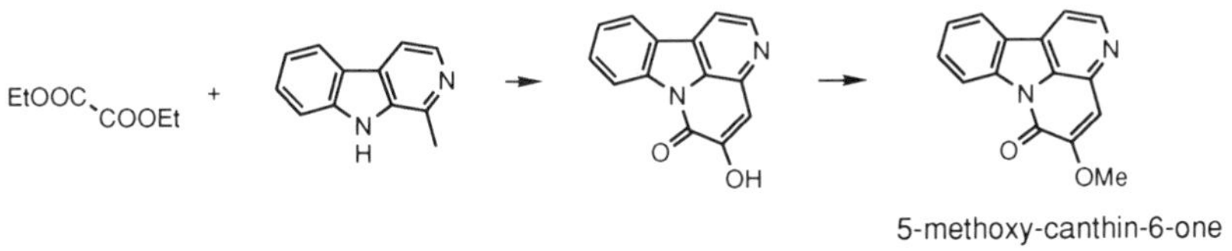

5-methoxy-canthin-6-one

The use of diethyl oxalate as an oxalyl anion by conversion into the bisenoxy silane derivative [Reetz, 1984] may be mentioned.

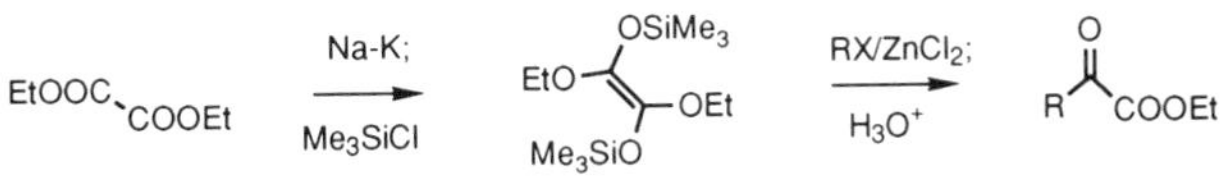

1,3-Dioxol-2-one undergoes certain Paterno–Büchi and Diels–Alder reactions. Syntheses of apiose [Araki, 1981] and conduritol-D [Criegee, 1957] took advantage of this property.

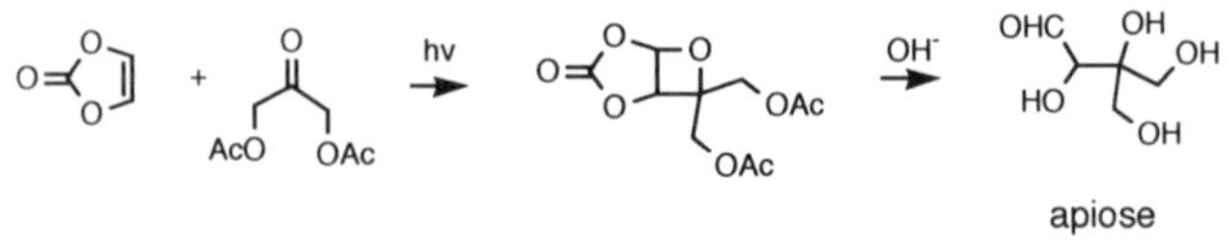

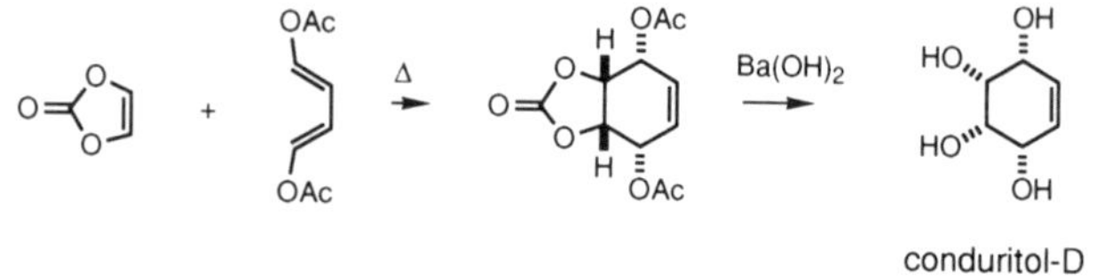

apiose

conduritol-D

1,2-Bisphenylsulfonylethene also behaves as a dienophile. Thus in a synthesis of ibogamine [Herdeis, 1989, 1991] desymmetrization started in the cycloaddition and the subsequently modified bicyclic intermediate underwent elimination of phenylsulfinic acid to enable introduction of the ethyl group.

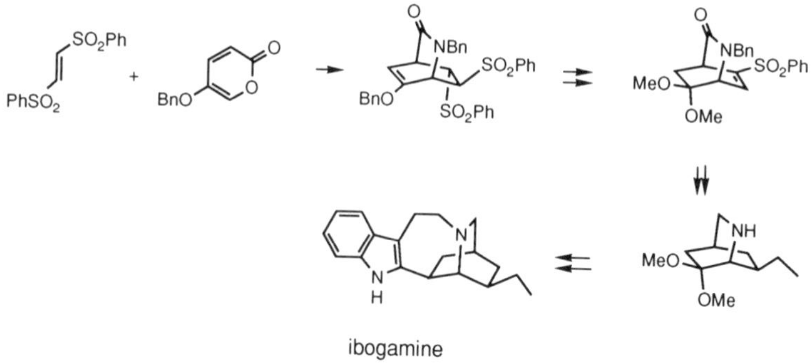

ibogamine

Very elegant synthetic approaches to rapamycin [Nicolaou, 1993] and the ABC-ring system of dynemicin-A [Shair, 1994] were based on coupling of bis(iodoalkene)s with (E)-1,2-bis(tri-n-butylstannyl)ethene and (Z)-1,2-bis(trimethylstannyl)ethene, respectively.

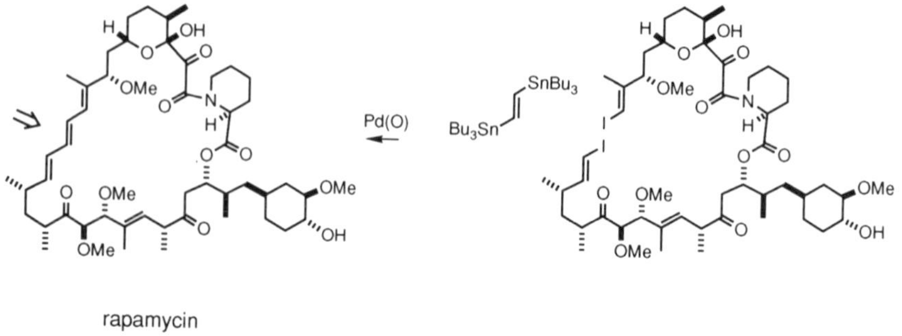

rapamycin

dynemicin-A

3.2. THREE-CARBON CHAINS

Building blocks comprising a three-carbon chain have been under rapid development in recent times, particularly those having one or more substituents at the central carbon and identical end groups which are differentiated by simple manipulations. Interchange of the end groups of such *meso* compounds results in their enantiomeric forms. Chiral compounds may be synthesized from the specified isomers.

Cursory examination of many compounds is sufficient to identify whether local symmetry affects their synthesis which would benefit from choosing symmetrical building blocks. The following delineates some applications of those containing a three-carbon chain.

3.2.1. 2-Methyl-1,3-propanediol: Its Analogs and Synthetic Implications of Desymmetrization

With respect to synthetic utility the most thoroughly explored three-carbon main chain system is the desymmetrized 2-methyl-1,3-propanediol [Banfi, 1993a]. The possibility of enantiomeric interconversion of this family of compounds by relatively simple protecting group manipulations is a particularly attractive feature for their application. Another attribute is the availability of the two enantiomeric monobenzyl ethers from a combination of biotransformation and chemical reactions in a few steps from isobutyric acid [Branca, 1977; K. Mori, 1985b]. Alternatively, desymmetrization of the *meso* compounds, such as that realized by acetylation of the diol with lipases (e.g., of *Pseudomonas fluorescens*) and lipase-catalyzed hydrolysis of its diesters, preferably the dibutanoate, gives rise to enantiomeric monoesters [Santaniello, 1990; Wirz, 1990].

A purely chemical method for accessing the (R)-form of 2-methyl-1,3-propanediol monotrityl ether is by acetalization of (−)-menthone with 2-hydroxymethyl-2-propenol, hydrogenation of the resulting acetal, opening of the heterocycle, tritylation, and reductive cleavage of the C—O bond [Harada, 1987a].

Synthons derived from modification of the unprotected hydroxyl group into sulfonate or halide undergo cross-coupling with various nucleophiles, enabling incorporation of the chiral isobutylene unit into synthetic intermediates of lacrimin-A [Takle, 1990], the milbemycins [Kocienski, 1987; Ley, 1989], ionophore A-23187 (calcimycin) [Evans, 1979], and the immunosuppressant FK-506 [A.B. Smith, 1989].

In a route to (−)-muscone [Z.F. Xie, 1988] the sulfonate was first converted into a nitrile and the initially protected hydroxyl group (as THP ether) was transformed into the electrophilic halide. (See Suginome [1987] or Dowd [1991] for an alternative ring expansion protocol.)

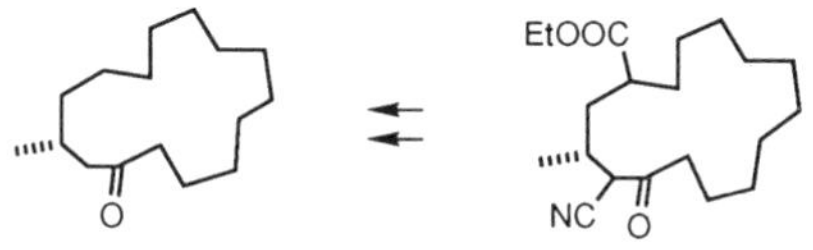

(-)-muscone

With the availability of the halides, umpoled organometallic species are readily formed. The applications include synthesis of the side chain of α-tocopherol [Cohen, 1976; Hird, 1989; K. Mori, 1991a, b], and the C-7/C-9 fragment of erythronolide-A [Kinoshita, 1986].

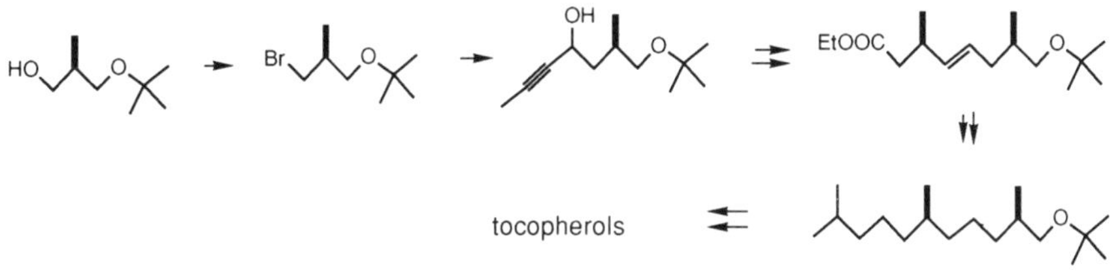

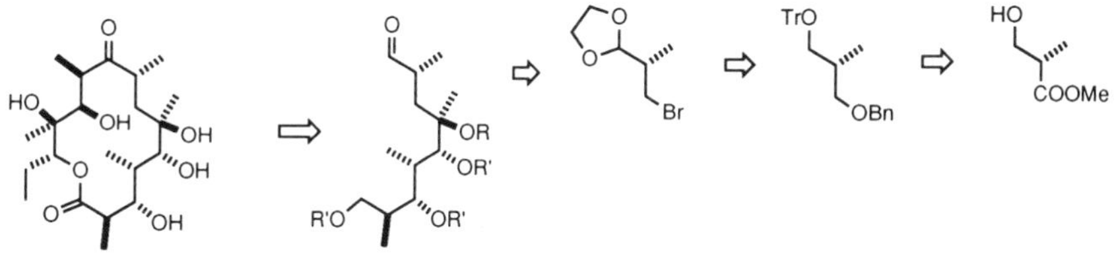

tocopherols

erythronolide-A

A Grignard reagent has been used in the displacement of a vinylic sulfone with retention of the double bond configuration during a synthesis of (+)-milbemycin-β₃ [Kocienski, 1987], while the coupling of a borane coupled with a vinylic bromide (Suzuki reaction) furnished the C-10/C-19 fragment of amphidinolide-A [O'Connor, 1989].

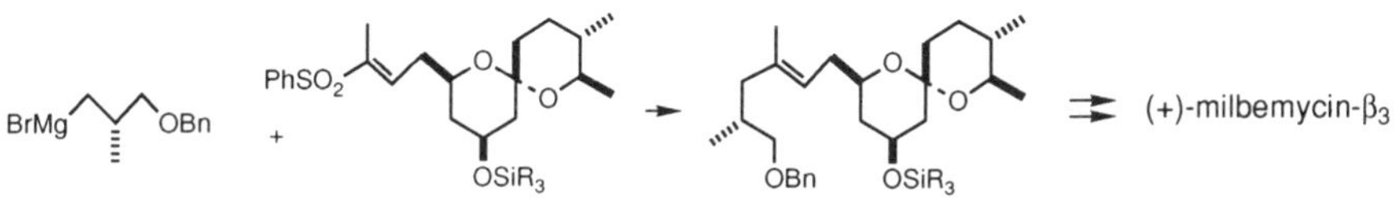

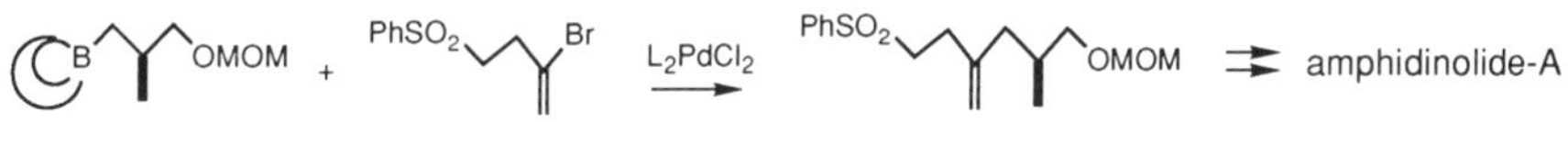

(+)-milbemycin-β₃

amphidinolide-A

Reactions of the chiral organometallic derivatives with oxygenated compounds including hydroxamate, aldehyde, and epoxides were exploited in the construction of portions of complex molecules such as palytoxin [Klein, 1982], FK-506 [D.R. Williams, 1988; T.K. Jones, 1990], and tylonolide [Marshall, 1992].

Proper manipulation of desymmetrized 2-methyl-1,3-propanediols enables construction of unsaturated carbon chain segments with a distinct methyl substituent in the allylic position, using Wittig, Julia, and related reactions. Such methods are very valuable for approaches to (+)-paliclavine [Kozikowski, 1984a], irumamycin [Akita, 1988], aplysiatoxin [Ireland, 1988], Prelog–Djerassi lactone [Yokoyama, 1991], the ansa bridge of rifamycin-S [Kishi, 1981; Nagaoka, 1981], and the botryococcenes [Hird, 1989]. It is of interest to note that the aldol reaction with the tetrahydropyranyl ether of (R)-β-hydroxyisobutryaldehyde was used in the A-ring construction of monensin [Collum, 1980].

After desymmetrization by conversion of one of the terminal carbon atoms to the aldehyde level, chain extension involving reaction with allylmetals (especially crotylmetals) can lead to homoallylic alcohols of defined stereochemistry. Syntheses based on this methodology include that of phyllanthocin [Trost, 1991], calcimycin [Boeckman, 1991; D.R. Williams, 1989], rifamycin-S [Roush, 1990], rapamycin [Fisher, 1991], and tirandamycin-B [Shimshock, 1991].

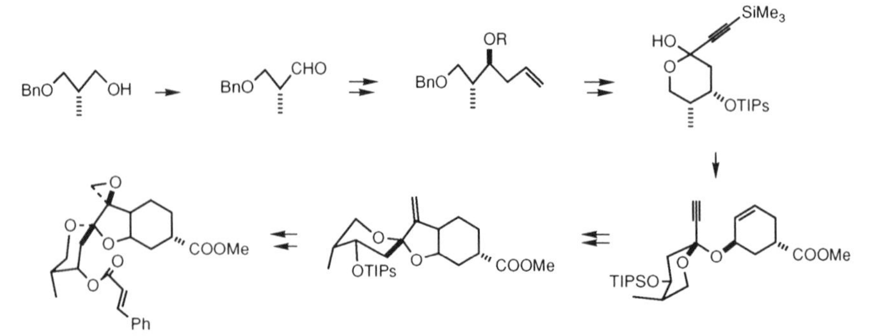

(+)-phyllanthocin

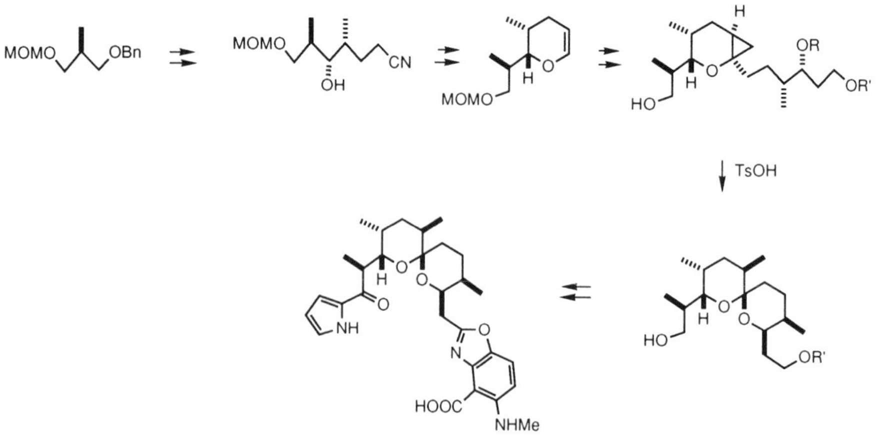

calcimycin

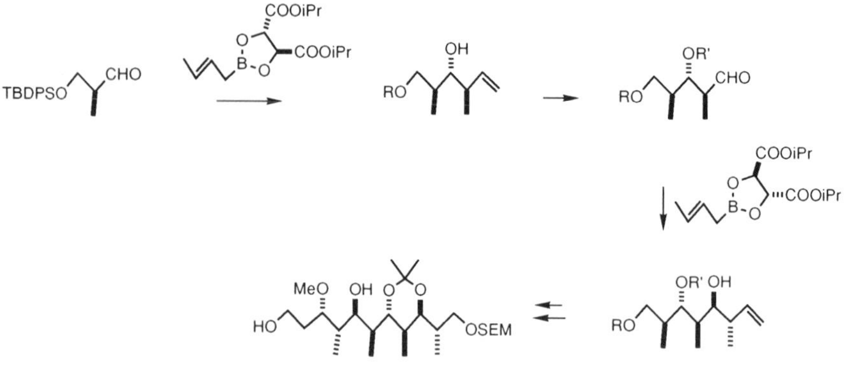

rifamycin-S
C-19/C-29 synthon

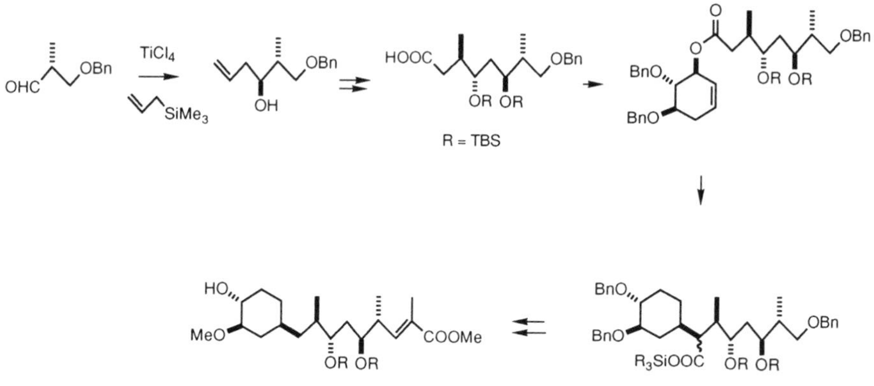

rapamycin
C-28/C-49 synthon

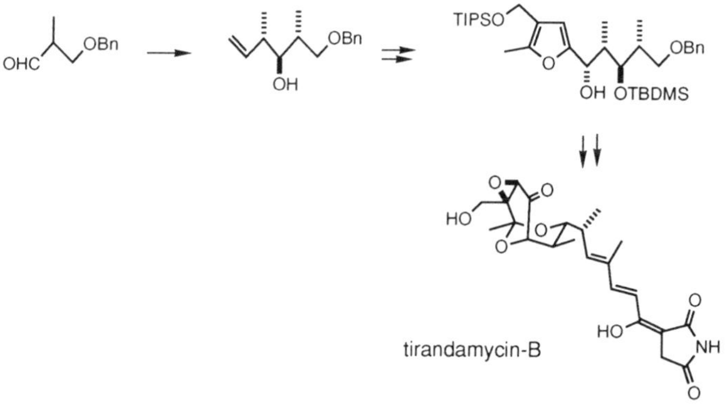

tirandamycin-B

Polypropionate chains having defined stereochemistry have also been assembled by cyclocondensation of the chiral aldehyde(s) with activated dienes, as illustrated in a route to rifamycin-S [Danishefsky, 1987a].

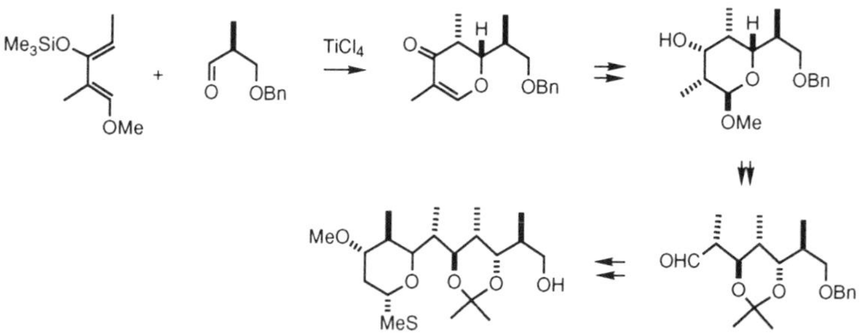

rifamycin-S
C-19/C-29 synthon

2-Ethyl-1,3-propanediol has been used in a synthesis of dihydrocorynantheol [Ihara, 1987]. Its desymmetrized chiral form, obtained from a monomenthyl malonate, was similarly converted into protoemetinol [Ihara, 1988]. The tetrahydropyranyl ether precursor was formed by an intramolecular free radical cyclization.

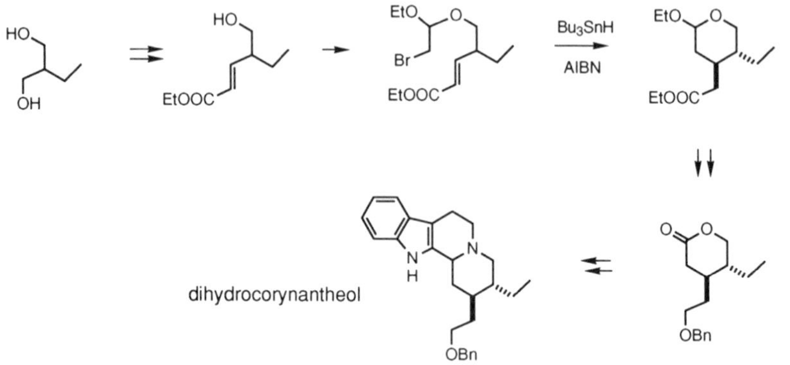

dihydrocorynantheol

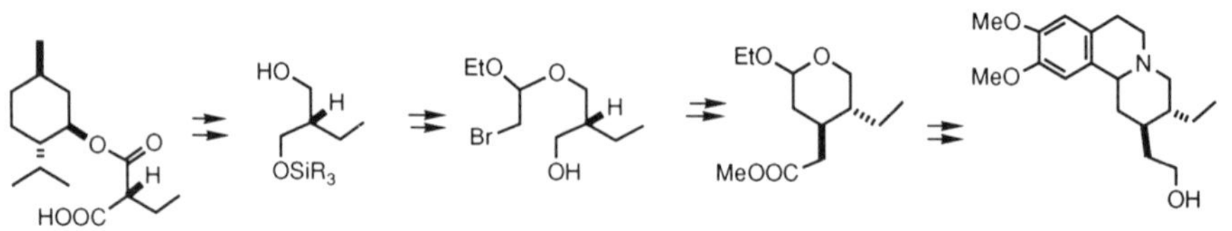

protoemetinol

A few other 2-substituted 1,3-propanediols have been exploited in synthesis. Usually these compounds undergo desymmetrization by lipase-catalyzed hydrolysis of the diacetates [Guanti, 1989, 1991c]. One such compound was employed in the elaboration of paraconic acid [K. Mori, 1989a] and the microbial growth factor, (−)-A-factor [Y.-F. Wang, 1984a], another in the construction of the C-11/C-17 fragment of tylonolide [Guanti, 1991a], and a key intermediate of carbapenems [Banfi, 1993b] via elaboration of the isopentenyl substituent into a desymmetrized tris(hydroxymethyl)methane.

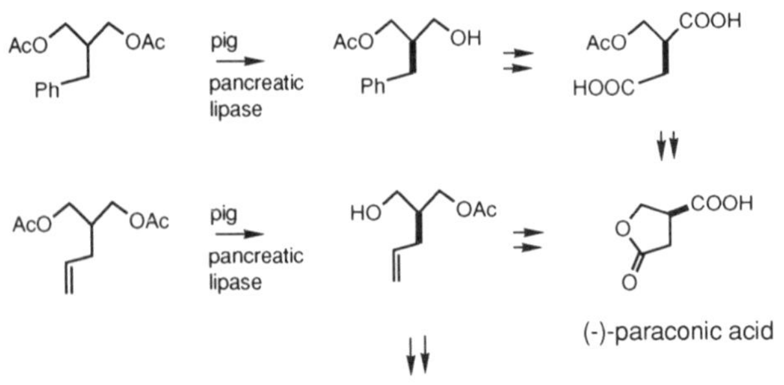

(-)-paraconic acid

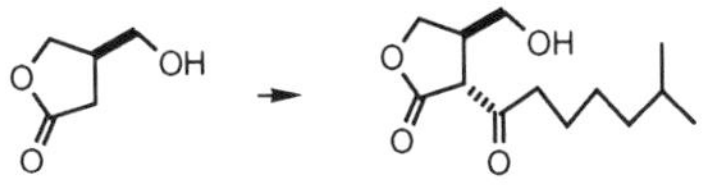

(-)-A factor

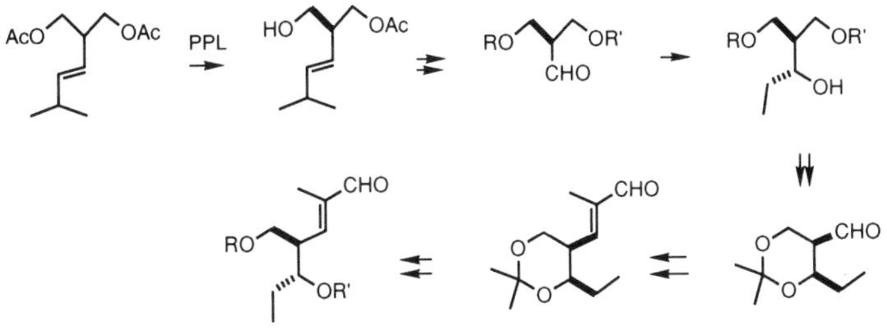

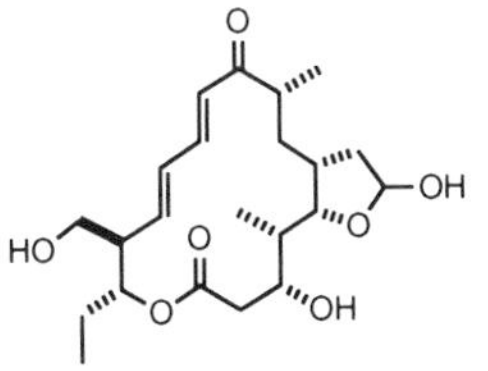

tylonolide

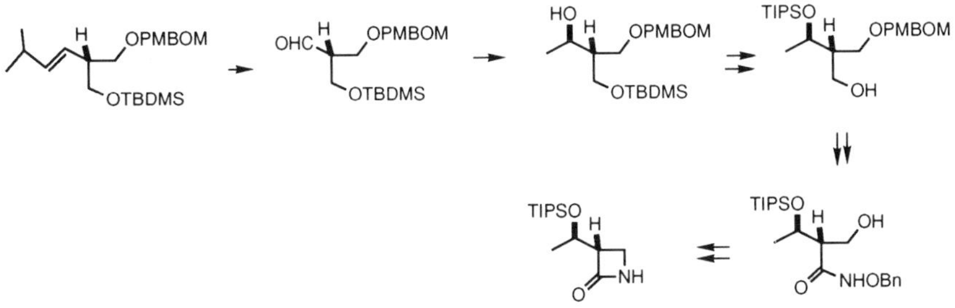

It is of interest to note that allylation of suitably protected β,β'-dihydroxyisobutyraldehydes exhibits diastereoselection as controlled by the protecting group(s) [Guanti, 1991b].

Underlying a synthetic approach to talaromycin-B [Kay, 1984] is desymmetrization of tris(hydroxymethyl)methane via acetalization, elaboration of the unprotected hydroxyl group, and subsequent C—O bond cleavage.

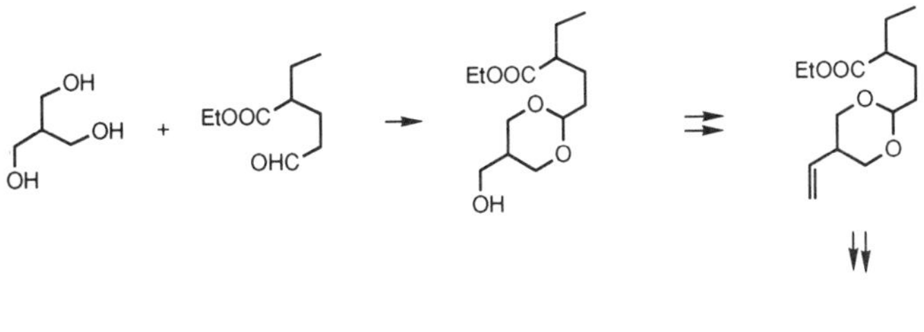

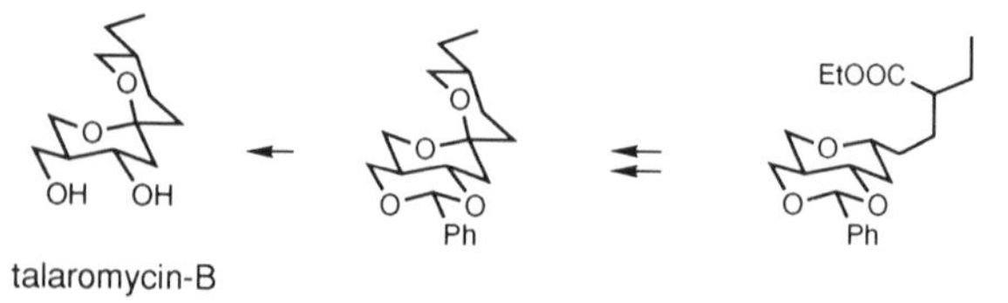

talaromycin-B

Interestingly, two symmetrical acetals of 2-substituted 1,3-propanediols were united by a 1,3-dipolar cycloaddition to give an intermediate of talaromycin-B [Kozikowski, 1984b]. When the ketopentaol was liberated, desymmetrization occurred by functional interactions which resulted in the most stable isomer.

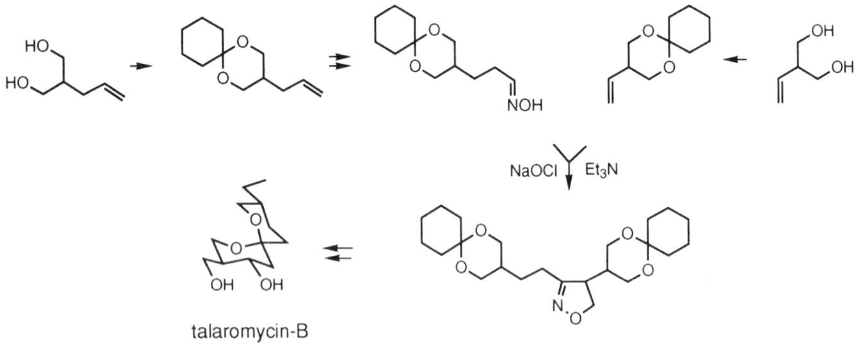

NaOCl / Et₃N

talaromycin-B

Formation and ring cleavage of 2-phenyl-1,3-dioxanes constitute a convenient way for desymmetrization of 1,3-diols, as the monobenzyl derivatives are easily cleaved after the desired transformation of the exposed alcohol. Thus a building block for monensin [T. Fukuyama, 1979] was obtained via such manipulation.

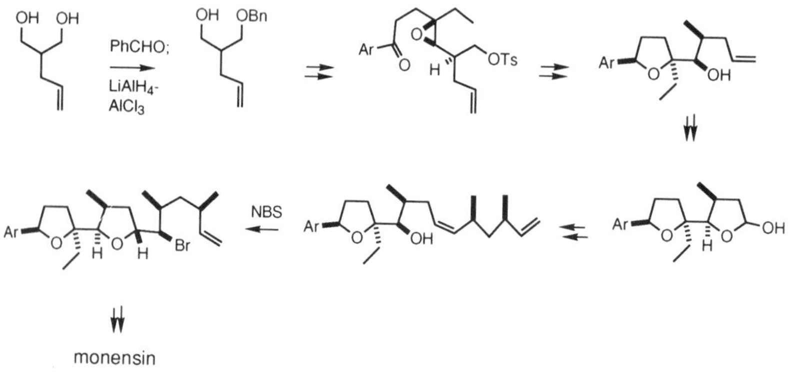

monensin

A less expedient desymmetrization method for 2-allyl-1,3-propanediol involves iodoetherification in the presence of a chiral tartrate ester [Kitagawa, 1992]. After benzylation of the free hydroxyl group, reversal of the first reaction serves to generate the monobenzyl ether.

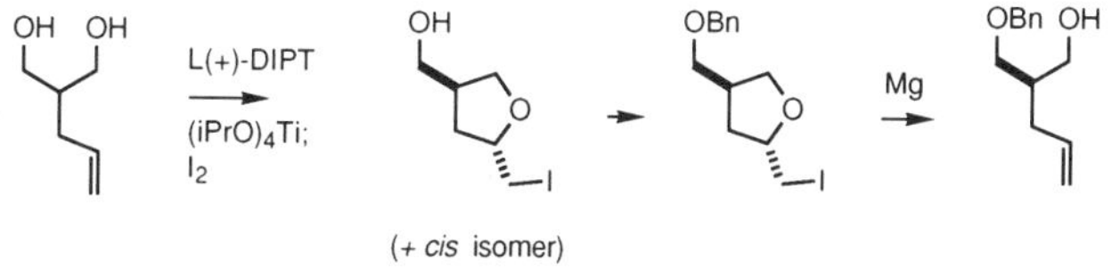

(+ cis isomer)

Less common 2-substituted 1,3-propanediols that have been considered and explored as building blocks include the aryl derivative for duocarmycin-A [Fukuda, 1990], and a steroidal component that lacks the terminal C_5 unit in the side chain of (20S)-25-hydroxycholesterol [Kurek-Tyrlik, 1988]. Functionalized long-chain analogs for the synthesis of invictolide and a fragment of ionomycin [Schreiber, 1985] were desymmetrized, taking advantage of spiroacetal formation involving one of the hydroxyl groups. This again illustrates internal protection giving rise to diastereotopic selection at a prochiral carbon center.

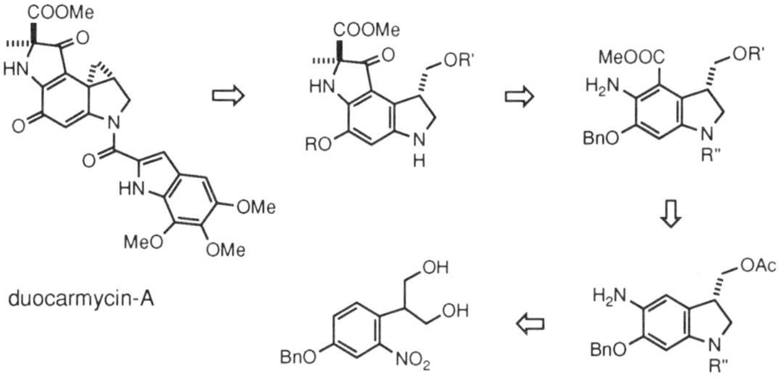

duocarmycin-A

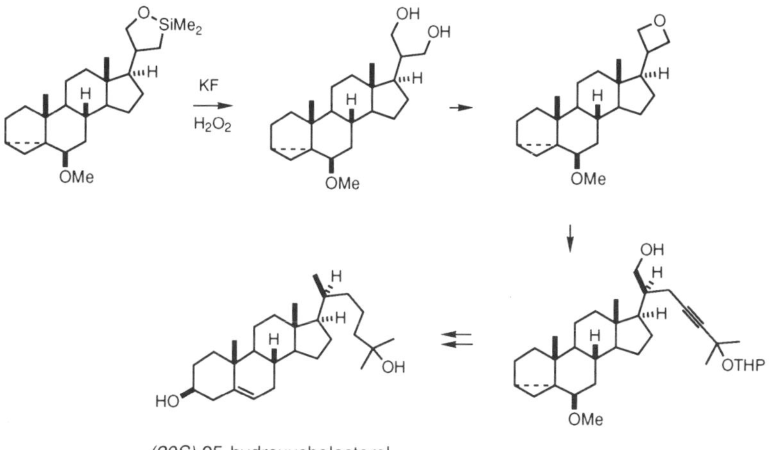

(20S)-25-hydroxycholesterol

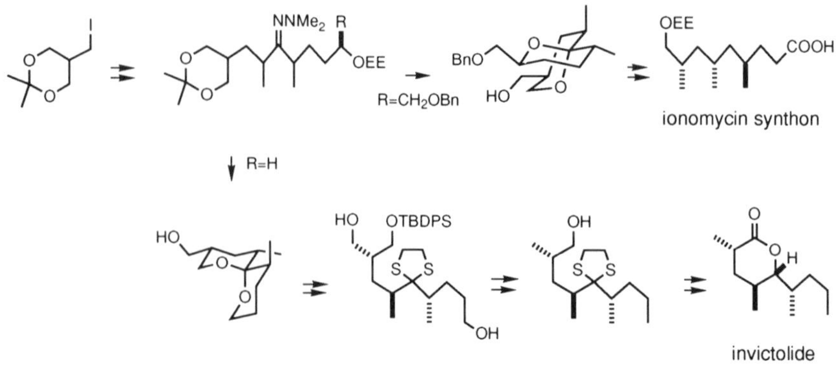

ionomycin synthon

R=H

invictolide

The bicyclic acetal component of (−)-secodaphniphylline has been synthesized [Stafford, 1990] from the acetonide of 2,2-bishydroxymethylpropionic acid by conversion into a chiral acetylenic alcohol. Hydration of the triple bond, which was accompanied by transacetalization, led to two bicyclic acetals (ratio 5:1); the major stereoisomer was converted into the required acid chloride.

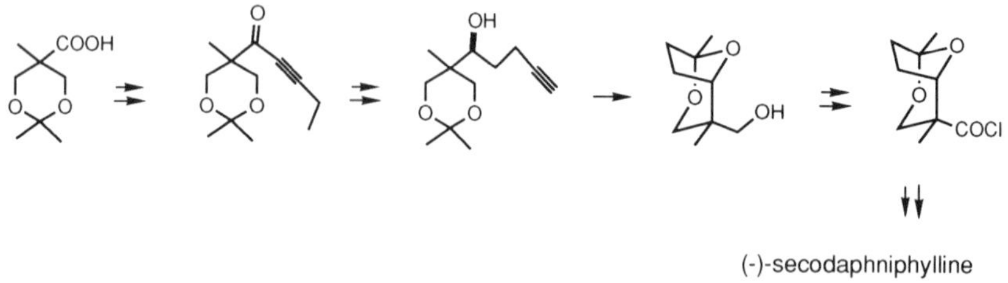

(-)-secodaphniphylline

Glycerol is readily desymmetrized via a stannane derivative of the 2-benzyl ether [M.H. Marx, 1985] by reaction with benzoyl chloride. The product was used in a synthesis of platelet activating factor. 1-O-Benzyl glycerol has been investigated as a promising precursor of fecapentaene-12 [Gunatilaka, 1983]. Enzymatic hydrolysis of the 1,3-diacetate of 2-O-benzylglycerol affords useful chiral building blocks [Breitgoff, 1986; Kerscher, 1987].

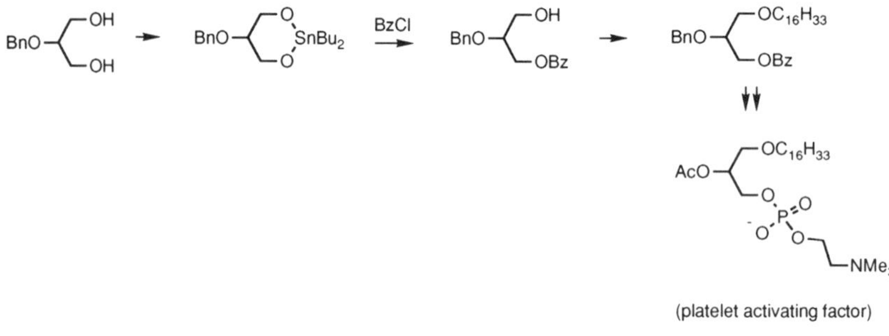

(platelet activating factor)

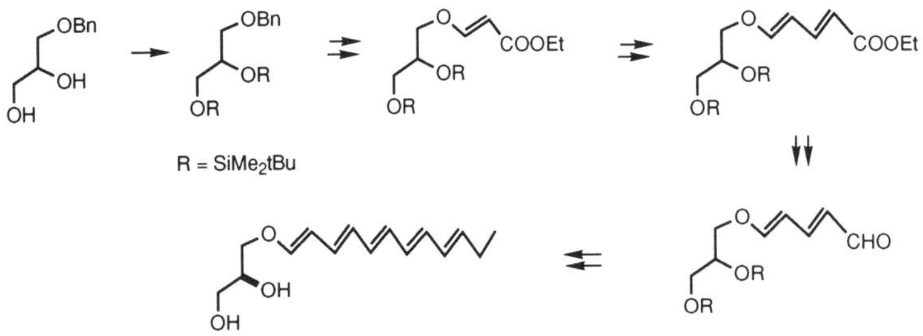

R = SiMe₂tBu

fecapentaene-12

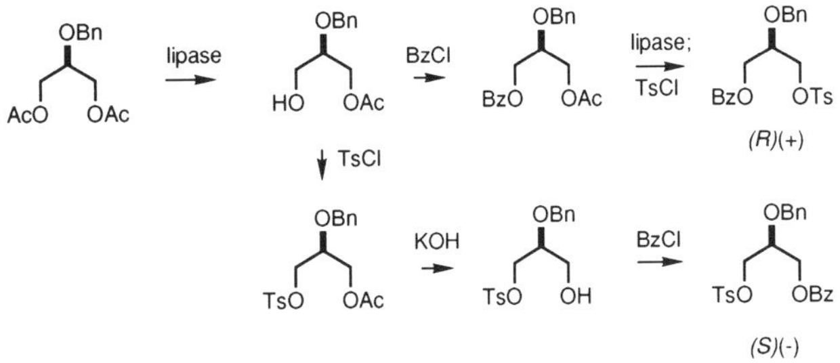

By far the most versatile building blocks representing desymmetrized glycerol are (S)-2,2-dimethyl-1,3-dioxolane-4-methanol [Takano, 1987d] and O-benzylglycidol [Takano, 1989g, 1990b, 1991b; Hanson, 1991]. Both enantiomers of the latter are available by simple transformations of the former and from L-serine. Ultimately, asymmetric epoxidation of allyl alcohol [Hanson, 1986] may prove to be the most economical method for the access of chiral glycidol.

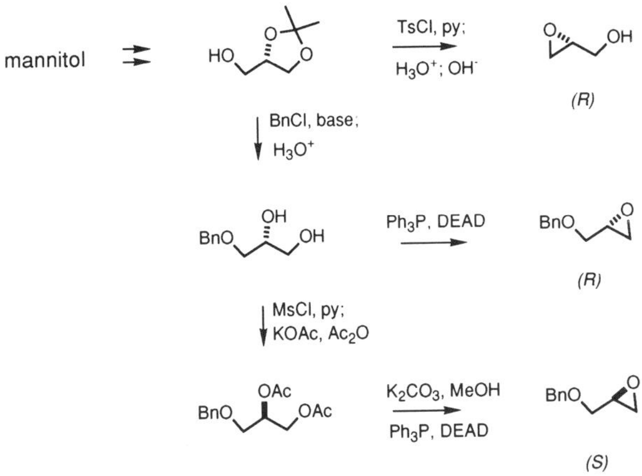

Some of the synthetic utilities of O-benzylglycidol are clearly shown below. From the (R)-isomer: mesembrine [Takano, 1981a], (−)-physostigmine [Takano, 1982a], (+)-nuciferol [Takano, 1982b], (+)-citronellal [Takano, 1987a], clavulone-II [Nagaoka, 1984], *ent*-latifine [Takano, 1985a], monoterpenes such as santolinatriene, lavandulol, rothrockene, and chrysanthemol [Takano, 1985b], (−)-kainic acid [Takano, 1988a], acromelic acid-A [Takano, 1987c], acromelic acid-B [Takano, 1989h], 4-hydroxyproline [Takano, 1988b], (−)-gloeosporone [Takano, 1988f], fragments of amphotericin and mycoticin [Takano, 1988e], protomycinolide-IV [Takano, 1992b], sulcatol [Takano, 1983a], (+)-muscarine [Takano, 1989c]. From the (*S*)-isomer: insect pheromones including serricornin and the spiroacetal portion of the milbemycins [Takano, 1989d], lipoxin-A_5 [Nicolaou, 1987], a three-carbon unit in the side chain of (−)-desmosterol [Takano, 1983c], (−)-anisomycin [Takano, 1989b], (+)-ramulosin [Takano, 1989e], the lactone portion of compactin [Takano, 1989f], polyol fragments of polyene antibiotics [Lipshutz, 1984] and amphotericin-B [Solladié, 1987a]. The utilization of both enantiomers in the synthesis of a molecule is especially satisfying, such as in the assemblage of the C-9/C-27 segment of milbemycin-K [Takano, 1992a]. Other notable achievements include the synthesis of (−)-vincadifformine [Kuehne, 1985] and an approach to leukotriene LTB_4 [Nicolaou, 1984] starting from other chiral glycidol derivatives.

(-)-mesembrine

(-)-physostigmine

(+)-nuciferol

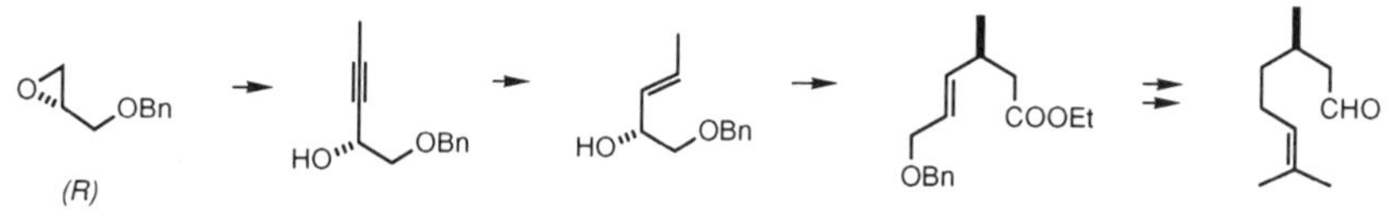

(+)-citronellal

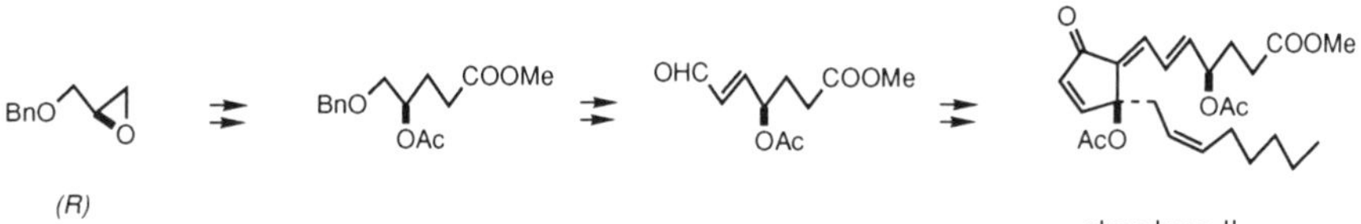

clavulone-II

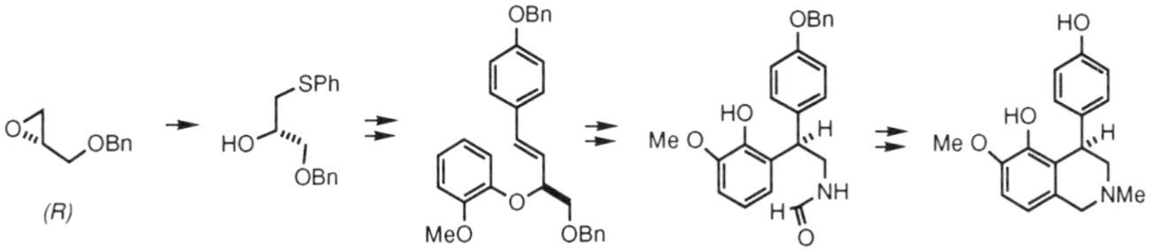

ent-latifine

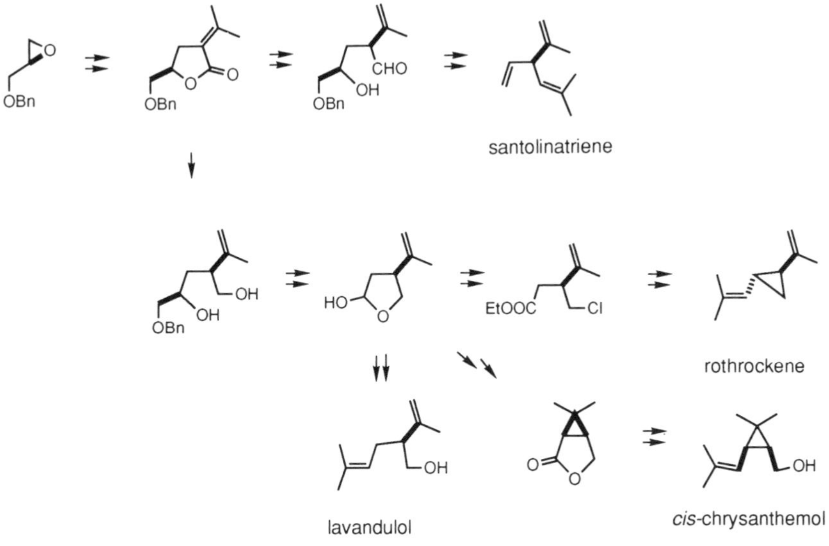

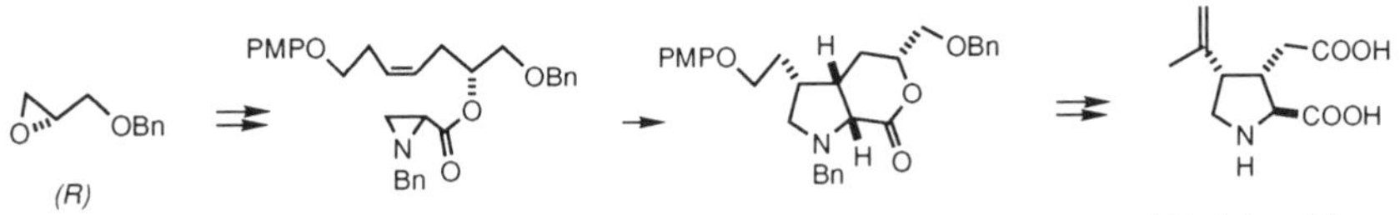

santolinatriene

rothrockene

lavandulol

cis-chrysanthemol

(-)-kainic acid

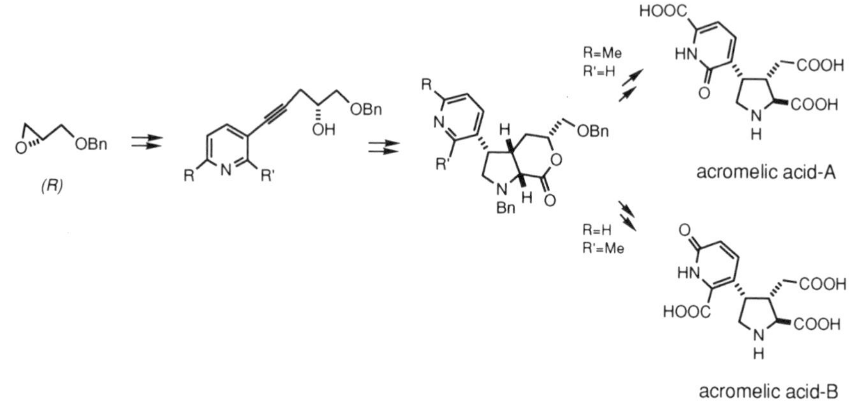
(R)
R=Me
R'=H
R=H
R'=Me
acromelic acid-A
acromelic acid-B

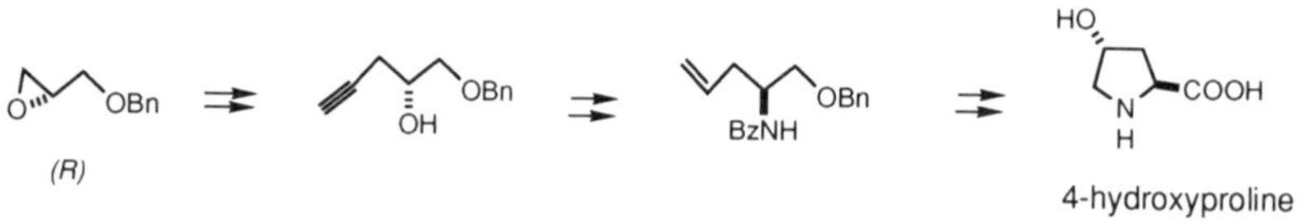
(R)
4-hydroxyproline

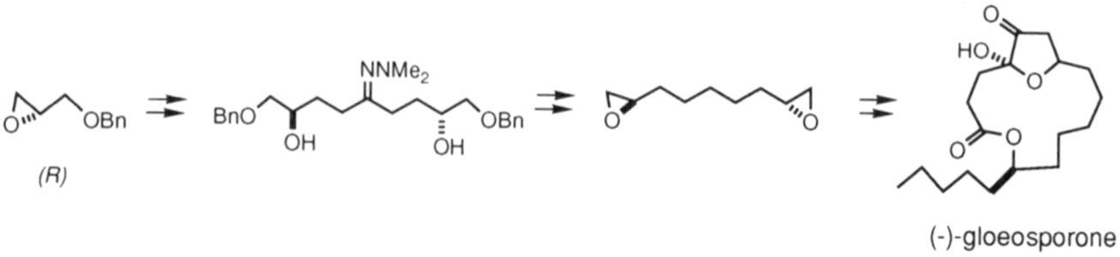
(R)
(-)-gloeosporone

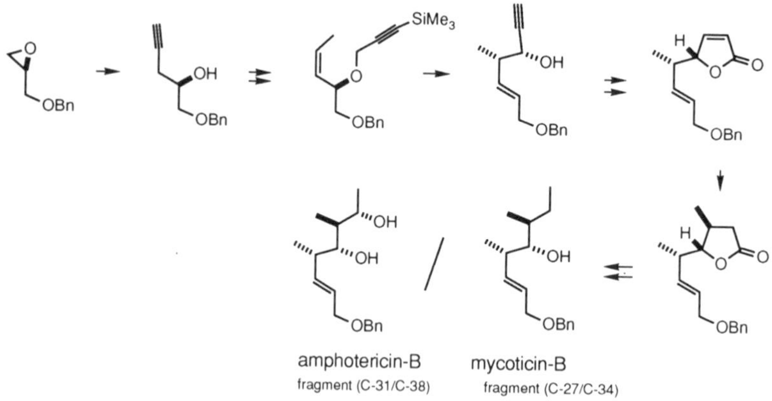
amphotericin-B
fragment (C-31/C-38)
mycoticin-B
fragment (C-27/C-34)

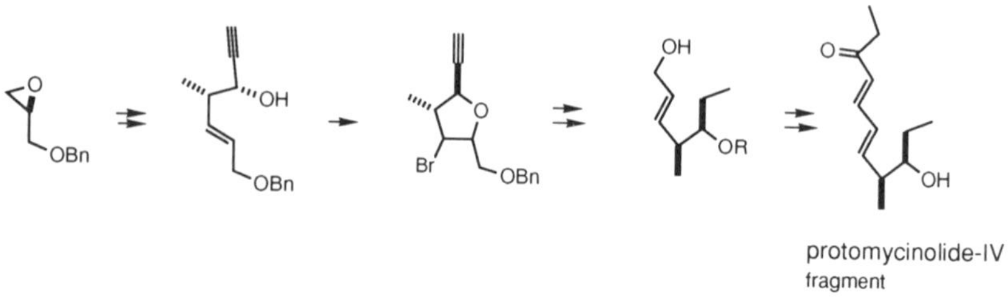
protomycinolide-IV
fragment

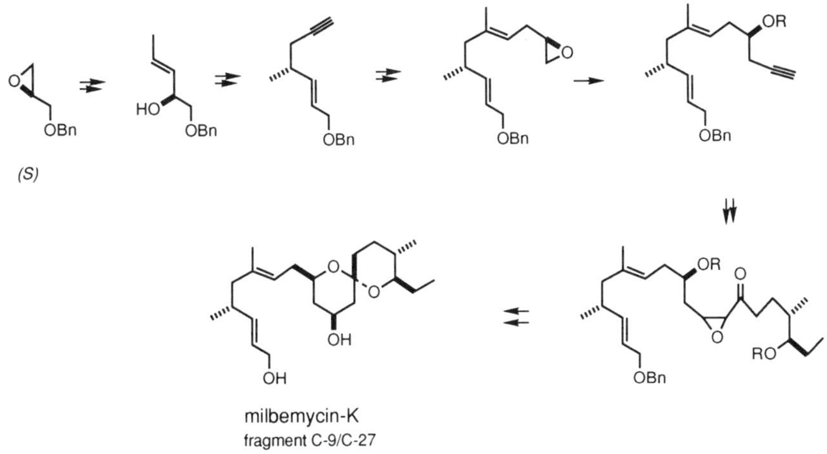

milbemycin-K
fragment C-9/C-27

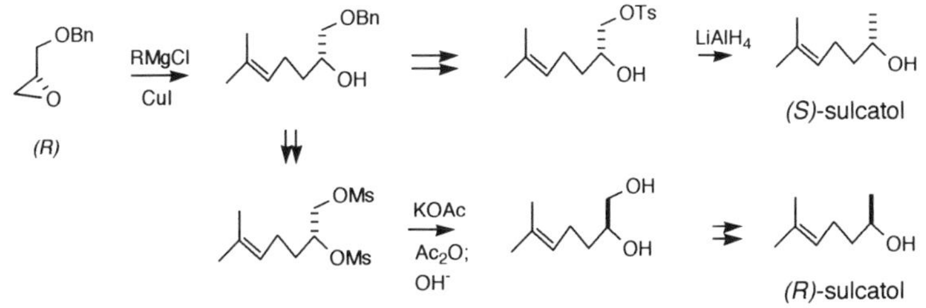

(S)-sulcatol

(R)-sulcatol

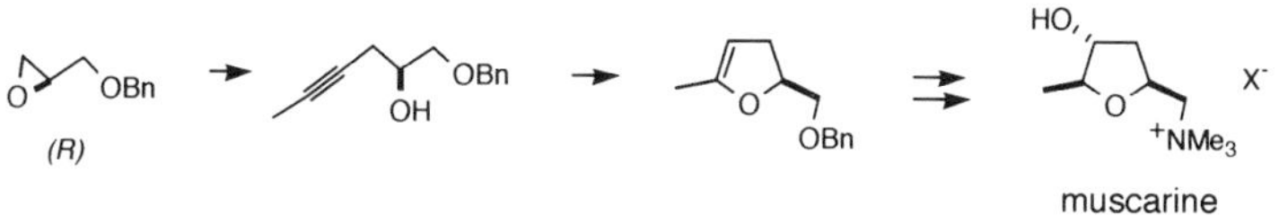

muscarine

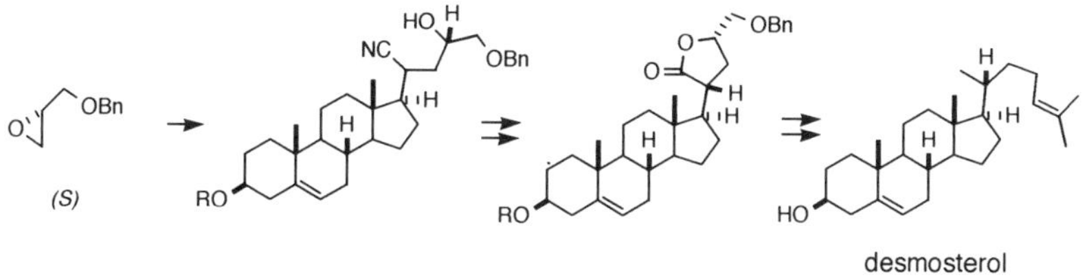

desmosterol

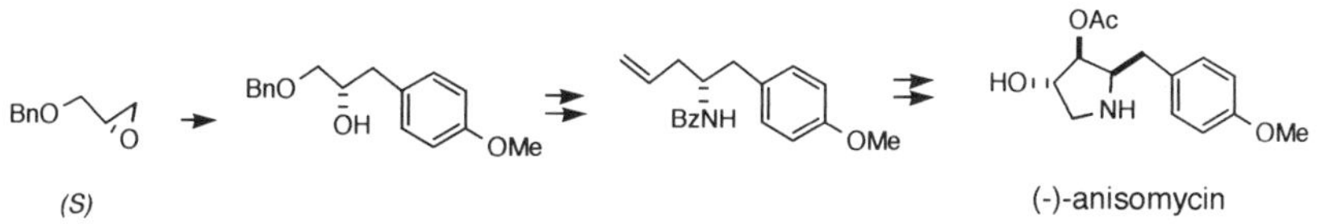

(-)-anisomycin

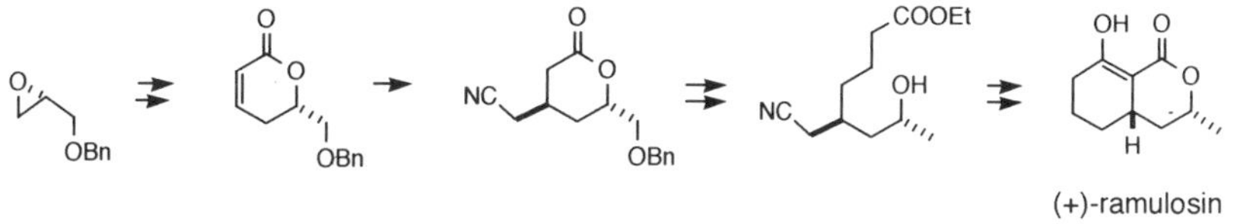

(+)-ramulosin

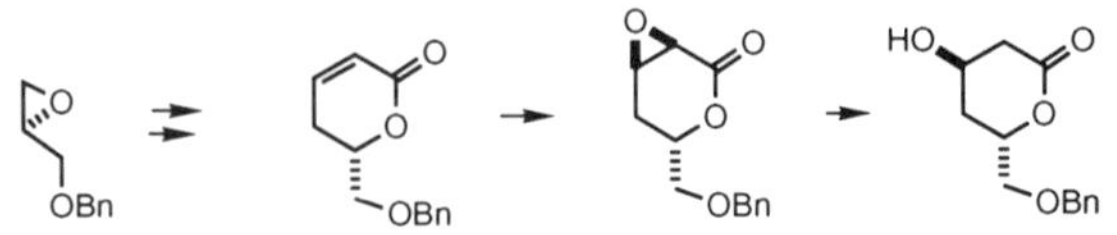

for compactin

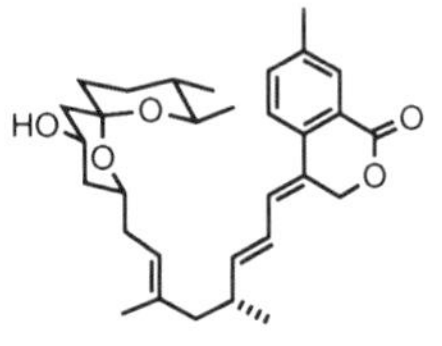

lacramin-A

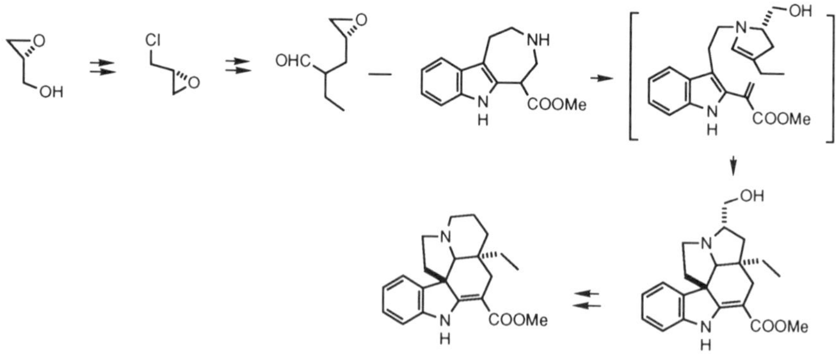

(-)-vincadifformine

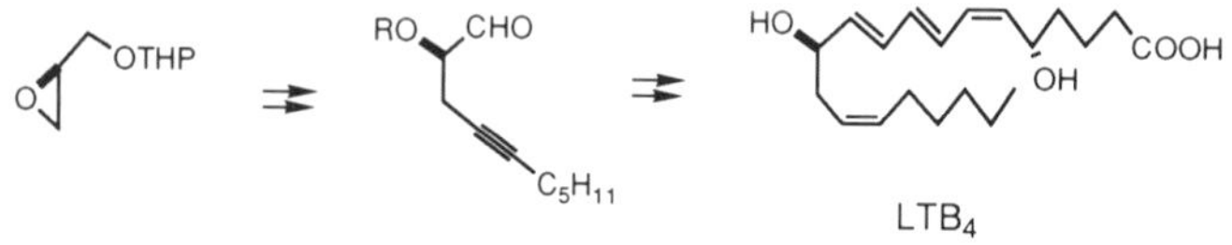

LTB₄

The convenient source of (S)-2,2-dimethyl-1,3-dioxolane-4-methanol is the corresponding aldehyde, also known as (R)-2,3-O-isopropylideneglyceraldehyde [Jurczak, 1986]. This chiral aldehyde, first obtained effectively from D-mannitol [Baer, 1939], is itself a valuable building block for synthesis. In fact its uses are far more extensive than the alcohol. The (S)-aldehyde is not as readily available, but it can be prepared from ascorbic acid [Jung, 1980]. However, the 2-benzyl ether of (S)-glyceraldehyde has been obtained from diethyl L(+)-tartrate [Jäger, 1989] and it can be used in a *syn*-selective synthesis of 1,2-diamino alcohols [Franz, 1994] via chelation-control Grignard reactions on the benzylaldimine.

Even more diverse synthetic potential of the aldehyde, as compared with the alcohol, is expected because of possible direct chain extension by aldol, organometallic, and Wittig reactions. Actually the most valued attribute of the aldehyde is the asymmetric induction at the formation of the new stereogenic center from the prochiral carbonyl group—the emphatic feature of many syntheses. The reactions with organometallic reagents (showing some *erythro-*selectivity) have been exploited in a great number of natural product syntheses, including 2-deoxy-D-ribose [T. Harada, 1981], (+)-disparlure [Pikul, 1987], *exo* and *endo*-brevicomin [Mulzer, 1986a], δ-multistriatin [Mulzer, 1986c], two chain-segments of erythronolide-B [Mulzer, 1987b], (+)-muscarine [Mulzer, 1987a], D-α-amino acids [Mulzer, 1986b], (3S,4S)-statine [Mulzer, 1988], *ent*-pestalotin [K. Mori, 1976], leukotriene LTA$_4$ [Y. Kobayashi, 1986], and (R)-11-HETE [Corey, 1981].

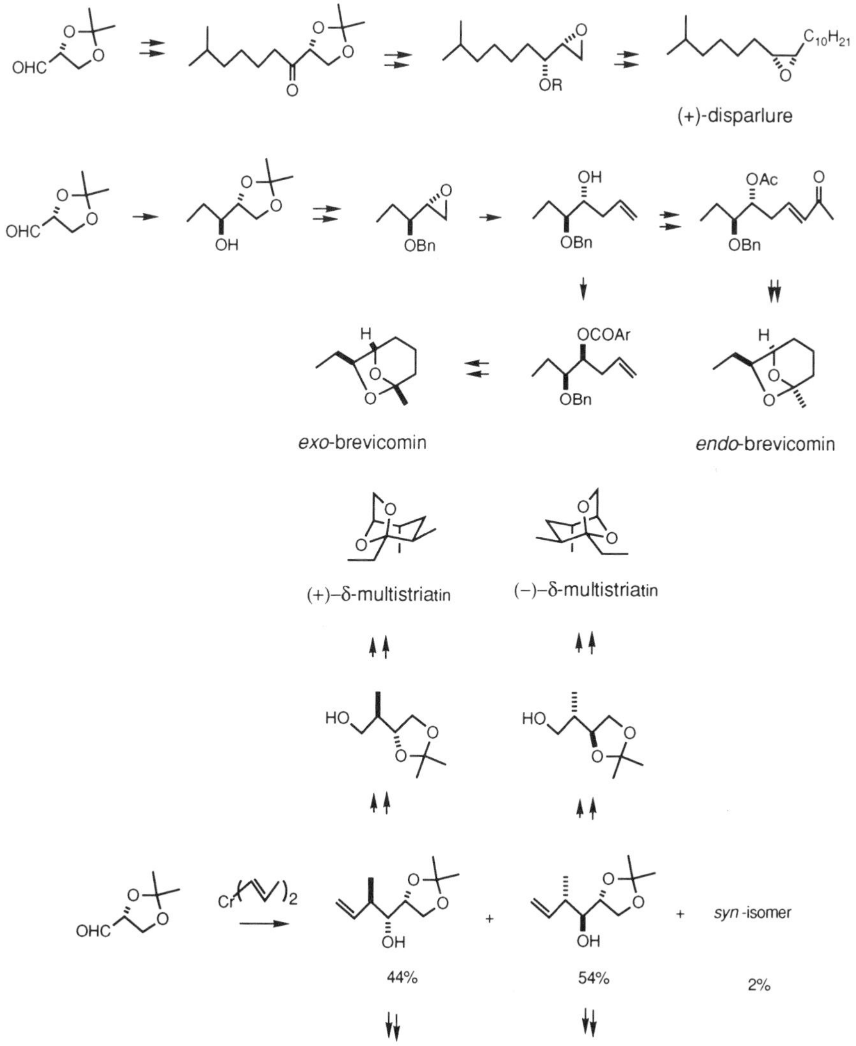

(+)-disparlure

exo-brevicomin

endo-brevicomin

(+)-δ-multistriatin

(−)-δ-multistriatin

44%   54%   *syn*-isomer

2%

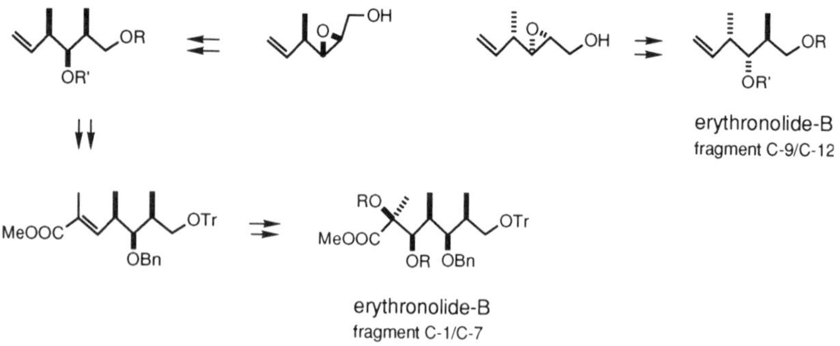
OR
OR'
OH
OR
OR'
erythronolide-B
fragment C-9/C-12
MeOOC
OTr
OBn
RO
MeOOC
OTr
OR OBn
erythronolide-B
fragment C-1/C-7

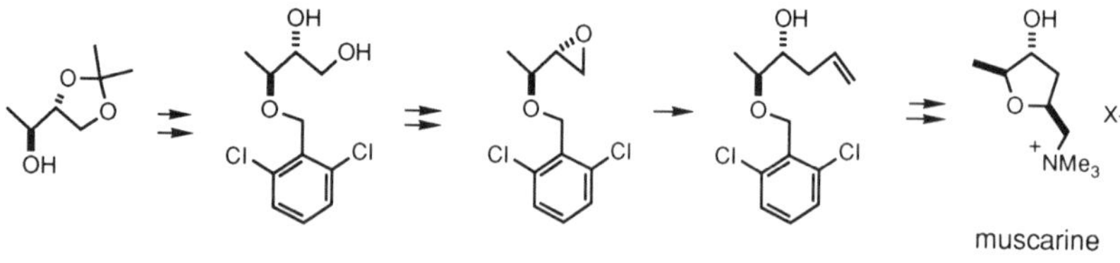
OH
OH
Cl
Cl
X-
NMe3
muscarine

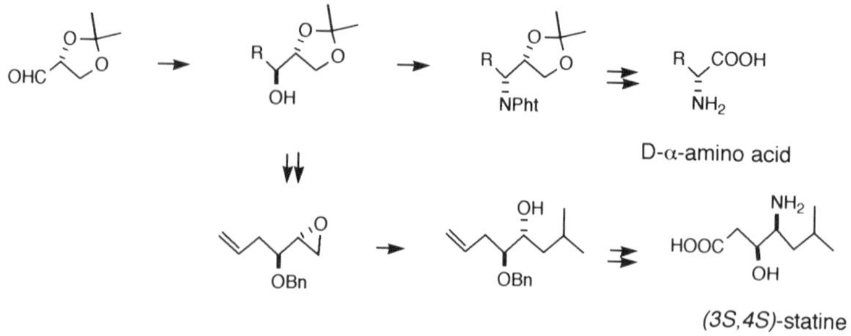
OHC
R
OH
R
NPht
R COOH
NH2
D-α-amino acid
OBn
OH
OBn
NH2
HOOC
OH
(3S,4S)-statine

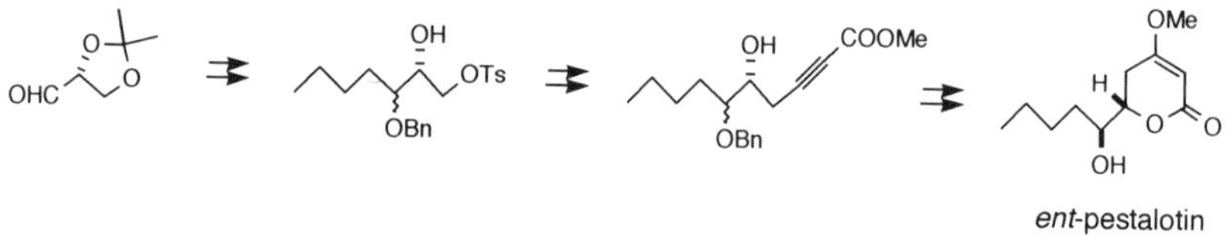
OHC
OH
OTs
OBn
OH
COOMe
OBn
OMe
H
OH
ent-pestalotin

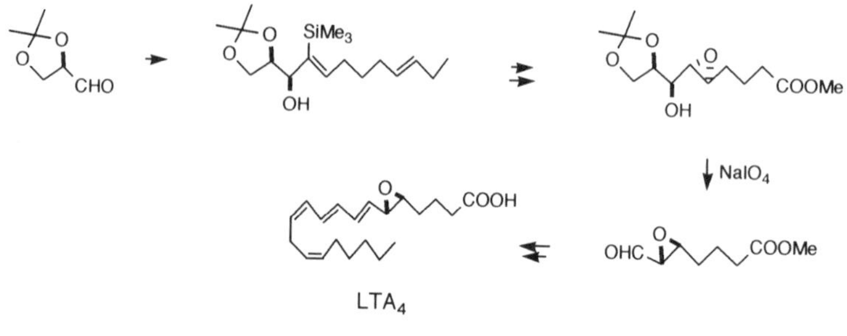
CHO
SiMe3
OH
COOMe
OH
NaIO4
COOH
OHC
COOMe
LTA4

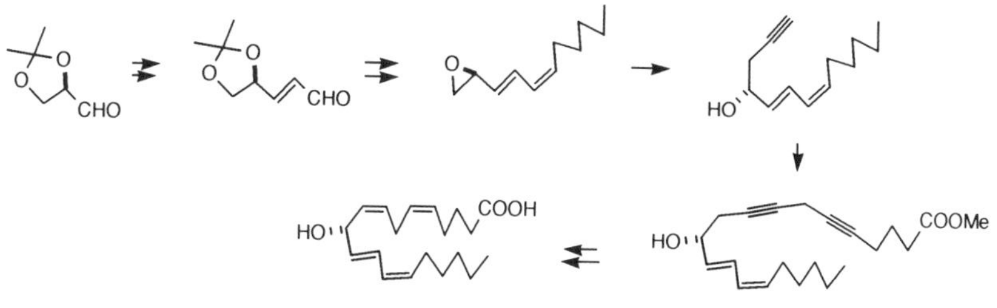

(R)-11-HETE

A series of publications described the elaboration of a chiral cyclopentanone which underwent controlled ring cleavage to provide intermediates for (−)-dihydrocorynantheol [T. Suzuki, 1985a], (−)-antirhine [T. Suzuki, 1985b], and (−)-chokol-A [T. Suzuki, 1988]. Syntheses involving chain extension by Wittig reaction and subsequent transformations are also quite numerous, as in the approaches to ipsdienol [K. Mori, 1979b], leukotriene LTA_4 [Rokach, 1981], *trans*-chrysanthemic acid [Mulzer, 1983], ikarugamycin [Boeckman, 1983, 1989], brefeldin-A [Trost, 1986b], the tetrahydropyran portion of compactin [Kozikowski, 1985], fuscol [Iwashima, 1992].

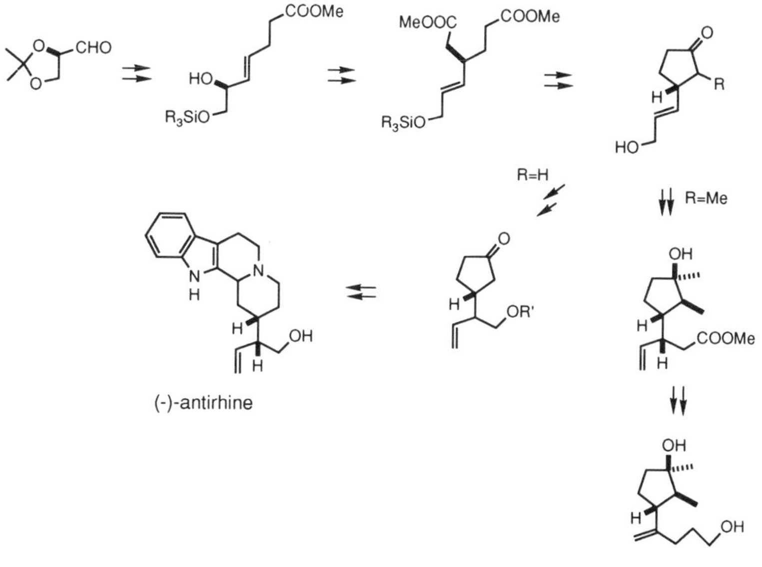

(-)-antirhine

(-)-chokol-A

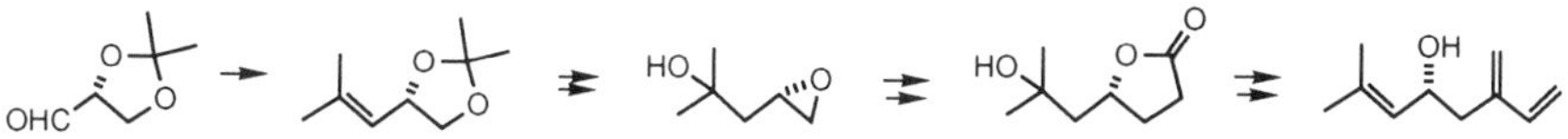

ipsdienol

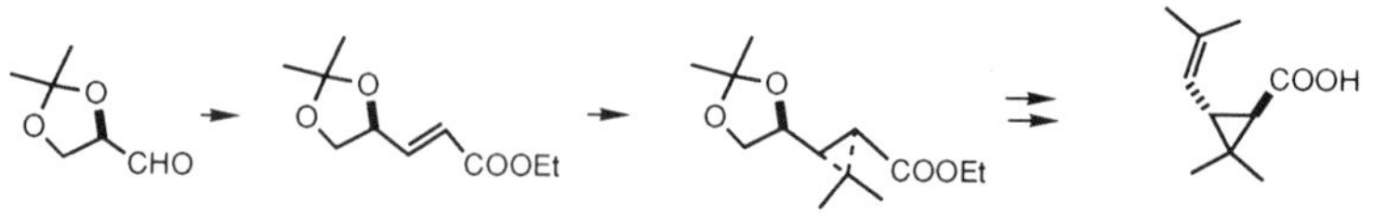

trans-chrysanthemic acid

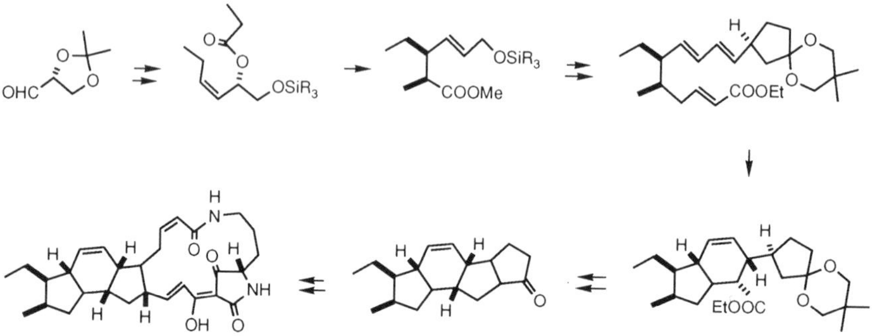

(+)-ikarugamycin

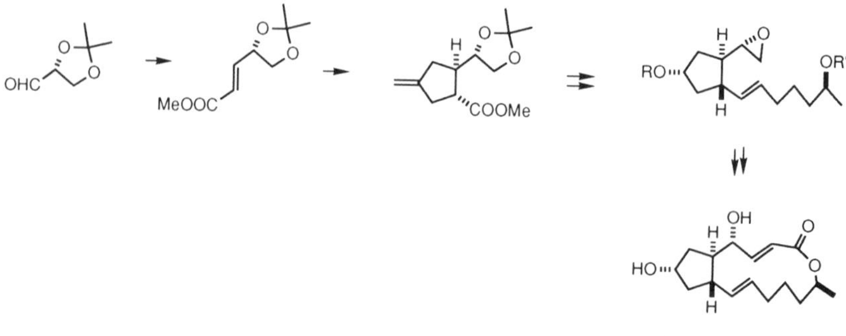

(+)-brefeldin-A

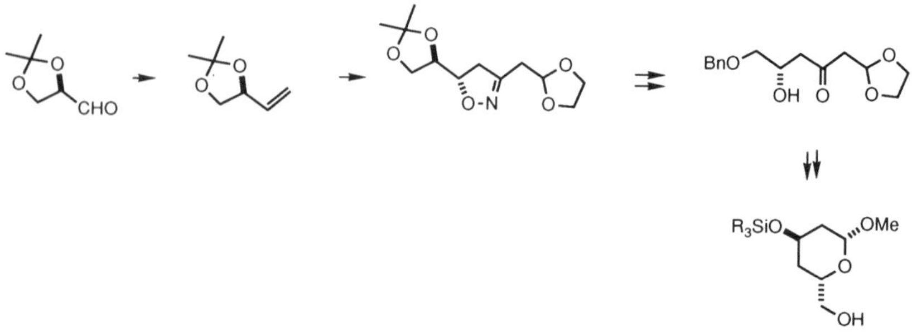

for compactin

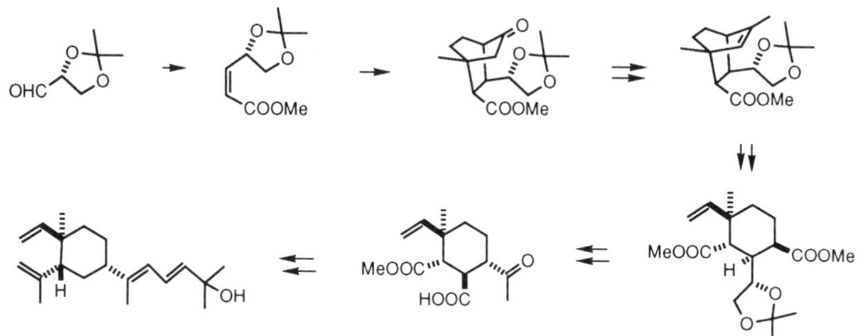

fuscol

By using other reactions, such as aldol condensation or conversion into a chiral alkylating agent, synthetic routes to prostaglandin-E₁ [Stork, 1977], brefeldin-A [Kitahara, 1979], milbemycin-β₃ [D.R. Williams, 1982], and tulipaline-B [A. Tanaka, 1980] were realized.

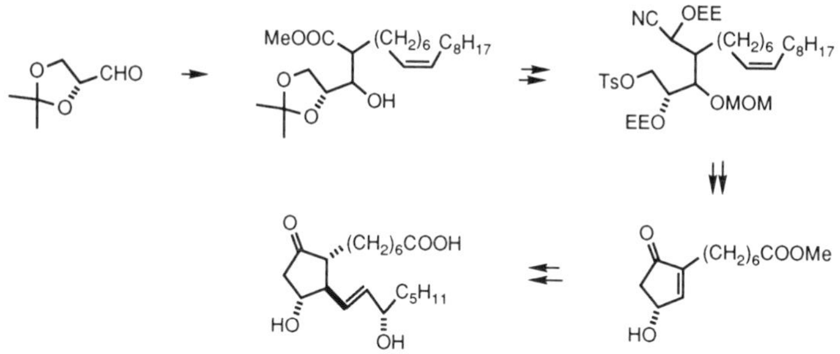

PGE₁

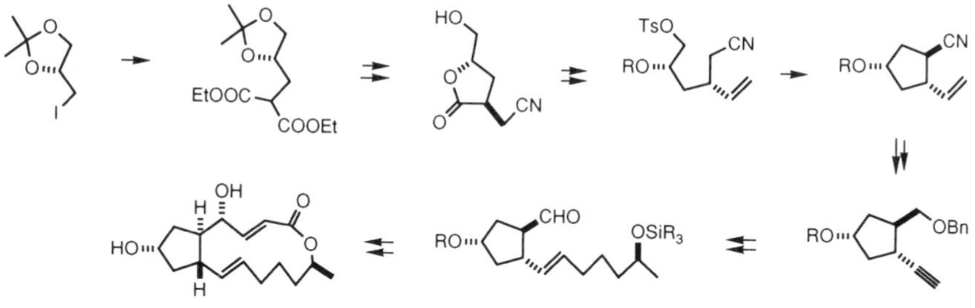

(+)-brefeldin-A

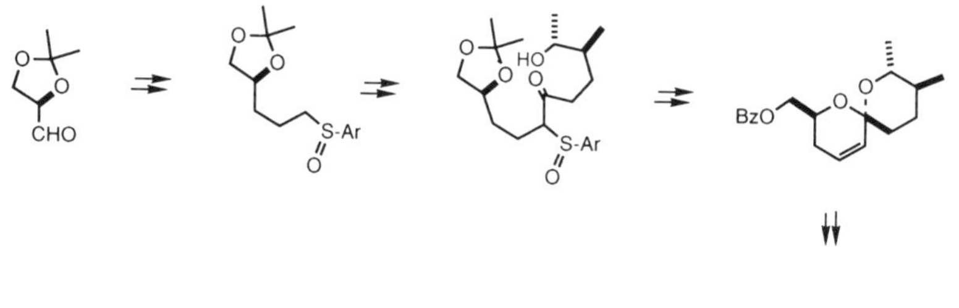

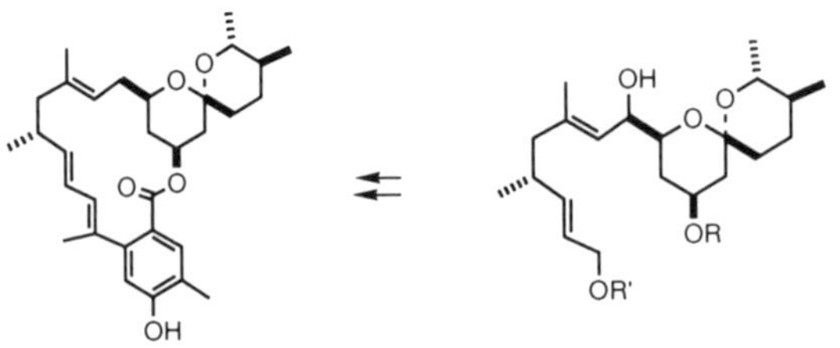

milbemycin-β₃

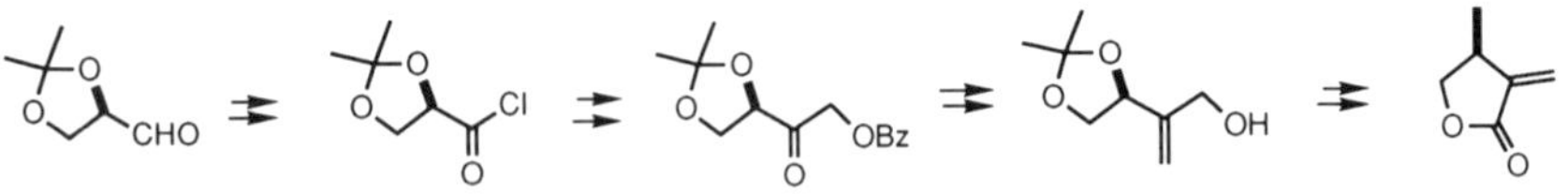

(-)-tulipaline-B

2,2-Dimethyl-1,3-propanediol is rarely used as a source of a 2,2-dimethylpropylene unit in a complex natural product. However, it was the starting material in a synthesis of moenocinol [Stumpp, 1986].

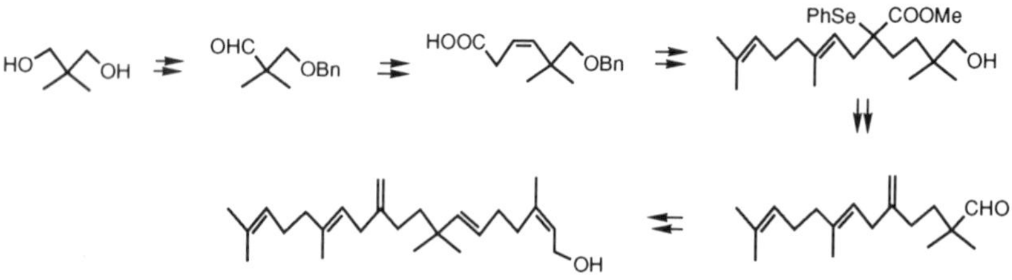

moenocinol

3.2.2. Other Symmetrical 1,3-Difunctional Compounds

Syntheses involving desymmetrizing operations of malonic esters are numerous. Those relying on enzymatic hydrolysis of one ester group are particularly worthy of note [Björkling, 1985; Luyten, 1987]. For example, both (R)- and (S)-isomers of β-cuparenone are readily acquired from the same intermediate [Canet, 1992].

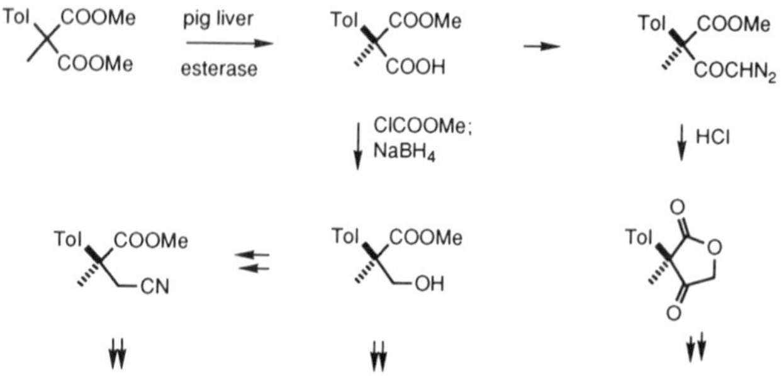

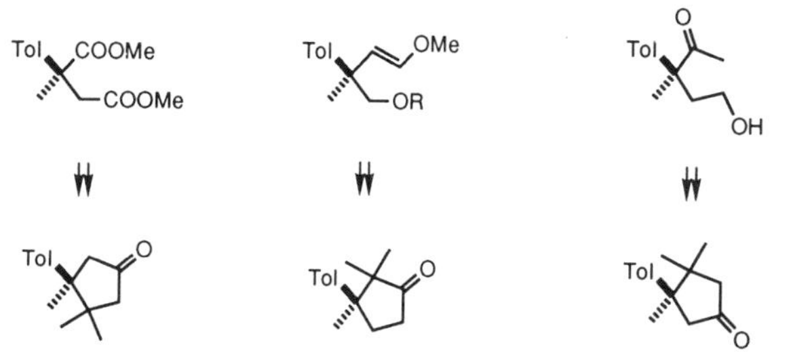

(-)-β-cuparenone   (+)-α-cuparenone   (+)-β-cuparenone

The use of symmetrical 1,3-difunctional compounds in syntheses does not always involve preliminary desymmetrization of substrates. While the applications of malonic esters, even with the restriction of desymmetrization during the synthetic processes, are too numerous to assess, an arbitrary example is the synthetic approach to neooxazolomycin [Kende, 1988], which featured a dianion cyclocondensation.

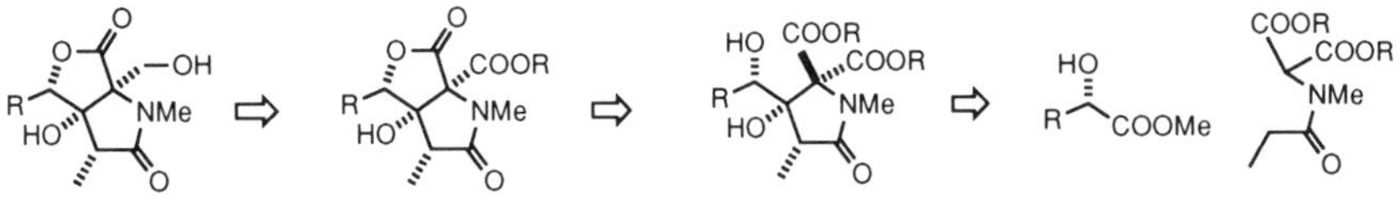

neooxazolomycin

Two 2,2-ethano malonate esters and their synthetic utility may be mentioned. One has served as an intermediate of eburnamonine [Hakam, 1987], the other having a simpler structure was prepared directly in a desymmetrized state (from ethyl cyanoacetate); imine formation and rearrangement paved the way to a formal synthesis of aspidospermine [Stevens, 1971].

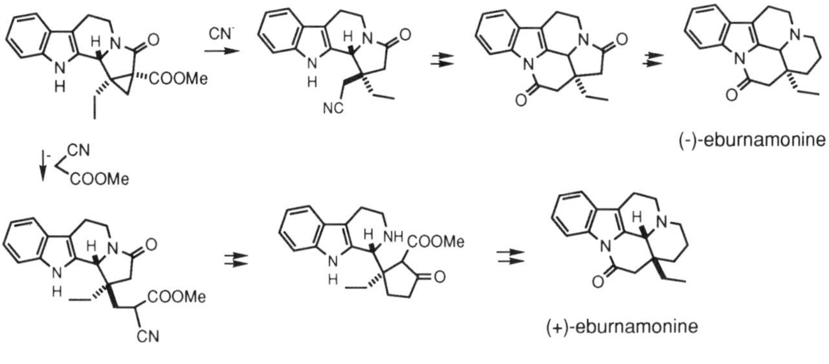

(-)-eburnamonine

(+)-eburnamonine

aspidospermine

In the cephalosporin-C synthesis [Woodward, 1966] a 2-glyoxylidenemalon-dialdehyde was employed as a Michael acceptor during construction of the thiazine ring. The locally symmetrical intermediate underwent cyclization to give only one product. 2-Ethyl-1,3-dibromopropane acted as the electrophile in double alkylation of dihydroharman to provide a precursor of flavopereirine [Danieli, 1980]. On the other hand, the reaction of 2-iodomethyl-3-iodopropene with 1,5-dilithio-1,5-bis(benzenesulfonyl)pentane gave rise to a methylenecyclooctane derivative [Lansbury, 1990], readily transformed into pseudopelletierine. The latter reaction sequence passed through symmetrical and unsymmetrical intermediates.

cephalosporin-C

pseudopelletierine

flavopereirine

1,3-Dihalopropane building blocks, besides serving as electrophiles, may be converted into nucleophilic species. An example is the preparation of 1,3-bis(dimethylphosphono)propanone from 3-iodo-2-iodomethylpropene and its use in a synthesis of muscone [Büchi, 1979b].

muscone

The homologation of 1-bromo-2-bromomethyl-3-methyl-2-butene to the symmetrical pimelic ester, followed by Dieckmann cyclization set the stage for completion of an isonootkatone synthesis [Marshall, 1967]. An earlier desymmetrization using homologation by an allylic and a diene unit at the chain termini was involved in an assemblage of selina-3,7(11)-diene [Chou, 1989].

isonootkatone

selina-3,7(11)-diene

A particularly interesting molecule is 2-bromomethyl-1,4-dibromo-2-butene, obtainable by 1,4-bromination of isoprene followed by allylic bromination. Isoprenylation of carbonyl compounds is achieved on treatment of the mixture with zinc [Tokuda, 1993].

Reactivity desymmetrization of 1,3-difunctionalized 2-alkylidenepropanes deserves and has received attention. For example, the conversion of 2-hydroxymethyl-3-methyl-2-butenol into a monobenzyl ether and phenyl sulfide gives it both nucleophilic and electrophilic roles to play in realizing a design of spiroannulation [McCurry, 1973].

β-vetivone

The diol is usually obtained from a propylidenemalonic ester by lithium aluminum hydride reduction. The Meldrum's acid derivative, which is a cyclic form of the diester, has been employed as the dienophile in a Diels–Alder approach to δ-damascone [Dauben, 1975].

δ-damascone

The simple diol forms acetals with carbonyl compounds, therefore furnishing a new method of protection for the latter species [Corey, 1975c]. These acetals undergo double isomerization to the highly acid-sensitive cyclic vinyl ethers with a Rh(I) complex.

Relevant chemistry involves reduction and hydroboration of the cyclic acetal derived from menthone [T. Harada, 1989], and enantioselection in deoxygenation of 1,3-diols to provide useful synthons for α-tocopherol [T. Harada, 1987a]. Enantiodifferentiating functionalization [T. Harada, 1987b] is also possible.

2-Oxyallyl zwitterion is a valuable synthetic intermediate for many purposes. The parent species cannot be generated by dehalogenation of 1,3-dihaloacetone, but the problem can be circumvented by using polyhalo analogs with a subsequent dehalogenation step on the product to achieve the transformation. Usually such circumstances deal with [3 + 2]- and [3 + 4]-cycloadditions. When a situation demands an acetone synthon as electrophile and then nucleophile, protection of the ketone group as an enol derivative is in order. Several such reagents are available, in the context of the present discussion the conversion of 1,3-dichloroacetone into 3-chloro-1-propen-2-yl diethylphosphate [Welch, 1986] by reaction with triethyl phosphite may be mentioned. The transformation achieved structural as well as reactivity desymmetrization (umpolung) at one of the two identical sites of 1,3-dichloroacetone.

Metal complexes of trimethylenemethane can often be generated in situ to serve as 1,3-dipolar (conjunctive) reagents for cycloadditions [Trost, 1986a, 1988]. Precursors such as 2-trimethylsilylprop-2-enyl acetate undergo symmetrization in the process. 1,3-Dication synthons with differentiable reactivity are also valuable in synthesis. 2-Bromomethylprop-2-enyl acetate, readily prepared by dehydrobromination of a 1,3-dibromide [Magnusson, 1990], is such a synthetic equivalent.

Somewhat related to these above compounds is β,β'-(bisphenylthio)isobutyraldehyde. On exposure to an organolithium reagent it is converted into a cyclopropylcarbinol [K. Tanaka, 1991].

Finally, a desymmetrized and uniformly oxygenated isobutane derivative has been prepared [Seu, 1994]. Grignard reaction on the epoxide served to assemble a carbon chain from which either enantiomer of frontalin could be obtained by deoxygenation of the proper terminus.

4

FOUR-CARBON CHAIN BUILDING BLOCKS

4.1. TARTARIC ACID AND DERIVATIVES

Among tartaric acid's many attributes as a chiral building block, the availability of both enantiomeric forms at reasonable cost and the presence of two types of functionality are its most valuable [Seebach, 1980]. The L(+)-tartaric acid which has the (2R,3R)-configuration is obtained in large quantities from potassium hydrogen tartrate (known as tartar or cream of tartar) which is a waste product from the wine industry, whereas the D(−)-isomer ((2S,3S)-isomer) occurs in *Bauhinia* plants. As (−)-tartaric acid is commercially produced by efficient resolution methods, it is not excessively expensive as a research chemical.

$(R,R)(+)-$ $(R,S)-$ $(S,S)(-)-$

meso-

III   tartaric acid   III

The optically active tartaric acids have a C_2 axis of symmetry. The four functionalized carbon atoms are pairwise homotopic, therefore a reaction at either pair of the functional groups gives rise to one product only. When reaction

130

occurs only at one site, one compound having four constitutionally different functional groups results. A great variety of chiral structures can be derived from simple transformations, when care is taken to avoid equilibration. These operations include inversion of configuration, deoxygenation, chain branching, chain elongation, and chain shortening.

Certain symmetrical intermediates are readily constructed by directional homologation of tartaric acid derivatives. For example, furaneol, a volatile flavor component of various fruits and roasted coffee, has been synthesized [Briggs, 1985]. In a preparation of a tetraoxygenated C_8 diene [Saito, 1988, 1990] simultaneous suprafacial allylic rearrangement of two acetoxy groups by Pd(II) catalysis translated the original chirality to the product.

furaneol

A diene diester derived from tartaric acid through chain elongation underwent cyclopropanation on exposure to isopropylidenetriphenylphosphorane and the product could be converted into two molecules of *trans*-chrysanthemic acid [Krief, 1988].

It is interesting to note that the same diester of the (*S,S*)-series is accessible from D-mannitol. With respect to synthetic applications involving functionalization of the conjugated double bond, an approach to loganin [Takano, 1987c] may be mentioned.

trans-chrysanthemic
acid

(3 : 2)

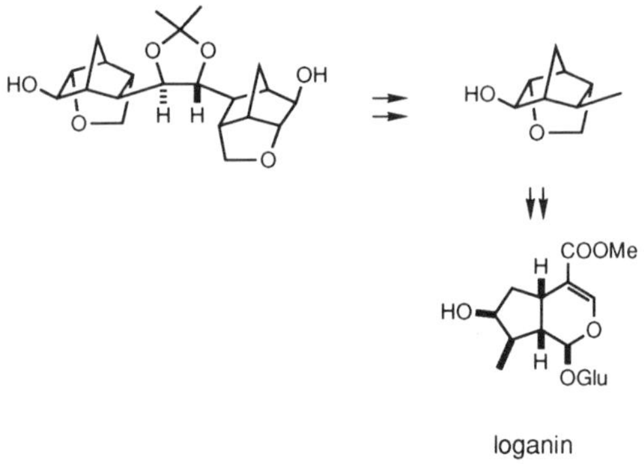

loganin

In both examples asymmetric induction was based on a template effect.

4.1.1. Syntheses via Initial Internal Desymmetrization

Alkylation of the dimethyl tartrate acetonide was the first step in a synthesis of (+)-malyngolide [Tokunaga, 1986]. Unfortunately the required stereoisomer was the minor alkylation product in a 1:4 mixture, due to steric factors.

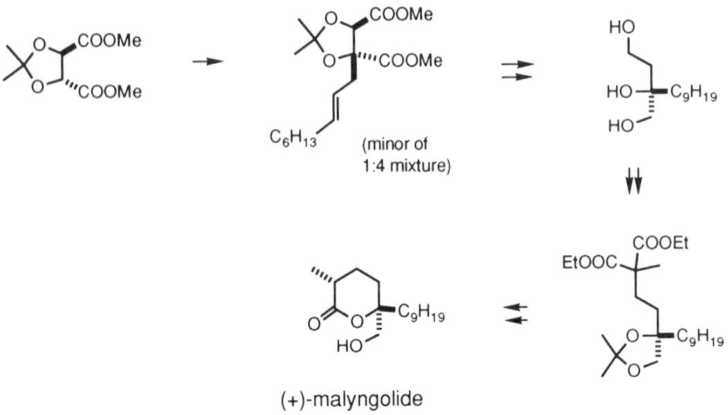

(+)-malyngolide

A route to (4R)-hydroxy-2-cyclopentenone, which is a valuable building block for synthesis of natural prostaglandins and other substances including clavulone-II [Nagaoka, 1984], has been developed based on elaboration of the (S,S)-tartaric acid acetonide into a diiodide which is a double alkylating agent [Ogura, 1976]. When the cyclopentanone derivative was generated under acidic conditions the concomitant β-elimination (desymmetrization) led to only one product, due to C_2-symmetry of the acetonide.

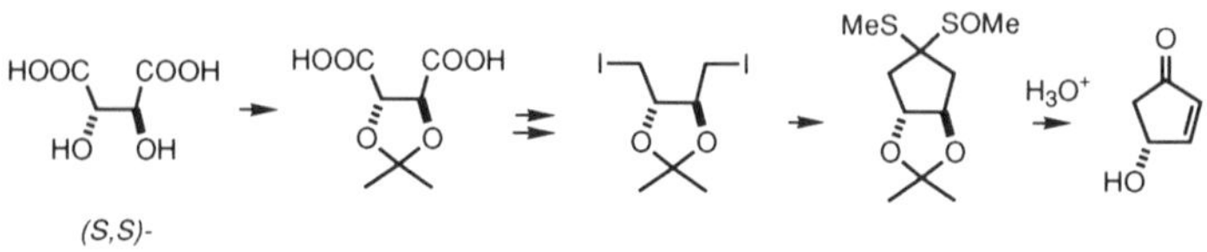

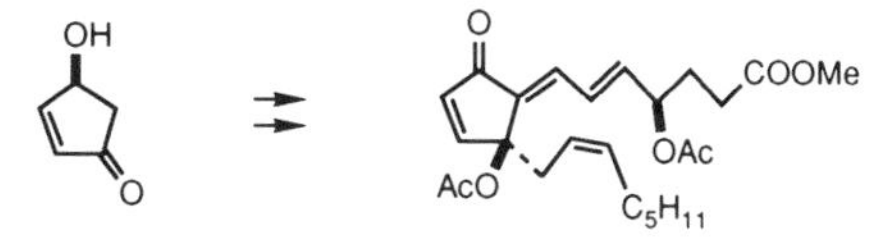

clavulone-II

Tartaric esters can be converted into 2,3-epoxybutanedioates which undergo stereospecific ring opening with cuprate reagents. A synthesis of δ-multistriatin [K. Mori, 1980] was based on this method. Interestingly, the third asymmetric center was induced via equlibration during the intramolecular acetalization.

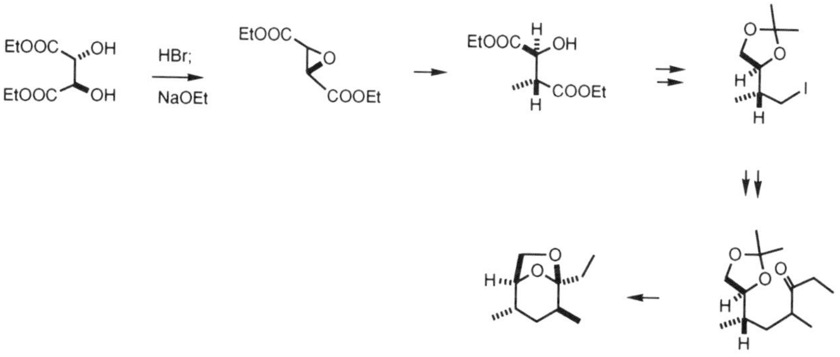

δ-multistriatin

Two molecules of (S,S)-tartaric acid were fashioned and combined to form the tetrahydropyran unit of the antibiotic X-14547A [Nicolaou, 1981a, b, 1985], as synopsized in the scheme.

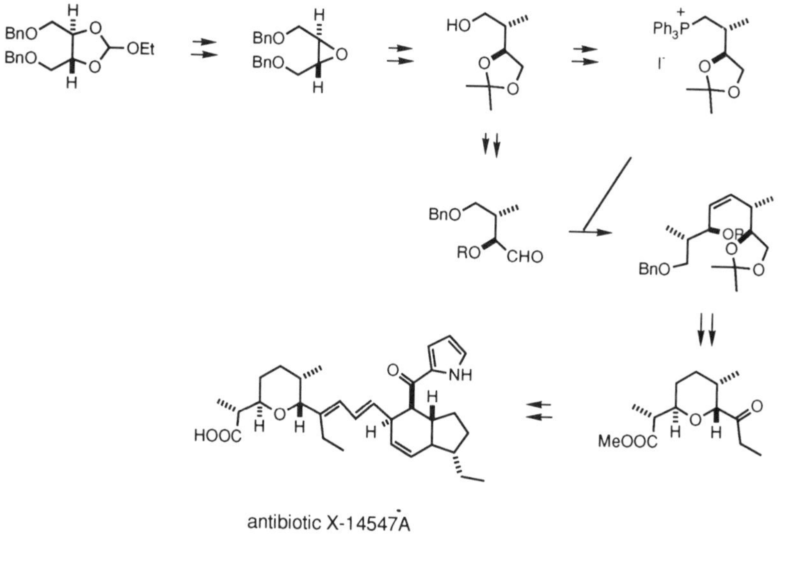

antibiotic X-14547A

In a synthetic effort toward debromoaplysiatoxin [P.-U. Park, 1987] one of the five chirons which defined the two chiral secondary carbon atoms of the spiroacetal subunit was prepared from diethyl tartrate.

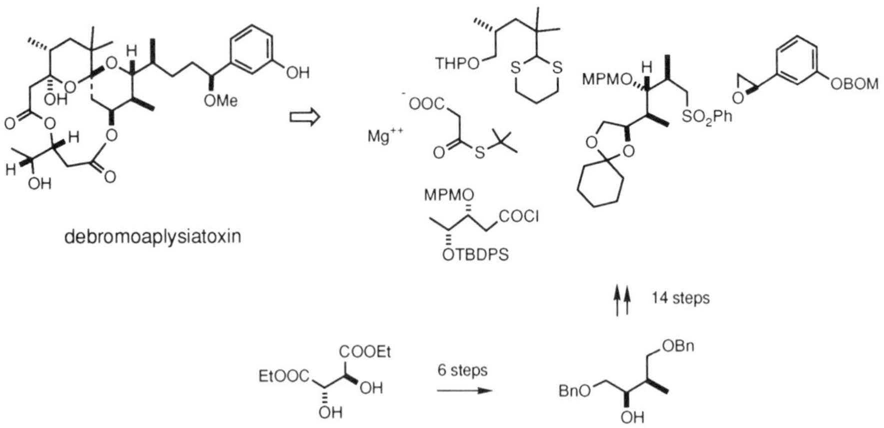

debromoaplysiatoxin

Another report described a synthesis of the unusual amino acid component of cyclosporine [Wenger, 1985].

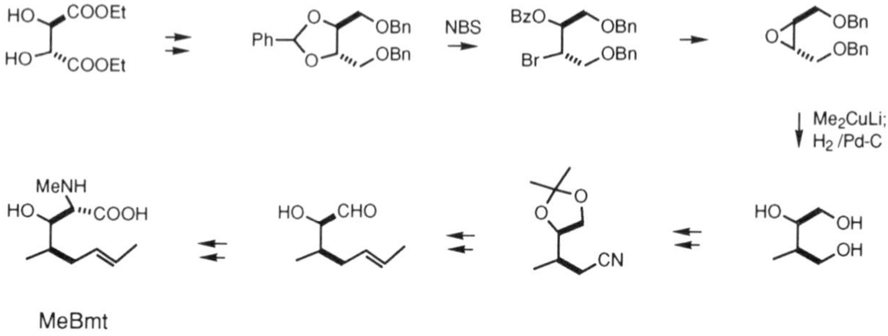

MeBmt

β-Substituted aspartic acid derivatives can be obtained by manipulation of 2,3-epoxybutanedioic esters through reaction with azide anion and the formation of the C_2-symmetric aziridines [Tanner, 1990].

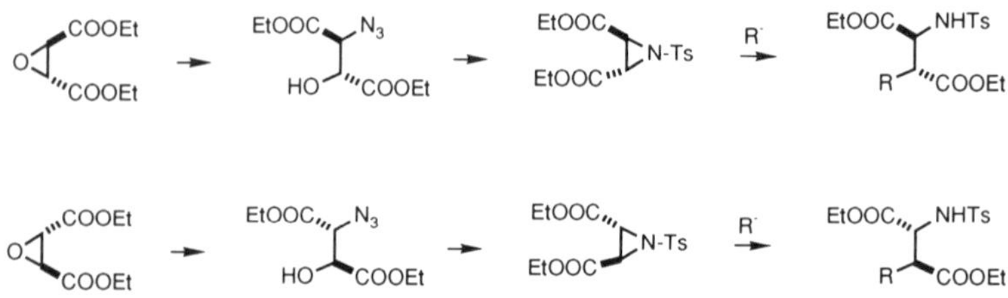

The use of diethyl tartrate in the construction of the polyene unit of the calyculins [Yokokawa, 1993] needs retention of only one carbinolic center.

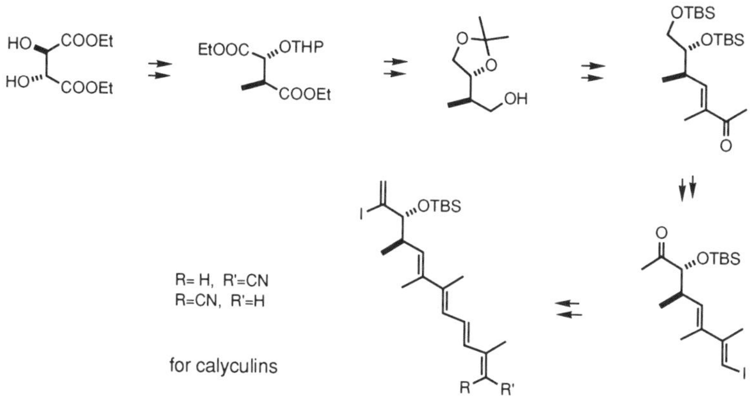

for calyculins

R= H, R'=CN
R=CN, R'=H

The cyclic acetal formed by condensation of a tartaric ester with benzaldehyde can be cleaved to give the tartaric ester monobenzyl ether. Synthetic use of this desymmetrized product includes formation of a building block for the cyclohexane moiety of the immunosuppressant compounds FK-506 and rapamycin [Kotsuki, 1992]. Even more interestingly, a chiral β-lactam has been constructed [Gateau-Olesker, 1986] via catalyzed hydrolysis of the ester group adjacent to the benzyloxylated carbon atom by pig liver esterase catalysis.

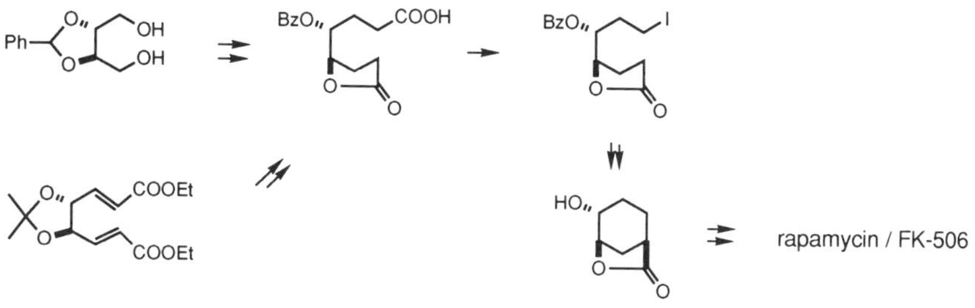

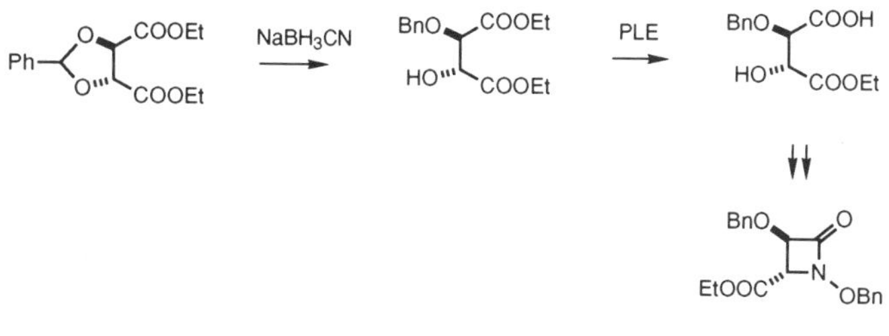

rapamycin / FK-506

Many different combinations of reactions on the monobenzyl ether have been explored, leading to useful synthetic intermediates. Examples are chiral cyclopentanones [Barriere, 1985], and the unsaturated lactone synthon for olguine [Valverde, 1985].

olguine

(S,S)-1,2:3,4-Diepoxybutane is available from the (R,R)-tartaric esters via isopropylidenation, reduction, tosylation, acid hydrolysis, and intramolecular displacement [Seebach, 1977]. By reaction of this diepoxide with Grignard reagents in the presence of copper(I) iodide, numerous C_2-symmetric glycols can be prepared [Devine, 1991]. Such a diynic glycol has been converted into the 2-phenyl-1,3-dioxolane, partially hydrogenated and desymmetrized by ring cleavage in a synthetic approach to the diol of a brown alga, *Notheia anomala* [Hatakeyama, 1985a].

Cleavage of the cyclic acetal by N-bromosuccinimide followed by debromination with tributyltin hydride and debenzoylation furnished the unnatural and expensive malic ester, which has been employed in many syntheses

[Seebach, 1980]. A more recent method of D-malic ester synthesis involves reductive cleavage of the cyclic sulfite of an L-tartaric ester [Gao, 1991].

L-tartrate esters   D-malate esters

A short route to the 2,8-dioxabicyclo[3.2.1]octane system of the squalestatins [Aggarwal, 1994] involves an aldol reaction using the acetonide of dimethyl tartrate and subsequent transacetalization.

4.1.2. Synthesis via External Desymmetrization

Tartrate esters in which the hydroxyl groups are properly protected afforded a convenient intermediate for (+)-terrein [Altenbach, 1990] on reaction with diethyl lithiomethylphosphonate. The desymmetrization occurred in the intramolecular Horner–Wadsworth–Emmons reaction step. Another strategy for assembling chiral polyhydroxylated cyclopentenes, such as that used in a syntesis of (−)-neplanocin-A [Bestmann, 1990], involves semihydrolysis of tartrate ester derivatives.

(+)-terrein

(-)-neplanocin-A

The value of L-tartaric acid as a building block for carbohydrate synthesis has been recognized. The diastereoselective addition of (γ-alkoxyallyl)stannanes to aldehydes derived from L(+)-diethyl tartrate provides a convenient method of chain homologation [Marshall, 1991].

Synthesis of chiral syringolide-1 and synringolide-2 [Wood, 1995] involving chain extension at one end of selectively protected threitols derived from tartaric acid has been accomplished.

n=5   (-)-syringolide-1
n=7   (-)-syringolide-2

Chain elongation from both ends of tartaric acid after proper protection and modification of its functional groups for the synthesis of chiral natural products is a well-documented and reliable strategy. Thus the elaboration of *exo*-brevicomin [K. Mori, 1974, 1986c; Meyer, 1977] and disparlure [K. Mori, 1979a] are some of the examples.

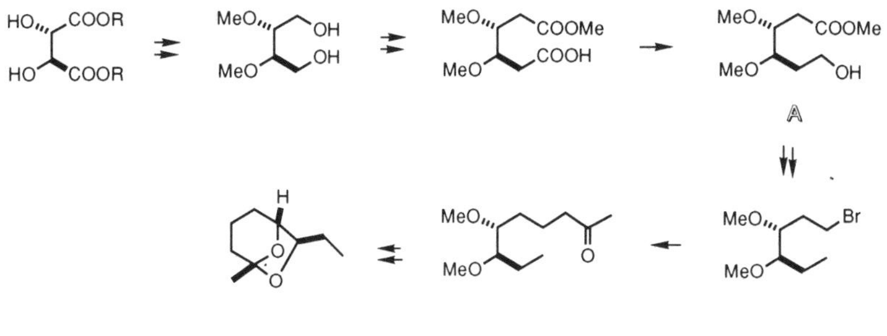

(+)-*exo*-brevicomin

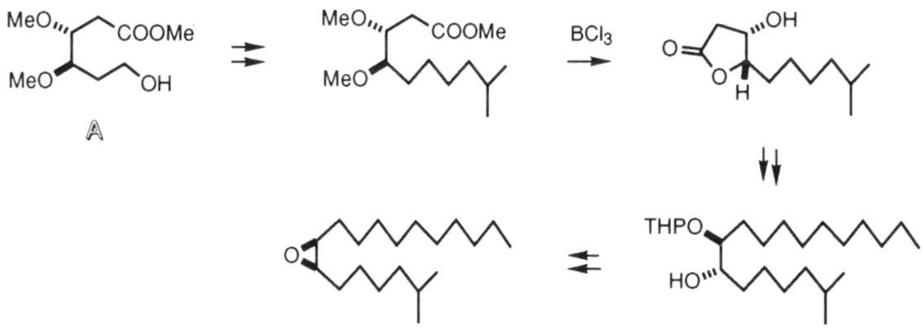

(+)-disparlure

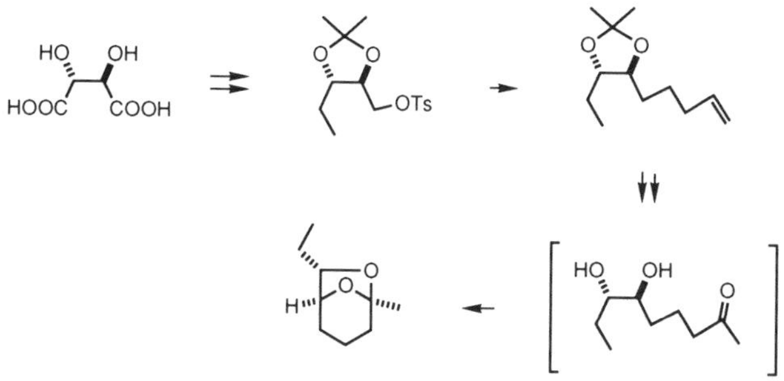

(-)-*exo*-brevicomin

The popularity in transforming tartaric acid into optically active *exo*-brevicomin is noteworthy. In one route [Achmatowicz, 1988] the insertion of an oxidoreduction sequence permitted the access to (−)-*endo*-brevicomin. The enantiomeric (−)-*exo*-brevicomin was obtained by several other research groups [Seu, 1986; Giese, 1988; Yadav, 1988b].

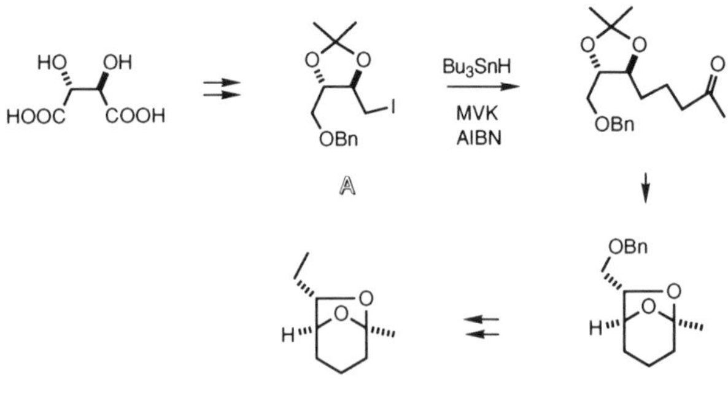

(-)-*exo*-brevicomin

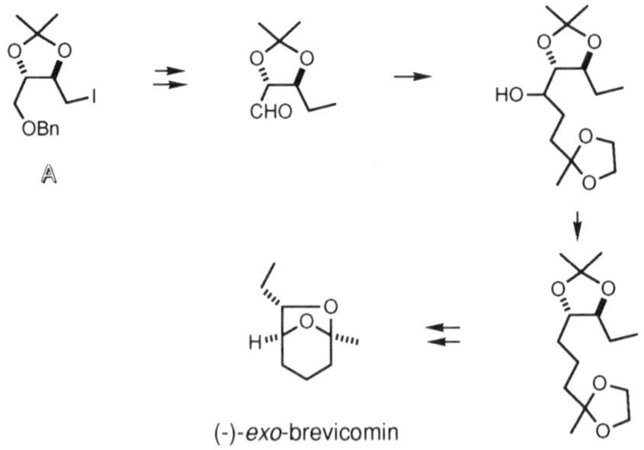

(-)-*exo*-brevicomin

Another enlightening synthesis of (+)-*exo*-brevicomin from D-tartaric acid [Kotsuki, 1990] was based on the different reactivities of tosylate and triflate toward nucleophiles. Thus the 2,3-acetonide of 1,2,3,4-butanediol derived from D-tartaric acid was monotosylated and then triflated, and the mixed disulfonate was subjected to a Kharasch reaction and a cuprate reaction to give the protected brevicomin. Isolation of intermediate was unnecessary.

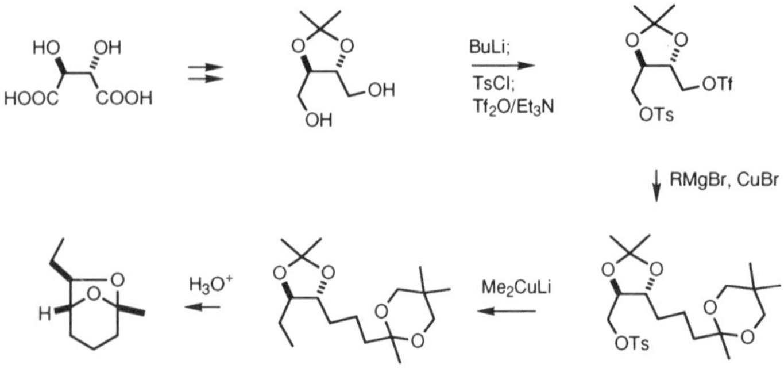

(+)-*exo*-brevicomin

The fungal metabolite (+)-LLP-880β and its epimer have been obtained [H. Meyer, 1975] from reaction of a tartaric acid-derived chiral aldehyde with methyl acetoacetate dianion, followed by lactonization and O-methylation.

LLP-880β

Structural correlation between (+)-colletodiol and (−)-tartaric acid was made, and a synthesis [Schnurrenberger, 1984] featuring bidirectional chain extension of the latter compound was accomplished. The other chiral hydroxy acid was derived from ethyl (R)-3-hydroxybutanoate.

(+)-colletodiol

(6S,7S)-*trans*-Laurediol also has a vicinal secondary alcohol unit which readily invokes the chain elongation strategy based on tartaric acid [Fukuzawa, 1986]. The required desymmetrized intermediate is available [Hungerbühler, 1979].

(6S,7S)-*trans*-laurediol

It should also be mentioned that the ω-chain of prostaglandins [Takano, 1990d], chiral 3-butyn-1,2-diol derivatives [Yadav, 1988a], and synthons for leukotriene-B$_4$ and lipoxin-B [Cohen, 1980; Rama Rao, 1987b] are readily prepared from L(+)-tartaric acid.

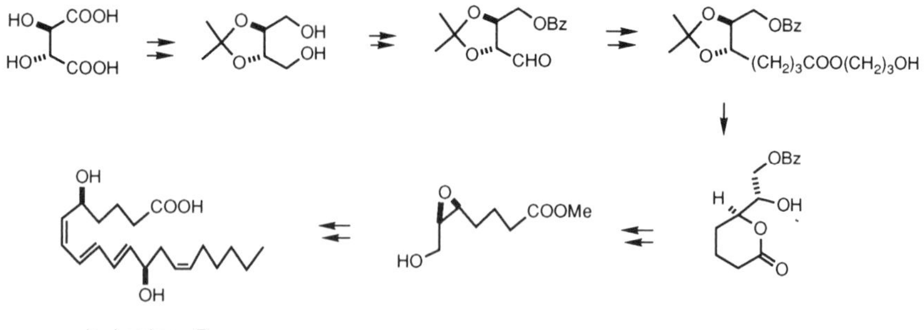

leukotriene-B₄

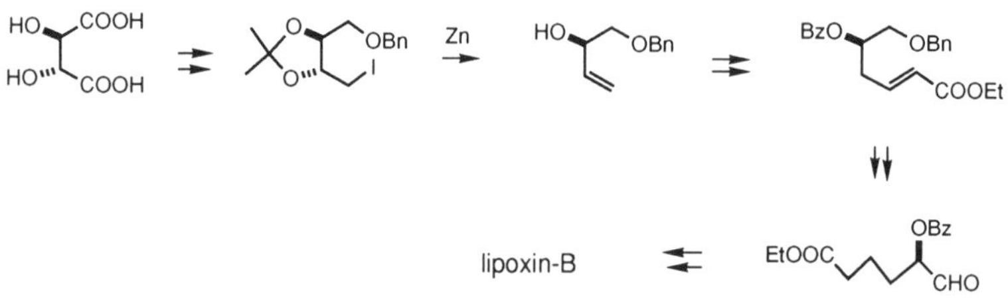

lipoxin-B

L-Tartaric acid has also been converted into 4-O-t-butyldimethyl- and 4-O-benzyl-2,3-O-isopropylidene-L-threose, a useful building block for synthesis of rare sugars such as (+)-norjirimycin [Iida, 1987], related substances including (+)-galactostatin [Aoyagi, 1991] and polyoxin-J [Mukaiyama, 1990], and the α-chain synthon of punaglandin-4 [Sasai, 1987; K. Mori, 1988c].

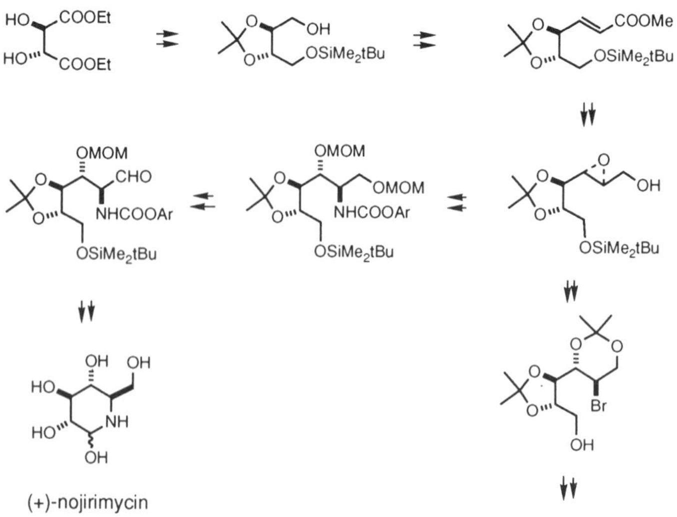

(+)-nojirimycin

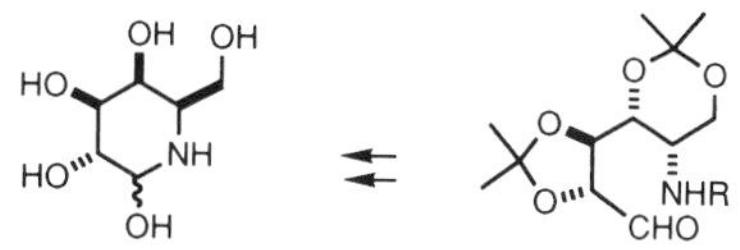

(+)-galactostatin

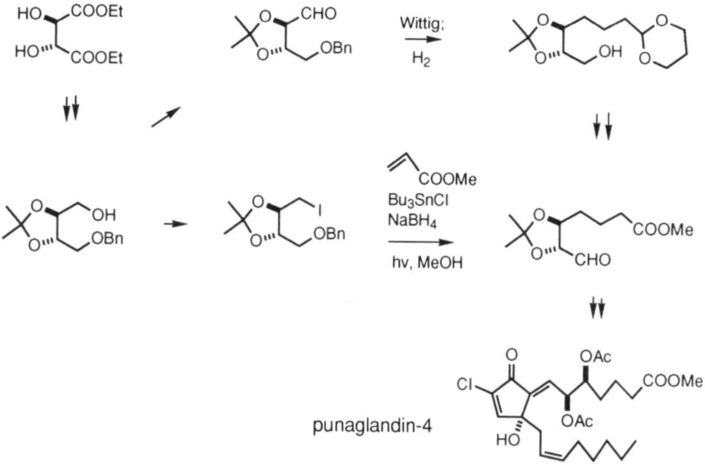

punaglandin-4

Analogous threose derivatives, especially those with the internal hydroxyl groups protected as the methoxymethyl ethers, have found extensive applications to synthesis of natural products, for example, *ent*-codonopsinine [Iida, 1985b], *N*-benzoyl-L-daunosamine [Iida, 1986b], and (−)-anisomycin [Iida, 1986a]. (Note that a previous approach to (+)-anisomycin [Wong, 1969] from *N*-benzyltartrimide was initiated by reaction with anisylmagnesium chloride; and also a route to lentiginosine [Yoda, 1993].) By a slight modification of intermediate stereochemistry, enantiodivergent synthesis may be developed, for example, of pinidine [Yamazaki, 1989].

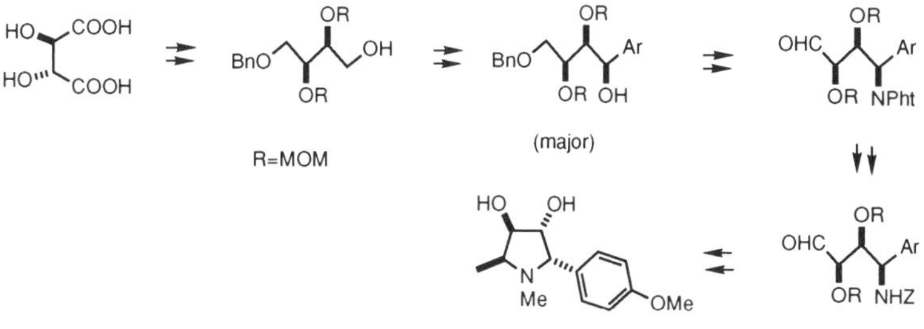

ent-codonopsinine

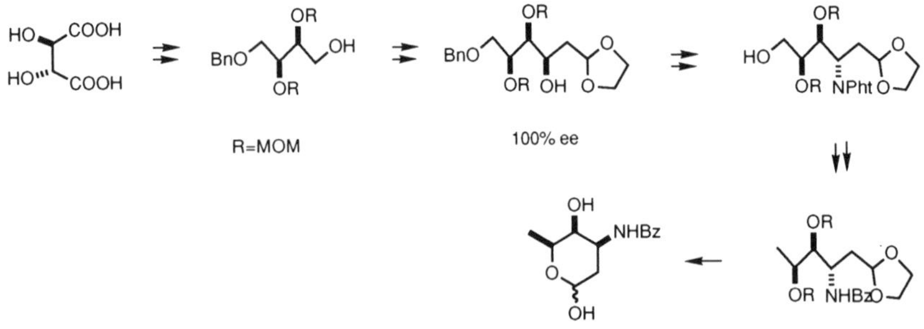
R=MOM
100% ee

N-benzoyl-L-daunosamine

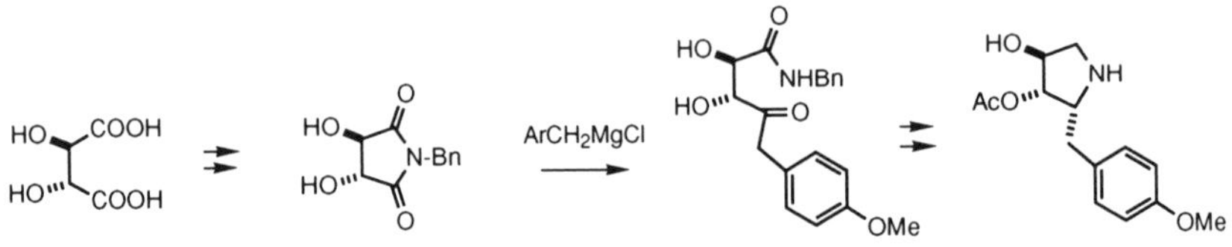
ArCH2MgCl

(+)-anisomycin

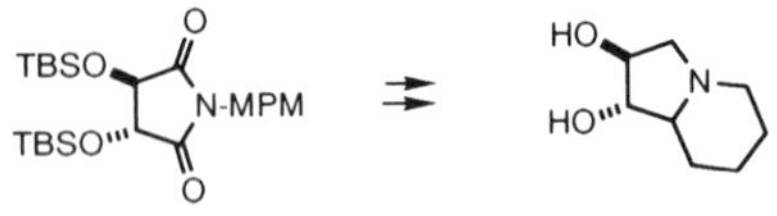

lentiginosine

For the synthesis of 15,16,19,20,13,24-hexaepiuvaricin [Hoye, 1991] the acetonide of (+)-diethyl tartrate was reduced and a four-carbon unit was added at each terminus before differentiation by monotosylation and further homologation.

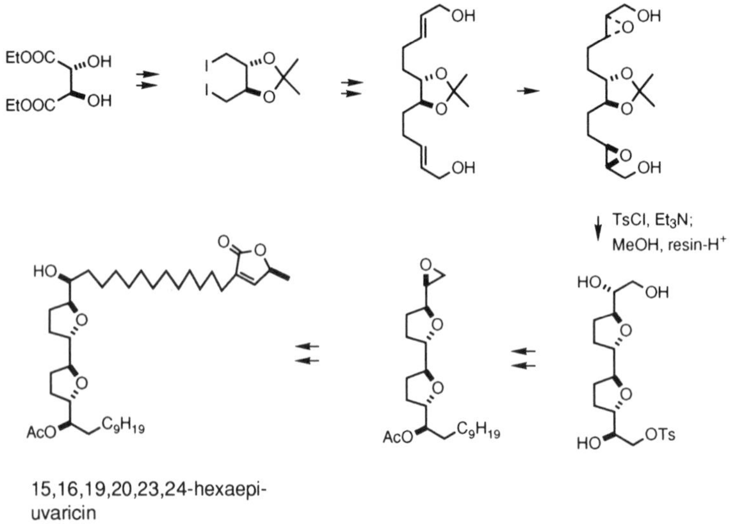
TsCl, Et3N;
MeOH, resin-H+

15,16,19,20,23,24-hexaepi-uvaricin

Corossolone has also been synthesized [Yao, 1995] from D(−)-tartaric acid.

corossolone

The very effective strategy of bidirectional chain elongation has been applied to a synthesis of hikizimycin [Ikemoto, 1990]. Desymmetrization was performed when the octanedioic ester containing six alkoxy and siloxy substituents in a *syn–anti–syn–anti–syn* stereochemistry was established.

hikizimycin

To plan a synthesis of (+)-*exo*-brevicomin [Masaki, 1982] it is imperative to extend both tartaric ester groups separately and reductively. The chemoselectivity problem associated with this need was neatly solved by adopting an intramolecular alkylation scheme in which the alkylating agent is one of the tosylate groups derived from the diester. The C_2-symmetry of the tartaric acid and various derivatives with intact asymmetric centers is essential to the concise synthesis that avoided protection–deprotection maneuvers.

A 2-methyl-2-(2-phenylsulfonylethyl)-1,3-dioxolane proved most useful for the synthetic endeavor as the intramolecular alkylation established directly the skeleton of *exo*-brevicomin except for the ethyl chain. The remaining tosylate can then be converted into the ethyl group by reaction with lithium dimethylcuprate. The final step was reductive removal of the sulfonyl residue.

(+)-*exo*-brevicomin

This intriguing strategy of constructing chiral heterocycles from tartaric esters has been successfully exploited in the elaboration of (+)-disparlure [Masaki, 1986], (−)-pestalotin [Masaki, 1984], (+)-asperlin [Masaki, 1992], a mosquito oviposition pheromone [Masaki, 1983a], a constituent of the glandular secretion of civet cat [Masaki, 1983b], and conhydrine [Masaki, 1989]. In these syntheses the bridged ring systems of the intermediates were severed.

An epoxy alcohol prepared from diethyl L-tartrate was converted into an α-amino aldehyde by an S_N2 displacement of one of the internal hydroxyl groups and degradation of the remote terminal carbon atom. By means of a hetero-Diels–Alder reaction a precursor of (−)-kainic acid was assembled [Takano, 1988g].

(-)-kainic acid

(S,S)-1,2:3,4-Diepoxybutane, available from the (R,R)-tartaric esters, was incorporated into mycinolide-IV as C-14, 15, 16 and the hydroxymethyl side chain [K. Suzuki, 1987] by a series of transformations featuring a stereospecific pinacol rearrangement.

mycinolide-IV

Partial utilization of tartaric acid in synthesis is illustrated in the elaboration of a C_{16}-platelet aggregation factor [Fujita, 1982]. Those involving incorporation and degradation of tartaric acid are usually keyed to an initial asymmetric induction by the existing center(s) of chirality, as shown in the assemblage of (+)-homolaudanosine [Czarnocki, 1987] and quinocarcin [Katoh, 1993].

An interesting synthesis of furo[3,4-*d*]pyran derivatives, accompanied by the induction of three contiguous asymmetric centers by an intramolecular hetero-Diels–Alder reaction, is based on the availability of chiral α-allyloxy aldehydes. These aldehydes are obtained from dialkyl tartrates, and their condensation with Meldrum's acid is usually followed by the Diels–Alder reaction. The bicyclic heterocycles are useful intermediates of acuminatolide, sesamin, and sesamolin [Takano, 1988c], ajmalicine and tetrahydroalstonine [Takano, 1988d].

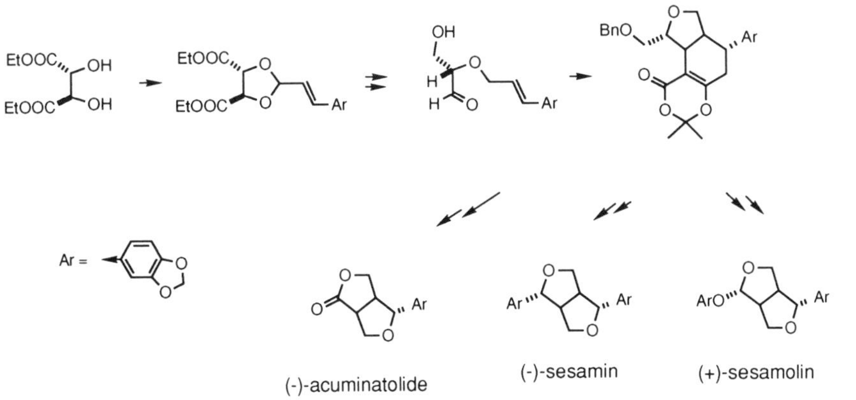

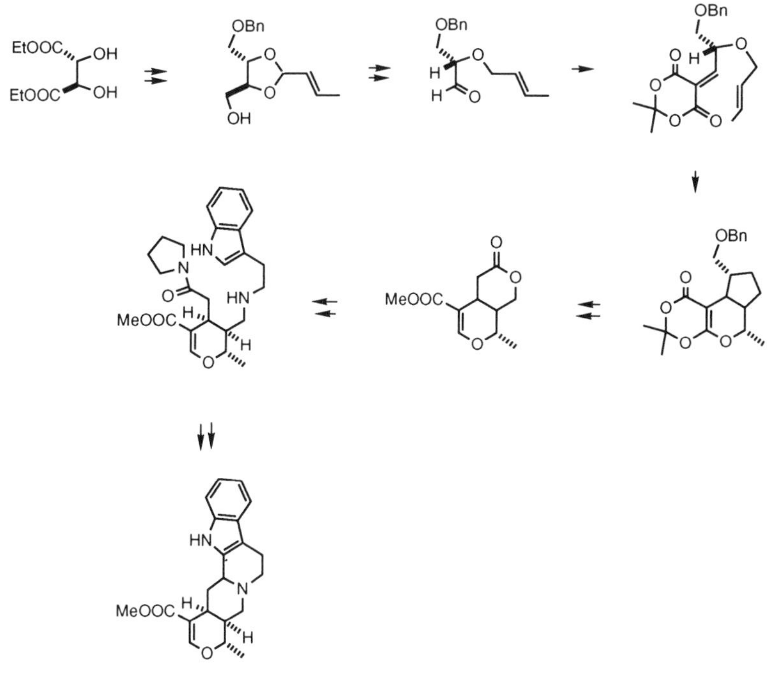

Ar =

(-)-acuminatolide   (-)-sesamin   (+)-sesamolin

(-)-tetrahydroalstonine

4.2. 2-BUTENE-1,4-DIOLS AND DERIVATIVES

In principle, 2,3-epoxy-1,4-butanediol in a specific stereochemical form can be made from tartaric acid. Alternatively, a monoether of a 2-butene-1,4-diol may undergo Sharpless asymmetric epoxidation using a tartaric ester as catalyst. Investigations in the use of these epoxides are contuining.

A useful building block for *ar*-bisabolane sesquiterpenes [Takano, 1989f], phytol [Takano, 1991c), and the C-11/C-17 segment of mycinamicin-III and -IV [Takano, 1992b] is the chiral 1-benzyloxy-3-butyn-2-ol, available in three steps from (*E*)-4-benzyloxy-2-butenol.

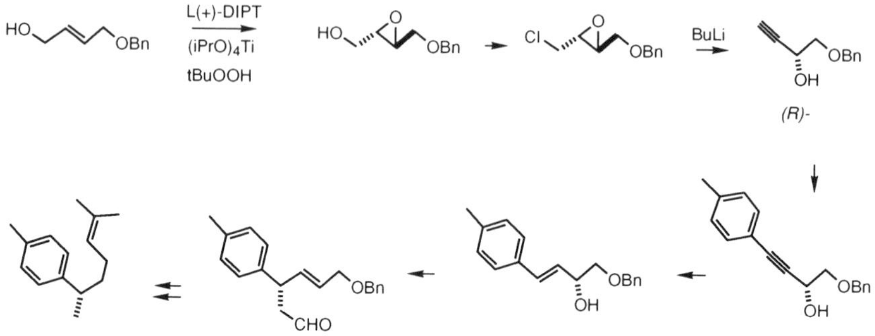

α-curcumene

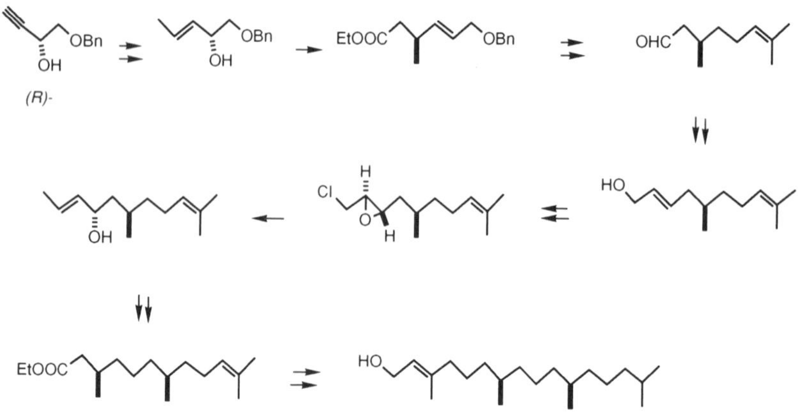

phytol

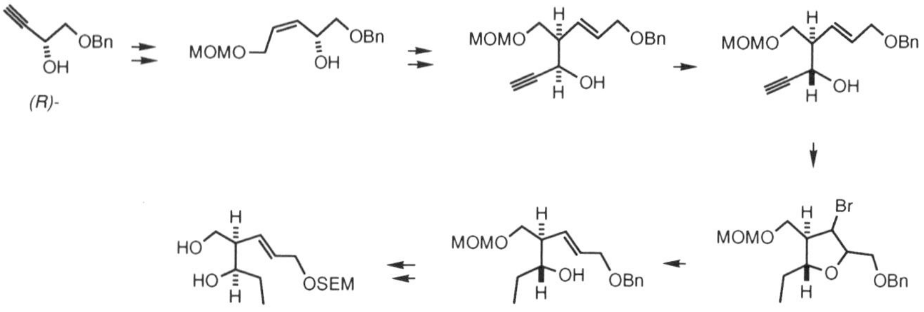

mycinamicins

A synthon for okadaic acid [Ichikawa, 1987] was also derived from the same epoxide.

1-Benzyloxy-2,3-epoxy-4-dodecanone, obtained from the monobenzyl ether of (S,S)-2,3-epoxy-1,4-butanediol, reacted with vinylmagnesium bromide to provide an intermediate of avenaciolide [K. Suzuki, 1986]. Interestingly, when the vinylmagnesium bromide reagent was silylated at the α-position, the reduction of the β-ketol derived by rearrangement of the Grignard reaction product pursued a different stereochemical course, affording predominantly the anti-1,3-diol. The latter compound was converted into isoavenaciolide.

(ds 98:2)   avenaciolide   isoavenaciolide   (ds >99:1)

From the monobenzyl ether of (E)-2-butene-1,4-diol the oxidation and chain elongation generated an α-acetoxy ketone. On treatment with fermenting baker's yeast the compound afforded a chiral *anti*-diol, which could be readily converted into (+)-*endo*-brevicomin [Pedrocchi-Fantoni, 1991].

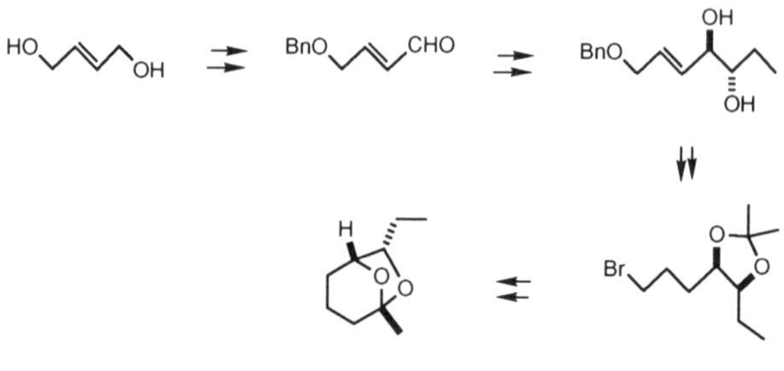

(+)-*endo*-brevicomin

A building block for the tetrahydropyran subunit of compactin has been prepared from (Z)-2-butene-1,4-diol by initial desymmetrization via monobenzylation and epoxidation [Prasad, 1984b].

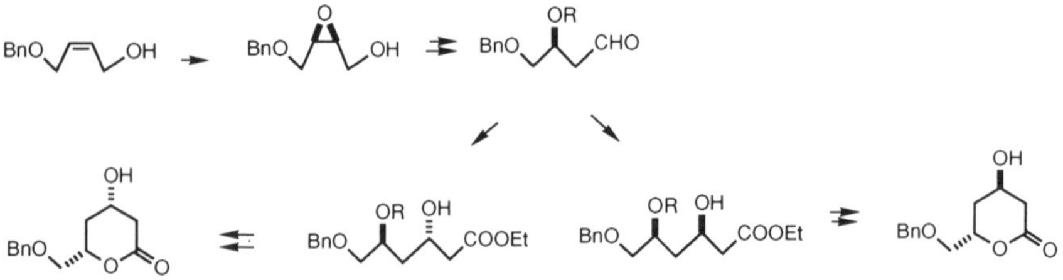

for compactin

A reiterative method for elaboration of carbohydrates via systematic homologation with complete stereocontrol is based on the Sharpless asymmetric epoxidation and Payne rearrangement [Katsuki, 1982; Ko, 1990]. An aldehyde group is regenerated at the chain terminus by a Pummerer rearrangement and it can be extended to a new allylic alcohol unit by the Wittig reaction.

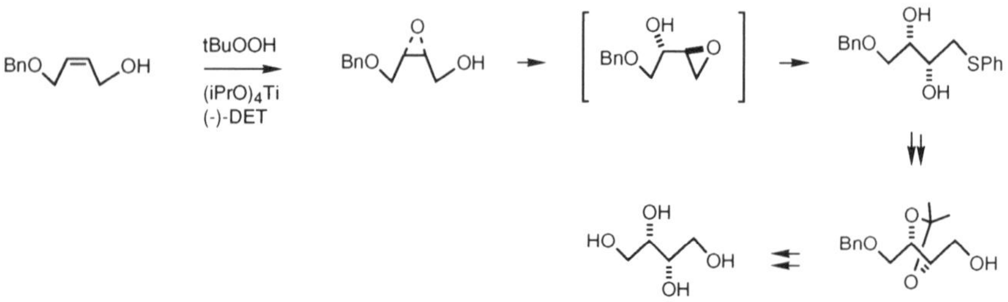

threitol

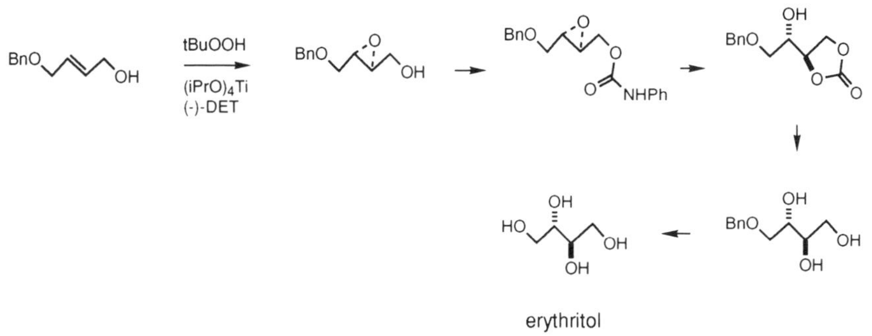

erythritol

For a general synthesis of 2-deoxyhexoses [Roush, 1991] the key process is asymmetric reaction of the epoxy aldehydes, obtained via asymmetric epoxidation of the monoprotected butenediols and oxidation, with allylboronates.

The preparation of the β-hydroxymethyl-β-lactam [Takano, 1983b] from a chiral 4-benzyloxy-2,3-epoxybutanol is quite convenient.

The acetonide of cis-2-butene-1,4-diol has been elaborated into the C-5 to C-8 fragment of N-methylmaysenine [Corey, 1975a]. The reaction sequence involved epoxidation, reaction with lithium dimethylcuprate, and transacetalization which provided a compound with suitably differentiated chain termini for further manipulation. The compound is also useful for the assemblage of the tetrahydropyranyl component of antibiotic X-14547A [P.-T. Ho, 1982], and the isopropyl homolog constitutes a key intermediate of thunbergol [Astles, 1989].

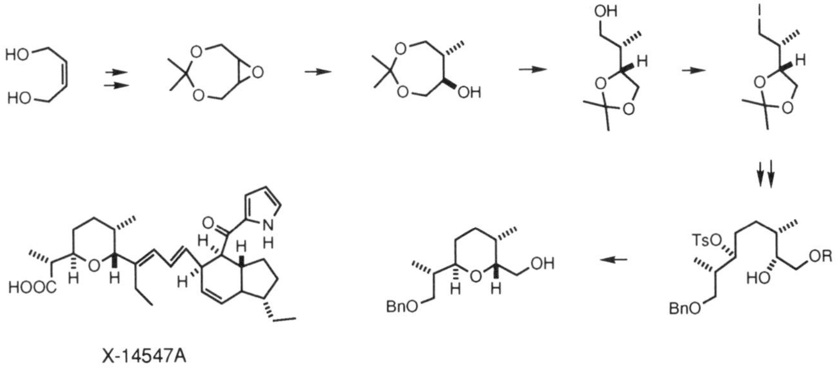

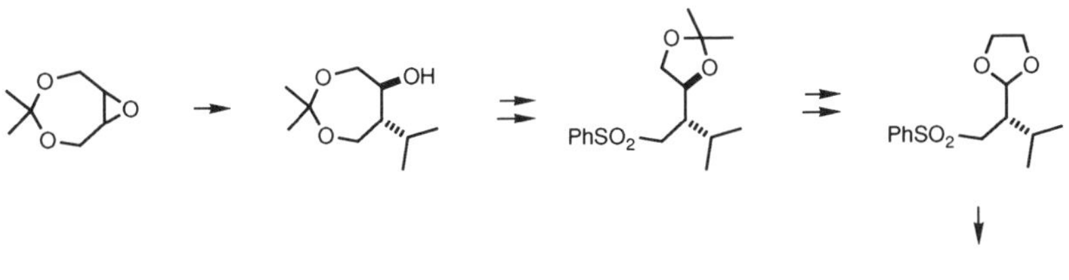

X-14547A

thunbergol

cis-2,3-Epoxy-1,4-butanediol is readily desymmetrized by enzymatic hydrolysis of the diester. The dibutanoate appears to be the substrate of choice [Grandjean, 1991] since it is less sensitive to pH and temperature variations. However, the highest enantiomeric excess was obtained in media with pH 6.0–6.5. The aziridine analogs [Fuji, 1990a] may find use in the synthesis of mitomycins.

The palladium-catalyzed displacement of allylic esters and carbonates is subject to asymmetric induction when the catalyst contains a chiral ligand. Optically active dimethyl 2-vinyl-1,1-cyclopropanedicarboxylate is obtainable by reaction of a (Z)-2-butene-1,4-diol dicarbonate [T. Hayashi, 1988b].

Extending the intramolecular version of the above S_N2' reaction to the dicarbamates delivers precursors of 2-amino-3-butenols [T. Hayashi, 1988a]. Acivicin has been synthesized [Vyas, 1984] from 2-trifluoroacetamino-3-butenol, although in racemic form.

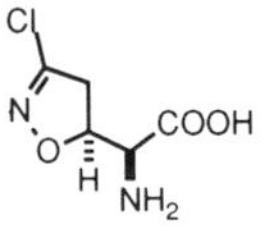

acivicin

2-Butene-1,4-diols have many uses in synthesis. A few examples involving desymmetrizing transformations are shown below. Thus a butenolide synthon for furoventalene [Kido, 1981] was prepared via a Claisen rearrangement. The monobenzyl ether of the (*E*)-isomer was one of two building blocks for (+)-retronecine and (+)-turneforcidine [Ohsawa, 1982], the union of the two building blocks was also effected by a Claisen rearrangement.

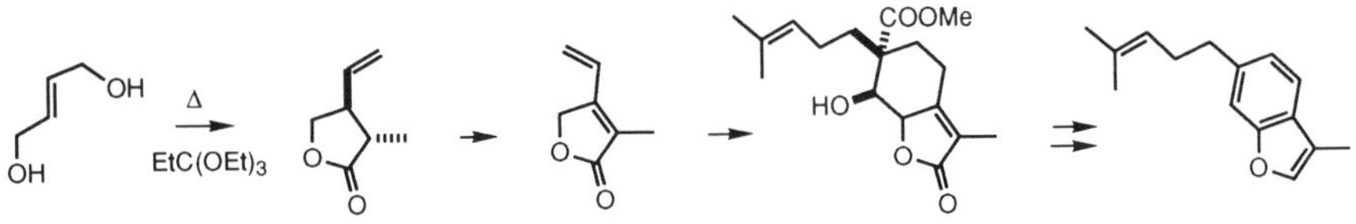

furoventalene

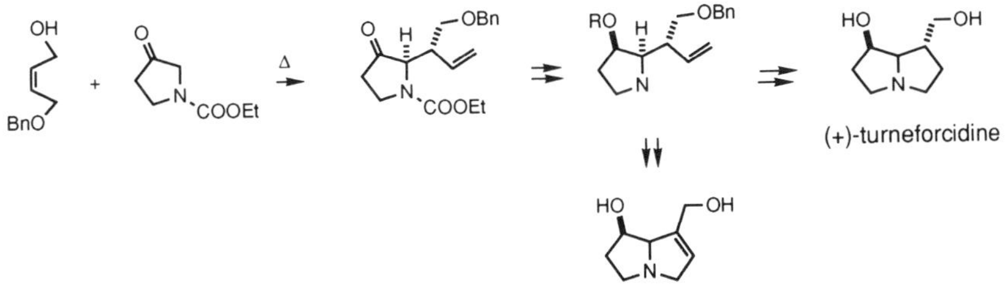

(+)-turneforcidine

(+)-retronecine

The cyclic acetal derived from (*Z*)-2-butene-1,4-diol and pivalaldehyde has been used in a Heck reaction to unite with a *p*-tolyl group. The product, having two oxygenated branches of different length, was readily converted into several sesquiterpenes, including *ar*-turmerone and α-curcumene [Takano, 1993c].

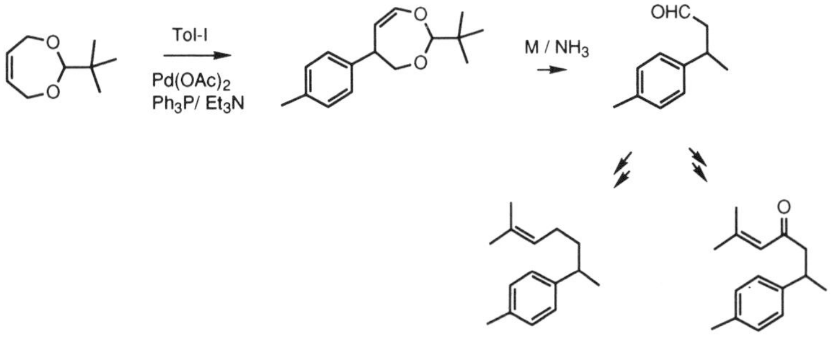

α-curcumene

ar-turmerone

During a study of polyene cyclization a synthesis of progesterone [van Tamelen, 1981] was accomplished. The unsymmetrical substrate was assembled by alkylation with (E)-1,4-dichloro-2-butene. In a photochemical route to crocetin dimethyl ester [Quinkert, 1977] the four-carbon central unit was derived from (E)-1,4-dibromo-2-butene.

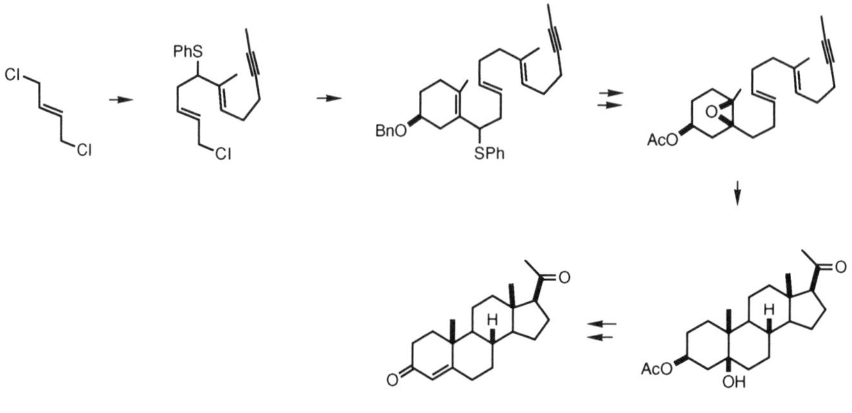

progesterone

For access to an (E)-4-amino-2-butenol derivative as substrate for asymmetric epoxidation and further elaboration into swainsonine [C.E. Adams, 1985] the 1,4-dichloro-2-butene was processed. Actually the asymmetric epoxidation was employed on two different occasions along the synthetic pathway.

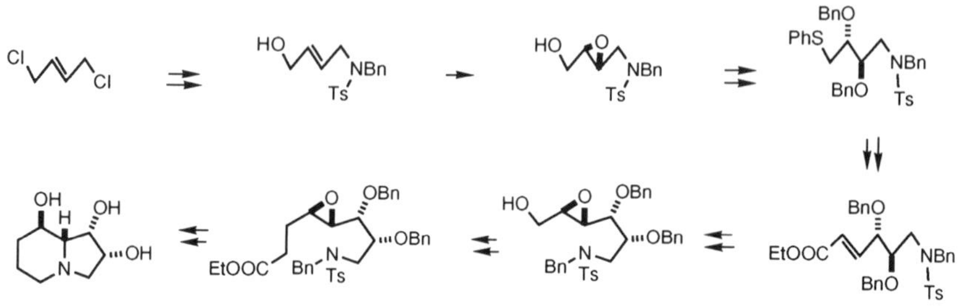

swainsonine

A chemical synthesis of (R)(+)-aspartic acid [Vigneron, 1968] from dimethyl butynedioate with a chiral auxiliary was based on an elegant concept. A template effect was demonstrated, but unfortunately the auxiliary was destroyed.

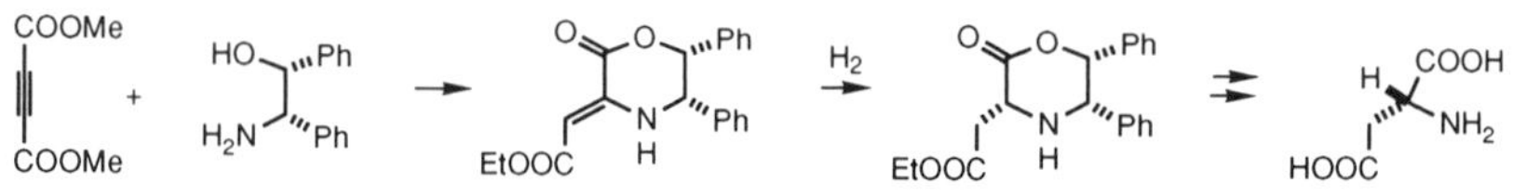

Related C_4 building blocks whose use involves desymmetrization include 1,4-dichoro-2-butyne in the synthesis of the pheromones of *Hyphantria cunea* [K. Mori, 1989d], and its conversion to 2-trimethylsilyl-1,3-butadiene [Batt, 1978; Franck-Neumann, 1986]. 2,3-Bis(3,4-dimethoxyphenyl)-1,3-butadiene suffered desymmetrization in a 1,3-dipolar cycloaddition en route to septicine [Iwashita, 1980].

R= Me, Pr   *Hyphantria cunea*
               pheromones

Ar = 3,4-(MeO)₂C₆H₃   septicine

4.3. SUCCINALDEHYDE AND SUCCINIC ACID DERIVATIVES

The abundance of succinaldehydes and succinic acid derivatives and the presence of functional groups within those molecules make them very useful synthetic building blocks. The highly reactive dialdehyde, available in the form of a cyclic acetal (2,5-dialkoxytetrahydrofuran), is usually generated in situ during a reaction, as illustrated in a tandem Horner–Wadsworth–Emmons reaction/aldol cyclization process for the preparation of a cyclopentenecarboxylic ester en route to mitsugashiwalactone [Amri, 1990].

mitsugashiwalactone

Succinaldehyde itself provided the four central carbon units of squalene [W.S. Johnson, 1970] at the starting point of an elegant approach. The two-directional extension of the chain involved a reiterative sequence of organometallic reaction, Claisen rearrangement, and reduction to the aldehyde level. The synthesis was highly efficient, not only because the carbon chain doubled in each unit introduced in one relevant operation, due to the symmetrical character of the target molecule, but also because reiterative steps could be employed.

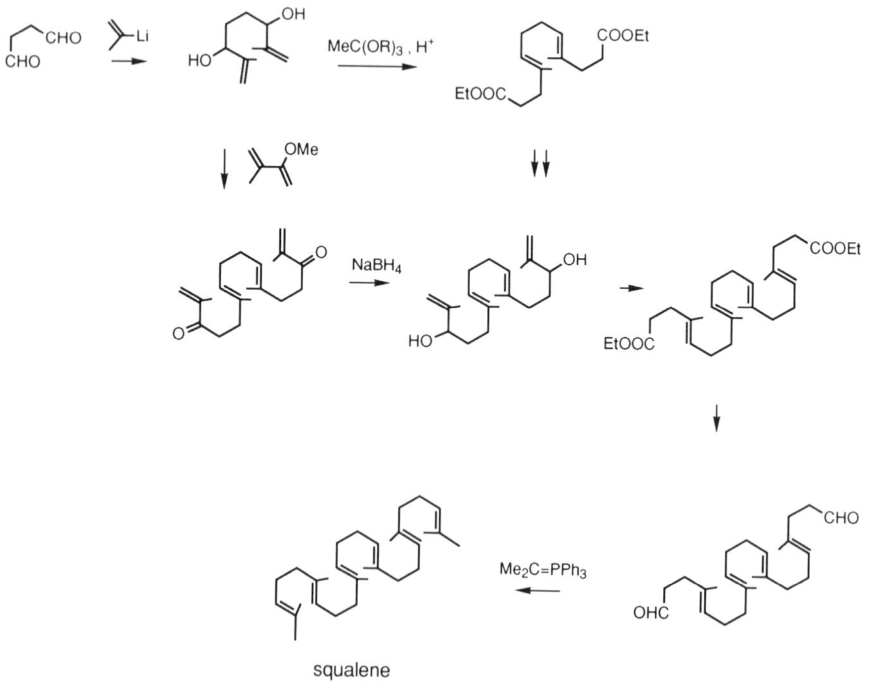

squalene

In the form of a bisdithiane derivative of succinaldehyde the stepwise alkylation formed the basis of a route to cis-jasmone [Ellison, 1972].

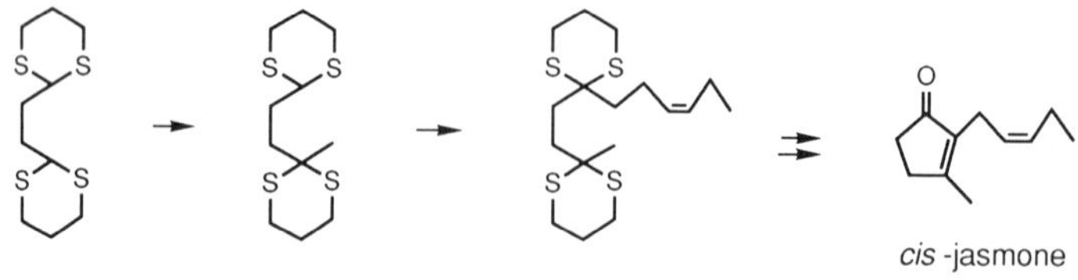

cis -jasmone

A chiral bicyclic oxazolidine derived from the protected succinaldehyde [Royer, 1987] is a valuable intermediate for pyrrolidine alkaloids including δ-coniceine [Arseniyadis, 1988a], and components of ant venoms [Arseniyadis, 1988b, c].

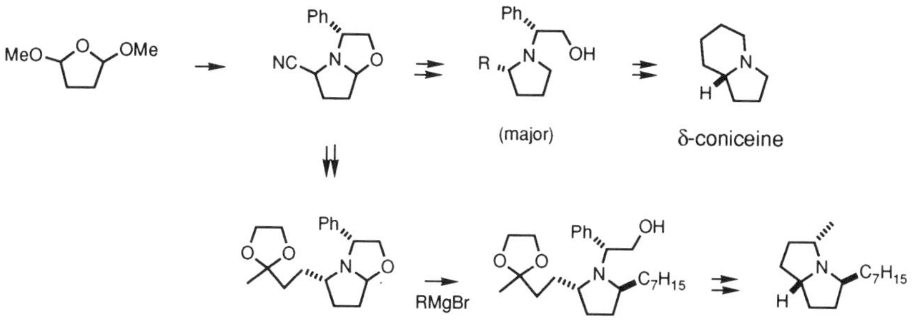

Succinyl dichloride serving as acylating agent to monomethyl malonate enabled elaboration of dimethyl 3,6-dioxooctanedioate which was desymmetrized on aldol cyclization. The resulting cyclopentenone was used in a synthesis of methyl jasmonate [F.J. Johnson, 1982].

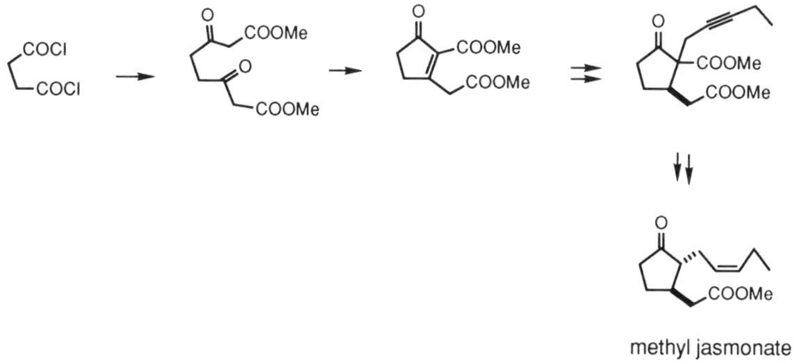

methyl jasmonate

Certain succinic esters and derivatives can be deprotonated to give α,α'-dianions [Misumi, 1984; Furuta, 1984]. Interestingly, a C_2-symmetric dimenthyl 1,2-cyclopropanedicarboxylate was obtained in 99% ee by alkylation with bromochloromethane [Misumi, 1985]. Extension of this chemistry led to a synthesis of (+)-ambruticin [Kende, 1990].

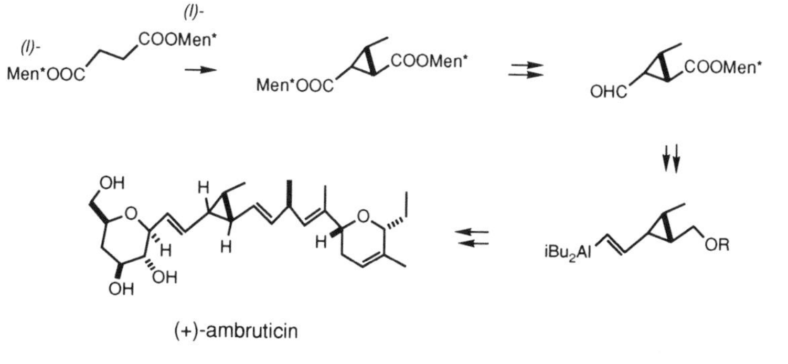

(+)-ambruticin

5

SYMMETRICAL ALIPHATIC COMPOUNDS: FIVE-CARBON AND LONGER CHAINS

In this chapter the synthetic applications of many symmetrical bifunctional aliphatic compounds which have five or more skeletal atoms are discussed. These include substances with all-carbon chains and those containing heteroatoms. A rough division of the various compounds is according to length of the main chains except for those containing framework heteroatom(s).

5.1 FIVE-CARBON UNITS

Intriguing synthetic chemistry has been developed which involves functionalization of 1,4-pentadiene derivatives. Illustrated below are several examples [Wilson, 1978, 1980; Eilbracht, 1985; Ghosez, 1986] concerning chemical operations which also perturb molecular symmetry. The main reason for adopting the symmetrical starting materials in each case is evident; formation of the products was uncomplicated due to identicalness of the two reactive groups. Generally the unreacted moiety is a latent function for another purpose.

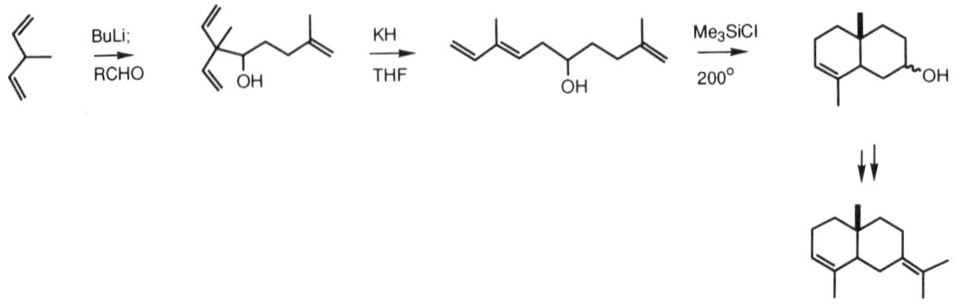

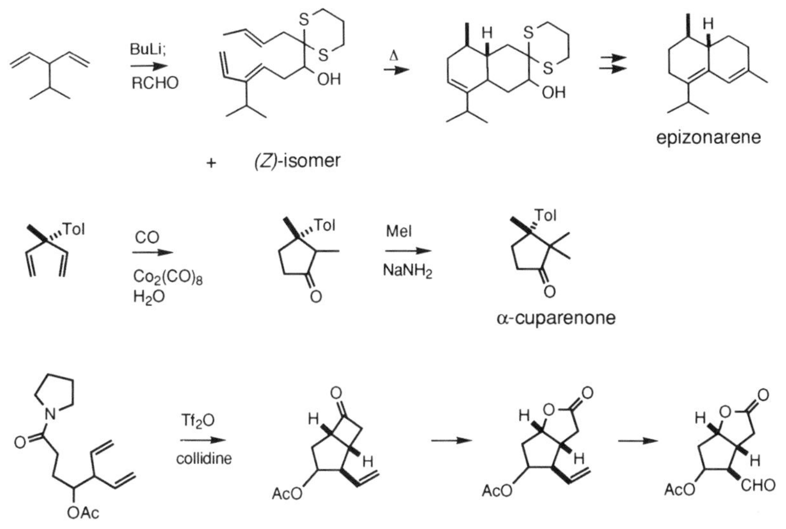

epizonarene

(Z)-isomer

α-cuparenone

The availability of many methods for differentiation of the two double bonds of 1,4-pentadien-3-ol makes this compound (and derivatives) quite versatile in synthetic applications. For example, thermal rearrangements of proper derivatives of 1,4-pentadien-3-ol provided conjugate dienes which acted as substrates for intramolecular Diels–Alder reactions employed in the synthesis of δ-coniceine [Weinreb, 1979] and valerenal [Bohlmann, 1980], respectively.

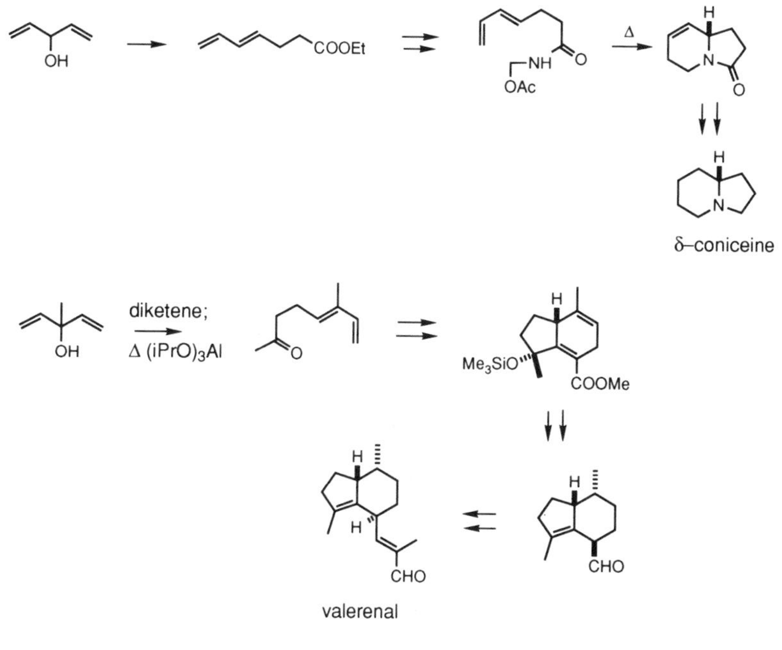

δ–coniceine

valerenal

3-Trimethylsiloxy-1,4-pentadiene forms a delocalized anion which undergoes alkylation at the terminal carbon atom. To demonstrate the synthetic utility of this desymmetrizing behavior as a methyl vinyl ketone β'-anion synthon, a preparation of friedosnyderane [Oppolzer, 1980a] was accomplished.

friedosnyderane

The limitations to one reaction site (double bond) of the diene by rendering the reaction intramolecular is crucial for the desymmetrization. This tactic has further been exploited in hydrosilylation [Tamao, 1990] and hydroboration [T. Harada, 1990, 1992].

	anti, anti	syn, anti	syn, syn
9-BBN	13	1	-
ThexBH$_2$	1	15	1

	anti, anti	syn, anti	syn, syn
9-BBN	>100	1	1
ThexBH$_2$	1	24	0

With chemical processes such as the Sharpless asymmetric epoxidation the occurrence of kinetic resolution of a prochiral alcohol permits the generation of one major product, even the reaction is intramolecular. Achievement of both diastereofacial and enantiotopic selectivity is possible in epoxidation of the dienol [Hatakeyama, 1986]. One particular compound thus obtained has been transformed into (−)-*exo*-brevicomin and (+)-*endo*-brevicomin [Hatakeyama, 1985b]. To acquire the other pair of enantiomers it would require a change of the tartrate ester catalyst.

Prostaglandin intermediates are readily acquired from the same epoxy alcohol [Okamoto, 1988]. (Note the epoxy alcohol obtained from 3,5-dimethyl-2,5-heptadien-4-ol was a synthetic precursor of (+)-verrucosidin [Hatakeyama, 1988].)

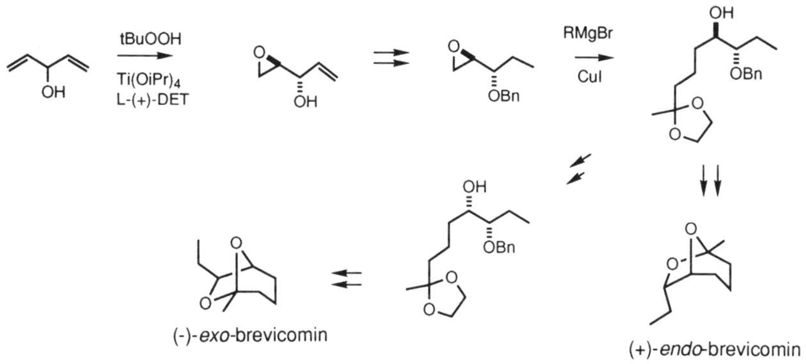

(-)-*exo*-brevicomin

(+)-*endo*-brevicomin

Actually the enantiomeric epoxy alcohols have been used in the synthesis of several rare sugars: digitoxose, cymarose, olivose, and oleandrose [Hatakeyama, 1986].

The construction of the chiral synthon containing the cyclohexane ring of the immunosuppressive macrolide FK-506 [M. Nakatsuka, 1990] has been delegated to a process initiated by asymmetric epoxidation of 1,4-pentadien-3-ol. A Claisen rearrangement was also involved.

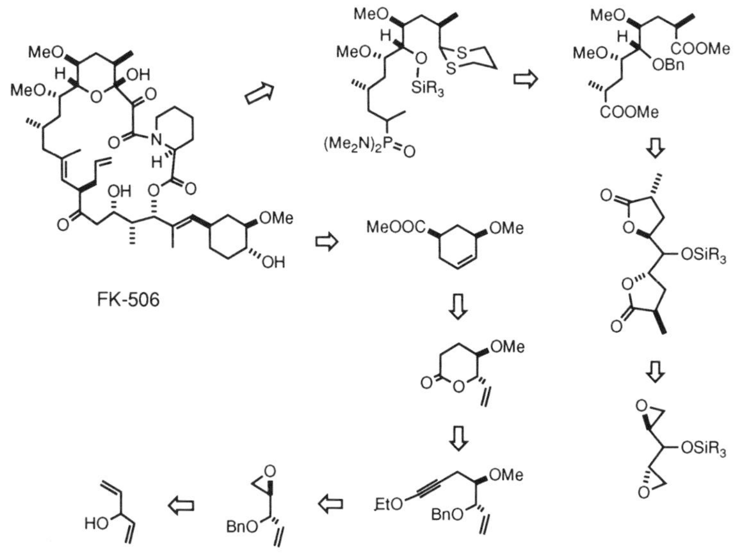

FK-506

The chiral diepoxide is a related compound employed in the synthesis. However, this building block was more conveniently obtained from L-arabinitol.

The Sharpless asymmetric epoxidation was used to desymmetrize 1,5-bis(trimethylsilyl)-1,4-pentadien-3-ol in a formal synthesis of lipoxin-B [Y. Kobayashi, 1988].

Analogous to the asymmetric epoxidation is the ene reaction of 2,4-dimethyl-1,4-pentadien-3-yl t-butyldimethylsilyl ether with methyl glyoxylate in the presence of a chiral titanium catalyst [Mikami, 1992]. Both diastereoselectivity and enantioselectivity are optimal.

A conceptually elegant approach to seychellene [Hagiwara, 1985] involved bisannulation by a reflexive Michael reaction. It is interesting to note that after incorporation of 1,4-pentadien-3-one into an unsymmetrical environment the resulting diketone was also readily differentiated in that one site was converted to a methyl-bearing tertiary carbon atom whereas the more hindered ketone was replaced by an exocyclic methylene group.

1,4-Pentadien-3-one has been employed in the construction of linear tricyclic ketones [Winkler, 1994] by tandem Diels–Alder reaction with sulfolenes which bear a terminal conjugated diene unit at the α-position. The transformation exhibits high levels of stereoselectivity.

Because glutaraldehyde (in aqueous solution) is a readily available chemical and its derivatives are implicated as biosynthetic intermediates, many efforts have been devoted to the use of glutaraldehyde in synthesis, sometimes with explicit biomimetic pretension. Thus over sixty years ago a four-component condensation involving glutaraldehyde as one of them was already designed to produce lobelanine [Schöpf, 1935] in one step. More recently 1,4-dihydropyridine was generated in situ from glutaraldehyde and ammonia in a process for nicotine synthesis [Leete, 1972]. When ammonia was replaced with tryptamine, a tetracyclic base was formed and this was readily reduced to give an alkaloid from *Dractontomelum magniferum* [Gribble, 1972].

A versatile scheme for piperidine alkaloid synthesis consists of Strecker reaction of glutaraldehyde with chiral amino alcohols at the starting point. Such adducts contain a piperidine ring fused to an oxazolidine and they possess ambiphilic potential. Firstly, the dual reactivity of the cyano group is exploitable in that its stabilization of an α-carbanion enables alkylation at the captodative site, while under the influence of silver ion its departure renders the Strecker adducts electrophilic. Removal of the cyano group in any of the synthetic intermediates is achievable by reaction with complex metal hydride reagents. Furthermore, the oxazolidine ring suffers cleavage on exposure to these reducing agents or Grignard reagents, therefore it is possible to generate either 2-substituted or 2,6-disubstituted piperidines as desired. Most importantly the bicyclic oxazolidines are chiral; asymmetric induction accompanies the nucleophilic or electrophilic alkylation.

The effectiveness of the methodology is readily demonstrated in the synthesis of both enantiomers of coniine and dihydropinidine [Guerrier, 1983]. Other applications include preparation of (+)-conhydrine [Ratovelomanana, 1985], (+)-solenopsin-A [Grierson, 1986], (−)-monomorine-I [Royer, 1985a], (−)-gephyrotoxin-223AB [Royer, 1985b], and precoccinelline [Yue, 1994].

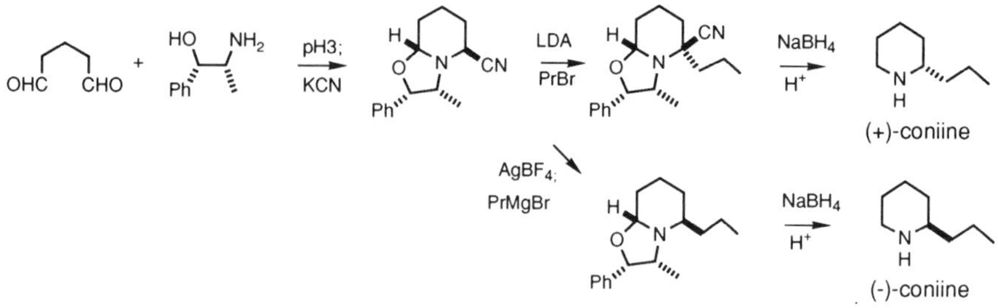

(+)-coniine

(-)-coniine

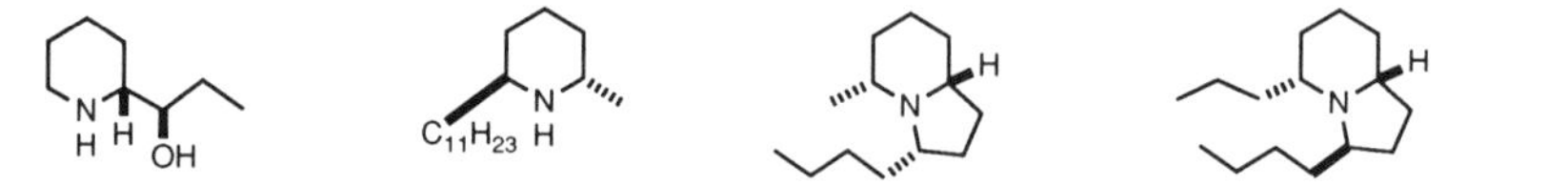

(+)-β-conhydrine   (+)-solenopsin-A   (-)-monomorine-I   (-)-gephyrotoxin-223-AB

Interestingly, (+)-tetraponerine-8 which contains a tricyclic perhydropyrimidine core was synthesized using a chiral piperidine derivative accessed in the same manner and via another Strecker reaction [Yue, 1990]. Two new asymmetric centers were induced by the latter intramolecular reaction.

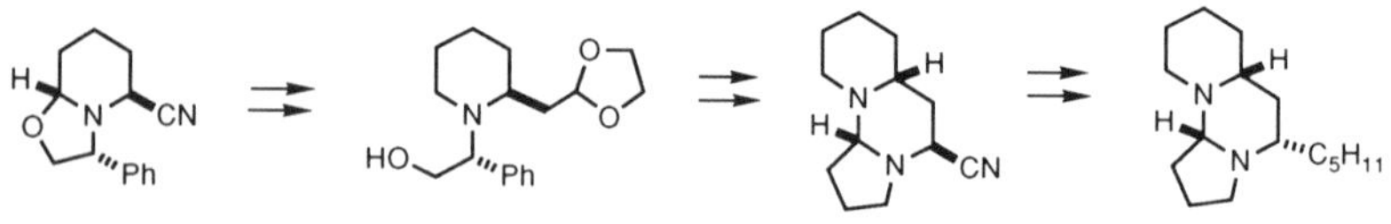

(+)-tetraponerine-8

It should also be noted that spirocyclic systems may be constructed from appropriate intermediates of this series, as shown by the elaboration of (+)- and (−)-isonitramine [Quiron, 1988] and (−)-perhydrohistrionicotoxin [Zhu, 1991].

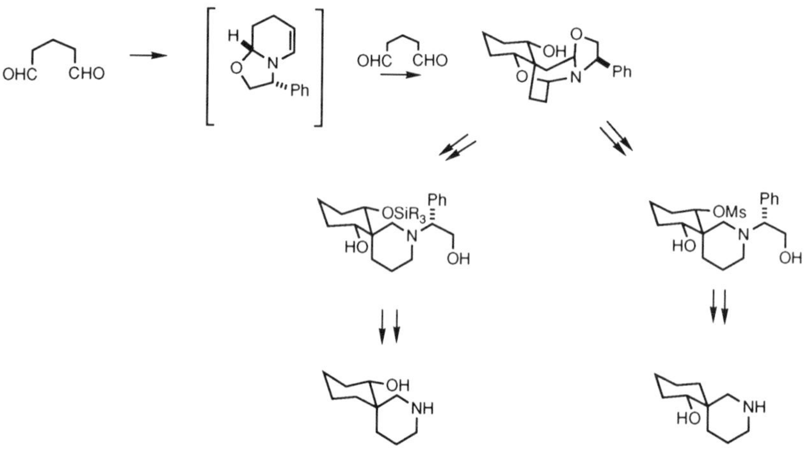

(+)-isonitramine   (-)-isonitramine

The reactivity differentiation of two identical aldehyde groups of prochiral glutaraldehydes by methods such as the above is exceptional. Synthetic chemists often resort to the monoprotection tactic on the corresponding primary alcohols to achieve the formal desymmetrization [Sarkar, 1990], or de novo generation of the monoprotected dialdehydes [Boeckman, 1980].

hirsutene

Dicarboxylic acids are very useful synthetically owing to many possible ways of desymmetrization. The desymmetrization is particularly easy by monoesterification of succinic and glutaric acids by simple alcoholysis of cyclic anhydrides. Interestingly, *t*-butanolysis of 3-substituted glutaric anhydride under catalysis of Amano-P lipase afforded the (*R*)-half ester in 60–90% ee [K. Yamamoto, 1988]. The free carboxyl group of such monoesters may be activated as an acid halide or aldehyde to undergo selective homologation. In this regard the preparation of an acetylenic isocyanate from monomethyl succinate for a [2+2+2]-cycloaddition approach to camptothecin [Earl, 1983] may be mentioned.

camptothecin

Among the uses of glutaraldehydic esters is a contribution of the five-carbon terminal unit of leukotriene LTA$_4$ methyl ester [Gleason, 1980].

LTA$_4$ methyl ester

Many synthetic applications involve other transformations of glutaric anhydrides/esters. For example, the two termini of 3-methylglutaric anhydride differentiated by aminolysis en route to paniculide-A [Jacobi, 1987] were transformed into an oxazole ring and an ynone. An intramolecular hetero–Diels–Alder reaction/elimination process accomplished the assemblage of a bicyclic intermediate.

paniculide-A

A low enantiomeric excess was observed when 3-substituted glutaric anhydrides were hydrogenated in the presence of a Ru(II) catalyst which contained chiral ligands [Osakada, 1981]. The products were the δ-valerolactones.

cis-2,4-Dimethylglutaric anhydride is a very useful starting material for the synthesis of natural products. In many cases a half-ester of the diacid was prepared to allow chemoselective manipulations at the two termini without disturbing the two stereocenters. It is not surprising that syntheses of α-multistriatin [P.A. Bartlett, 1979; Walba, 1982b] and the EF-ring component of nigericin [P.A. Bartlett, 1985] were pursued on this basis in view of the presence of a *cis*-3,5-dimethyltetrahydropyran subunit in the targets.

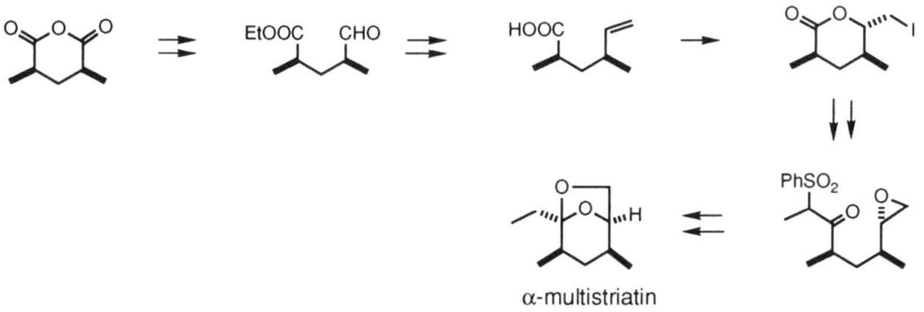

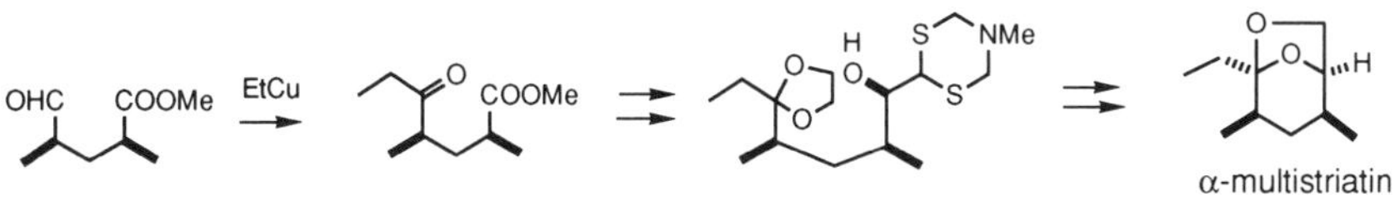

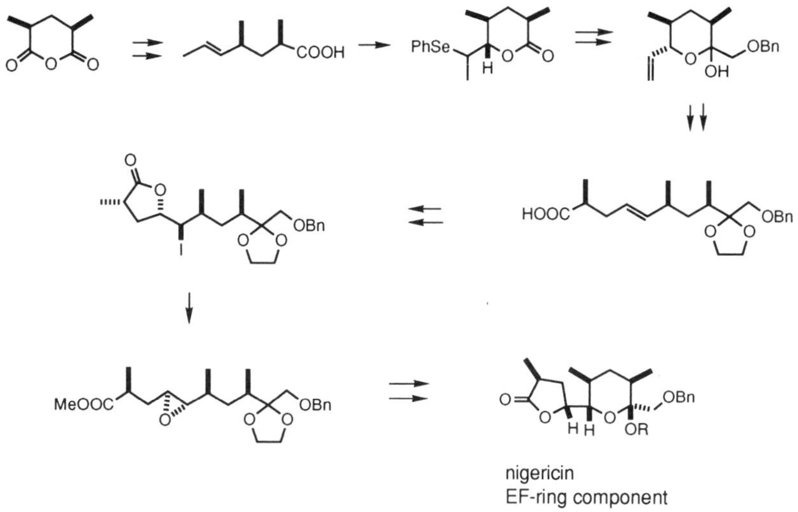

α-multistriatin

α-multistriatin

nigericin
EF-ring component

Note that despite the presence of a plane of symmetry in fulvinic anhydride and the apparently uncomplicated reconstitution of fulvine from it via direct acylation, the expedient protocol involved formation of a half-ester [Vedejs, 1984].

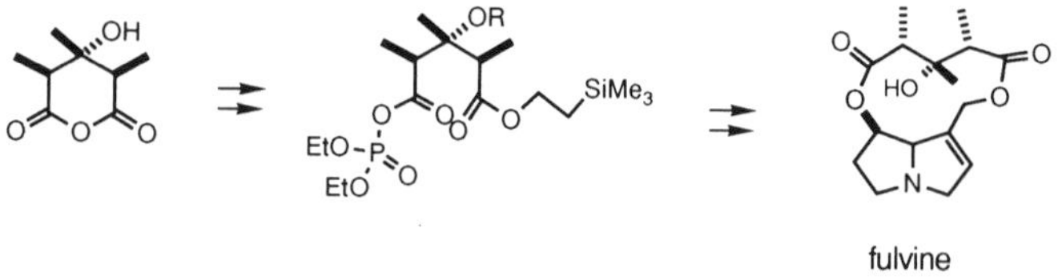

fulvine

While many approaches to the Prelog–Djerassi lactone delayed the methyl group introduction at the α-position of the lactone, the methylation step can be obviated by using cis-2,4-dimethylglutaric anhydride in the synthesis [Santelli-Rouvier, 1984]. Desymmetrized intermediates may undergo reactions such as lactonization [P.A. Bartlett, 1980], matched aldol reaction [Masamune, 1980, 1981], and crotylmetal reactions [Maruyama, 1981; R.W. Hoffmann, 1982, 1989]. It is also possible to use the diol to achieve the synthesis [M. Honda, 1984].

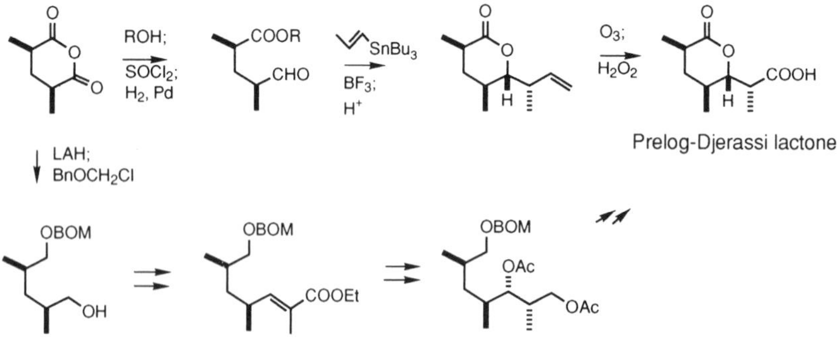

Prelog-Djerassi lactone

(Note the nonstereoselective syntheses via C-acylation and reduction [Bergel'son, 1963; Nakano, 1979].)

A convergent synthesis of the putative triene precursor of monensin-A [Patel, 1986] consisted of three-fragment coupling, the terminal subunit being derived from dimethyl meso-2,4-dimethylglutarate.

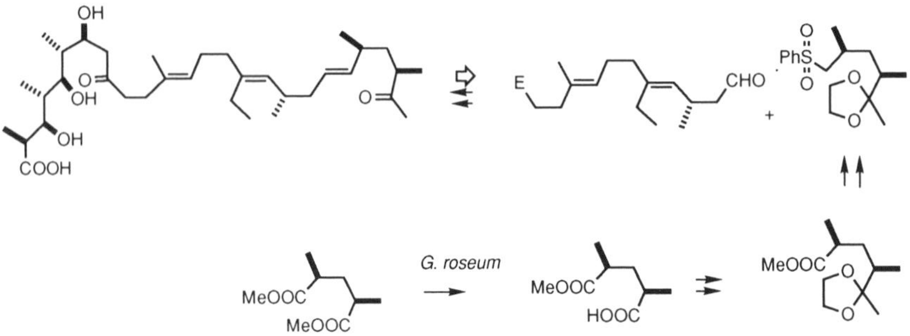

G. roseum

Chiral compounds are accessible by reaction of *meso*-anhydrides with chiral nucleophiles. For example, in a synthesis of (+)-compactin [Rosen, 1985] a ketophosphonate reagent that comprised the pyran portion was conveniently prepared by alcoholysis of a 3-*t*-butyldimethylsiloxyglutaric anhydride with (*R*)-1-phenylethanol. A surprsingly high (8:1) asymmetric induction in the alcoholysis step was observed (cf. aminolysis method [Karanewsky, 1991]).

(+)-compactin

The anhydride of *trans*-2,4-dimethylglutaric acid has a C_2-symmetry. Its use in the elaboration of the hydronaphthalene subunit of kijanolide has been demonstrated [Marshall, 1987, 1990].

kijanolide

Chemoselectivity is observed in Grignard reactions of cyclic anhydrides; this enabled a synthesis of catalpalactone [Marx, 1982]. Friedel–Crafts acylations also exhibit chemoselectivity and one such reaction was used to prepare 5-oxo-6-heptenoic acid for synthesis of a bicyclic intermediate of strigol [Dolby, 1976].

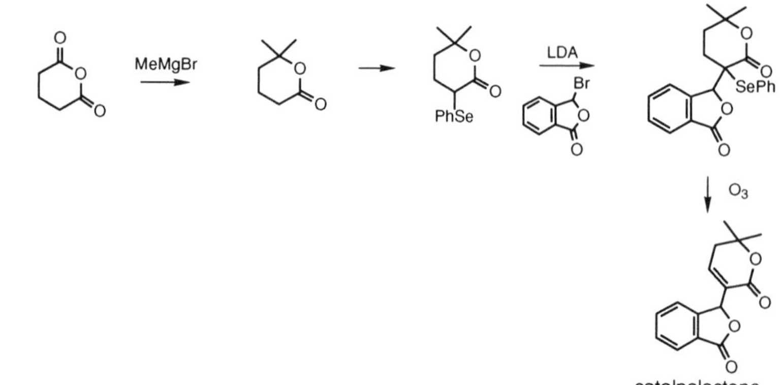

catalpalactone

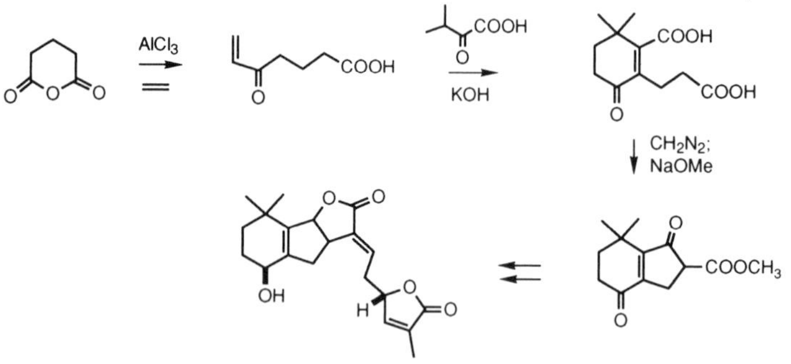

strigol

An anionic version of the *C*-acylation with 3-methylglutaric anhydride was involved in annulation of a furan ring en route to a synthesis of gnididione [Buttery, 1990].

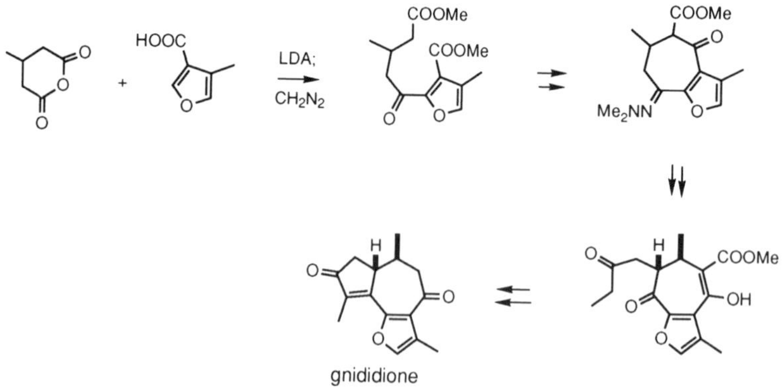

gnididione

Temporary setting of two identical termini of a carbon chain in different oxidation levels represents another frequently employed strategy for desymmetrization. In a synthetic scheme for hirsutene [Little, 1979] the precursor of a fulvene intermediate was identified as 3,3-dimethylglutaraldehyde. However, reactivity problems dictated a surrogate in the form of the hydroxy aldehyde.

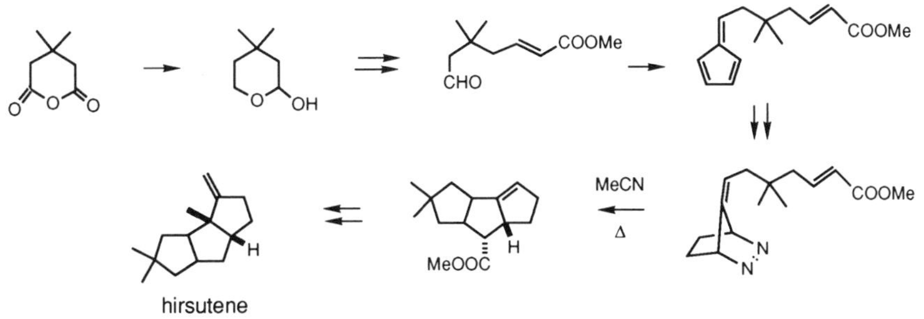

hirsutene

It is interesting to note that quadrone has been synthesized [Sowell, 1990] from dimethyl 3,3-dimethylglutarate via monoallylation. The first cyclization made use of an electrochemically promoted intramolecular hydroacylation on the mono-Horner–Wadsworth–Emmons adduct of the glutaraldehyde.

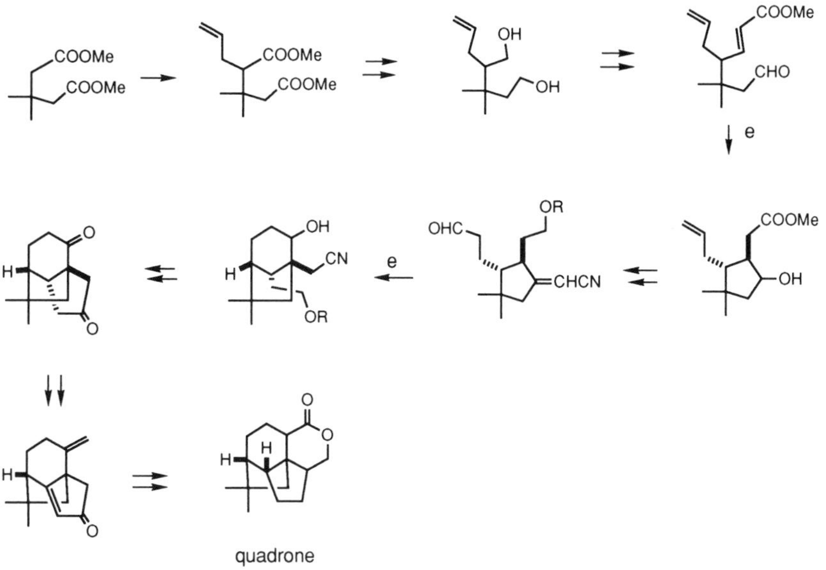

quadrone

The aldol reaction of the tin(II) enolates of diamides derived from 3-substituted glutaric acids and (4S)-isopropyl-1,3-thiazolidine-2-thione with aldehydes is under chelation control [Nagao, 1992]. Thus δ-lactone products bearing three asymmetric centers are readily prepared by this method. A somewhat different protocol for group differentiation is shown in a synthesis of the methyl ester of the (+)-Prelog–Djerassi lactone [Nagao, 1985].

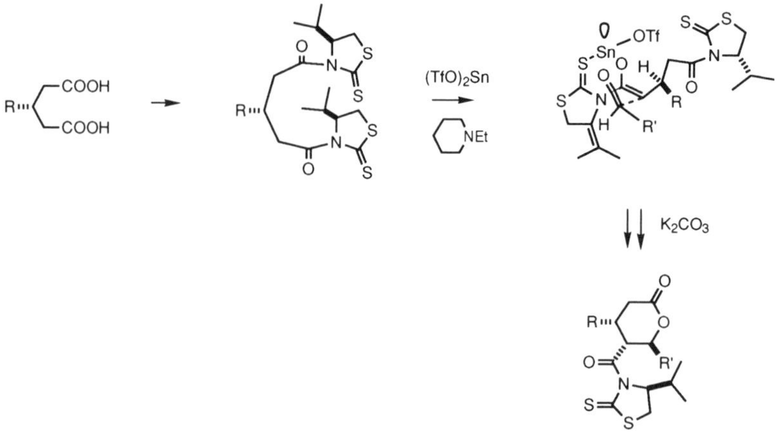

Polyketide chain formation is efficiently achieved using dianion acylation with 3-substituted glutaric esters. Biomimetic cyclization of the adducts leads to aromatic compounds. Thus chrysophanol and related substances are readily accessible [Harris, 1976] (see also Yamaguchi [1986]).

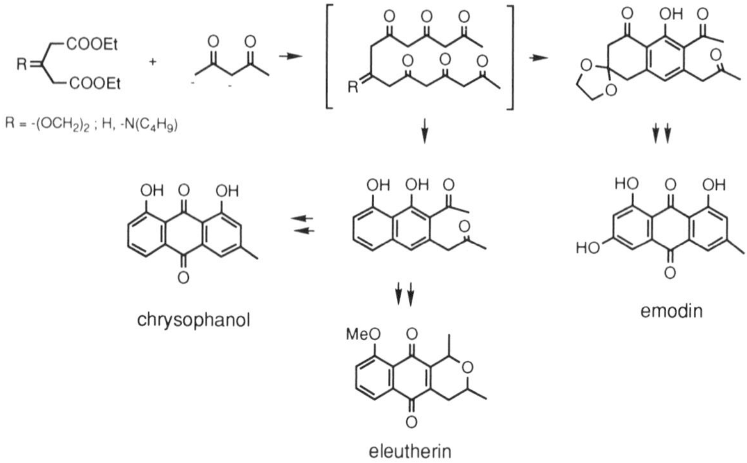

chrysophanol

eleutherin

emodin

Tricarballylic acid is glutaric acid with an additional carboxyl group at C-3. Its condensation with anhydrides leads to symmetrical fused dilactones. Two members of these lactones have been employed in the synthesis of avenaciolide [W.L. Parker, 1973], and methylenomycin-B [Strunz, 1982].

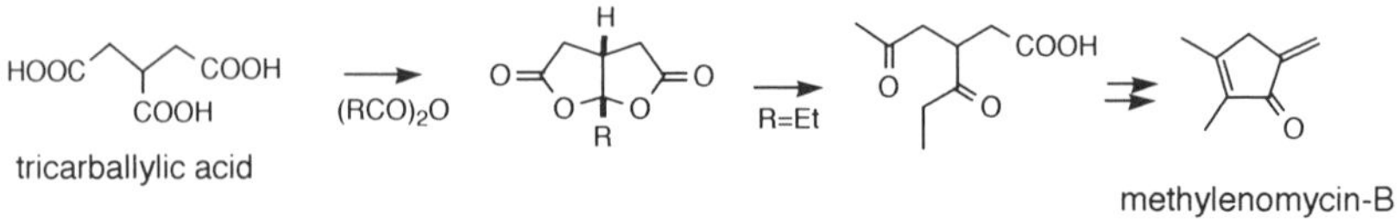

tricarballylic acid

methylenomycin-B

avenaciolide

By far the most common method for the preparation of chiral synthetic intermediates from *meso*-glutaric acid derivatives is by enzymatic hydrolysis of the diesters. Pig liver esterase (PLE) has been used to advantage, for example, in the synthesis of (+)-faranal [Poppe, 1986], the yellow scale pheromone [E. Alvarez, 1988], and the chiral hydroxy acid portion of the nonterpene moiety of verrucarin-A [Mohr, 1982]. The C-1/C-7 segment of 14-membered ring macrolide antibiotics (oleandomycin, lankamycin, etc.) has been synthesized from a more highly substituted diester [Born, 1989].

(+)-faranal

yellow scale
sex pheromone

A much less obvious application of the half-ester is its incorporation into (−)-retigeranic acid [Wender, 1990], initially in the side chain of an aromatic precursor. Transformation of the aromatic compound into a tetracyclic olefin containing the CDE-ring portion of the target molecule was achieved by an intramolecular arene *meta*-photocycloaddition.

retigeranic acid

It is interesting that the (S)(+)-monoester obtained by PLE-catalyzed semihydrolysis of dimethyl 3-hydroxy-3-methylglutarate was converted into both (S)(−)- and (R)(−)-mevalonolactone, respectively, via selective reduction of the carboxylic acid with diborane and of the ester group by lithium borohydride [F.-C. Huang, 1975].

The PLE-catalyzed hydrolysis of dimethyl 3-N-benzyloxycarboxamidoglutarate followed by deprotection of the amine provided a building block for synthesis of carpetimycin-A [Iimori, 1983] and thienamycin [Kurihara, 1985].

Enzymatic hydrolysis of a 3-substituted glutaric ester was also the first step toward preparation of the C-1/C-9 fragment of rhizoxin [S. Kobayshi, 1992]. It is of interest to note that the two-carbon unit on the ethereal side of the intermediary δ-lactone was destined to become the highly oxidized branch (CH$_2$CO) of the corresponding lactone subunit of rhizoxin.

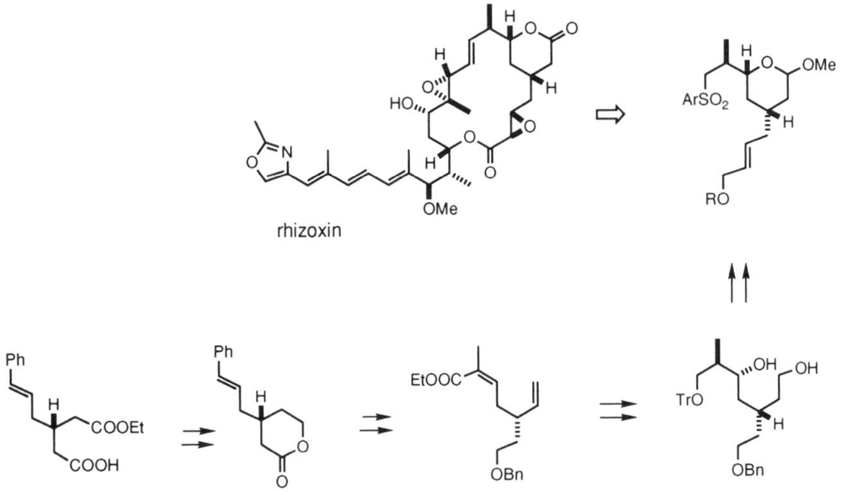

rhizoxin

The synthetic applications of other 3-monosubstituted and 3,3-disubstituted glutaric acid derivatives are numerous. After examination of the several examples listed below one should be impressed by the ingenuity of the chemists who formulated and executed such work. These synthesis include those of camphor [Komppa, 1903], dihydrocorynantheine [van Tamelen, 1969b], ajmalicine [van Tamelen, 1969c], and corynantheine [van Tamelen, 1969d]. The 3-monosubstituted glutaric esters employed in the indole alkaloid syntheses were readily prepared by Michael reactions on dimethyl glutaconate.

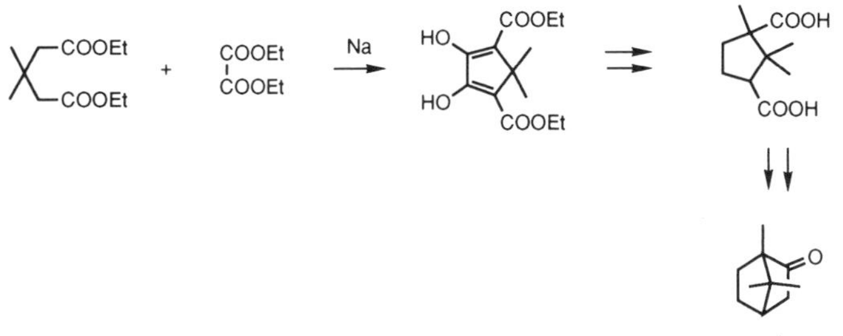

camphor

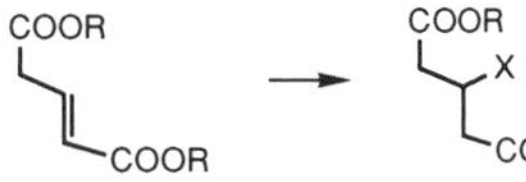

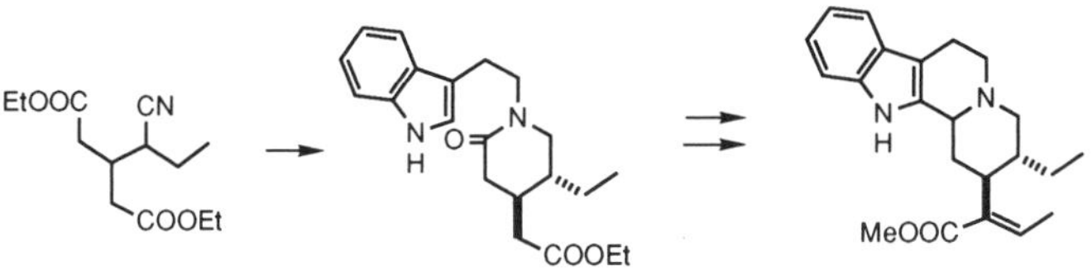

dihydrocorynantheine

corynantheine   ajmalicine

It is pertinent to compare the above to an approach to camptothecin [Kametani, 1981]. The pairwise predifferentiation of the tetraester to better serve the synthetic operation for camptothecin should be noted.

camptothecin

The use of 3-oxoglutaric acid in the synthesis of emetine [Chapman, 1962; Clark, 1962a, b] was based on an excellent exploitation of hidden symmetry. Thus a twofold Mannich condensation with 6,7-dimethoxy-3,4-dihydroisoquinoline led to a ketone which could be controlled in the form of either the *meso* or the racemic modification by choosing conditions to ensure crystallization of a particular diastereomeric salt. The *meso* compound was desired and after aldol ring closure of its 1:2 adduct with methyl vinyl ketone, only functional group adjustments (including reduction) remained to be made to complete the synthesis.

The most critical aspect of this concise approach concerned desymmetrization of the prochiral ketone by cyclization. The *meso* intermediate dictated the generation of only one product. It is of interest to note that one of the methyl vinyl ketone units introduced into the diisoquinolinyl acetone acted as a protecting group for the secondary amine, thereby saving relevant manipulation.

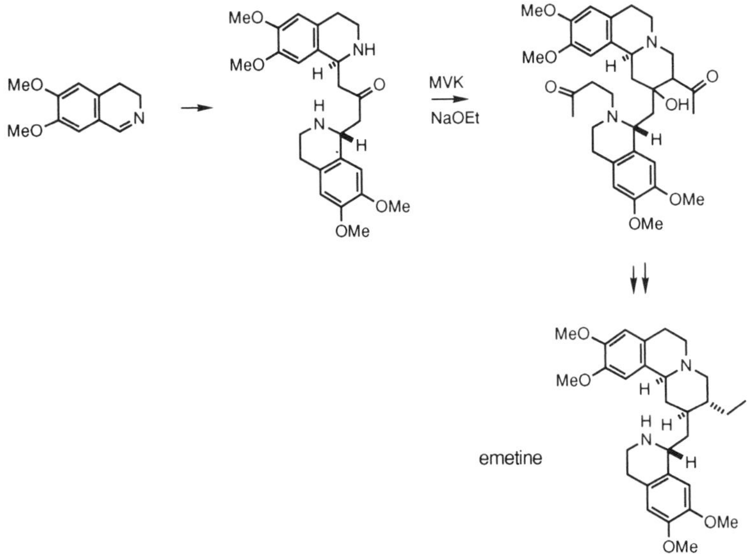

emetine

A route to camptothecin [Curran, 1992] based on [4 + 1]-radical annulation required an α-pyridone. The substitution pattern of this precursor indicated a convenient source from the Doebner condensation product of dimethyl 3-oxoglutarate and cyanoacetic acid.

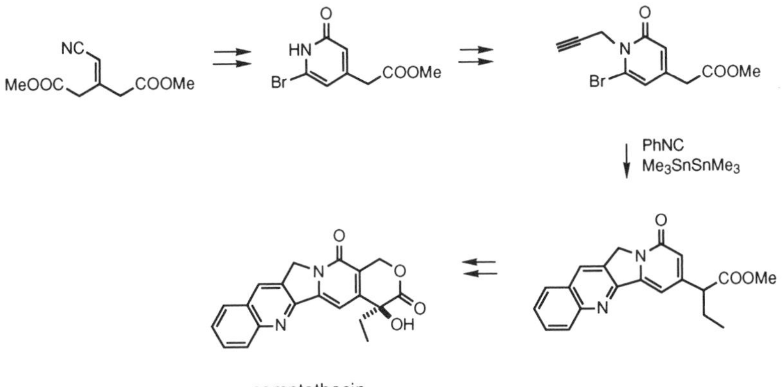

camptothecin

The discovery of the reproducible conditions for the synthesis of *cis*-bicyclo[3.3.0]octane-3,7-diones by condensation of a dialkyl 3-oxoglutarate with α-dicarbonyl compounds [Weiss, 1968] has opened a new avenue towards many intriguing molecules. The benefit conferred upon synthesis of natural products, particularly terpenes, from the ready acquisition of the di- and polyquinane intermediates is evident. For example, the parent diketone was used in hirsutic acid-C [H. Hashimoto, 1974; Z.-F. Xie, 1987], pentalenene [Piers, 1989], quadrone [Piers, 1985], and a precursor of the heteroyohimbe alkaloids [Leonard, 1990]. The 1,5-dimethyl derivative found service in gymnomitrol [Coates, 1979; Paquette, 1981] and bifurcarenone [K. Mori, 1989e], whereas a propellanedione derived from 1,2-cyclopentanedione was verified as an ideal starting material for modhephene [Wrobel, 1983].

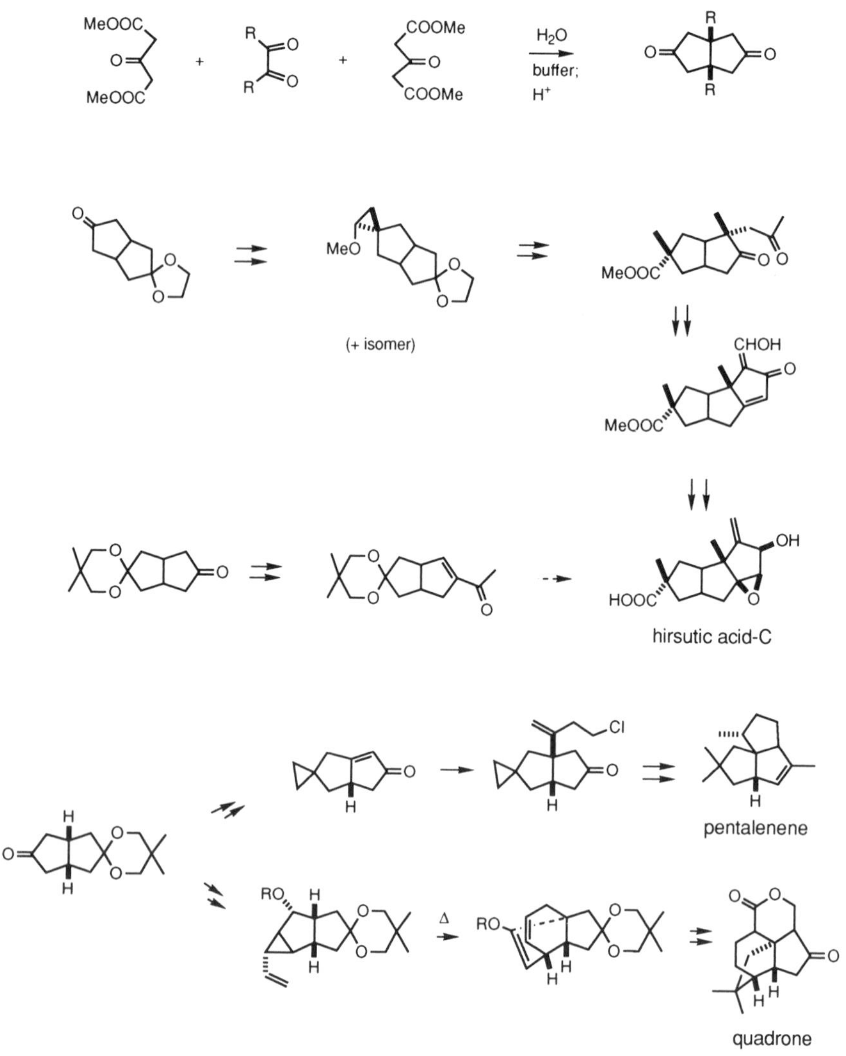

(+ isomer)

hirsutic acid-C

pentalenene

quadrone

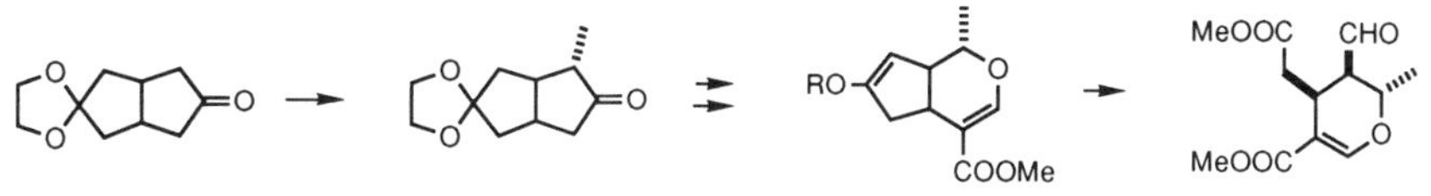

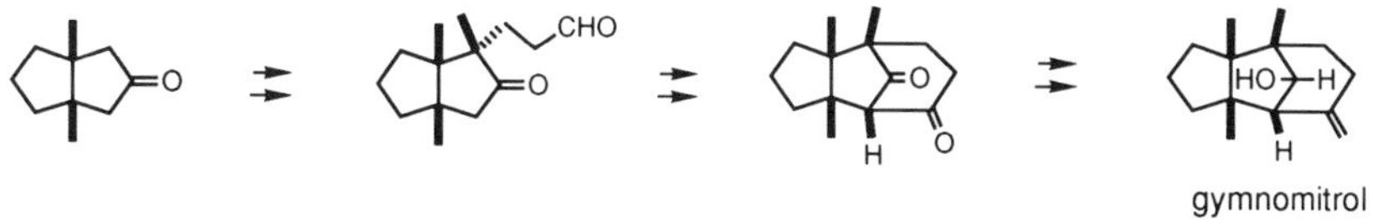

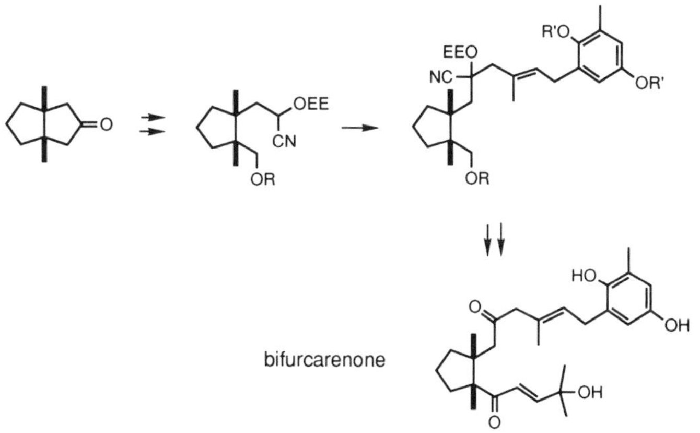

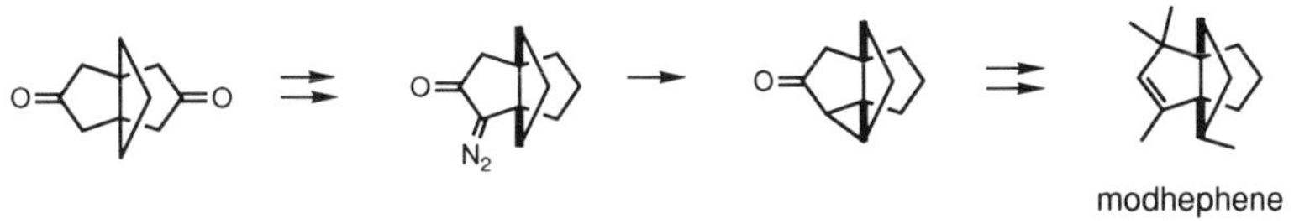

gymnomitrol

bifurcarenone

modhephene

The desymmetrization step can sometimes be effected by different methods, such as in approaches to loganin [Caille, 1984; Garlaschelli, 1992]. It is noteworthy that the Baeyer–Villiger reaction could be effected without protection of one of the ketone groups.

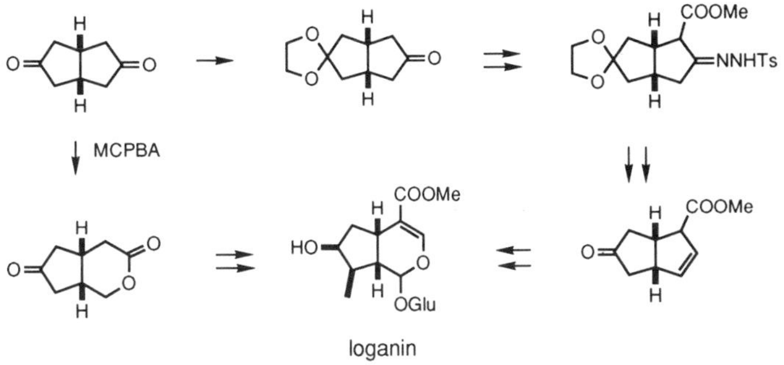

loganin

All the above target molecules are unsymmetrical but the synthetic approaches took advantage of the possible derivation of the various skeletons from a symmetrical core. Interestingly, a slightly less symmetrical diquinane, which has a plane of symmetry only through the ring junction, was employed in the construction of isocomene [Dauben, 1981]. The desymmetrization involving monoacetalization to permit one of the two identical carbonyl groups to be removed was not affected by the lower symmetry.

These diquinanediones also constitute useful building blocks for many theoretically significant structures, including substituted semibullvalene [L.S. Miller, 1981; Quast, 1982], triquinacene [Bertz, 1985], and staurane-2,5,8,11-tetraene [Mitschka, 1978; Deshpande, 1985; Kubiak, 1990].

triquinacene

While the Weiss reaction consists of a series of aldol and Michael reactions, simpler annulation processes which exploited the enolizability of 3-oxoglutaric esters are also known. Thus the assemblage of pyrazomycin was via alkylation

[De Bernardo, 1976], the precursor of endocrocin by a double aldol reaction [Steglich, 1970] and that of terramycin by the tandem Michael–Dieckmann cyclization [Muxfeldt, 1968] which involved a desymmetrized half-ester/amide.

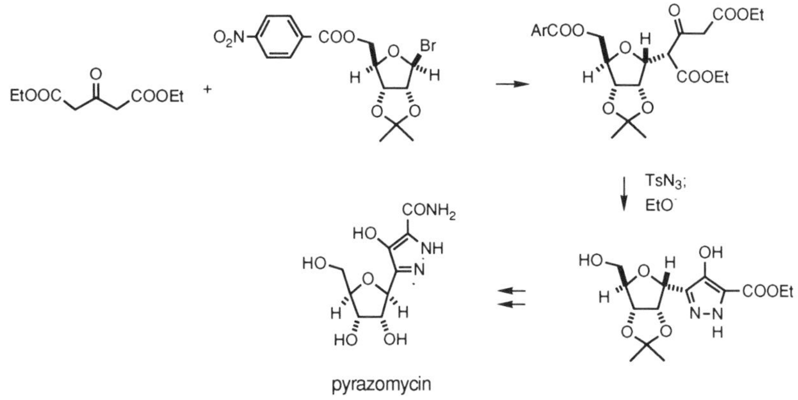

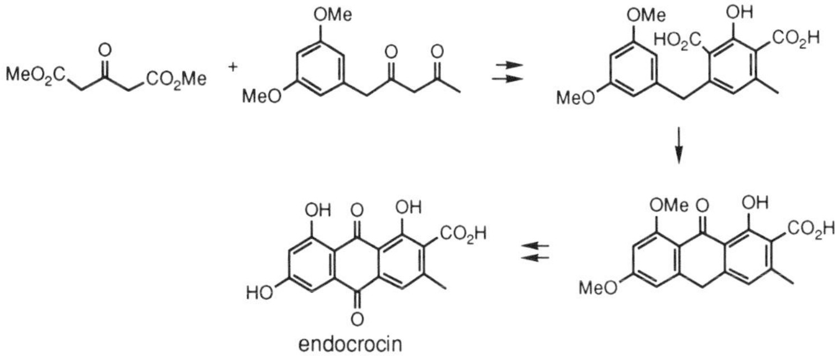

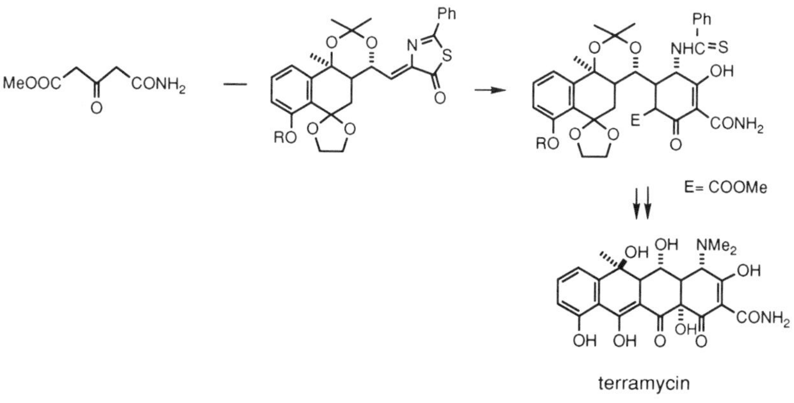

pyrazomycin

endocrocin

E= COOMe

terramycin

The 2,3-disubstituted furan formed by reaction of dimethyl 3-oxoglutarate with chloroacetaldehyde is easily converted into a 4-pyranone, and ultimate patulin [Tada, 1994].

It is also interesting that a building block for the C-terminal dipeptide of vancomycin was formed via the self-condensation (aldol–Dieckmann cyclization) product of dimethyl 3-oxoglutarate [Stone, 1991].

The ethylenedioxy derivative of diethyl 3-oxoglutarate contributed the central 1,3,5-trioxopentane unit of the polyketide precursor for a biomimetic synthesis of emodin [Harris, 1976]. The symmetry of the polyketide was responsible for the generation of only one tetralin which was ideal for emodin. The expedient preparation of the complete acyclic precursor was relevant to such structural patterns.

A convergent route to (+)-thienamycin in which the ethylenedioxy derivative of diethyl 3-oxoglutarate provided all the carbon atoms of the dehydroproline moiety has been developed [Kametani, 1985]. The key intermediate was assembled by a 1,3-dipolar cycloaddition between benzyl crotonate and a nitrone derived from the monoaldehyde which was obtained from the diester.

(+)-thienamycin

In a synthesis of (−)-talaromycin-A and -B [K. Mori, 1987a] dimethyl 3-oxoglutarate was reduced to the triol. Acetonide formation served to differentiate the two terminal groups. Subsequently a chiral tetrahydropyran derivative was prepared to act as a template for spiroannulation.

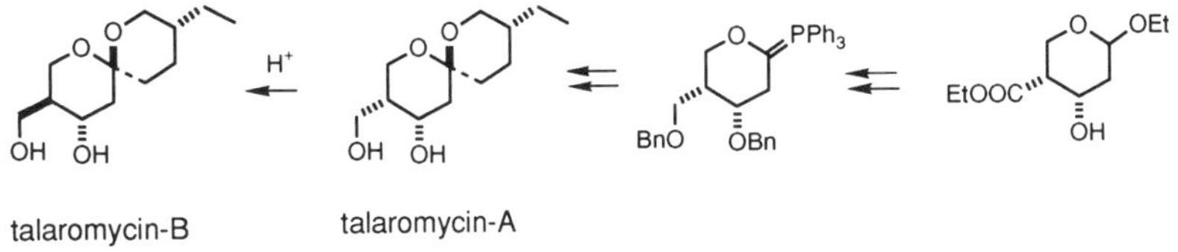

talaromycin-B   talaromycin-A

The incorporation of dimethyl 3-oxoglutarate into the framework of aspidosperma alkaloids has been contemplated [Troin, 1991]. However, experimental findings indicated dimethyl 2,3-pentadienoate to be a superior reaction partner for *N*-substituted anilines. The approach based on the following scheme is a viable alternative to prior methods. The spirocyclic intermediate has been obtained.

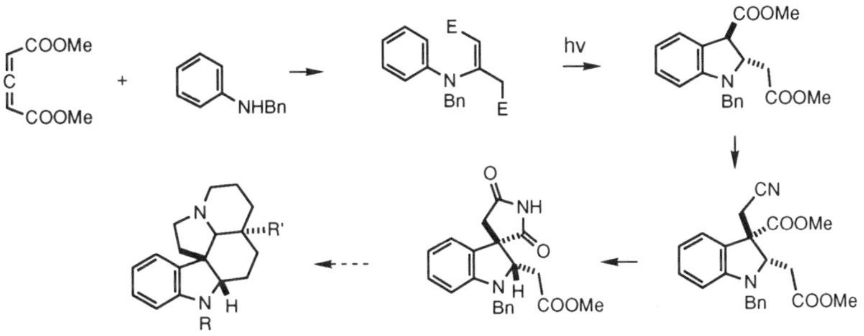

The intrinsic symmetry of 2,3-pentadienoic acid and esters plays an important role as dienophile during a Diels–Alder reaction, for only one regioisomer is expected for such a reaction. A unique construction of the homophthalic ester precursor for the synthesis of lunularic acid [Arai, 1973] was based on this method.

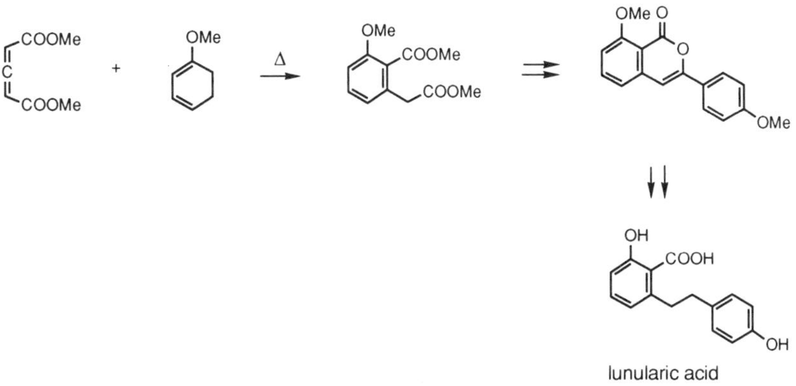

lunularic acid

A synthesis of fulvoplumierin [Büchi, 1969] started from a Diels–Alder reaction of the allene diester with 1,3-butadiene. After fashioning the two ester groups of the adduct into the α-pyrone unit the six-membered ring was contracted via cleavage and intramolecular aldol condensation.

In the construction of the pyridone unit of camptothecin [Danishefsky, 1971; Volkmann, 1971] the diester was employed as a Michael acceptor.

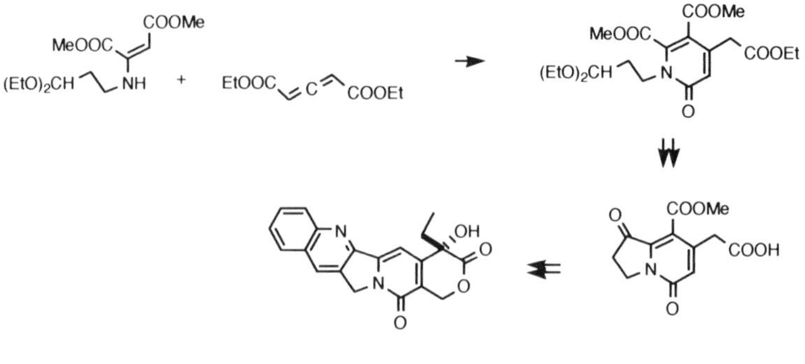

camptothecin

The use of 2,4-pentanedione and its derivatives as building blocks is most interesting when desymmetrization is involved. One such process, due to mechanistic variation in accordance with change of reaction conditions [Meyers, 1987b], has enormous synthetic implications.

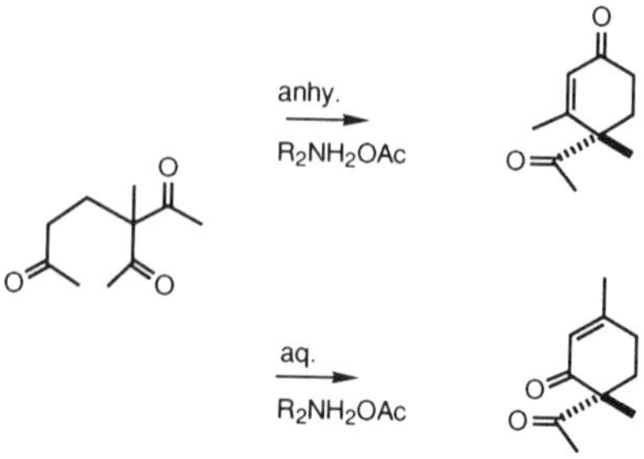

Differentiation of the two identical ketone groups by monoprotection is a general practice. Examples shown below involved an enol ether in a route to *cis*-cinerolone [Büchi, 1971b] and incorporation of one ketone into an oxazole ring in a synthesis of triacetyldaunosamine [Wong, 1975].

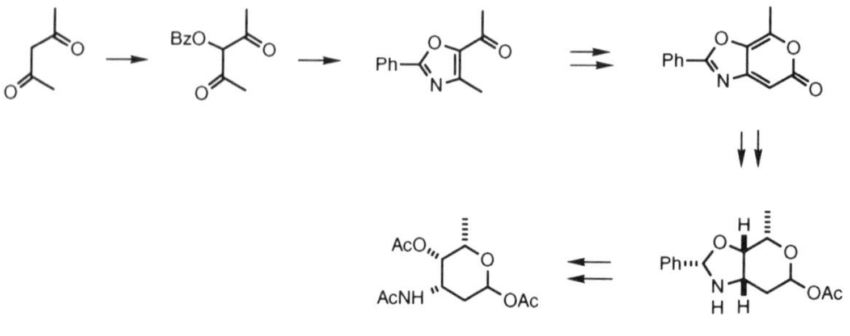

triacetyldaunosamine

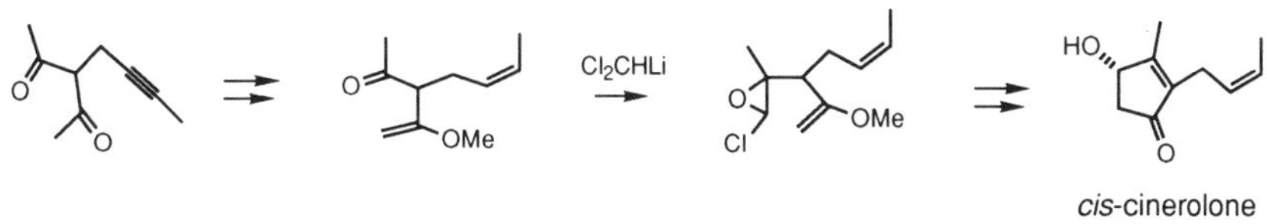

cis-cinerolone

A valuable latent and desymmetrized form of 2,4-pentanedione is 3,5-dimethylisoxazole. The 4-chloromethyl derivative is an excellent equivalent of methyl vinyl ketone [Stork, 1967a, b]. Actually both acetyl groups may be retained in synthetic intermediates, as shown in a synthesis of ferruginol [Ohashi, 1968].

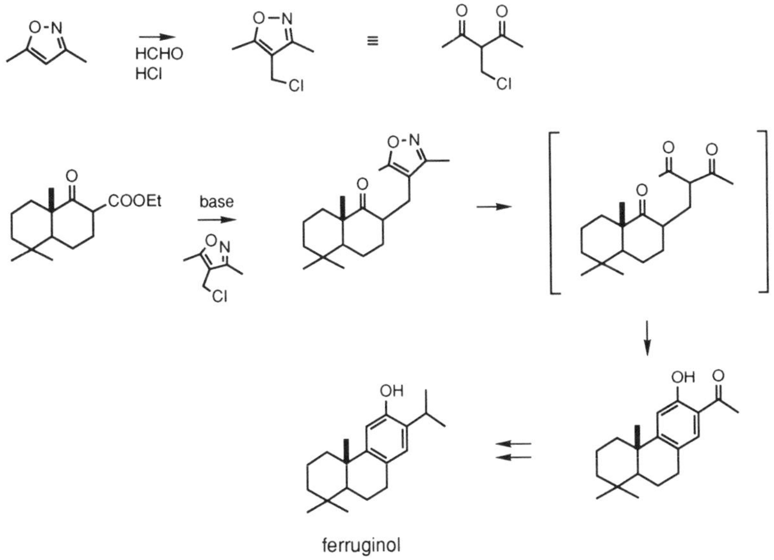

ferruginol

The facile formation of dianions from β-diketones (and β-ketoesters) enables chain elongation to afford unsymmetrical intermediates for synthesis. The method particularly suits assemblage of bridged and spiroacetals [Kongkathip, 1984; Ardisson, 1987; Ferezou, 1988], and other 1,3-dioxygenated substances, including nonactin [Gerlach, 1974, 1975].

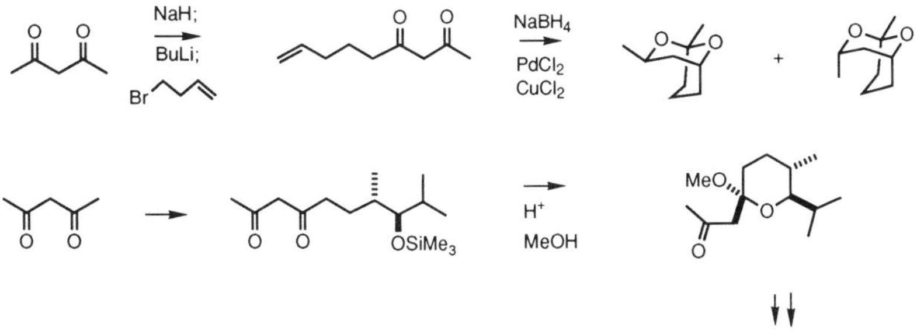

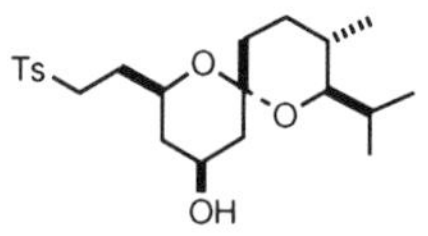

spiroacetal synthon of
22,23-dihydroavermectin-B$_{1b}$

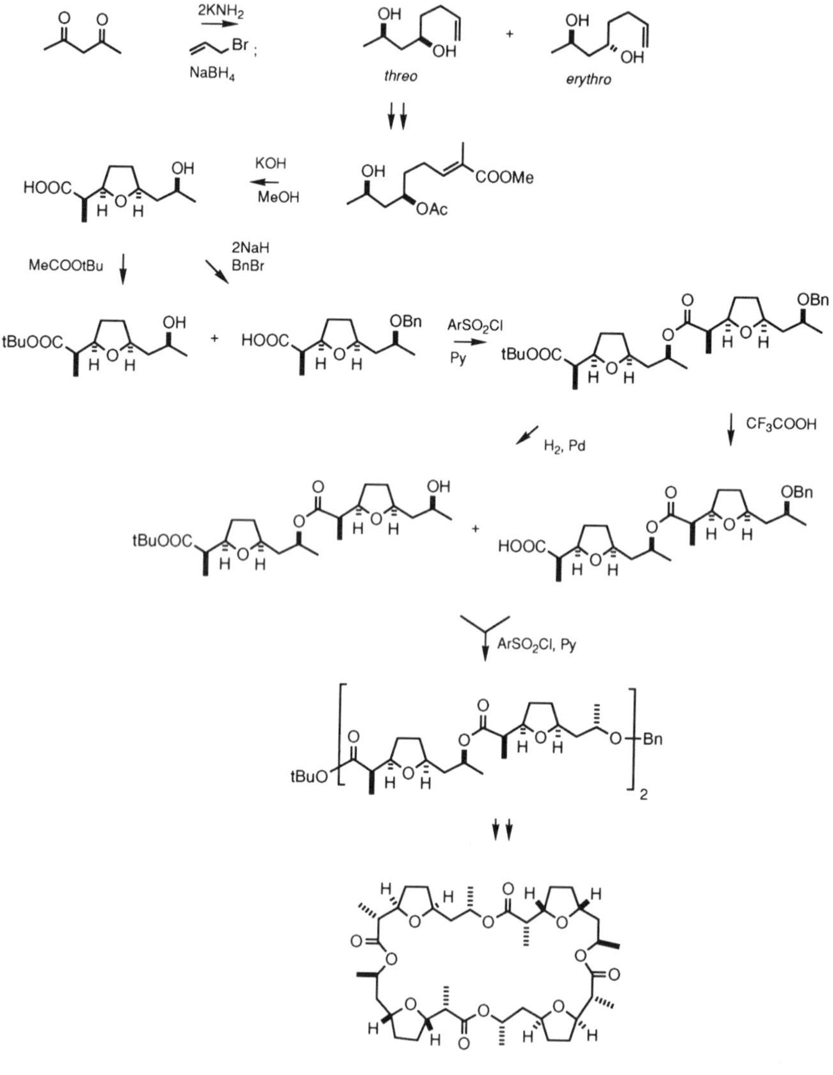

threo

erythro

nonactin

A less obvious application of 2,4-pentanedione through an unsymmetrical alkylation was in the construction of the chroman portion of α-tocopherol [Olson, 1980]. The C$_5$ unit became part of the aromatic ring.

The 2,3,5-trimethylisoxazolium ion has been employed as an equivalent of 2,4-pentanedione 1,5-dianion, to complete a facile synthesis of curcumin [Kashima, 1977].

Chrysophanol and elutherin have been synthesized [Harris, 1986] by combining two molecules of 2,4-pentanedione and a glutaric ester. Desymmetrization occurred during cyclization.

It should also be mentioned that expediency may demand desymmetrization such as monoenol ether formation when using 2,4-pentanedione at the start of a synthesis. Thus in a route to prostaglandin-$E_{2\alpha}$ [Stork, 1976] the reaction sequence must follow the steps dictated by the functional groups.

An aldol process is probably the most convenient method for the synthesis of hypacrone [Y. Hayashi, 1975], as it is simplified by the identicalness of the two acceptor groups.

hypacrone

A steroid synthesis was formulated on the basis of ketone stitching via a cyclic borane [Bryson, 1980]. The D-ring synthon required for this approach was prepared from an annulation involving a base-promoted condensation of 2,4-pentanedione and 1-carbethoxycyclopropyl triphenylphosphonium tetrafluoroborate.

estrone methyl ether

The frequent appearance of 1,3-dioxygenated carbon chain units in natural products have stimulated development of synthetic methods for their assemblage. It is interesting to note that the very valuable (R,R)-1,2:4,5-diepoxypentane and its bischlorohydrin precursor have been prepared from 2,4-pentanedione and used in an approach to roxaticin [Rychnovsky, 1991]. The polyhydroxy unit of the proposed roflamycoin structure has also been elaborated [Rychnovsky, 1994] from (S,S)-1,2:4,5-diepoxypentane. The latter compound, available from γ-ribonolactone [Attwood, 1984], has also been employed in the assemblage of the spiroacetal subunit of (+)-milbemycin-β_1 [Ley, 1989].

An elegant approach to all-*syn* skipped polyol systems is based on two-directional chain extension starting from (*S*,*R*)-1,2:4,5-diepoxypentane [Schreiber, 1987]. Both odd-numbered and even-numbered chain polyol chains are obtainable. The most important aspect of this work concerns differentiation of chain termini at a proper time by application of reactions such as asymmetric epoxidation of alkene units, which results in kinetic resolution. Either enantiomer of a chiral product is produced from a *meso* precursor.

5.2. SIX- AND SEVEN-CARBON UNITS

(*R*,*R*)- and (*S*,*S*)-1,2:5,6-diepoxyhexanes are easily obtained from D-mannitol. Synthetic applications of such molecules may be illustrated by a total synthesis of (+)- and (−)-pyrrolidine-197B [Machinaga, 1991a] which involved chain homologation in both directions by epoxide opening with a benzyloxyalkylmagnesium halide followed by termini differentiation of the tetraol intermediates. After careful protection of one of the two primary hydroxyl groups the synthetic operations became rather straightforward. Although at the late stage a cyclic sulfate intermediate suffered cleavage at the two carbinyl centers equally easily on reaction with an azide ion, both products eventually converged to the same chiral pyrrolidine. Slight variation of the same strategy provided an option to (−)-indolizidine-239CD [Machinaga, 1991b].

Desymmetrization may be effected in the epoxide opening, as shown in an approach to (+)-decarestrictine [Machinaga, 1993]. The third and last asymmetric center was established by an intramolecular Michael reaction.

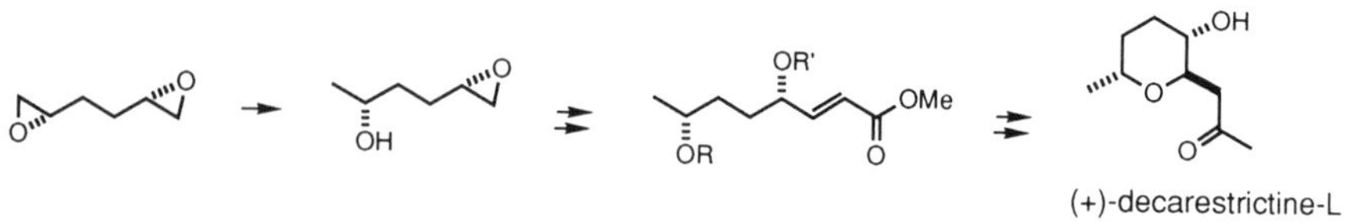

(+)-decarestrictine-L

(R,R)-1,5-Hexadiene-3,4-diol obtained from D-mannitol can be monobenzylated. Claisen rearrangement of the resulting monohydroxyl compound paved the way to α-lipoic acid [Rama Rao, 1987a].

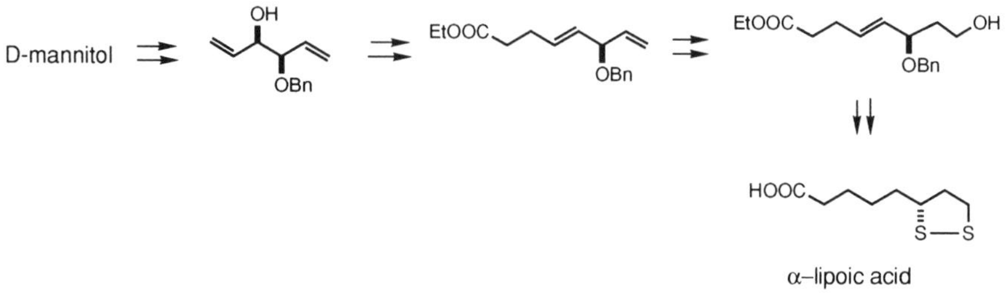

α–lipoic acid

The Sharpless asymmetric epoxidation of allylic alcohols is now well known. An anomalous version of this reaction on 1,5-hexadien-3,4-diols has been observed [Takano, 1991a]. Accordingly the results were exploited in the synthesis of L-*erythro*- and D-*threo*-sphingosine.

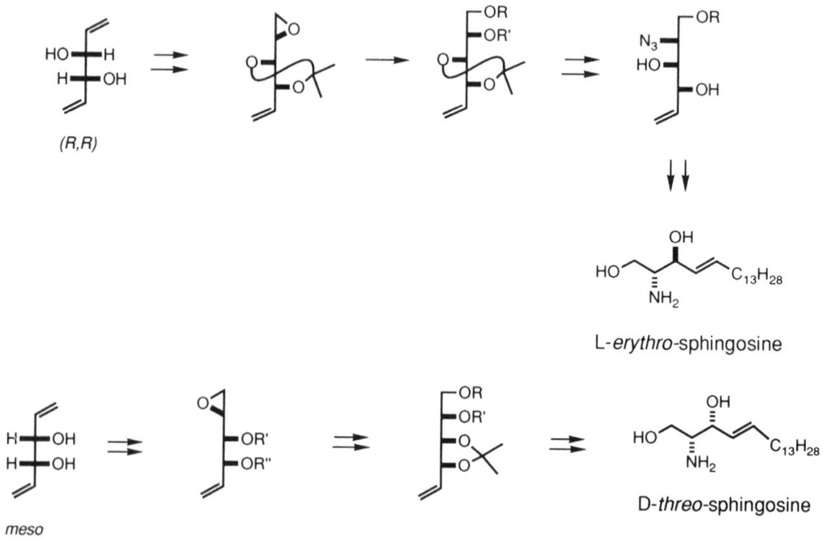

L-*erythro*-sphingosine

D-*threo*-sphingosine

2,5-Dimethyl-2,4-hexadiene is a versatile starting material for the synthesis of *trans*-chrysanthemic acid. In one of the approaches the diene was monoepoxidized and converted into pyrocine [Devos, 1978], another method involved dihydrofuran formation and photochemical rearrangement [Ohkata, 1978]. The most straightforward scheme consisting of reaction of the diene with a diazoacetic ester is that due to Staudinger, but technical improvements have led to the synthesis of a chiral product [Aratani, 1977].

pyrocine

chrysanthemic acid

90% ee

(Z)-2,5-Dimethyl-3-hexene-2,5-diol has also been used in a synthesis of *trans*-chrysanthemic acid [Gênet, 1980] via an allylic displacement reaction.

A functional equivalent of the six-carbon segment which constitutes the nonquinone portion of the mitomycins is (2Z,4Z)-hexadiene-1,6-diol [Naruta, 1986]. Elaboration of a tricyclic intermediate was successful by means of a nitrene route. Only one cycloadduct was produced owing to the identicalness of the two double bonds in the diol. However, the terminal units were reactively and sterically set apart at the formation of the indoloquinone chromophore, which could lead only to the desired skeleton. This is an elegant exploitation of hidden elements of molecular symmetry.

mitomycin-A

In a synthesis of the leukotriene LTA$_4$ methyl ester [Corey, 1979a] the (E,E)-isomer of 2,4-hexadiene-1,6-diol was monosilylated and oxidized to an aldehyde for chain elongation by a Wittig reaction.

leukotriene LTA$_4$ Me ester

2,5-Hexanedione has been used to advantage in a synthesis of ellipticine [Cranwell, 1962]. Its condensation with indole provided 1,4-dimethylcarbazole to which the terminal pyridine ring was further annexed. Note that the diketone was symmetrically incorporated, but the product was endowed with reactivity dissymmetry. A distinct use of 2,6-heptanedione is in a synthesis of frontalin [Sato, 1980] involving a photoinduced, titanium tetrachloride-catalyzed addition of methanol to one of the carbonyl groups (C—C bond formation).

ellipticine

Dimethyl *anti*-3,4-bis(p-tolyldimethylsilyl)hexanedioate, readily obtained by reductive coupling of the (Z)-acrylic ester, is convertible into the 3,4-bissilylcyclopentanone. Owing to its σ-symmetry Baeyer–Villiger reaction of the ketone led to only one lactone, which could be elaborated into deoxyribonolactone diacetate and arabonolactone triacetate [Fleming, 1992].

A more elaborate application of the *meso*-diester is in a synthesis of nonactin [Fleming, 1994].

benzyl (-)-nonactate

methyl (+)-nonactate

The fused dilactone of 3,4-dihydroxyhexanedioic acid has been transformed into the Geissman-Waiss lactone [Hizuka, 1989], and an intermediate of necine bases [Geissman, 1962; Narasaka, 1984] such as retronecine and platynecine, via ring opening to give a butenolide in which the two functional groups are different. Selective reduction of the side chain and conversion to a primary amine preceded the closure of the pyrrolidine ring by a Michael reaction.

Geissman-Waiss lactone

platynecine

Facilitation of a synthetic approach to pseudodomonic acid [Snider, 1982] by the generation of a terminally differentiated (3*E*,5*E*)-octadien-1,8-diol derivative and its mediation of an intramolecular hetero-Diels–Alder reaction is notable. Interestingly, the intermediate was prepared initially by a Lewis acid-promoted Prins reaction from the symmetrical 1,5-hexadiene.

Hydroboration of 2,5-dimethyl-1,5-hexadiene and 2,6-dimethyl-1,6-heptadiene has been shown to exhibit asymmetric induction in the intramolecular stage [Still, 1980a], resulting in product ratios in the range of 15–20:1 in favor of the *meso* compound. The two hydroxyl groups of the major product (*meso*:DL ratio = 15:1) from the heptadiene were differentiated to permit elaboration into a building block for the side chain of vitamin-E.

A trioxygenated diene similarly underwent *threo*-selective hydroboration and the major product was used in the construction of the ansa bridge of rifamycin-S [Still, 1983a].

The symmetrical 1,7-dibenzyl ether of (E,E)-2,5-heptadiene-1,4,7-triol undergoes Sharpless asymmetric epoxidation [Hatakeyama, 1993] to afford useful compounds for synthesis.

The cleavage of homoallylic alcohols by treatment of potassium hydride to generate ketones has been applied to symmetrical 4-hydroxylated 2,6-dimethyl-1,6-heptadienes. This method represents a technically expedient way to convert esters into enones such as ar-turmerone [Snowden, 1987]. Despite sacrifice of one of the alkenyl residues in the process it does not require special attention in the Grignard reaction during preparation of the homoallylic alcohols. Of course the usefulness of the method is linked to the identicalness of the two allylic residues.

ar-turmerone

Mevalonolactone has been synthesized [Feitzon, 1975] from 4-methyl-1,6-heptadien-4-ol via ozonolysis, lithium aluminum hydride reduction, and oxidation with silver carbonate on Celite. Desymmetrization was achieved during the final oxidation step.

A 4,4-disubstituted 1,6-heptadiene was used in a synthesis of ptaquilosin [Kigoshi, 1989b]. The intramolecular aldol reaction of the diketone obtained from a Wacker oxidation was the desymmetrizing process. Cleavage of the less hindered primary silyl ether during the Wacker oxidation actually facilitated the synthetic process.

ptaquilosin

One way of deriving the δ-lactone subunit of rhizoxin starts from 3-allyl-5-hexenecarbaldehyde [Keck, 1993]. This symmetrical compound acted as an acceptor for an asymmetrical aldol reaction to lengthen the carbon chain to include C-9 of the macrolide prior to an oxidative degradation of the two vinyl groups and formation of a lactol precursor. This work was obviously based on symmetry considerations.

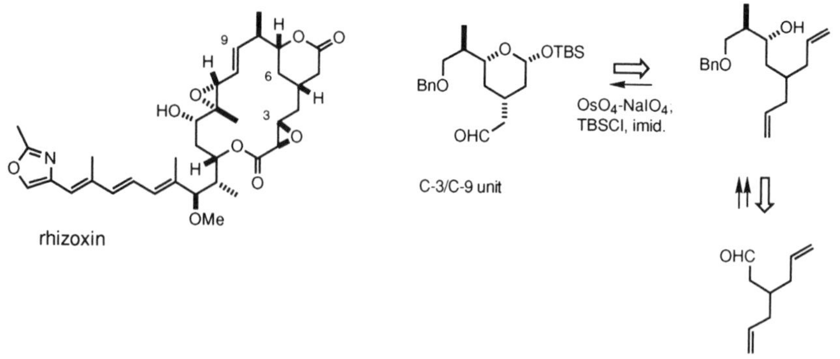

rhizoxin

C-3/C-9 unit

OsO$_4$-NaIO$_4$, TBSCl, imid.

2-Allyl-2-ethyl-4-pentenoic acid has been converted into a γ-lactone en route to quebrachamine [Takano, 1981b]. The two unsaturations were individually functionalized or degraded to become the skeletal elements of the nontryptamine portion of the molecule. Note that D-mannitol has also been elaborated into a related chiral lactone and used in the synthesis of quebrachamine in both enantiomeric forms [Takano, 1981d].

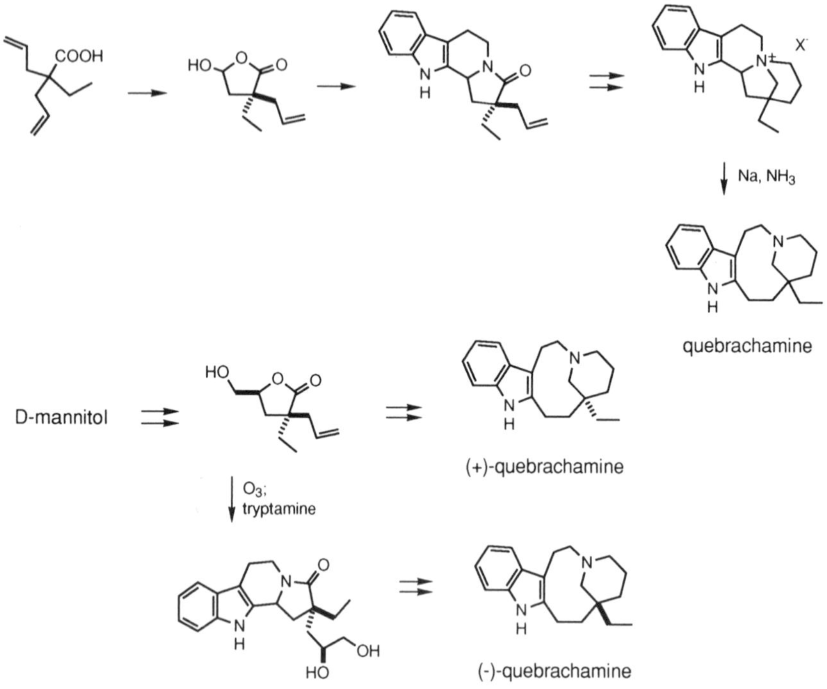

Na, NH$_3$

quebrachamine

D-mannitol

O$_3$;
tryptamine

(+)-quebrachamine

(−)-quebrachamine

4-Oxoheptanedioic acid is readily available from furfural via Perkin reaction and acid-catalyzed hydrolysis, and the diesters on further esterification [Lukes, 1952]. These symmetrical compounds are very useful synthetic intermediates owing to the presence of three functional groups. The acid itself forms a spirodilactone which can undergo selective reaction with alkyllithium reagents at only one of the carbonyl groups, such as demonstrated in a formal synthesis of *cis*-jasmone [Strunz, 1983]. On the other hand, organometallic reactions can be effected at the ketone site of the diethyl ester, which was the first step towards synthesis of rhazinilam [Ratcliffe, 1973].

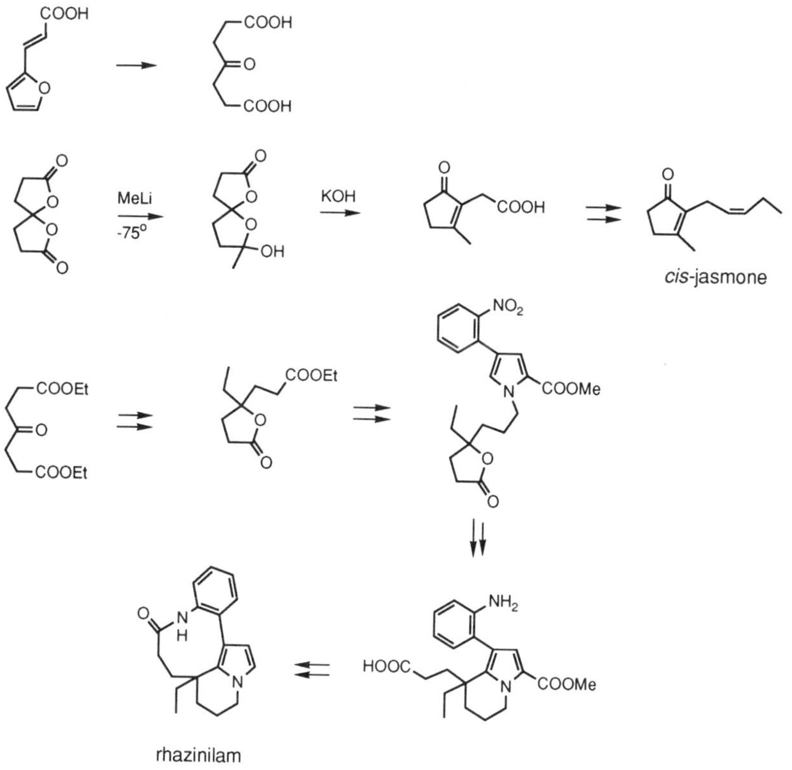

rhazinilam

The use of diethyl 4-oxoheptanedioate for construction of porphobilinogen [M.I. Jones, 1976] is most advantageous when considering the two carboxylic acid side chains together with C-3 and C-4 of the pyrrole ring as a unit. Retrosynthetic analysis involving disconnection at the tetrasubstituted cyclic double bond and functional group interchange readily reveals a simple monoalkylation approach starting from the keto diester.

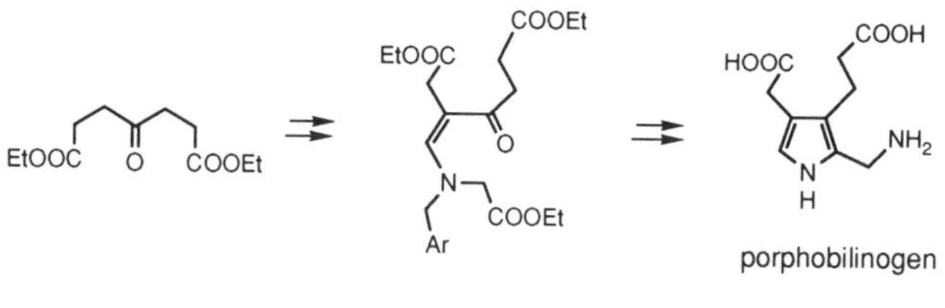

porphobilinogen

A synthesis of jasmine ketolactone [Shimizu, 1992] was initiated from an acetal of diethyl 4-oxoheptanedioate. After converting the ester groups in two different stages to the aldehydes to effect Wittig and Horner–Wadsworth–Emmons reactions, macrolactonization was effected by an intramolecular Michael reaction to partition the ring into a cyclopentanone and a smaller lactone.

jasmine ketolactone

Dieckmann cyclization of a dialkyl 4-oxoheptanedioate acetal leads to an unsymmetrical product in which one of the α-positions of the cyclohexanone is activated, permitting selective functionalization prior to removal of the ester group. In other words, dissymmetry, wherever desired, may be maintained. Thus the desymmetrized 1,4-cyclohexanedione 4-mono(ethyleneacetal) which contains a methyl group at C-2 has been exploited in syntheses of nootkatone [McGuire, 1974], isopetasol [Yamakawa, 1974], isotelekin [R.B. Miller, 1974], yomogin [Caine, 1975], cauhtemone [Goldsmith, 1976], and β-elemenone [Majetich, 1977].

isotelekin

nootkatone

yomogin

cauhtemone

β–elemenone

Variations have been reported such as retaining the ester group in the Dieckmann cyclization products, for example in syntheses of siccanin [Kato, 1981], the upper portion of tetronolide [Matsuda, 1989], and nootkatone [Marshall, 1971]. The chiral synthon of tetronolide was derived from a bioreduction of the cyclic keto ester [Kitahara, 1985], and in the nootkatone synthesis the Dieckmann cyclization was accomplished concomitant to a Wittig reaction. (Compare with isonootkatone synthesis [Marshall, 1967] in which dimethyl 4-isopropylideneheptanedioate was prepared for the Dieckmann cyclization, and also the transformation of diethyl 4-methylheptanedioate into anhydrolycopodine [S. Kim, 1978].)

siccanin

nootkatone

When the ethyleneacetal of dimethyl 4-oxoheptanedioate was reduced with diisobutylaluminum hydride and treated briefly with sodium hydroxide, a cyclohexenealdehyde was produced. This profoundly desymmetrized compound was elaborated into (+)-ivalin [Tomioka, 1989].

(+)-ivalin

It is interesting that several indole alkaloids (vincamine [Kuehne, 1964], aspidospermidine [Laronze, 1974], vincatine [E. Ali, 1982], and aspidofractinine [Cartier, 1989]) were constructed from tryptamine and 4,4-disubstituted heptanedioic esters, the latter species supplying the monoterpene portion. The perhydroquinoline unit of aspidospermine was constructed from an analogous C_{11} dinitrile [Ban, 1965; Kuehne, 1966]. Note that one of the nitrogen atoms of the dinitrile was incorporated into the alkaloid.

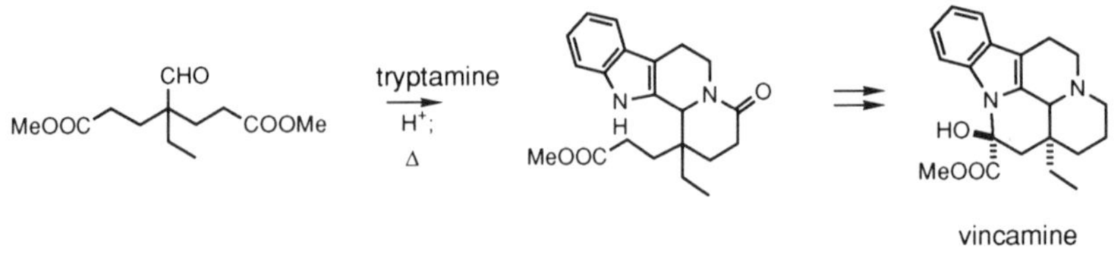

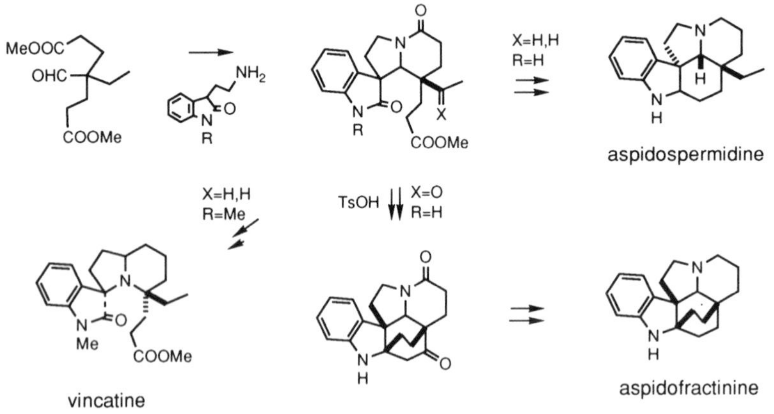

vincamine

aspidospermidine

vincatine

aspidofractinine

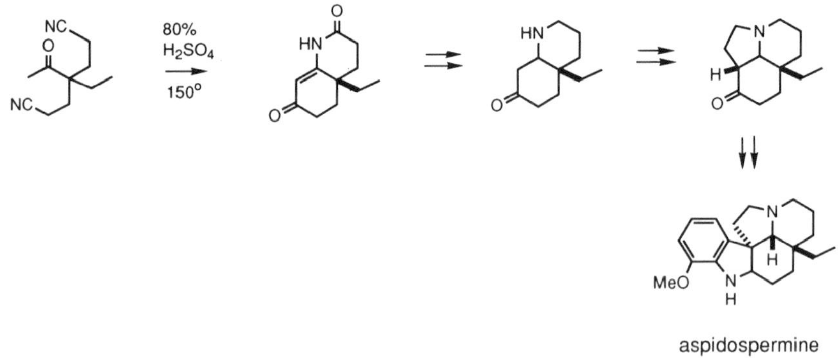

aspidospermine

Condensation of two desymmetrized components from a tetraester and a diol provided all the skeletal carbon atoms of periplanone-J (T. Harada, 1992b) in an acyclic precursor. An intramolecular alkylation served to close the ten-membered ring.

periplanone-J

An elegant stereocontrolled method for the introduction of *syn*-1,3-diol units and chain extension is based on a reaction sequence consisting of semihydrogenation, epoxidation, benzylic deoxygenation, Birch reduction, ozonolysis, and ketone reduction of 1,7-bis(3-methoxyphenyl)-1,6-heptadiyn-4-ol [Z. Wang, 1992]. Thus enantiodivergent synthesis from a *meso* compound was developed by subsequent manipulations which involved terminal differentiation via diastereotopic group-selective reaction.

5.3. EIGHT-CARBON AND LONGER CHAINS

1,7-Octadiene has been found to be a convenient starting material for synthesis of *endo*-brevicomin [Kshibashi, 1987]. Thiomethylation added a one-carbon unit to the end of the chain and accomplished the shift of the double bond to the position which was required for dihydroxylation.

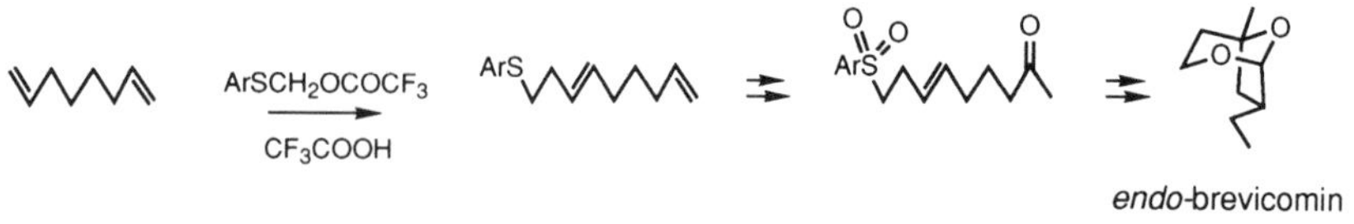

endo-brevicomin

2,7-Dimethyl-2,6-octadiene, the symmetrical head-to-head dimer of isoprene, has served as the central unit for a synthesis of squalene [Bhalerao, 1971], by way of chain extension at two termini after proper activation.

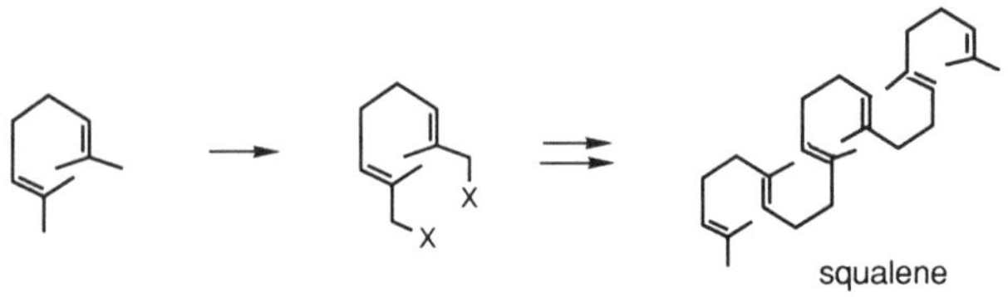

squalene

The head-to-head dimer of ethyl acetoacetate contains a 1,4-diketone unit which undergoes aldol reaction to give only one cyclopentenone, due to symmetry. The cyclopentenone derivative is a suitable intermediate for the synthesis of dehydroiridodiol [Kimura, 1982].

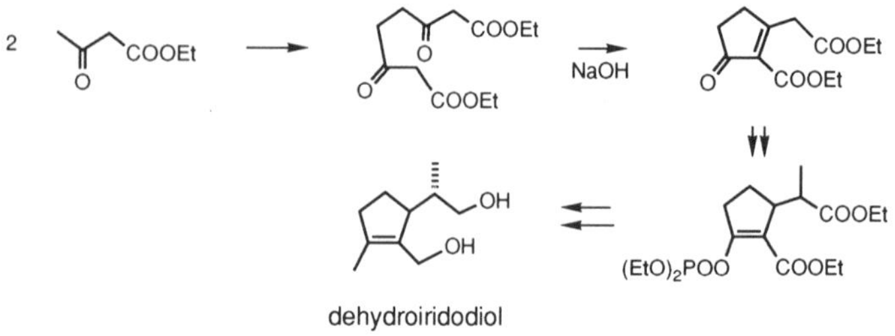

dehydroiridodiol

(S)-O-Benzylglycidol is an excellent chiral building block derived from D-mannitol. Through a chain extension process a 1,2:8,9-diepoxynonane was readily acquired as intermediate for (−)-gloeosporone [Takano, 1988f]. Desymmetrization involved epoxide ring opening by different nucleophiles at two stages.

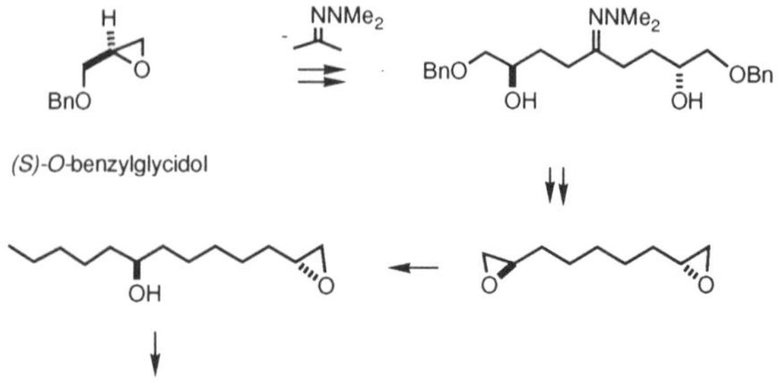

(S)-O-benzylglycidol

(-)-gloeosporone

It is of interest to examine the structural requirements attending synthesis of sterpurene. While the terpene molecule lacks symmetry elements, retrosynthetic analysis reveals that an enone with a hydrindane skeleton is a suitable intermediate. This enone can in turn be derived from an acyloin which possesses C_2-symmetry in the enediol form. Thus an acyloin condensation from a *vic*-cyclopentanediacetic ester is readily identified [Möens, 1986]. The necessary diester was prepared by electrochemical reductive cyclization of dimethyl 5,5-dimethyl-2,7-nonadiendioate.

sterpurene

An excellent synthetic approach to the ladybug defense alkaloids, precoccinelline and coccinelline [Stevens, 1979], began with reductive amination of 5-oxononanedialdehyde bis(dimethylacetal). The dialdehyde was released at low pH and intercepted by dimethyl 3-oxoglutarate in a Robinson–Schöpf condensation when the acidity of the medium was diminished. The azaphenalene system was produced in one step.

cocinelline

The tricyclic tryptamine obtained from degradation of the Fischer indolization product of 5-oxononanedioic acid is an extremely versatile intermediate for the synthesis of many complex indole alkaloids [Ban, 1983]. A key step of the synthesis is a photoinduced rearrangement.

tubifoline   condyfoline

A symmetrical dimethylnonadienediol, readily prepared from 1,6-hepta-diyne, underwent transformation into a monoether and then a homologated acid for lactonization to give suspensolide [Hidalgo-Del Vecchio, 1994].

suspensolide

The recognition of a hidden symmetrical 4,6-dimethyl-3,5,7-nonanetrione segment in the sex pheromone of the female drugstore beetle facilitated its elaboration [Kuwahara, 1978]. Essentially it requires an aldol reaction of the trione with acetaldehyde to construct the complete skeleton.

Both enantiomers of 1,7-dioxaspiro[5.5]undecane (olive fly pheromone) have been assembled via a C_2-symmetric nine-carbon chain [K. Mori, 1984]. The spirocyclic system permitted configuration inversion of two carbinol centers (equatorial to axial) and thereby prompted a conformational change which led to the enantiomeric series.

From the acyloin condensation product of a levulinic ester a synthesis of *cis*-jasmone [Wakamatsu, 1974] has been developed. The symmetrical nature of the bistrimethylsiloxyalkene removed any problem of multiple product formation from the alkylation. In fact this method is a sacrificial operation to ensure chemoselectivity, one molecule of levulinic ester being lost. It must also be emphasized that this concept has been extended to synthesis of other compounds which include cyclic analogs, and the C_{10} chain is only one of the many cases.

cis-jasmone

Disconnection of manzamine-C for synthetic considerations is straightforward. The less familiar 1-aza-6-cycloundecene was acquired from 5-decyne-1,10-diol in a conventional manner [Torisawa, 1991].

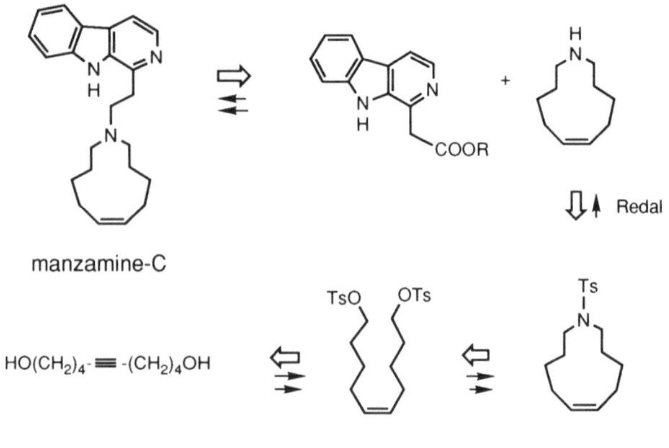

manzamine-C

$HO(CH_2)_4$-≡-$(CH_2)_4OH$

The slight difference at the two termini of the ten-carbon chain of a toxic mushroom constituent has made its synthesis [Takano, 1991d] relatively simple in that a tetraol was formed and a selective deoxygenation was carried out at the final stage.

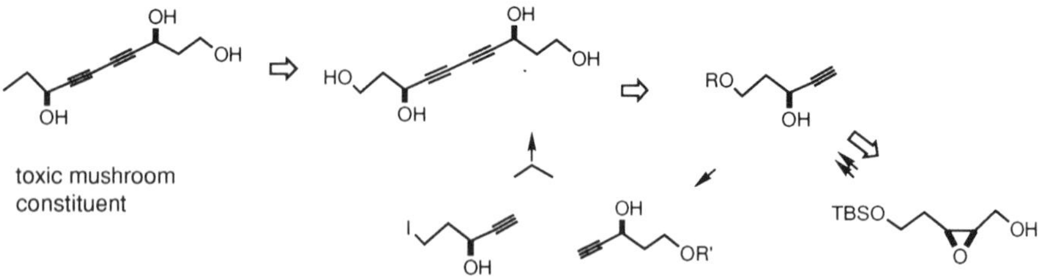

toxic mushroom constituent

The power of retrosynthetic analysis and biogenetic considerations is amply expressed in a description of a synthesis of endiandric acid-A [Nicolaou, 1982]. Semihydrogenation of a symmetrical C_{10} dienediynediol directly generated a bicyclic diol which was efficiently elaborated to endiandric acid-A.

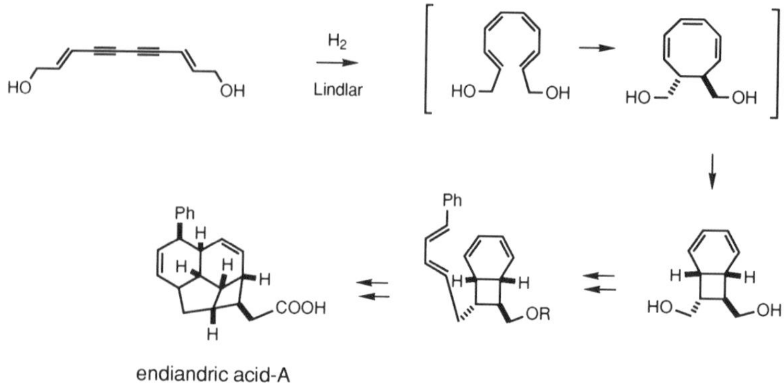

endiandric acid-A

Apparently the simplest method for the synthesis of 13-hydroxy-10-sesquigeran-1,12-olide [Simonet, 1993] is via iodolactonization of a symmetrical precursor.

For a synthesis of porantherine [Ryckman, 1987] that featured two Mannich reactions, a 2,2-disubstituted piperidine was prepared by addition of 5-lithio-2-pentanone 2'2'-dimethylpropyleneacetal to O-methylvalerolactam. This diacetal is a symmetrical 2,10-undecanedione derivative, but it was soon desymmetrized upon condensation of the amine with one of the ketone groups.

porantherine

An intramolecular Glaser coupling of 1,14-pentadecadiyne and hydration of the cyclic product were featured in a muscone synthesis [Fliri, 1979]. The symmetrical 1,3-cyclopentadecanedione is easily processed into the target molecule.

muscone

The telomers of butadiene with acetic acid have been coupled to give a symmetrical C_{16}-tetraene. After Wacker oxidation of the two terminal alkenes and hydrogenation of the two internal saturations, 2,15-hexadecanedione was produced [Tsuji, 1976]; the latter is a precursor of muscone [Tsuji, 1979b].

muscone

A symmetrical C_{19} triketone was elaborated in a protected form en route to perhydrohistrionicotoxin [Glanzmann, 1982]. After the trione was released, two cyclohexenones were generated by aldol reaction. One of the isomers was used to complete the synthesis.

perhydrohistrionicotoxin

5.4. HETEROATOM-INSERTED CARBON CHAINS

Heteroatom-inserted carbon chains are generally bifunctional molecules used in heterocycle synthesis, and those leading to unsymmetrical products are mentioned here. There are instances when one of the two identical chain segments is involved in bond formation with another functionality, such as that shown in an approach to actinidine [Cossy, 1988]. It must be emphasized that only one N-allyl group was needed but steps were saved with the second group serving as a protective device.

actinidine

The functional groups on the two chain segments may engage in reaction with one another, as in syntheses of arecoline [Wohl, 1907; Dankova, 1941] and trachelanthamidine [Takano, 1981c]. Interestingly, in an approach to epilupinine [van Tamelen, 1969a] a pseudosymmetrical acyloin (as considered in the enediol form) was generated from a diester, yet it was transformed through a diol and a dialdehyde, and using a Mannich reaction to reach the quinolizidine skeleton.

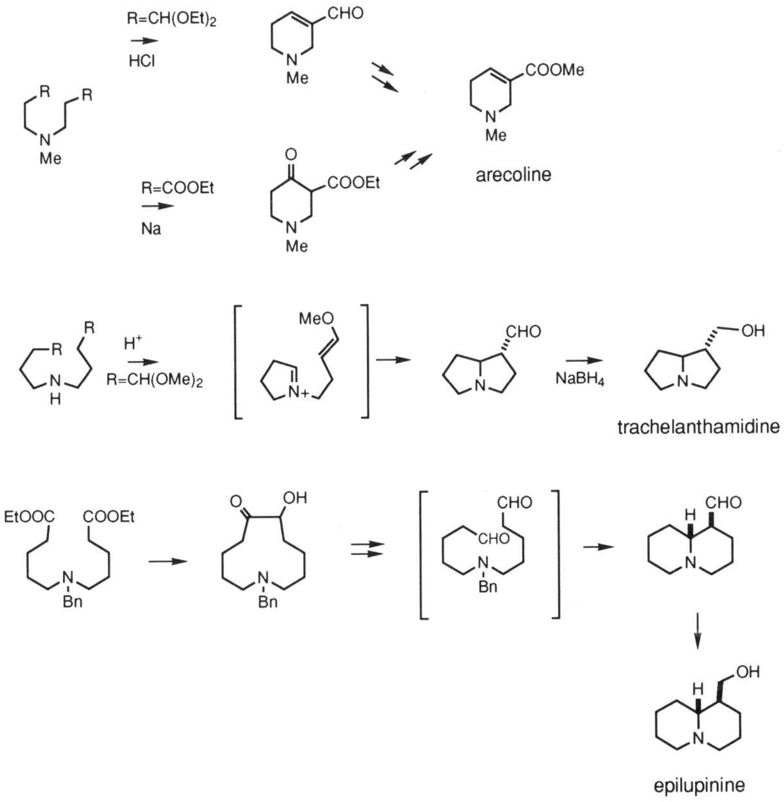

arecoline

trachelanthamidine

epilupinine

Isoalchorneine, a natural pseudosymmetrical bicyclic guanidine, has been elaborated from acyclic precursors which are symmetrical [Büchi, 1979]. Desymmetrization accompanied bicyclization which was effected by intramolecular S_N2' displacement reactions.

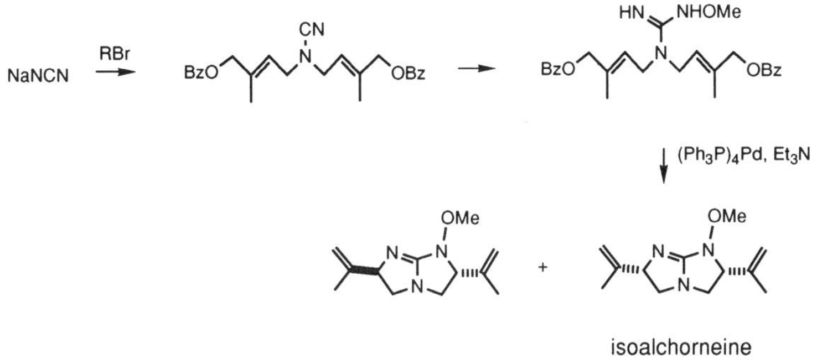

isoalchorneine

Intramolecular translactamization was featured in a synthesis of $(-)$-homaline [Wasserman, 1982] from a symmetrical diamine.

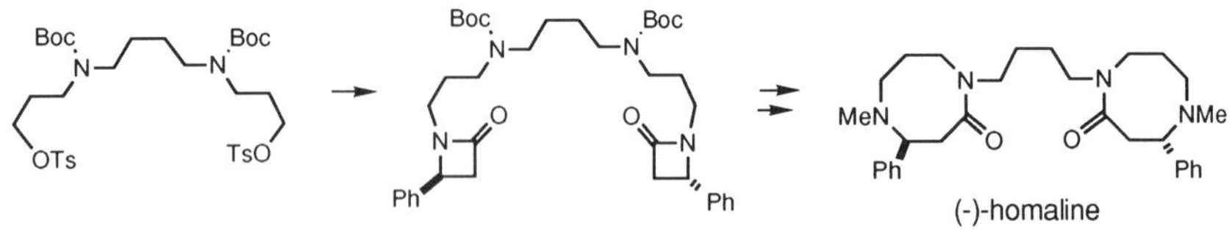

(-)-homaline

A synthesis of pentalenolactone-G [Pirrung, 1987] featured an intramolecular photocycloaddition of a precursor which contained two allenyl groups in a side chain. Note that one of the butadienyl residues played no active role, but its presence had the purpose of expediency to simplify preparation.

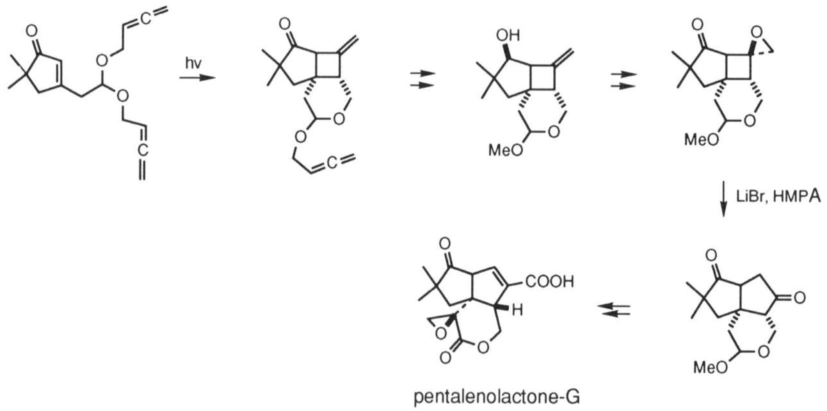

pentalenolactone-G

The cobalt-mediated cotrimerization of alkynes can be extended to the synthesis of pyridines by using nitriles as one of the addends. The versatility of the method is demonstrated by the efficient preparation of a precursor of pyridoxine [Geiger, 1984] from bis(trimethylsilylpropargyl) ether and acetonitrile.

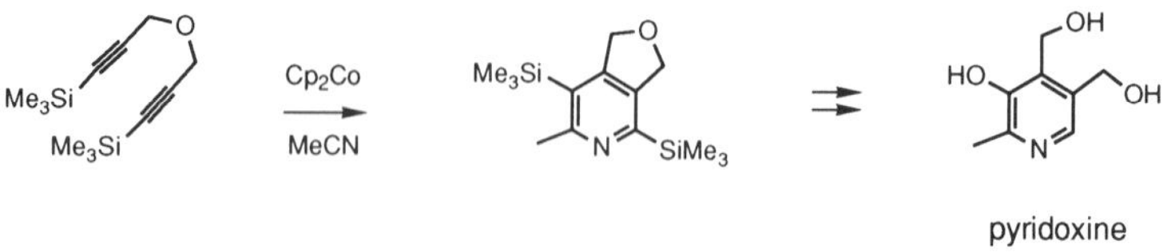

pyridoxine

6

FIVE-MEMBERED RING SYNTHONS

6.1. SIMPLE CYCLOPENTANE DERIVATIVES

Generally, it is most expedient to pursue the synthesis of a complex cyclopentanoid molecule starting from a precursor already containing the five-membered ring, due to limitation of cyclopentannulation methodologies. Of course the access to polyquinanes from lower quinanes still requires construction of the five-membered ring units. The following examples highlight various cyclopentane derivatives which served as building blocks.

From cyclopentene a diquinane intermediate for the synthesis of modhephene [Oppolzer, 1981] was rapidly assembled by acylation and Nazarov cyclization. Subsequently, an intramolecular ene reaction served to convene the propellane skeleton.

modhephene

1,2-Dimethylcyclopentene is a convenient building block for cuparene [Ishibashi, 1988]. Thus a Lewis acid-catalyzed reaction with chloromethyl *m*-tolyl sulfide gave a thiochromane derivative which on desulfurization afforded the sesquiterpene.

cuparene

Among the reported syntheses of gymnomitrol the most concise route is that involving cycloaddition of 1,2-dimethylcyclopentene with a *p*-benzoquinone monoacetal [Büchi, 1979a]. In one step a tricyclic intermediate was formed; this intermediate contained all the functionalities for elaboration of the sesquiterpene.

The cycloaddition combining symmetrical and unsymmetrical moieties resulted in an unsymmetrical product. Such desymmetrization is a necessity according to the target. The critical aspect is that recognition of local symmetry in the cyclopentane portion of the molecule and exploitation of this structural feature proved very beneficial.

gymnomitrol

Within the tricyclic framework of protoillud-7-ene the cyclopentane ring bearing a *gem*-dimethyl group is locally symmetrical. Consequently, synthetic schemes based on reductive 1,2-dialkylation of 4,4-dimethylcyclopentene should hold certain advantages over other routes. Particularly well suited for the pursuit seems to be the deMayo reaction with 2,4-pentanedione [Takeshita, 1979] from which a diketone is easily obtained. Intramolecular aldolization and a photocycloaddition with ethylene are then needed to elaborate the molecular skeleton of the sesquiterpene.

protoillud-7-ene

Similar symmetry arguments suggest that synthetic approaches to hirsutene based on annulation of 4,4-dimethylcyclopentene (cf. α-caryophyllene alcohol synthesis [Corey, 1964]) must be entertained seriously. Indeed, a convenient route to hirsutene [Iyoda, 1986] consisting of photocycloaddition and cyclomutation of the 5:4:6-fused ring system by reaction of the symmetrical diketone with iodotrimethylsilane was developed.

A method for α-chlorocyclopentanone synthesis via [2+2]-cycloaddition involving chloroketene and ring expansion on reaction with diazomethane is endowed with good regioselectivity. An iterative application of this reaction sequence on 4,4-dimethylcyclopentene, with proper modification of various intermediates, resulted in hirsutene [Greene, 1980]. Hirsutic acid-C has also been synthesized in an analogous manner [Greene, 1985], although there was only moderate stereoselectivity (3:1) in the first annulation. In both cases the plane of molecular symmetry vanished upon the first cycloaddition step.

hirsutic acid-C

A synthesis of sesbanine [Wada, 1985] was initiated by coupling of the symmetrical methyl 3-cyclopentenecarboxylate with an activated methyl nicotinate. Functional group transformation of the resulting diester completed the work.

sesbanine

cis-1,2-Cyclopentanedimethanol undergoes HLADH-catalyzed oxidation to give a chiral γ-lactone. The latter compound has been employed in a synthesis of (+)-multifidene [Boland, 1983].

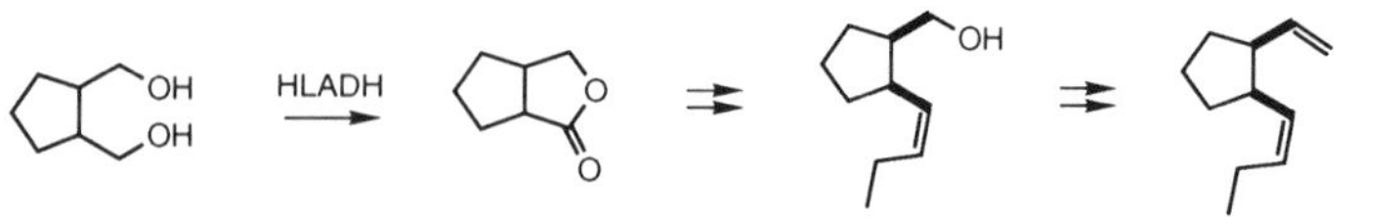

(+)-multifidene

6.3. CYCLOPENTENOLS AND CYCLOPENTENEDIOLS

Cyclopentadiene is a versatile source of various compounds containing a five-membered ring. For example, 3-cyclopenten-1-ol, readily obtained via hydroboration of cyclopentadiene, has been used as a starting material for many syntheses. Thus in a route to prezizaene [Piers, 1987], 3-cyclopentenol was functionalized to give a substrate for Cope rearrangement which led to the bridged ring portion of the target compound. The two side chains provided therein were so modified that the remaining cyclopentane unit could be closed by an intramolecular alkylation.

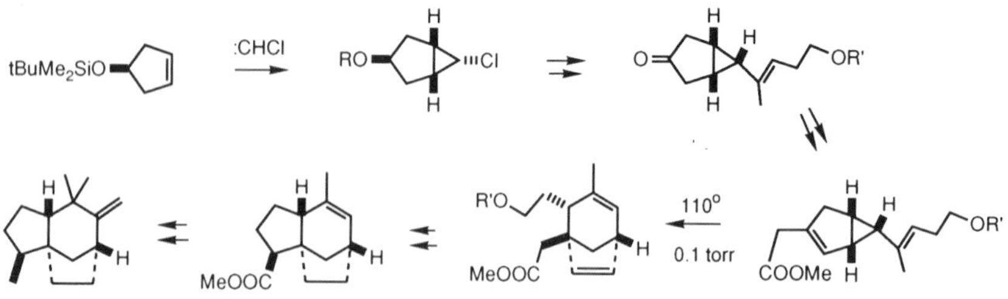

prezizaene

Among early applications of the deMayo reaction to natural product syntheses is that of loganin [Büchi, 1970] in which 3-cyclopentenol was the template for annexation of the pyran moiety. An intermediate for this synthesis has also been converted into sweroside aglucone methyl ether [Kinast, 1976].

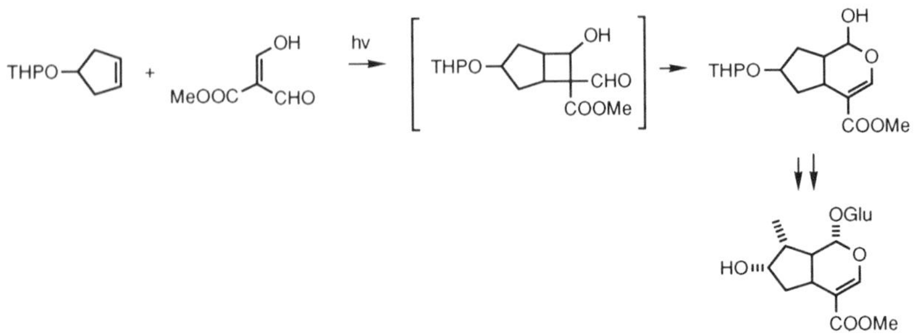

loganin

While the iridoid skeleton was assembled in one operation from the photoreaction of 3-cyclopenten-1-yl tetrahydropyranyl ether and 2-carbomethoxymalondialdehyde, many steps must be devoted to the establishment and adjustment of the substitution pattern of the carbocycle.

A modified version [Partridge, 1973a] used a chiral methylcyclopentenol. Comparison of the two routes makes it clear about the importance of symmetry with reference to synthetic needs. Basically the target is unsymmetrical; desymmetrization after construction of the molecular framework turned out to be far less efficient than desymmetrization of a reaction component.

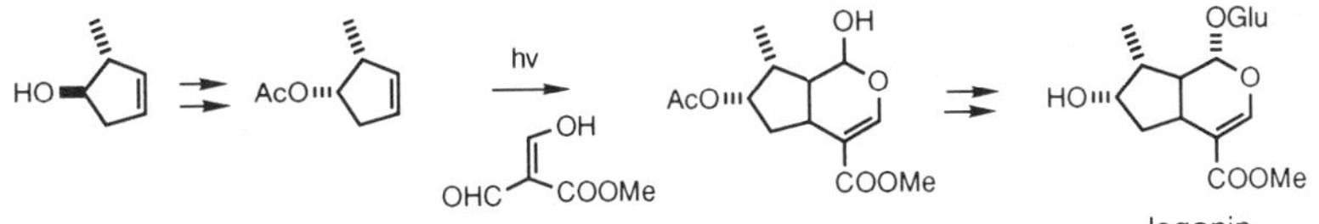

loganin

Because of the significant synthesis of the complete chiral segment of erythronolide-A [Stork, 1982] from the (−)-enone derived from (+)-2-methyl-3-cyclopenten-1-ol, a digression is made here to show the relationship between two building blocks and the enone.

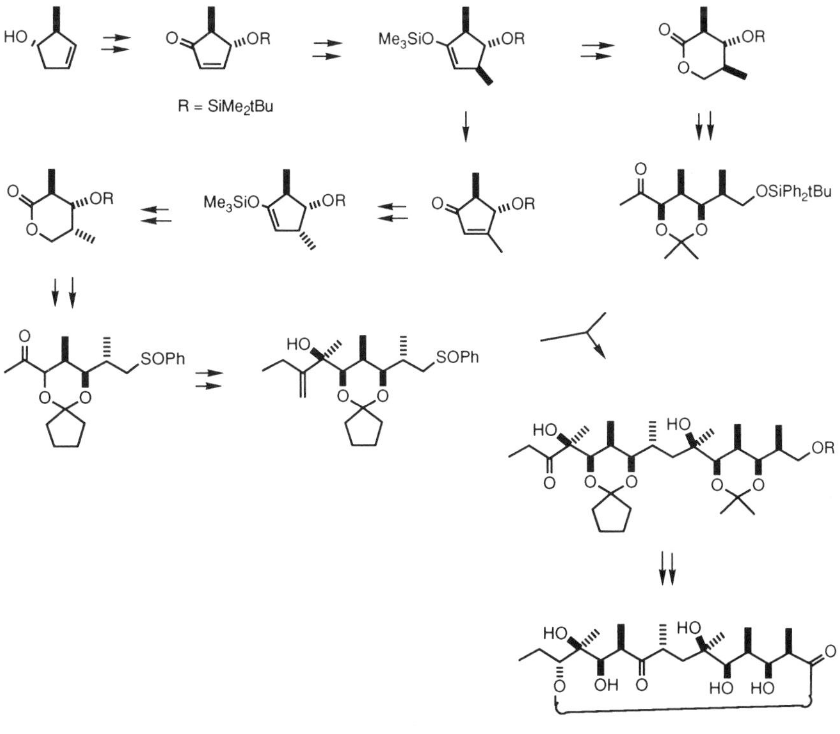

R = SiMe₂tBu

erythronolide-A

Note also that hydration of ethyl (1,3-cyclopentadien-5-yl)acetate via hydroboration with (+)-di-3-pinanylborane provided a chiral hydroxy ester [Partridge, 1973b]. Fried's prostaglandin intermediate (q.v.) could then be prepared via internal displacement of the mesylate which accompanied saponification of the ester to afford a bicyclic lactone.

The enantiomeric hydroxy ester is a valuable material for synthesis of pseudomonic acid-C [Barrish, 1987].

pseudomonic acid-C

The unsubstituted cyclopentenol THP ether was also employed in the synthesis of several prostaglandins [W.P. Schneider, 1968; Just, 1969] via C-6 substituted bicyclo[3.1.0]hexane derivatives. The molecular symmetry was broken upon alkylation of a ketone intermediate to introduce the "α-chain." Solvolytic opening of the cyclopropane provided the hydroxyl group to the five-membered ring and revealed the side chain containing the allylic alcohol system.

$PGE_{1\alpha}$

MCPBA

HCOOH;
Na_2CO_3

$PGF_{1\alpha}$

+ C-15 epimer

The major symmetrical adduct from cyclopropanation of the *t*-butyldimethylsilyl ether of 3-cyclopentenol with the carbene generated from diazo Meldrum's acid was converted into a potential precursor of brefeldin-A [Livinghouse, 1978]. The plane of symmetry in the adduct was destroyed upon opening of the three-membered ring.

brefeldin-A

The complex alkaloid ajmaline is a challenging synthetic target. Symmetry considerations figured prominently in two reported syntheses [Masamune, 1967; van Tamelen, 1970]. Guided by biogenetic speculation an indolic secoaldehyde was identified as the subtarget for synthesis. The configuration of the aldehyde is inconsequential due to its epimerizability.

Formulation of a synthetic plan for ajmaline [Masamune, 1967] was greatly simplified by further ignoring the *C*-ethyl group of the secoaldehyde. A consideration based on a Pictet–Spengler transform of this imaginary intermediate, with recognition of the amino aldehyde equivalency to a carbinolamine, would place C-15 in a pivotal position.

It is now apparent that a locally symmetrical 3-substituted glutaraldehyde would serve well as a central intermediate. The latent C-4 substituted cyclopentene is available by alkylation with 3-cyclopenten-1-yl tosylate.

ajmaline

The other biomimetic route to ajmaline [van Tamelen, 1970] relied on a Mannich reaction to form the bond between C-5 and C-16. The iminium species for the cyclization is structurally akin to tetracyclic yohimbe alkaloids such as corynantheidine.

Consider now the aldehyde corresponding to corynantheidine, and its Pictet–Spengler transform. The dialdehyde precursor is again a C-3 substituted glutaraldehyde, which may be released from a cyclopentene by oxidative cleavage. Recognition of the local symmetry of the dialdehyde and hence the cyclopentene unit facilitated the synthesis.

ajmaline

The approach led to 21-deoxyajmaline. Introduction of the missing hydroxyl group was made possible by the relative ease in oxidative cleavage of the C—N bond.

Two other salient features of this work are that tryptophan was employed to unite with a C_9 subunit. After the Pictet–Spengler cyclization, treatment of the tetracyclic amino acid with dicyclohexylcarbodiimide led to decarbonylative dehydration and the desired intramolecular Mannich reaction. The second critical point is that only deoxyajmalal (A/B) resulted. The Pictet–Spengler product with either 3α,15β,20β- or 3α,15α,20β-methine hydrogens would be reluctant to cyclize for steric reasons. Cyclization of the 3α,15α,20β-isomer requires a transition state in which two bulky substituents are pseudoaxial.

Related syntheses of geissoschizine [K. Yamada, 1974], rhyncophyllol [van Tamelen, 1969e], and a pentacyclic model of strychnine [van Tamelen, 1960a] have also been accomplished.

geissoschizine

rhyncophyllol

The dithiane derivative of a protected *cis*-3,4-dihydroxycyclopentane-carbaldehyde served in a synthesis of crassin [McMurry, 1989].

crassin

cis-4-Cyclopentene-1,3-diol is produced by reaction of cyclopentadiene with singlet oxygen followed by reductive cleavage of the endoperoxide. It is of interest to note that this *meso*-diol has been transformed into muscarine [Still 1980b].

The glutaraldehyde derived from the cyclic diol is still symmetrical. However, however, reaction of its diacetate (a tetrahydropyran) with excess methylmagnesium bromide is *threo*-selective as a result of chelation control. Formation of the tetrahydropyranoxide anion prevented overreaction. Internal etherification, after regeneration of the original alcohol groups, with inversion at C-5 delivered a precursor of muscarine having the correct stereochemistry.

muscarine

The C-22 to C-34 segment of halichondrin-B and -C has been synthesized from *meso*-2-cyclopentene-1,4-diol [Burke, 1991]. The significant feature of this work is the stereocontrol of four stereocenters by means of concurrent [3.3]sigmatropic dioxanone-to-dihydropyran rearrangements.

R = H   halichrondrin-B

R = OH   halichrondrin-C

An analog of neocarzinostatin has been obtained [Hirama, 1991] from elaboration of the diol monosilyl ether.

The optically active monoacetate of the diol, secured by enzymatic hydrolysis, served as starting material for a synthesis of (+)-brefeldin-A [Nokami, 1991].

Most syntheses of brefeldin-A are based on Michael reactions to introduce the side chains of the five-membered ring that constitute the macrolactone moiety. 4-Acetoxy-2-cyclopentenone is a valuable synthetic equivalent of the highly unstable cyclopentadienone in view of the fact that after its reaction with a Michael donor, in situ elimination of acetic acid may be effected to generate a new cyclopentenone. Thus completion of a synthetic route to brefeldin-A [Köksal, 1980] demonstrated the validity of using an unsymmetrical molecule (4-acetoxy-2-cyclopentenone) as a surrogate of a symmetrical one (cyclopentadienone).

(+)-brefeldin-A

From the molecular symmetry viewpoint the conversion of the all-*cis* 2,3-epoxycyclopentane-1,4-diol to terrein [Auerbach, 1974] via hydrolytic cleavage and oxidation–dehydration steps is of great interest. Reactions involving the 1,3-diol segment did not disturb the symmetry prerequisite of terrein.

terrein

The unsymmetrical allosamizoline, an aminocyclitol obtained from hydrolysis of the pseudotrisaccharide allosamidin, has been elaborated from a symmetrical cyclopentenediol [Trost, 1990]. A benzyloxymethyl group acted as stereocontrol element during singlet oxygenation to provide the required diol. Unilateral manipulation of one of the hydroxyl groups desymmetrized and redistributed the functionalities. For a synthesis of the related allosamidin [Griffith, 1991] the symmetrical cyclopentenediol was acetylated and hydrolyzed in the presence of acetylcholinesterase to allow selective carbamylation.

allosamizoline

With a nearly symmetrical structure and special stereochemical arrangement of functionalities (+)-mannostatin-A is amenable to synthesis from a C-5 substituted cyclopentadiene. The choice of 5-methylthiocyclopentadiene is most logical on symmetry grounds, as its hetero-Diels–Alder reaction would leave a double bond for *cis*-dihydroxylation. The employment of a nitroso dienophile fulfills all the structural and stereochemical requirements of the synthesis [S.B. King, 1991].

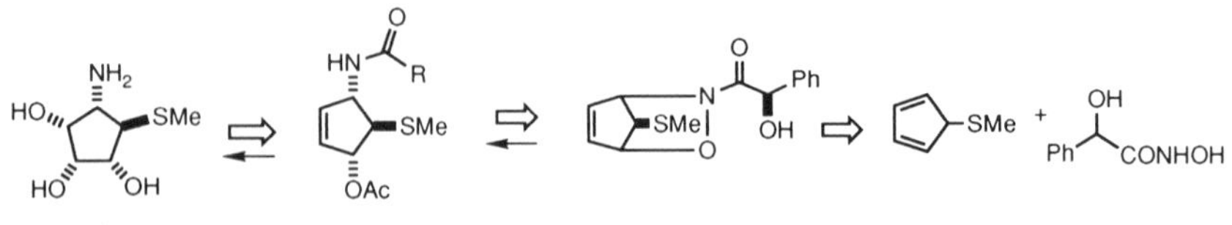

(+)-mannostatin-A

In the structure of funiculosin there is an all-*cis* pentasubstituted cyclopentane. The synthetic pathway to this subunit [Pattenden, 1990] involving intramolecular reactions to ensure stereochemical results is interesting. The choice of *cis,cis*-4-cyclopentene-1,2,3-triol [Cocu, 1970] as the building block was judicious.

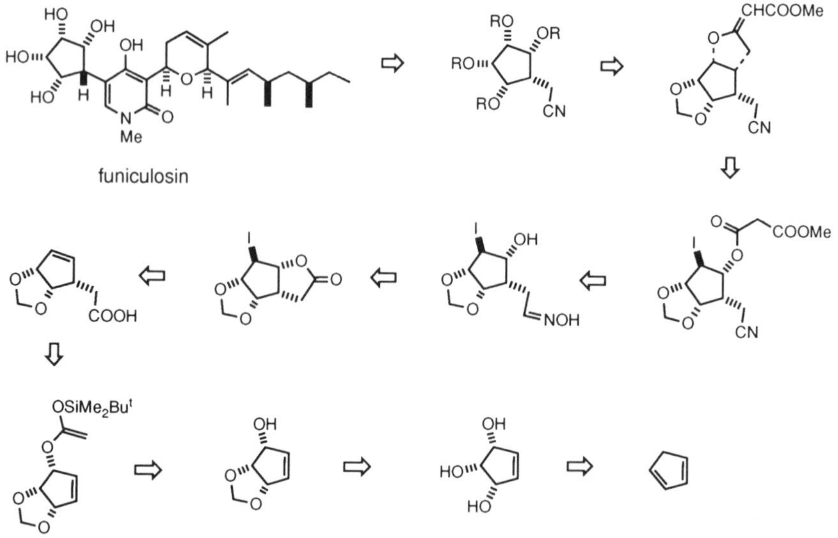

funiculosin

Retrosynthetic analysis of the anguidine problem [Ziegler, 1990] suggested the use of a locally symmetrical intermediate which contains a cyclopentenediol unit. This unit in turn can be derived from a 5,5-disubstituted cyclopentadiene.

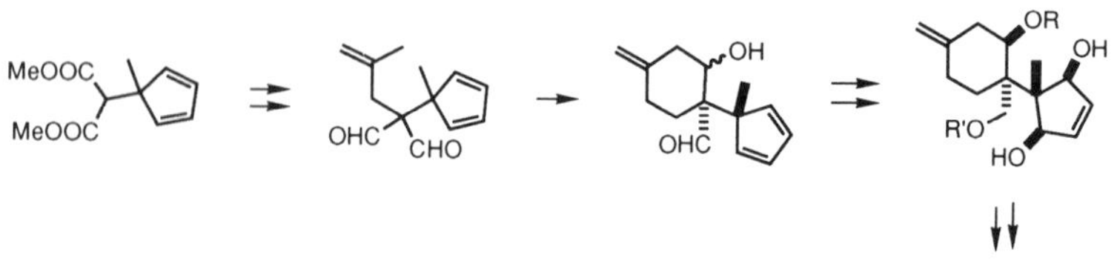

anguidine

The synthesis was thwarted by an unfortunate migration of an exocyclic double bond of the tricyclic intermediate to an undesirable position.

The use of *cis*-4-cyclopentene-1,3-diol in prostanoid (PG) synthesis is logical owing to the excellent attributes of the diol in terms of its availability and functional group correlation with the target molecules. In a synthesis of prostaglandin-$F_{2\alpha}$ [Fried, 1972] the epoxide of the dibenzyl ether was reacted with an allylcopper reagent. An unsymmetrically substituted epoxy cyclopentanol was elaborated. By leaving a hydroxyethyl chain unprotected, epoxide opening by attack with an alkynylaluminum was rendered regioselective. The product was a tetrasubstituted cyclopentane with stereochemistry corresponding to prostaglandin-F_2.

PGF$_{2\alpha}$

Desymmetrization of the cyclopentenediol by monoetherification permitted selective functionalization of the double bond. Cyclopropanation via intramolecular carbenoid interception followed by an organocopper reaction led to useful intermediates for the synthesis of various subclasses of prostaglandins [Corey, 1972a].

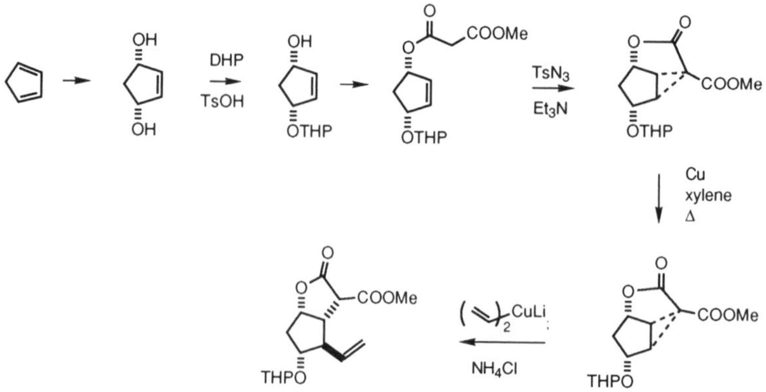

The cyclopentenediol diacetate can be partially hydrolyzed on exposure to *Bacillus subtilis* var. Niger. [Takano, 1976b]. A bicyclic lactone (35% optical purity) was obtained via a Claisen rearrangement. Reduction and epoxidation converted the lactone into Fried's intermediate for prostaglandins.

The same bicyclic lactone has been synthesized from both diastereomers of the monoprotected diol [Terashima, 1977], as shown below.

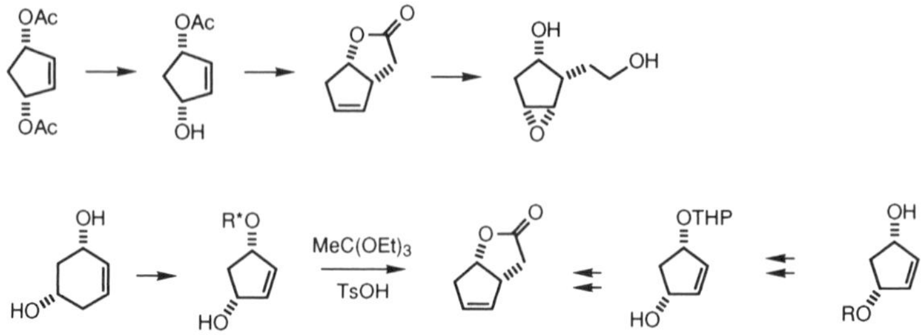

Both enantiomers of the cyclopentenediol monoacetate are now available in high purity by enzymatic hydrolysis of the diacetate (Laumen, 1984]. Pig liver esterase and cholinesterase have complementary selectivities.

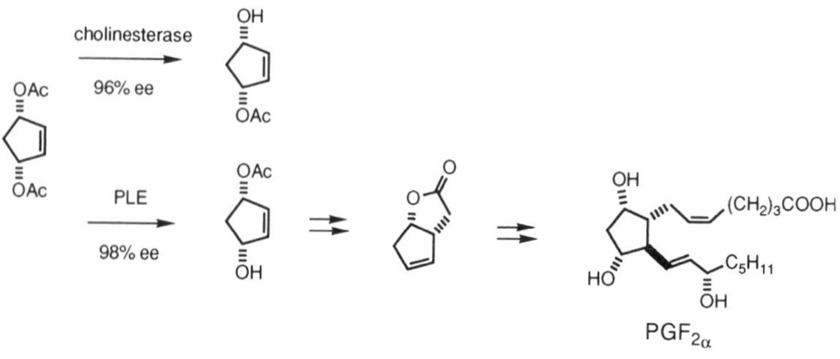

A nonenzymatic access to both enantiomers of the monoprotected *cis*-cyclopentenediol is by reaction of *cis*-3,4-epoxycyclopentyl ethers with a chiral amide base followed by functional group manipulations [Asami, 1985]. These compounds are readily transformed into chiral cyclopentenolones. Other methods for the preparation of the latter substances comprise of either diketone reduction with lithium trialkoxyaluminum hydride, in which two of the alkoxy groups are derived from a chiral 2,2'-binaphthol [Noyori, 1984], or resolution via a chiral sulfoxide adduct [Eschler, 1991].

An efficient synthesis of prostaglandin-$F_{2\alpha}$ based on chain attachment via free radical reactions has been achieved [Stork, 1986]. The starting material was the $(-)$-monoacetate of the cyclopentenediol. Magnificent stereoselectivity of the tandem cyclization–alkylation process was observed.

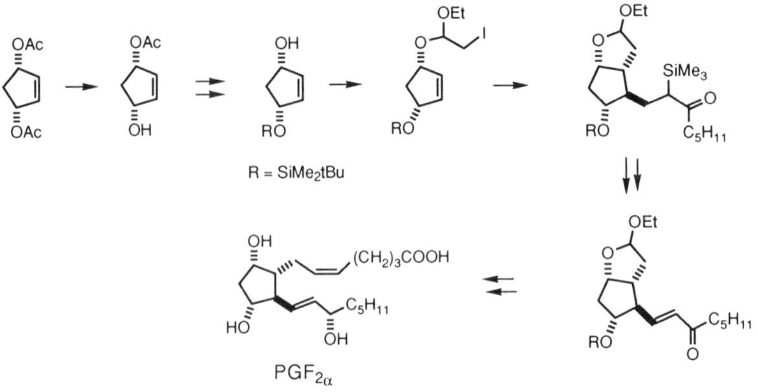

R = SiMe₂tBu

PGF₂ₐ

The free radical that initiated the cyclization can be generated from electrolytic decomposition of the proper carboxylic acid [Becking, 1988].

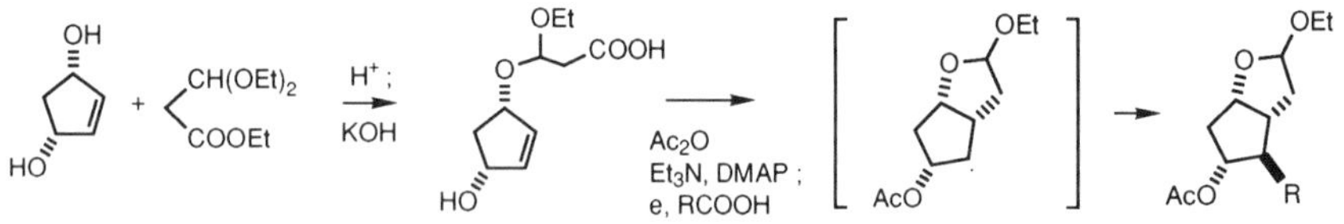

An alternative method for the *vic*-difunctionalization is via palladation [Larock, 1991].

Enantiocomplementary routes to the prostanoids via configuration inversion of 2,5-dihydroxycyclopent-3-enylacetic acid lactone derivatives [Tömösközi, 1985] are very appealing. In this work the critical pH effect on the preferential iodolactonization was discerned.

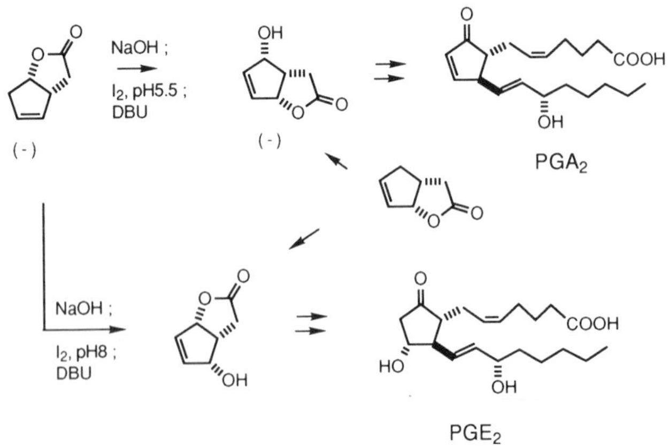

Pertaining to prostanoid synthesis, strategies involving desymmetrization of the cyclopentenediol system by oxidation have been very successful. An adequately protected cyclopentenolone is susceptible to conjugate addition by use of organocopper reagents to give enolate species which can be alkylated to furnish precursors of the prostaglandins [Stork, 1975].

Another promising method for desymmetrization of the diol is by abzyme-mediated hydrolysis of the diacetate [S. Ikeda, 1991]. The antibody was raised by inoculation of the diol monoacetate monophosphate to rabbit.

Prostanoid synthesis based on conjugate addition/trapping protocols has been refined to a state of unusually high efficiency and sophistication [Noyori, 1990; Danishefsky, 1989]. Recently, it has been found that a superior Michael reaction is mediated by organozinc species [M. Suzuki, 1989] thus obviating the need for $R_3P.CuI$ complex or Ph_3SnCl.

A convergent assembly of ($-$)-prostaglandin-E_2 methyl ester [C.R. Johnson, 1986] employed an alcohol secured from acetylcholinesterase-catalyzed hydrolysis. By virtue of the diquinane structure the tandem dialkylation of a dihydroxycyclopentenone acetonide was highly stereoselective. The extraneous oxygen α to the ketone function was removed in the final step of reduction.

The obvious potential of the dioxadiquinane for elaboration of neplanocin-A has been exploited [Medich, 1987].

(-)-neplanocin-A

A synthesis of didemnenones-A and -B has been reported [Forsyth, 1990]. A silyl ether of the cyclopentenolone was reacted with *t*-butoxymethyllithium and the ensuing tertiary alcohol was converted into a ketone. Cyclization was mediated by mercuration of the triple bond.

didemnenones

The C_2-symmetry of an oxocyclopentanediol acetonide (actually a methoxyimino derivative) has been exploited in a desymmetrizing formation of a prostaglandin synthon [Corey, 1986].

PGE$_2$

A segment of maytansine containing two stereocenters has been acquired [Samson, 1977] by the method which comprised conjugate addition of a methyl group to a protected cyclopentenolone and ring cleavage.

maytansine

Two Michael reactions between the symmetrical dimethyl acetonedicarboxylate and the inherently symmetrical 4-acetoxy-2-cyclopentenone, gave rise to a desymmetrized product. The differentiation of two ketone groups enabled a rapid synthesis of tetrahydroalstonine [Hölscher, 1990].

(+)-tetrahydroalstonine

In a synthesis of (−)-pentalenolactone-E methyl ester [K. Mori, 1988e] the same desymmetrized diquinanedione was modified at one of the ketone groups into *gem*-dimethyl substituents. The other carbonyl and the desymmetrizing ester were elaborated accordingly to provide the missing structural components.

pentalenolactone-E
methyl ester

One example of directly utilizing cyclopentadiene in the elaboration of aliphatic compounds is a synthesis of xylitol pentaacetate [Holland, 1982]. Photosensitized ring scission led to an epoxy aldehyde which was converted into the pentose derivative. Interestingly, both diastereomeric epoxides gave the same product.

xylitol pentaacetate   (major)

6.3. OTHER FUNCTIONALIZED CYCLOPENTANES

The first stage of another brefeldin-A synthesis [Corey, 1976] was to form a 4-substituted 2-cyclopentenone by a retro-Michael reaction. The substrate was a symmetrical 3-oxo-bicyclo[3.1.0]hexane-6,6-dicarboxylic ester.

brefeldin-A

Obviously the C_2-symmetric *trans*-4-oxocyclopentane-1,2-dicarboxylic acid is a useful starting material for the elaboration of brefeldin-A. It has been found that the corresponding hydroxycyclopentanemonocarboxylic acid can be obtained from the dimethyl ester by reduction and partial hydrolysis [M. Honda, 1981]. Because of C_2-symmetry, reduction of the ketone is not problematic, and the observed selective hydrolysis may have been due to participation of the hydroxyl group. The molecule with differentiated carbonyl groups can now be readily fashioned into brefeldin-A.

brefeldin-A

The lactone acid served equally well in the synthetic purpose [P.A. Bartlett, 1978] on account of its chemoselectivity toward α-sulfonyl carbanions. It is thus possible to append the "lower chain" to the cyclopentane ring without complication.

brefeldin-A

The keto diester has also found an application in a synthesis of prostaglandin-E$_1$ [Oda, 1975]. Desymmetrization was effected via reduction of the acetal diester and intramolecular acylation of the symmetrical mixed anhydrides. The condensation could occur only once to give an α-acyllactone which was readily alkylated. The route to the prostaglandin target became clear.

PGE$_1$

A rather unusual synthetic plan for ptaquilosin [Kigoshi, 1989a], the aglycon of a bracken carcinogen, specifies the methallylation of a *trans*-1,2-cyclopentanedicarboxylic acid monoester and the formation of the six-membered ring by an intramolecular Friedel–Crafts acylation. Degradation of the angular ester group was necessary.

The C_2-symmetry of the bis-(+)-menthyl ester has a significant bearing on the partial saponification and alkylation of the monoester. In other words, the chiral ester group controlled enantioselectivity of the alkylation. Once asymmetry was established at the quaternary angular carbon atom, the configuration at any of the tetrahedral centers followed automatically, as a result of equilibration or sterically controlling the approach of external reagents during rehybridization reactions.

It is interesting to note that the carbonyl group and the secondary methyl in the five-membered ring were introduced after spiroannulation. The final product was found to be (+)-ptaquilosin, the enantiomer of the aglycon derived from the natural compound.

ent- ptaquilosin

To solve the problem of differentiating the two aldehyde groups of a *trans*-1,2-cyclopentanedialdehyde and using them to synthesize (+)-brefeldin-C [Schreiber, 1988], a process was implemented to effect de novo formation of a mono(dithioacetal) via an intramolecular aldol condensation of an octenedialdehyde followed by degradation of a one-carbon unit.

(+)-brefeldin-C

1,2-Cyclopentenedicarboxylic anhydride has served as a dienophile for 1-methoxy-4-methyl-1,3-cyclohexadiene and the adduct was used in a synthesis of α-and β-barbatene, and gymnomitrol [Kodama, 1979]. Skeletal rearrangement of the bicyclo[2.2.2]octene nucleus was induced in the late stages of the process.

α- barbetene
β-

gymnomitrol

1,2-Bistrimethylsiloxycyclopentene is readily available from the modified acyloin condensation of a dialkyl glutarate. This compound provides a vicinal difunctionality to enable sequential nucleophilic and electrophilic reactions. Its many synthetic applications include the construction of the bicyclo[6.3.1.]dodecane framework [Trost, 1984].

Other examples are syntheses of punaglandin-4 [Sasai, 1987], clavulone-II [Shibasaki, 1985], and crinine [Overman, 1982]. The latter work involved a tandem cationic aza-Cope rearrangement/Mannich cyclization to establish a hydroindolone system.

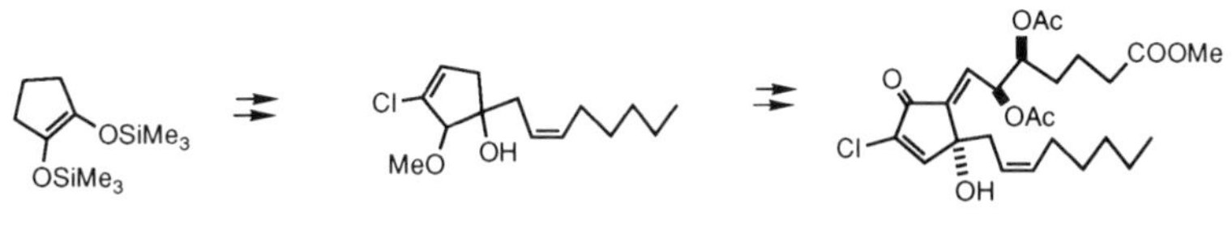

punaglandin-4

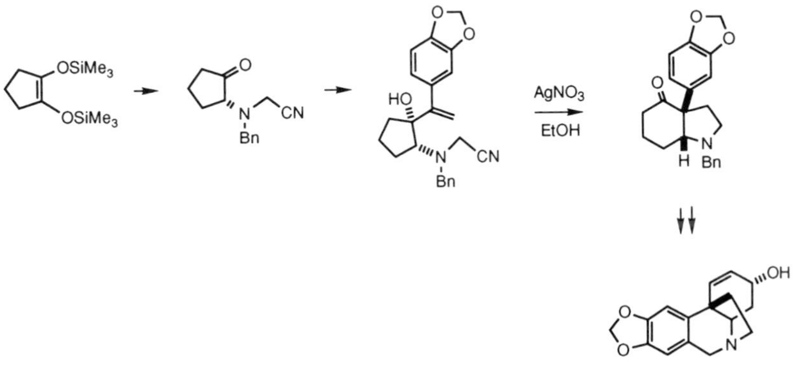

crinine

1,2-Cycloalkanediones are truly *meso* compounds, although most of them exist in the enol form. The desymmetrization can be rendered permanently through reactions such as with organometallic reagents (for example as the first step of a ring-enlarging annulation scheme for *trans*-kumausyne synthesis [M.J. Brown, 1991]) or by forming enol derivatives. The passage through 2-phenylthio-2-cyclopentenone toward aromaticin and confertin [Fang, 1982] is exemplary of the latter maneuver.

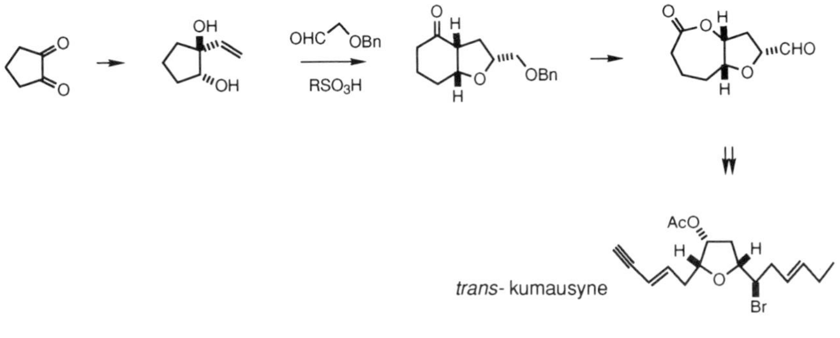

trans- kumausyne

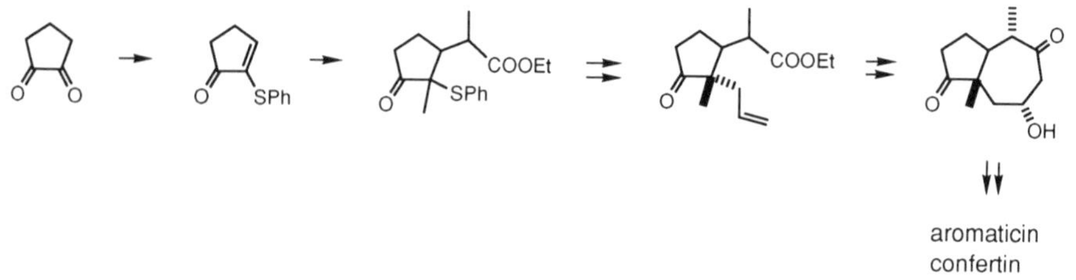

aromaticin
confertin

1,2-Cyclopentanedione has also been used in a Weiss condensation to provide a propellane intermediate for modhephene [Wrobel, 1983].

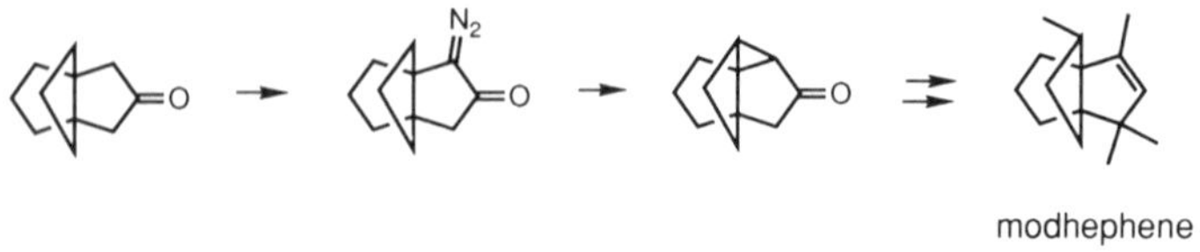

modhephene

A synthesis of aplysin and filiformin [Laronze, 1989] is representative of desymmetrizing assembly of more complex molecules from the symmetrical 2,6-dimethylcyclopentanone.

(+ isomers)

epiaplysin   aplysin

filiformin

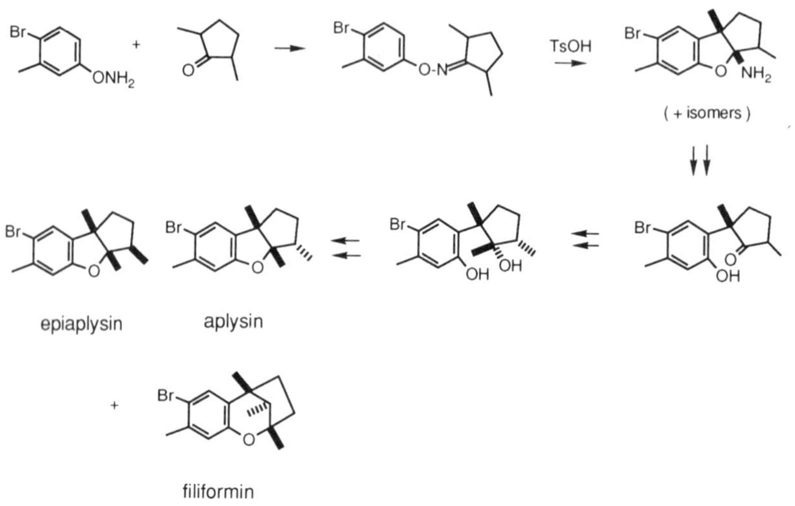

A retrosynthetic analysis of faranal by disconnection of the central double bond readily reveals a difunctional chain which can be correlated with cis-3,4-dimethylcyclopentanone. Accordingly, a synthetic process was pursued [R. Baker, 1981; K. Mori, 1982].

(3S,4R)-(+)-faranal

RCO₃H

1,2,3,4-Tetrachloro-5,5-dimethoxycyclopentadiene undergoes Diels–Alder reaction with electron-rich dienophiles. Such cycloadducts are useful precursors of 3,4-disubstituted cyclopentanones [Jung, 1977] based on a protocol involving dechlorination, double bond cleavage, and hydrolysis of the acetal followed by decarboxylation.

6.4. DIQUINANES

Two factors contributing to the rapid development of polyquinane chemistry are the identification of many physiologically active natural products with such a framework, and the recent evolution of efficient methods for construction of five-membered rings. The challenge of dodecahedrane and related molecules also supplies a continuing stimulus to the development.

A building block of considerable importance is the symmetrical perhydropentalenone. Some synthetic work described here serves to highlight the advantage of using the compound and its analogs. In the case of gymnomitrol the presence of an oxygen functionality in a position corresponding to the dimethylated bicyclic ketone makes it obvious to define a bridging synthetic pathway [Coates, 1979; Han, 1979]. The only caution pertains to the order of the alkylation steps prior to bridging the third ring.

gymnomitrol

Although loganin is devoid of symmetry elements, its synthesis can be contemplated in terms of desymmetrization of a diquinanedione [Caille, 1984].

Thus the transformation of one of the ketone groups into a β,γ-unsaturated ester unit paved the way to unraveling the hydropyran moiety, while the protected ketone was used to introduce the methyl group.

loganin aglucone

The diquinanedione monoacetal has featured in approaches to hirsutic acid-C [Sakan, 1971; H. Hashimoto, 1974]. However, the creation of a new stereocenter from the ketone was somewhat complicated for lack of control elements.

(isomers)

tBuOK

hirsutic acid-C

A chiral diquinane intermediate was obtained after desymmetrizing functionalization of the diketone monoacetal and enzymatic reduction of the regenerated ketone [Z.-F. Xie, 1987].

hirsutic acid

The fashioning of a symmetrical diquinane into the subunit present in quadrone [Piers, 1985b] is interesting. An oxy-Cope rearrangement led to the formation of the bridging six-membered ring, while most other syntheses of quadrone relied on intramolecular alkylation to accomplish this construction.

quadrone   K = ketal

The angular triquinane sesquiterpene, pentalenene, has also been synthesized from a symmetrical precursor [Piers, 1989]. On the other hand, a diquinanedione having only one plane of symmetry (along the ring juncture) was required to complete a synthesis of isocomene [Dauben, 1981].

pentalenene

isocomene

E = COOMe

Stauranetetraone, a fenestrane derivative, is readily obtained in a similar fashion [Deshpande, 1985; Kubiak, 1990].

In bifurcarenone only one cyclopentane ring is present. However, the functional group pattern suggests a diquinanone is synthetically serviceable [K. Mori, 1989e] as its desymmetrization by Baeyer–Villiger reaction and lactone cleavage would furnish two side chains of different length and oxidation level for convenient manipulation.

It has been asserted that symmetry may impede synthetic operations and the diquinanedione was taken as a specific example [Bertz, 1983]. However, a useful compound for selective functionalization at one of the rings is the acetal enone [Belletire, 1983]. The diketone also undergoes Norrish Type I photocleavage [Camps, 1986] to give predominantly an unsymmetrical cyclopentanone.

Another symmetrical diquinanedione is the heteronuclear 1,3-dione. On double annulation a seconor[5]peristylanedione was produced [Eaton, 1977].

[5]peristylane

Alkene units related by mirror symmetry have been discriminated by an ene reaction using enophiles which contain chiral auxiliaries. A diquinadiene is a member of the molecular family, and a kinetic selection of one enantiomer successfully led to a synthesis of (−)-xylomollin [Whitesell, 1986].

The isomeric 1,4-diene possesses a mirror plane, yet the two double bonds can be differentiated, as shown in an application of the process to a synthesis of (−)-specionin [Whitesell, 1988].

(-)-specionin

The unique carbocyclic portion of ikarugamycin is, except for one stereocenter at an angular position, pseudosymmetrical. Inversion of the configuration of the ring juncture and disconnection at the two double bonds of the macrolactam would lead to a symmetrical structure, if oxidation states of the substituents are ignored. Thus it seems possible to assemble the tricarbocyclic system in a modified form with efficiency, by taking advantage of the symmetry elements [Whitesell, 1987].

In fact the enantiomers of a single diquinanecarboxylic ester were used to create a substituted cyclopentenylmethylphosphonium salt and a bicyclic enal, respectively. The Wittig reaction of the two components, followed by photochemical electrocyclization, partial reduction, and adjustment of the position of the remaining double bond and the ring juncture by a series of oxidoreduction maneuvers, accomplished the synthetic objective.

ikarugamycin

It should be emphasized that synthetic schemes based on symmetry considerations for molecules like ikarugamycin are more amenable to access of the chiral form.

A diquinane aldehyde intermediate for hirsutene [Levine, 1988] has been elaborated from 2,2-dimethylindan via Birch reduction, dihydroxylation of the less hindered double bond, hydrogenation, ring cleavage, and aldol condensation. Molecular symmetry was broken in the last step.

hirsutene

A synthesis of illudinine [Woodward, 1977] was facilitated by the use of a symmetrical indacene derivative. But when symmetry of the molecule was destroyed upon conversion of the two bromine substituents into an ester and a methoxy group, cleavage of one of the five membered rings in the desired location could be implemented.

(major)

illudinine

6.5. NORBORNADIENE

Norbornadiene is a most readily available chemical and it has a rich chemistry. By virtue of the two nonconjugated but interactive double bonds not only various functional groups can be introduced, new carbon skeletons may be created. Accordingly, desymmetrization of norbornadiene leads to important building blocks.

The Prins reaction of norbornadiene affords a tricyclic skeleton which is readily converted into a keto acid by oxidation. As the cyclopropane ring of this ketone is cleaved on exposure to a hydrohalic acid, it is possible to secure versatile intermediates for the synthesis of cyclopentane derivatives in the form of a 2,5,7-trisubstituted bicyclo[2.2.1]heptane with three different functionalities. Thus the synthetic utility of norbornadiene relies on its ready desymmetrization.

A synthesis of sesquifenchene [Grieco, 1975] employing the norcamphor-carboxylic acid called for debromination, methylation at C-7 and conversion of the carboxyl function into the proper six-carbon side chain, and the ketone into an exocyclic methylene group. The *exo*-bromine atom at C-5 of the bromonorcamphorcarboxylic ester provided more effective steric hindrance to the approaching methylating agent than an oxygen atom of the ethyleneacetal unit, giving a 15:1 mixture of products which was enriched in the desired isomer.

sesquifenchene

The bromonorcamphorcarboxylic acid has been utilized in many other syntheses. Thus a simple approach to methyl dihydrojasmonate [T.-L. Ho, 1982a] was planned on the basis of a Baeyer–Villiger oxidation to generate percursorial functionalities for the ketone and the acetic ester side chain.

methyl dihydrojasmonate

The inversion of the configuration of C-7 set the stage for the transformation of the norcamphor ring system to boschnialactone, iridomyrmecin, and isoiridomyrmecin [T.-F. Wang, 1989]. The polar *syn*-7 substituent apparently intervened to ensure delivery of the peracid to the *exo*-face of the substrate [T.-L. Ho, 1982b]. A Henbest-type hydrogen bonding alignment can steer the reagent into the more hindered environ.

boschnialactone

iridomyrmecin   isoiridomyrmecin

A practical route has been developed for prostaglandin synthesis [Bindra, 1973; Peel, 1974] from norbornadiene. A normal Baeyer–Villiger reaction of the 5-*exo*-halo-2-oxobicyclo[2.2.1]heptane-7-*anti*-carboxylic acid furnished the lactone. Subsequently the halogen atom served as a leaving group in a translactonization step.

The Prins reaction of norbornadiene with chloral is an alternative method for the desymmetrizing functionalization which gave the same intermediate for prostaglandin synthesis [Takano, 1977].

Oxygen atom insertion between C-1 and C-2 of norborn-5-en-2-ones by reaction with alkaline hydrogen peroxide becomes feasible because of the increased strain of the carbon framework. The α-hydroxyhydroperoxide adducts may undergo heterolysis instead of a concerted rearrangement. Consequently, *syn*-7 substituted norborn-5-en-2-ones are useful precursors of cyclopent-2-enol-3-acetic acid derivatives, as demonstrated by a synthesis of iridomyrmecin [Grieco, 1981].

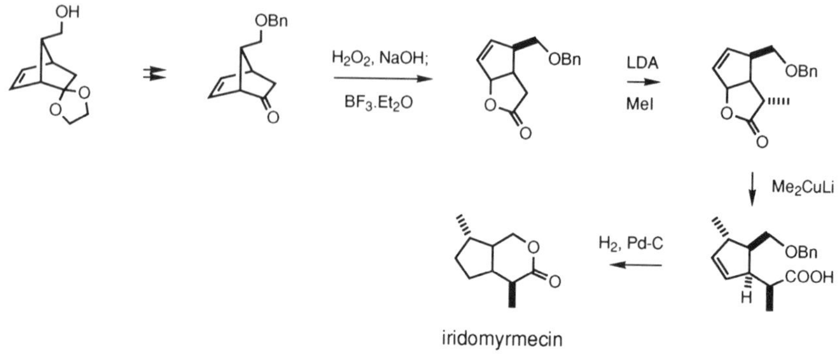

iridomyrmecin

Fused lactones of the type encountered in the above have been recognized as synthons for chain segments of certain polypropionate-derived antibiotics. This recognition led to the synthesis of the Prelog–Djerassi lactone [Grieco, 1979a], tylonolide hemiacetal [Grieco, 1982a], and calcimycin [Martinez, 1982] by proper modification of substituents at a stage before or after devolution of the condensed lactones.

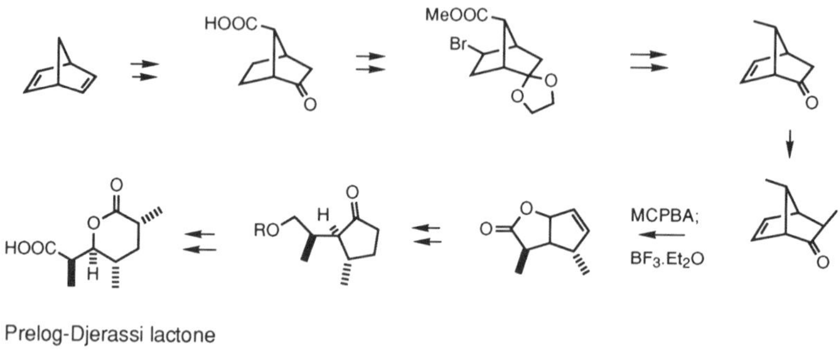

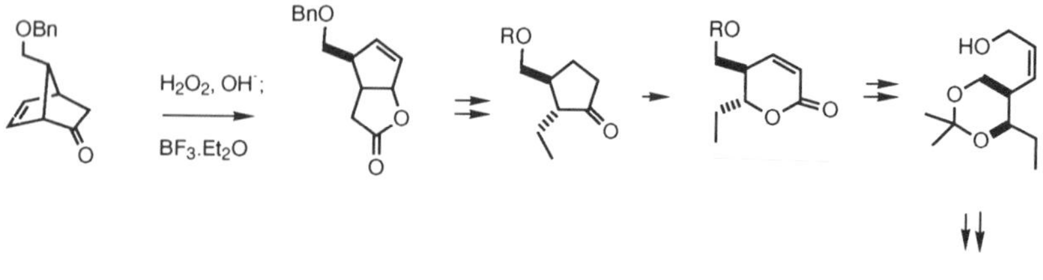

Prelog-Djerassi lactone

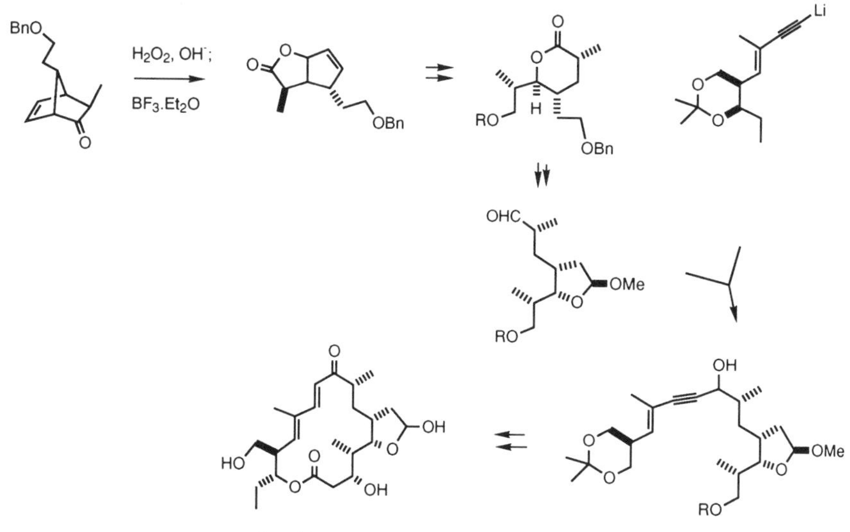

tylonolide hemiacetal

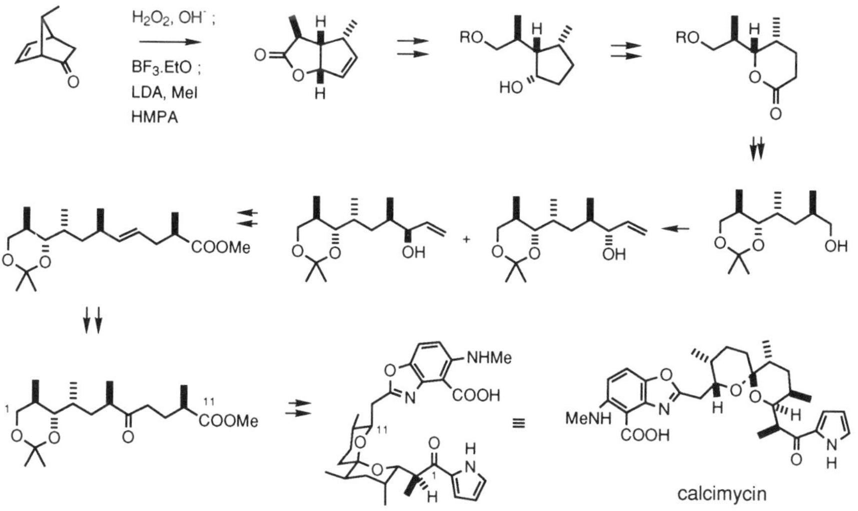

calcimycin

The carboxyl (or ester) group of the protected bromonorcamphor may be modified into a side chain containing an additional chirality center. Upon such a modification a β-lactam with all the essential functional groups in proper configurations for the elaboration of thienamycin may be established [Grieco, 1984]. Thus, it required insertion of a nitrogen atom instead of the oxygen atom indicated above, after the synthesis reached the cyclopentanone stage, and excision of a two-carbon-atom subunit corresponding to C-5 and C-6 of the norbornenone.

thienamycin

The presence of an oxygen substituent at C-4, a methyl group at C-5 of the pseudoguaianolides suggests their synthesis from analogous fused lactones. A retrosynthetic analysis of these sesquiterpenes specifies the need for prior methylation at C-3 and C-7 of the norcamphor precursors. As indicated in the sesquifechene synthesis during methylation of the 7-ester the presence of both bromo and acetal substituents favors attack of the electrophile from the *syn* side of the masked ketone, resulting in a relative configuration which corresponds to a *trans* ring fusion of the hydrazulene ring system. The full substitution of C-7 ensures *endo*-methylation at C-3, correlating with C-10 of damsin.

The cyclopentenol generated by cleavage of the modified norbornenone proved valuable not only for a synthesis of damsin [Grieco, 1977b], using a rather straightforward reaction sequence thereafter, but also the elaboration of helenalin [Ohfune, 1978] by proper reengagement of functionalities so as to allow annexation of a γ-lactone to the hydrazulene ring system. The critical issue of configuration inversion of C-10 was addressed by the formation and subsequent equilibration of a *trans*-fused δ-lactone in which the methyl group was axial.

damsin (= dihydroambrosin)
ambrosin

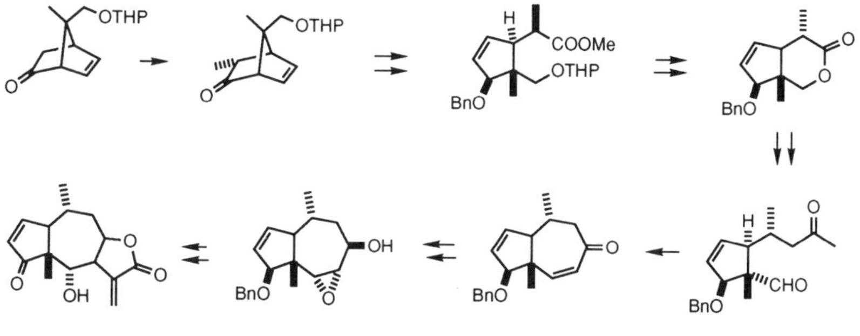

helenalin

The use of the norbornenone for steroid synthesis has also been conceived. By identifying the cyclopentene portion of the bridged ring system with the D-ring of steroids, the creation of three stereogenic centers (C-13, 17, 20) can be accomplished by methylation at C-7, reduction of the ester group, and methylation at C-3 [Grieco, 1979b; Trost, 1979a]. Introduction of the carbon chain for the C-ring construction took advantage of the allyloxy pattern of the transposed lactone which emerged from oxidative cleavage of the norbornenone and subsequent isomerization.

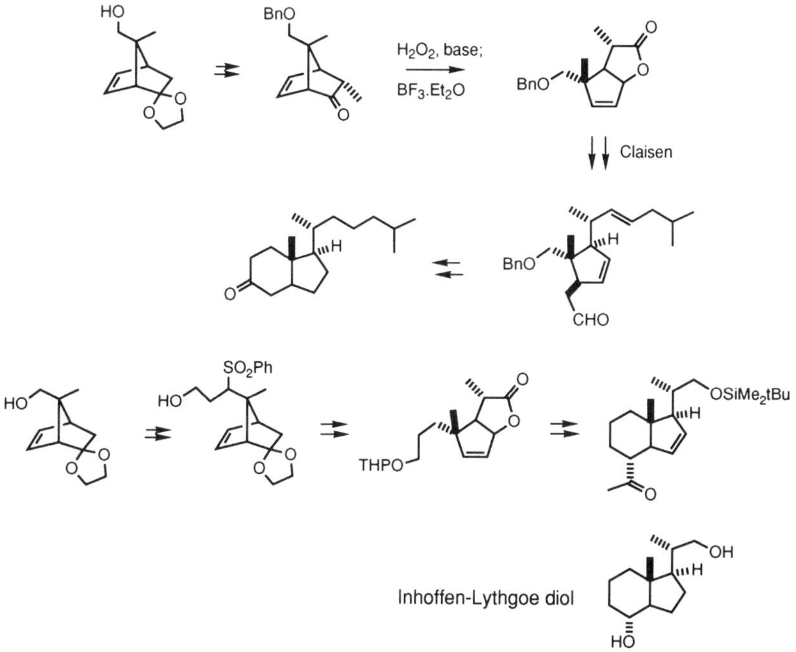

Inhoffen-Lythgoe diol

An adaptation of the *o*-quinodimethane route to estrone [Grieco, 1980b], by crafting the vinylcyclopentane moiety from the norbornenone, is particularly rewarding.

A biogenetically inspired approach to sordaricin [Mander, 1991] involved an alkylative assembly of the precursor from the norbornenone-7-carboxylic ester and a synthon derived from (+)-carvone.

Aristeromycin, a carbocyclic analog of adenosine, is an antibiotic which inhibits phytopathogenic organisms. The synthetic challenge posed by the cyclopentane moiety of aristeromycin was met by employing norbornadiene as starting material which permits dihydroxylation from the *exo* face, to establish a *trans* relationship with the oxymethyl and the nitrogen substituents. The latter two substituents were fashioned by oxidative cleavage of the residual double bond and appropriate manipulations [Shealy, 1966]. It should be noted that desymmetrization occurred after the scission of the norbornene. Aristeromycin in the chiral form can be obtained [M. Tanaka, 1992b] by enzymatic reaction of the *meso*-diol or its diacetate, produced by cleavage of the dihydroxylated norbornadiene.

aristeromycin

A route to echinosporin has been devised [Kinesella, 1990]. However, problems still exist in the introduction of the hydroxyl group into the iridoid skeleton.

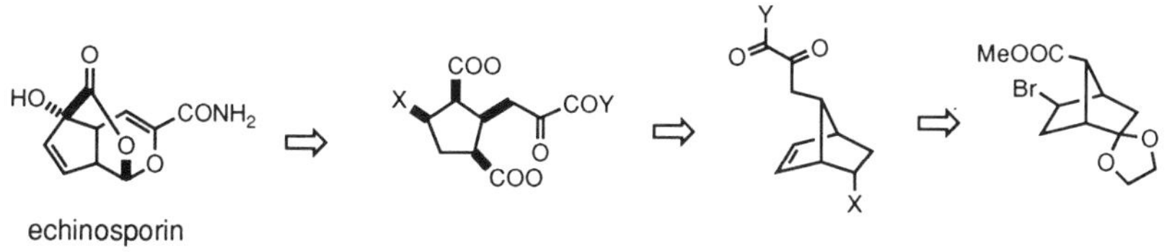

echinosporin

An imaginative approach to a spiroacetal intermediate of the milbemycins [Van Bac, 1988] consists of elaborating a symmetrical dilactone. This dilactone is readily obtained from nobornadiene.

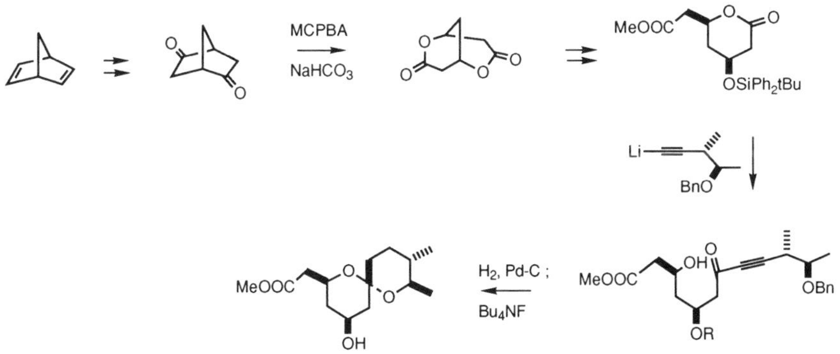

A desymmetrized norbornane-2,5-dione has found use in a synthesis of crassin acetate methyl ether [Dauben, 1990]. Both carbocycles suffered cleavage in the process.

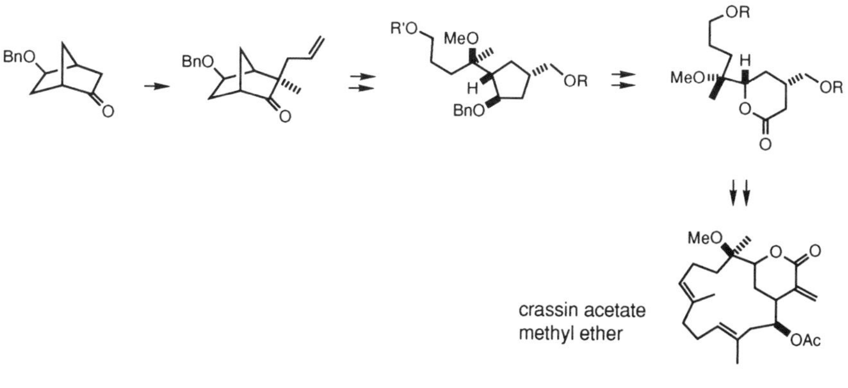

crassin acetate
methyl ether

A different scheme of utilizing norbornadiene for prostaglandin synthesis is that based on its facile conversion into 6-formylbicyclo[3.1.0]hex-2-ene on treatment with a peroxycarboxylic acid. The condensed ring system provided stereocontrol during introduction of the α-chain, and the cyclopropane moiety served as a masked unit for the hydroxyl group at C-11 and the ω-chain. The regioselectivity and stereoselectivity for unravelment of the latter two structural elements were in turn dictated by the α-chain [Axen, 1969; R.C. Kelly, 1973].

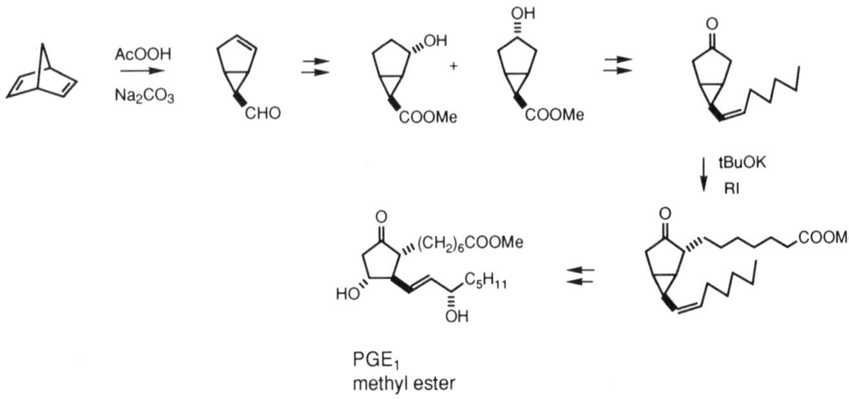

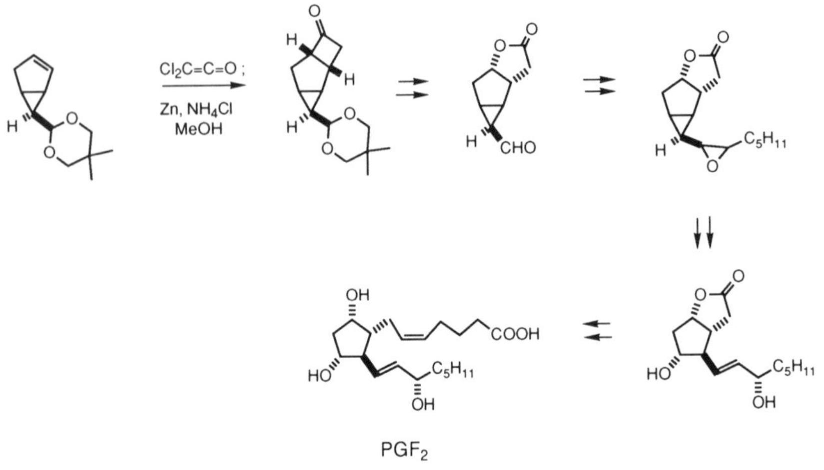

PGE₁
methyl ester

PGF₂

A synthesis of *trans*-4,5-dialkyl-2-cyclopentenones [Schore, 1979] employed norbornadiene as an ethylene equivalent. Thus the desired products were obtained via a Pauson–Khand reaction and Michael addition followed by a retro-Diels–Alder fission. Note that the reactivity of norbornadiene as the Pauson–Khand reaction substrate is due to its ring strain. The equivalency of the two double bonds removed chemoselectivity problems.

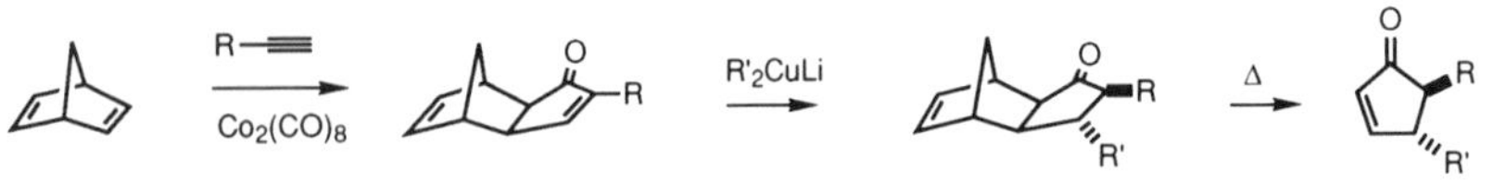

6.6. NORBORNENE DERIVATIVES

Besides norbornadiene other norbornanes are also versatile building blocks for synthesis. Norbornene itself is commercially available and inexpensive. In terms of its utility in the synthesis of unsymmetrical molecules, desymmetrization is achievable by functionalization at the trigonal carbon atoms. The simultaneous formation of C—C and C—O bonds at these sites is illustrated in a convenient synthesis of epi-β-santalene [Snowden, 1981].

epi-β-santalene

Many of substituted norbornenes are easily prepared from cyclopentadiene employing the Diels–Alder reaction. Interestingly, a ready preparation of cis,cis,cis,cis-1,2,3,4,5-pentakis(hydroxymethyl)cyclopentane [Tolbert, 1985] involved manipulation of the Diels–Alder adduct of 5-benzyloxycyclopentadiene and maleic anhydride.

An eminent example which shows the synthetic potential of this class of compounds pertains to the access to a common intermediate of all the primary prostaglandins [Corey, 1969c]. Accordingly, cyclopentadiene was alkoxymethylated and then submitted to the Diels–Alder reaction to gain a precursor of 7-anti-alkoxymethylbicyclo[2.2.1]hept-5-en-2-one. The norbornenone was cleaved oxidatively and the resulting cyclopentenolacetic acid was elaborated into a fused γ-lactone.

PGF$_2$

By a chain extension process the *meso* compound, 7-*anti*-methyl-7-*syn*-tosyloxymethyl-2-norbornene, has been converted into an allylic chloride which underwent an intramolecular magnesio-ene reaction. The product was reacted with carbon dioxide to complete the cyclic array of sinularene [Oppolzer, 1982]. Desymmetrization occurred in the cyclization step.

The asymmetry of albene is due to the double bond. Accordingly, synthesis of this sesquiterpene in the racemic form can take advantage of symmetry until the double bond is introduced. The [3+2]-cycloaddition involving 2-(trimethysily)methyl-2-propenyl acetate and dimethyl norbornene-2,3-dicarboxylate furnished an adduct with the correct skeleton [Trost, 1982b]. Another cyclopentannulation route to albene [Curran, 1987a] involved free radical transfer.

A more conventional approach to albene [J.E. Baldwin, 1981] used the minor Diels–Alder adduct of cyclopentadiene and dimethylmaleic anhydride as starting material.

While an *endo* transition state is preferred for the Diels–Alder reaction of cyclopentadiene and maleic anhydride, *exo*-cycloaddition predominates in the corresponding reaction involving a fulvene as the diene. The symmetrical cycloadduct of 6,6-diamethylfulvene and maleic anhydride [Alder, 1950] has been used in formal synthesis of longifolene [T.-L. Ho, 1992] with desymmetrization performed after lactonization.

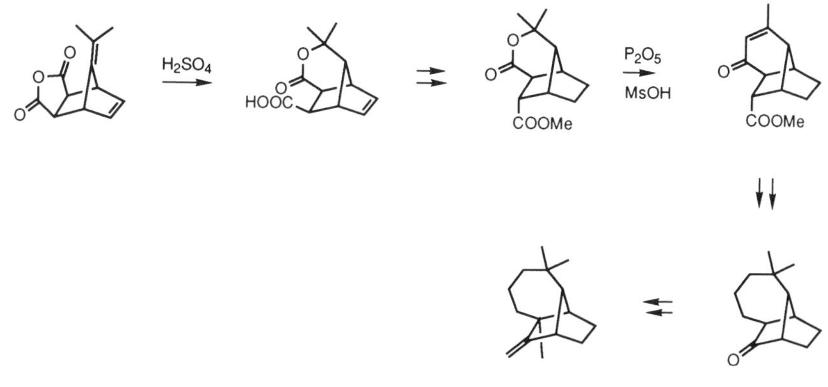

longifolene

The norbornenes obtained from furmaric and maleic acid derivatives are useful for elaboration of many natural products including prostaglandins [Jones, 1974; Fischli, 1975] and boschnialactone [Van der Eycken, 1985]. In these syntheses the cyclopentene ring was cleaved with or without further modification.

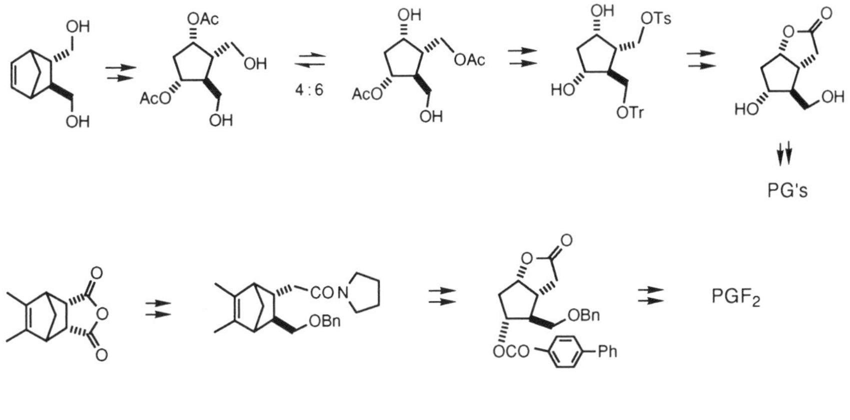

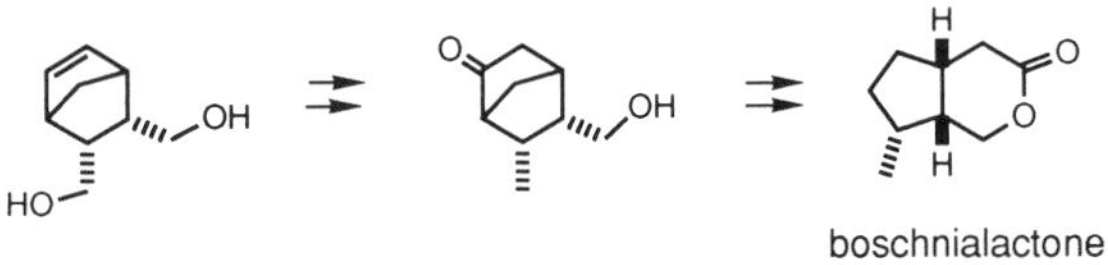

PG's

PGF$_2$

boschnialactone

An excellent synthesis of forsythide dimethyl ester [Furuichi, 1974] exploited the norbornane skeleton as a template for the construction of the *cis*-fused cyclopentane ring containing the carboxyl residue. The pyran moiety was crafted from the saturated five-membered ring of *endo*, *endo*-bicyclo[2.2.1]hept-5-ene-2,3-dicarboxylic acid.

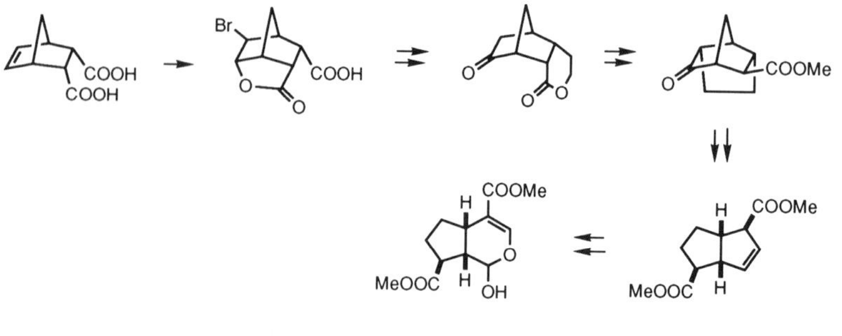

forsythide
dimethyl ester

Methanolysis of 3,6-methanophthalic anhydride served to differentiate the two carboxyl groups. Interestingly, anodic oxidation of the monoester led to decarboxylation and rearrangement. The 7-*anti*-carbomethoxy-2-*exo*-norbornyl acetate proved to be a convenient precursor of methyl dihydrojasmonate [Torii, 1975].

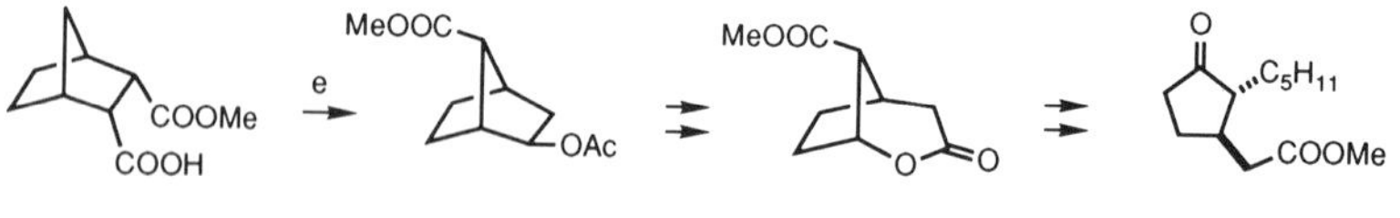

methyl
dihydrojasmonate

More highly functionalized norbornenes may be prepared from various Diels–Alder adducts. For example, dihydroxylation (OsO$_4$) of the unconjugated double bond of dimethyl bicyclo-[2.2.1]hepta-2,5-diene-2,3-dicarboxylate gave a suitable intermediate for the synthesis of (−)-aristeromycin, (−)-neplanocin-A [Arita, 1983] and pentalenolactone [Danishefsky, 1979]. In these syntheses a different five-membered ring of the bicyclic system was cleaved.

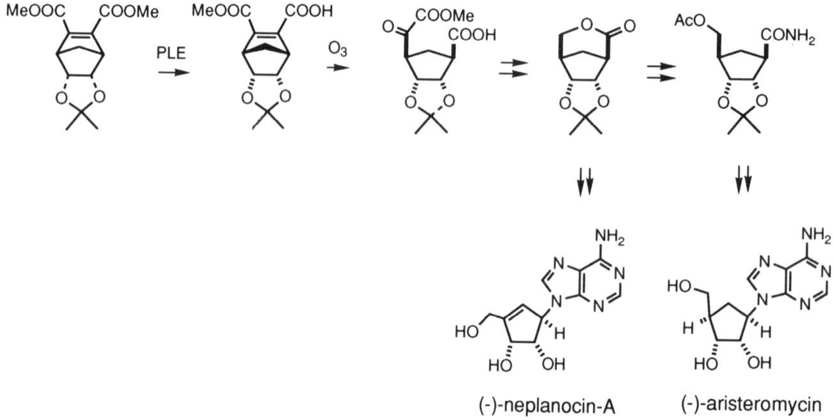

(-)-neplanocin-A   (-)-aristeromycin

pentalenolactone

The Diels–Alder adduct of cyclopentadiene and *p*-benzoquinone undergoes intramolecular photocycloaddition with great facility. On pyrolysis the cyclobutane ring is broken and a symmetrical triquinanedione is generated.

Desymmetrization with selective functionalization of this molecule is relatively easy to manage in view of the close proximity and parallelism of the two enone systems. In other words, the dipole–dipole interactions favor isometrization that removes them. Thus the semihydrogenation of the new bisenone is more readily achieved. This aspect was extremely beneficial to the execution of a precapnelladiene synthesis [Mehta, 1984].

precapnelladiene

Cyclopentadiene can provide a three-carbon unit to the central ring of the linear triquinane skeleton and *p*-benzoquinone is halved during the "metathetic" transformation. The application of this process for synthesis of linear triquinane sesquiterpenes needs to address the distribution of the methyl groups and to identify the methylation pattern for each terpene molecule. Thus, the synthesis of hirsutene [Mehta, 1986b] requires 2,5-dimethyl-*p*-benzoquinone as dienophile, while the synthesis of capnellene requires benzoquinone only (but the diene is methylcyclopentadiene).

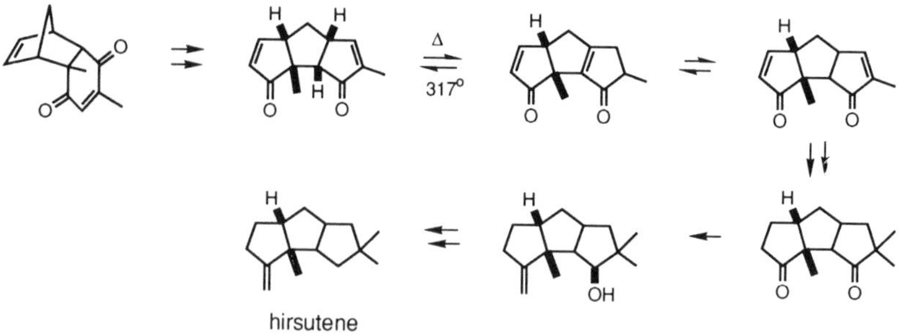

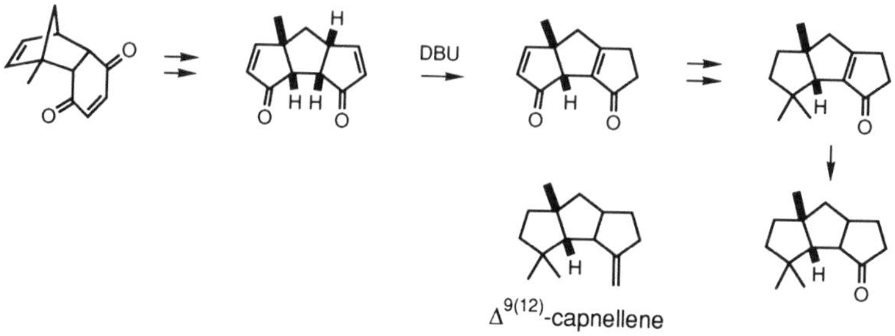

hirsutene

$\Delta^{9(12)}$-capnellene

The versatility of the triquinane derivatives has been further demonstrated in an elaboration of potential intermediates of crinipellin-A [Mehta, 1988a] and magellaninone (Mehta, 1987c).

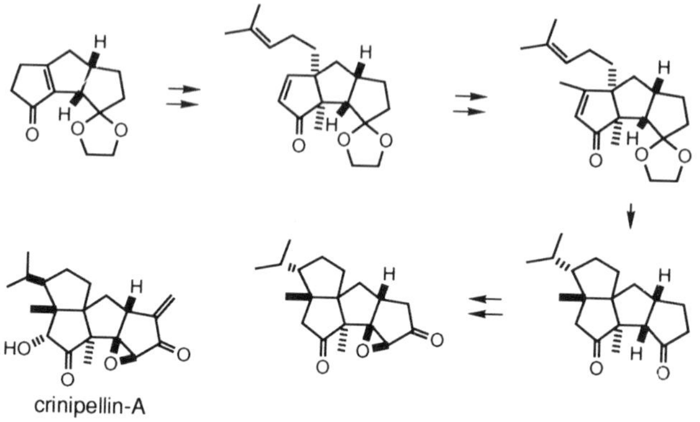

crinipellin-A

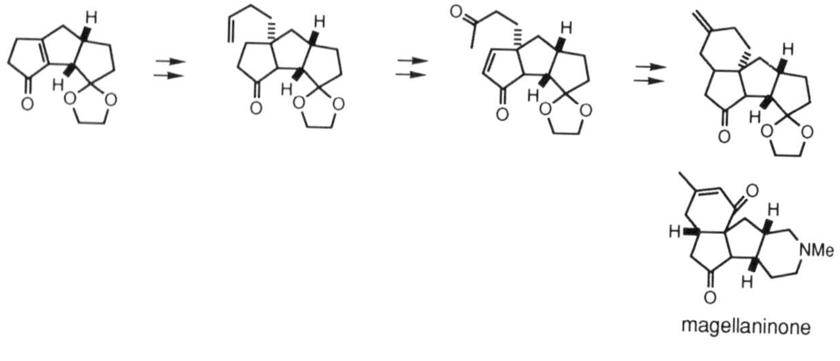

magellaninone

An intriguing role for a symmetrical norbornane-2,6-dione is its transformation into glycinoeclepin-A [K. Mori, 1989f]; one of the five-membered rings was destined to be part of the acid side chain, while the other served as the template for annulation. Asymmetric reduction by baker's yeast delivered a ketol which was then protected at the carbonyl site, allowing oxidation, methylation, epimerization, ring expansion, and at a later stage, ring cleavage to unravel the chiral side chain.

glycinoeclepin-A

6.7. 1,3-CYCLOPENTANEDIONES

Symmetrical 1,3-cyclopentanediones almost rival the 1,3-cyclohexanedione homologs in synthetic utility. According to structural changes necessary for various projects, examples are given below.

6.7.1. Bond Formation Involving C-1, C-2

Most applications of symmetrical 1,3-cyclopentanediones involve bond formation at C-1 and C-2, frequently via annulation. Thus a simple synthesis of dihydrojasmone [T.-L. Ho, 1971] made use of the symmetrical nature of 2-(*n*-pentyl)-1,3-cyclopentanedione which was protected as the enol methyl ether and reacted with methyllithium. On treatment of aqueous acid, dihydrojasmone was produced.

dihydrojasmone

One of the outstanding syntheses of modhephene [Oppolzer, 1981a] started from 1,3-cyclopentanedione, transited through a diquinane with a key step of an ene reaction to establish the propellane framework. In the intramolecular ene reaction the critical configuration of the tertiary carbon atom was fixed concurrently with generation of the methyl group attaching to it.

modhephene

The Diels–Alder reaction produces cyclohexene derivatives. Since there are many methods for ring contraction, sometimes it is more expedient to initiate a cyclopentane preparation by the Diels–Alder reaction. A case that illustrates the strategy is the synthesis of 3-oxosilphinene [Ihara, 1986a]; its advantage is control over four stereocenters during formation of the tricyclic ketone.

3-oxosilphinene

An unusual design for synthesis of verrucarol [Koreeda, 1988] consisted of an intramolecular Diels–Alder reaction to create the hydrochromane skeleton. While one of the original ketone groups became incorporated into a lactone at the intermediary stage, the other ketone was transformed into an exocyclic methylene residue for the generation of the epoxide. The diene required for the Diels–Alder reaction was attached to C-4 of the original cyclopentanedione via an oxygen atom, consequently the most crucial structural requirement of verrucarol was satisfied in the cycloadduct. It should be emphasized that in many other synthetic routes to the trichothecane sesquiterpenes the allylic ether is the most troublesome functionality to establish in a regioselective manner.

verrucarol

The application of a photocycloaddition to convene a precursor of hirsutene [Tatsuta, 1979] differed from most other routes as it gave a 6:4:5 fused ring system with inbuilt functionality to initiate the required skeletal rearrangement.

hirsutene

The preparation of a potential intermediate for the capnellane sesquiterpenes [Kagechika, 1991] featured Pd-catalyzed intramolecular cyclization of an enol triflate onto a cyclopentadiene. The substrate was obtained from 2-methyl-2-(3-oxobutyl)-1,3-cyclopentanedione and the cyclization can be made to generate a chiral product by using a palladium catalyst ligated to a 2,2'-bis(diphenylphosphino)-1,1'-binaphthyl unit (BINAP). [Lagechika 1993].

$\Delta^{9(12)}$-capnellene-3β-8β,10α-triol

Because many physiologically active and synthetically challenging molecules contain a polyquinane skeleton, functionalized diquinanes are important building blocks. A diquinenedione and its modified forms are frequently employed for the purpose [Trost, 1980].

2-Allyl-2-methyl-1,3-cyclopentanedione is convertible to the (2S,3S)-ketol in greater than 98% ee by reduction with baker's yeast. This hydroxy ketone served as a starting material for a synthesis of coriolin [Brooks, 1985] and an intermediate of anguidine [Brooks, 1982]. Note that in the latter approach a configuration inversion of the alcohol group was required.

coriolin

anguidine

An interesting assemblage of the hydrindan framework of coriamyrtin [K. Tanaka, 1982] relied on a tandem Michael reaction–aldol condensation. The substrate was prepared by a Michael reaction between the γ-methylenebutenolide acceptor and 2-methyl-1,3-cyclopentanedione donor. The carbocyclization step was initiated by a Kharasch reaction across the butenolide double bond.

coriamyrtin

Confertin has been synthesized [Schultz, 1982] via alkylation of 2-methyl-1,3-cyclopentanedione with a 3-iodomethylfuran. Direct aldol cyclization failed, but the hydrazulene skeleton could be convened after converting the 2,2-disubstituted cyclopentanedione into the enedione. It should be noted that in this synthesis both the cyclopentanedione and the furan are symmetrical. The *meso* intermediates were desymmetrized in the aldol step.

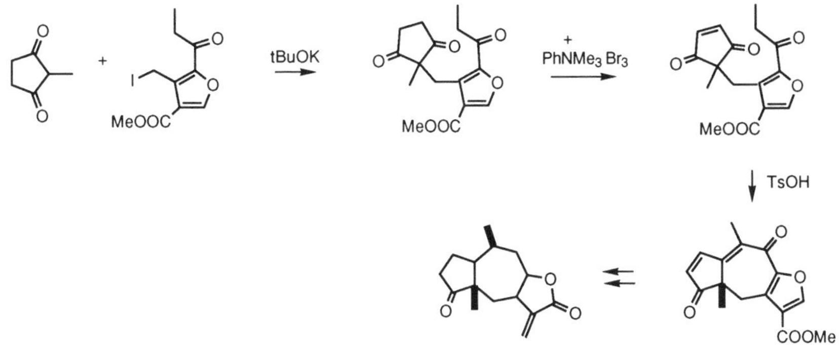

confertin

(There are several complex syntheses featuring the direct use of 4-cyclopentene-1,3-dione. One recent example is that of (+)-hitachimycin [A.B. Smith, 1992].)

(+)-hitachimycin

A less convergent route to another pseudoguaiane sesquiterpene, damsinic acid [Lansbury, 1978], involved intramolecular interception of an allyl cation by an alkyne linkage to form the seven-membered ring. In this intermediate the newly created ketone group furnished a handle for the attachment of the unsaturated acid side chain.

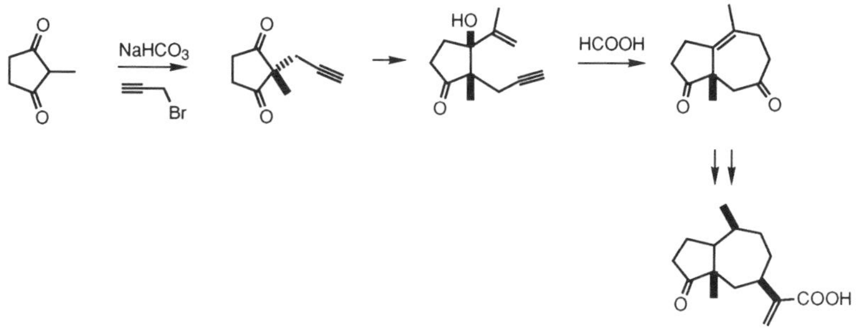

damsinic acid

Annulation of the seven-membered ring by means of a homo-Cope rearrangement provided an alternative pathway for the synthesis of damsinic acid [Wender, 1979]. The substrate was secured by reaction of 3-ethoxy-2-methyl-2-cyclopentenone with 1-methyl-2-vinylcyclopropyllithium and acid workup. The rearrangement step was efficiently conducted by a combination of photochemical and thermal treatments.

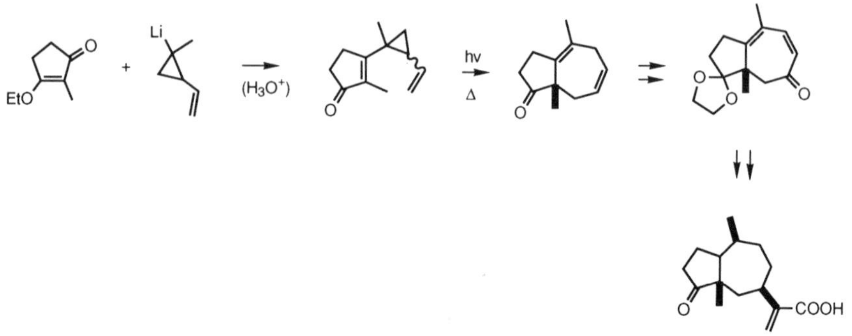

damsinic acid

There have been many synthetic routes to the steroids based on alkylation of 2-methyl-1,3-cyclopentanedione and subsequent cyclization. Generally, Michael reaction of the diketone with an enone affords a triketone which undergoes aldol condensation readily. Variations pertain to the nature of the enone, thus it is possible to choose a compound that permits modification and chain extension [Velluz, 1965; Bucourt, 1967; Saucy, 1971; Danishefsky, 1976].

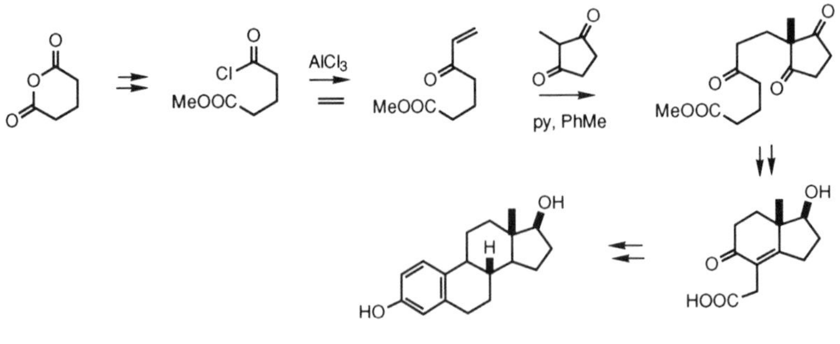

estradiol

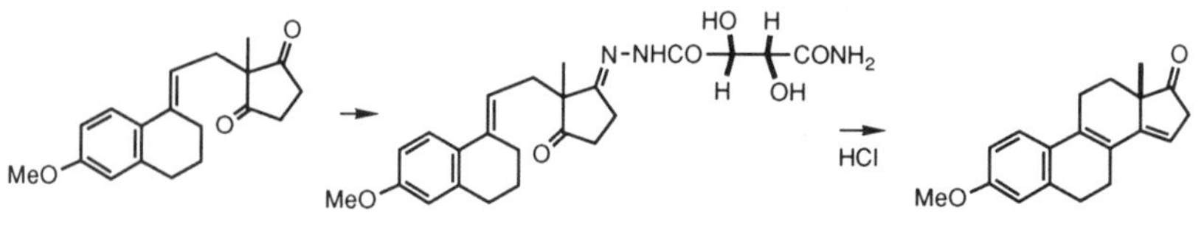

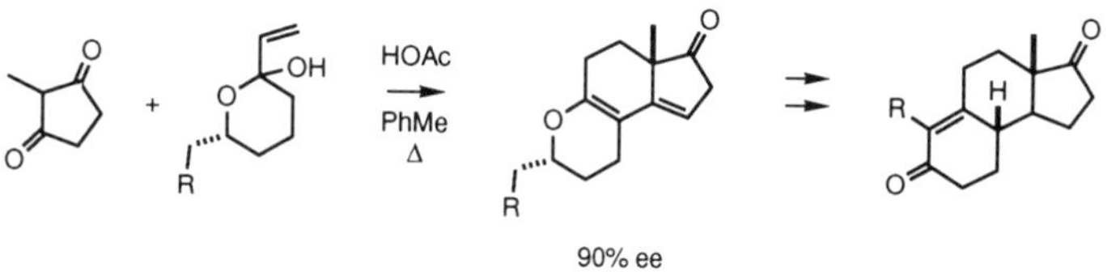

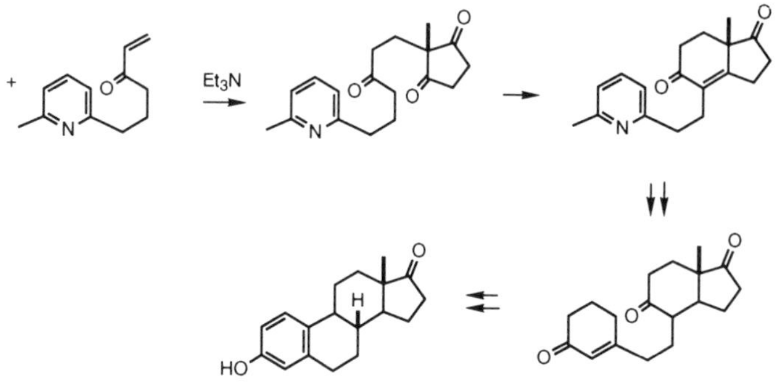

90% ee

estrone

Among variants of steroid synthesis employing 2-methyl-1,3-cyclopentanedione route [Daniewski, 1975] is one which an arylpropyl chain is coupled to a ketone by reaction of a borane and a diazoketone. The product underwent cyclization to form the C-ring.

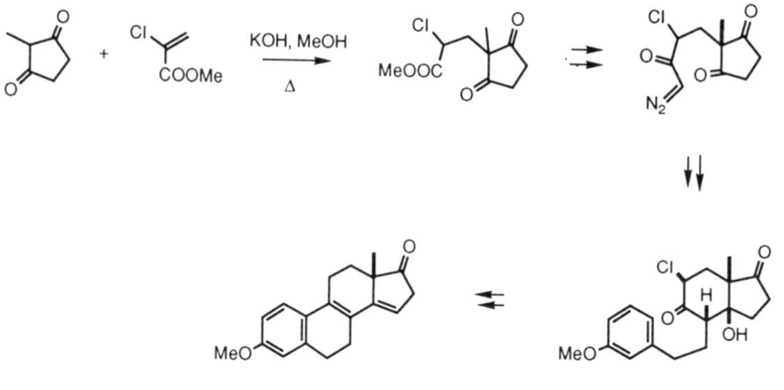

The interconvertibility between khusimone and isokhusimone validates retrosynthetic analysis based on one of the compounds. Disconnection of isokhusimone at the tetrasubstituted double bond (McMurry transform) generates a triketone which has a symmetrical segment. As 2,2-disubstituted 1,3-cyclopentanediones are accessible from reductive diacylation of 1,2-bis(trimethylsiloxy)cyclobutene and rearrangement, a route to khusimone [Y.-J. Wu, 1988] is readily constituted.

khusimone   isokhusimone

The hydrindan analog of Wieland–Miescher ketone, readily available from 2-methyl-1,3-cyclo-pentanedione, also enjoys great popularity as a building block for synthesis. Furthermore, both (R)- and (S)-isomers can be secured at will by using (S)(−)-proline and (S)(+)-homoproline, respectively, to promote the aldol cyclization [Buchschacher, 1977].

The hydrindandione lacks three methyl groups and a furan ring to match the structure of pinguisone, but the locations of such missing elements are well suited for remedy [Bernasconi, 1981].

pinguisone

The presence of a hydrindan subunit in (−)-punctatin-A augurs a synthesis [Paquette, 1986] via ketone reduction, alkylation, and [2.3]-Wittig–Still rearrangement to furnish a substrate for photocyclization to create the cyclobutane unit.

(-)-punctatin-A

A hydrindan segment is also embedded in dendrobine. A synthesis of dendrobine [Inubushi, 1974] started from addition of a cyano group to the saturated ketone which was to be used to complete the pyrrolidine ring.

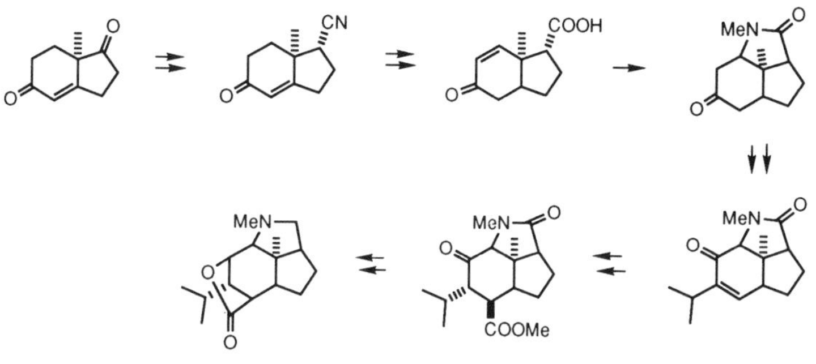

dendrobine

A route to isoclovene [Baraldi, 1983] consisted of reductive *gem*-dialkylation of the cyclohexanone moiety of the *cis*-hydrindan. An ester-Claisen rearrangement to deliver an acetic ester chain to the *exo*-face of the molecule was implemented to give a product containing two functionalized side chains. The third carbocycle was formed by an intramolecular Michael reaction.

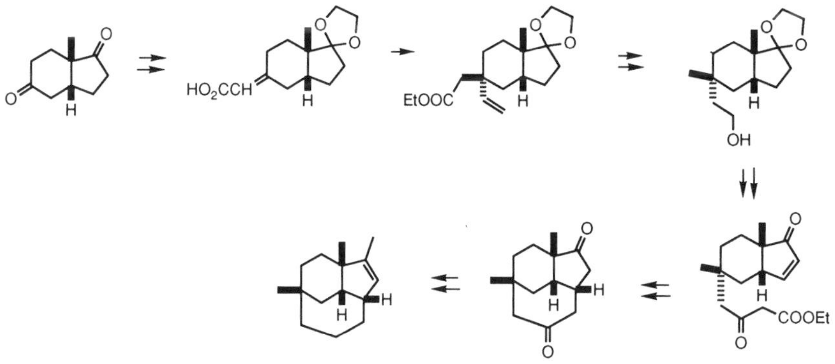

isoclovene

Estrone can be prepared from the hydrindandione [Groth, 1991] in a convergent manner which is clearly advantageous. A new route to testosterone [Ihara, 1990b] featured an intramolecular nitrile oxide-to-alkene cycloaddition to form the B-ring.

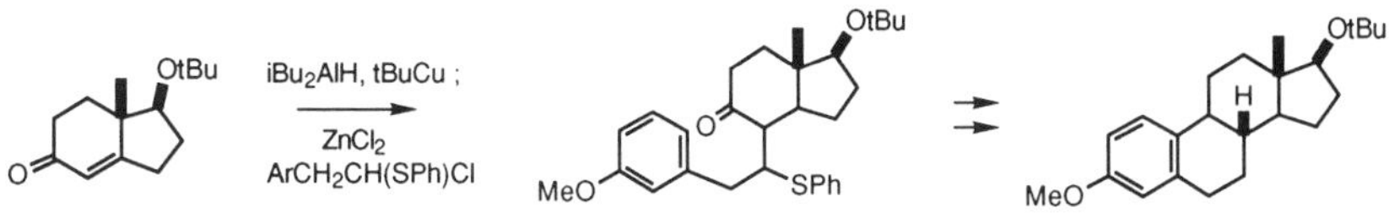

The synthesis of pseudoguaianolides can also take advantage of the substitution pattern of the hydrindandione, that is the presence of an oxygen function at the subangular position next to the methylated ring juncture carbon atom. Expansion of the six-membered ring is necessary.

Different schemes for the ring expansion was instituted in the access to damsin [Quallich, 1979] and helenalin [M.R. Roberts, 1979], and confertin [Marshall, 1976]. In the confertin work the regioselective and stereoselective formation of the ethereal O—C bond of the lactone was concurrent with the ring expansion via cleavage of a fused cyclopropane unit.

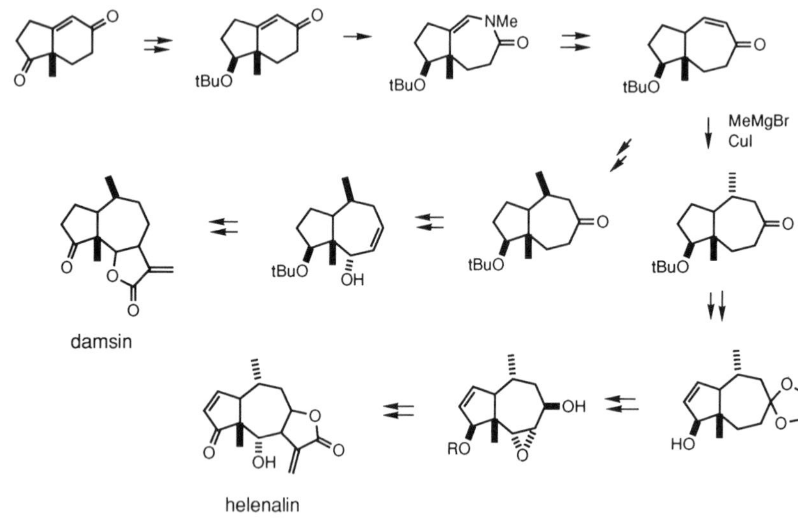

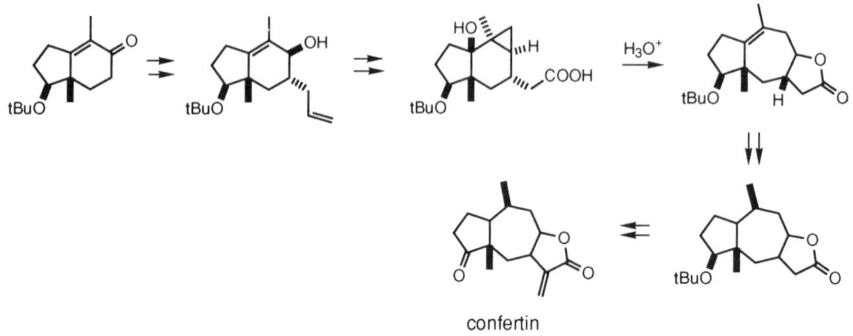

damsin

helenalin

confertin

A route to (+)-ceroplastol-I [Paquette, 1993] featured expansion of the six-membered ring by a Claisen rearrangement to afford an 8:5-cused intermediate.

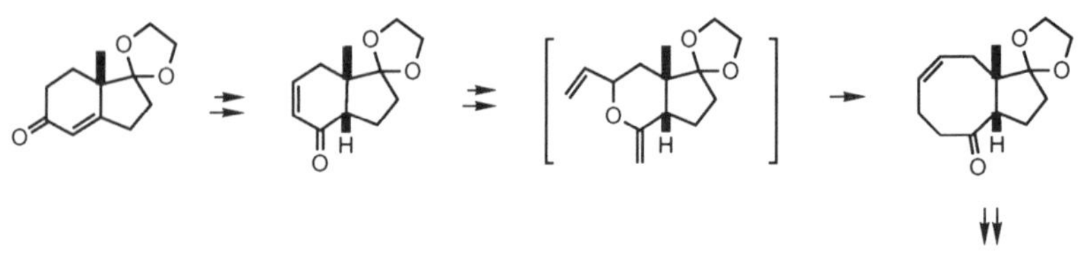

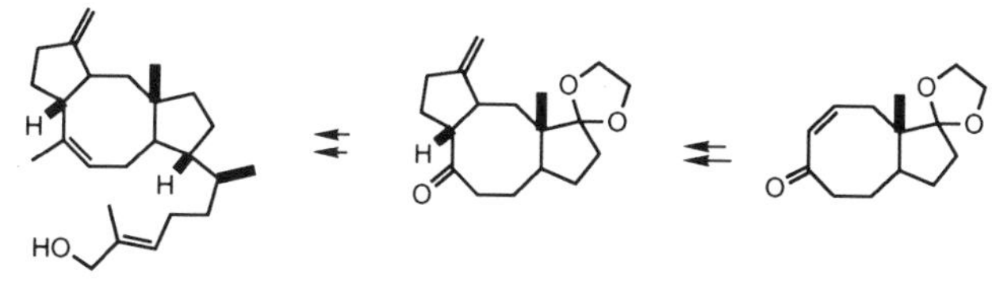

(+)-ceroplastol-I

In an approach to isocomene [Ihara, 1991] based on the same building block, contraction of the six-membered ring was involved.

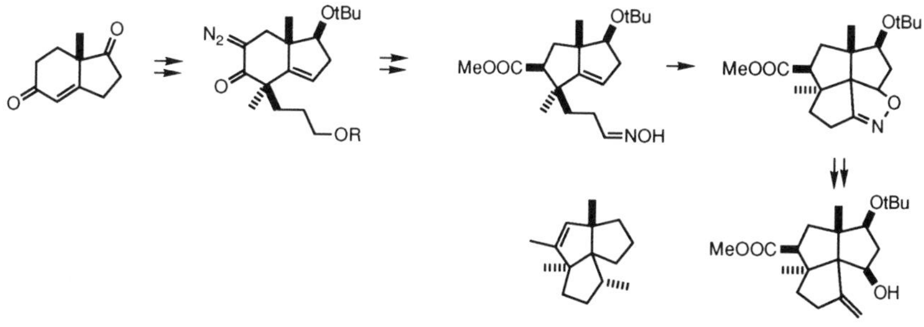

isocomene

Diels–Alder reactions of dienes derived from the cyclohexenone subunit of the hydrindenedione provided substrates for photochemical oxadi-π-methane rearrangements. Proper choice of reactions allowed the syntheses of (−)-silphiperfol-6-en-5-one [Demuth, 1988] and a precursor of hirsutene [Demuth, 1985].

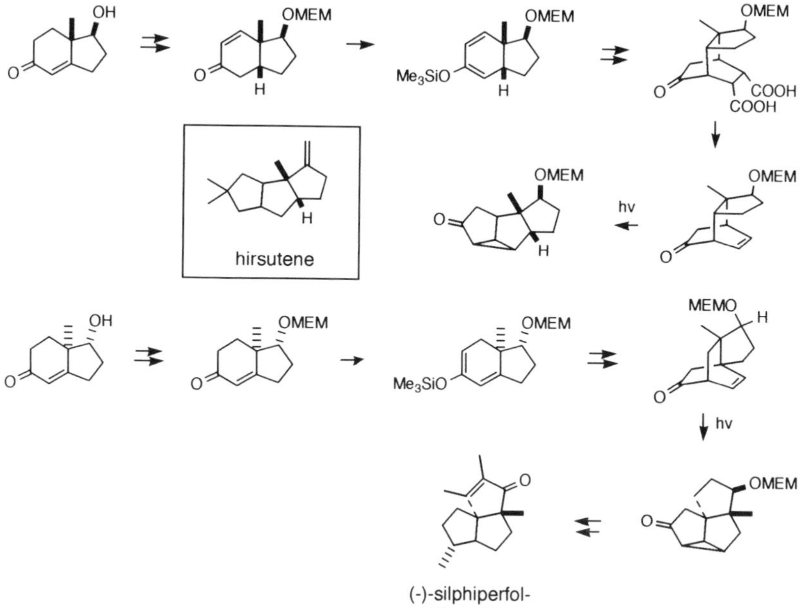

hirsutene

(-)-silphiperfol-
6-en-5-one

Ring cleavage of the bicyclic system was featured in an excellent method for the synthesis of vitamin-D$_3$ [Nemotot, 184, 1985]. After introduction of a C$_6$ pendant the six-membered ring was severed. Elongation of carbon chain adjacent to the ketone for assemblage of a new hydrindanone followed. Thus the steroidal side chain was established with complete stereocontrol at C-17 and C-20.

vitamin-D$_3$

A hydroxy cyclohexenecarbaldehyde suitable for the synthesis of vitamin-E has been prepared [Daniewski, 1984] by degradation of the hydrindendione.

The use of the hydrindandione to provide the six-membered ring and a three-carbon segment which terminates at the hydroxypyranone moiety of compactin [Daniewski, 1990] is also quite novel. A reductive alkylation at the α-position of the enone to initiate construction of the other six-membered ring preceded oxidative cleavage of the cyclopentanone unit.

compactin

A protected hydrazulenedione was considered as a potential precursor of radiatin [Lansbury, 1988a]. The functional group distribution of the bicyclic ketone was so exploited that attachment of an acetonitrile chain to the saturated ketone site, removal of the conjugated carbonyl group, and oxidative cleavage of the olefin resulted in a cyano ester. The latter compound was used in a hybrid Dieckmann–Thorpe cyclization.

radiatin

The synthetic strategy of preserving only the five-membered ring of the hydrindandione was also adopted in a synthesis of verrucarol [Schlessinger, 1982]. In the derived secondary alcohol, two adjacent stereogenic centers corresponded to those of the target molecule. Construction of the pyran ring required degradation of the cyclohexenone by two carbon units and rebridging with an oxygen atom to the cyclopentane. By establishing the remaining portion of verrucarol with a Diels–Alder reaction, two residual carbon atoms from the original six-membered ring ultimately appeared in an angular position with the pendent hydroxymethyl group.

verrucarol

A less complicated process was involved in the elaboration of the angularly ethylated hydrindandione into vincamine [Takano, 1986]. Degradative conversion of the enone to a δ-lactone was accompanied by translocation of the oxygenated site in the five-membered ring so that an α-glycol system could be revealed upon condensation with tryptamine.

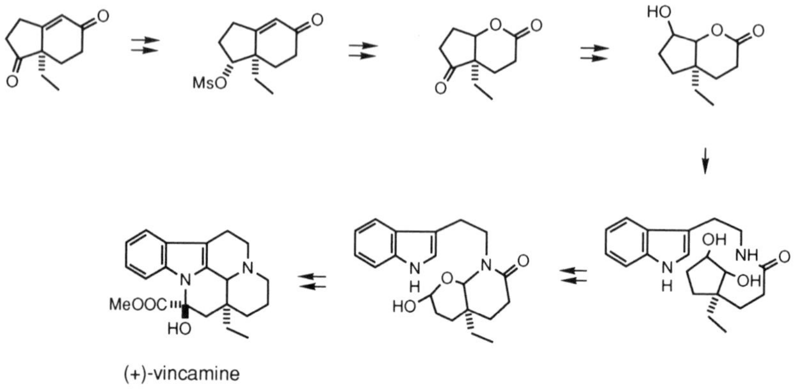

(+)-vincamine

A side chain homolog of the above hydrindandione was identified as an intermediate for synthesis of the cecropia juvenile hormone JH-I [Zurflüh, 1968]. In the process both rings were cleaved by Grob fragmentation. The advantage of this design is in the control of double bond stereochemistry.

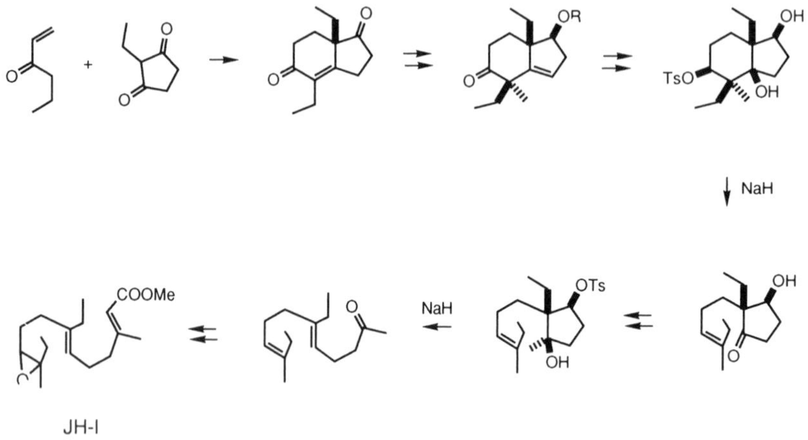

JH-I

A facile synthesis of tribenzo[3.3.3]propellane [Paisdor, 1991] embodied the treatment of 2,2-dibenzyl-1,3-inandione with polyphosphoric acid and removal of the ketone group of the product. The net process performed two intramolecular Friedel–Crafts alkylations.

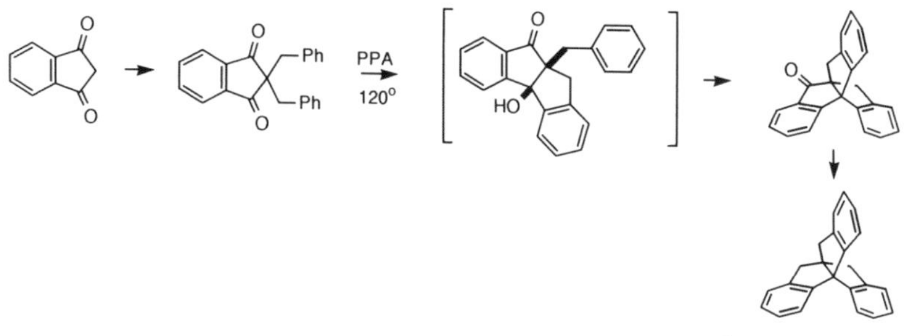

tribenzo[3.3.3]propellane

6.7.2. Bond Formation at C-1, C-2 with Subsequent Ring Modification

There are quite a few syntheses based on bond formation at C-1 and C-2 of 1,3-cyclopentanedione with subsequent changes of the ring. For example, parvifoline has been synthesized [Villagomez-Ibarra, 1994] via Grob fragmentation of a tricyclic intermediate. Heteroatom insertion with exocyclic movement of a ring carbon was involved in an approach to indolizidines [Ohnuma, 1983]. When an amino group was placed in a proper position of a side chain at C-2, $C \rightarrow N$ transacylation occurred. Reclosure of the mesocyclic keto lactam furnished the bicyclic structure.

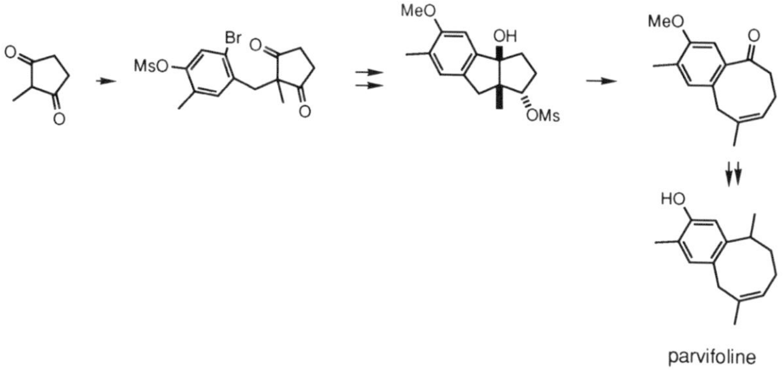

parvifoline

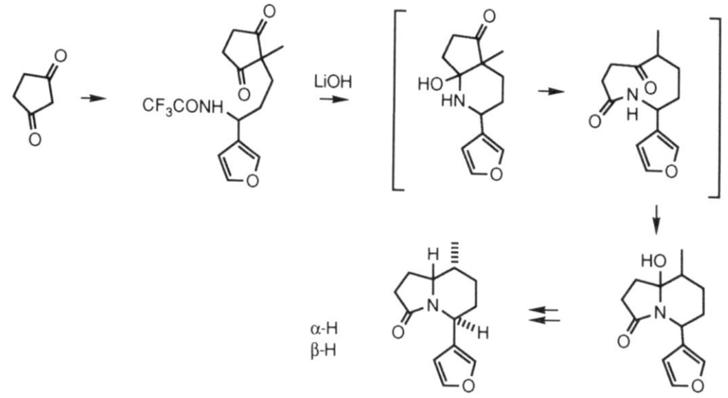

In its development into cedrol [Corey, 1973] 1,3-cyclopentanedione formally underwent a one-carbon insertion between C-1 and C-2 and additional C—C bond formation at these locations.

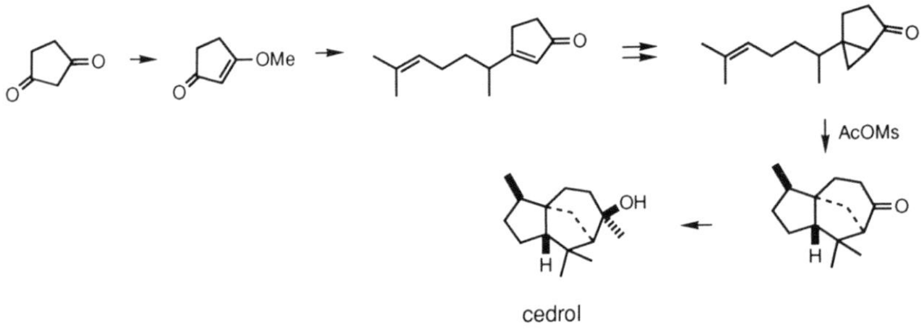

cedrol

A synthesis of β-himachalene [Challand, 1969] from 2-methyl-1,3-cyclopentanedione represents an early application of the deMayo reaction. The fragmentation of the photocycloadduct was implemented after modification in order to differentiate the functionalities of the intermediate. It is noted that the original five-membered ring was expanded by two carbon units.

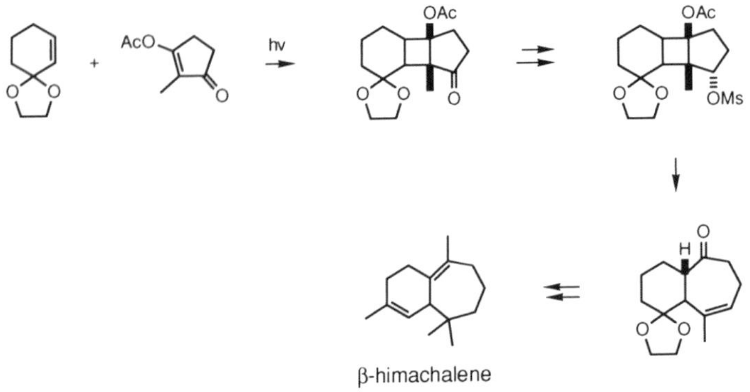

β-himachalene

β-Bulnesene and its epimer have been elaborated [Oppolzer, 1980b] by an intramolecular photocycloaddition followed by fragmentation of a tricyclic diol. Unfortunately, the photocycloaddition transition state leading to the desired product was unfavorable due to steric compression. Consequently the methyl epimer was the major product.

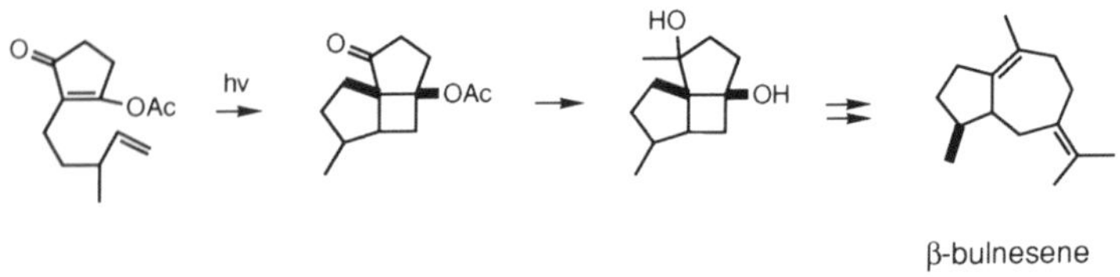

β-bulnesene

For incorporation into zizaene (Barker, 1983] the member atoms of 1,3-cyclopentanedione were redistributed in a more intricate manner. An intramolecular [2+2]-photocycloaddition generated a bridged ring system which was caused to fragment after suitable manipulation.

zizaene

The ring construction scheme for acquiring potential intermediates of the pseudoguaiane sesquiterpenes [Sampath, 1987] was initiated by photocycloaddition of 3-methoxy-2-methyl-2-cyclopentenone with dichloroethene. It also featured a Pauson–Khand reaction and scission of the original five-membered ring.

6.7.3. Bond Formation at C-1, C-3 or at C-1, C-4

These structural amendments of 1,3-cyclopentanediones are less frequently witnessed in the synthesis of complex molecules. A few cases are indicated below. When a 3-substituted 2-cyclopentenone containing an oxo side chain created from 1,3-cyclopentanedione was exposed to alkali, steric and entropic factors favored a tandem Michael–aldol condensation. Thus assemblage of a zizaene precursor [Alexakis, 1978] was most efficiently accomplished.

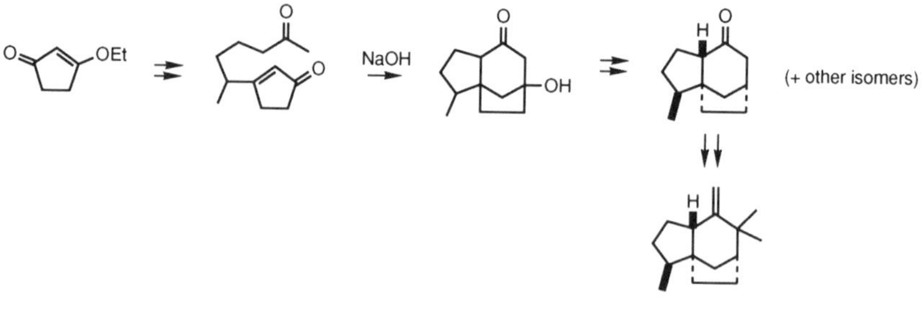

zizaene

The formation of a functionalized brexane by an intramolecular Diels–Alder reaction constituted a key step in a synthesis of sativene [Snowden, 1986]. The synthesis started by alkylation at C-4 of a 1,3-cyclopentanedione derivative, proceeded through a trimethylsiloxycyclopentadiene, which was induced to undergo an intramolecular Diels–Alder reaction with the dienophile in the side chain. Ring expansion of the brexanone and affixation of the isopropyl group followed.

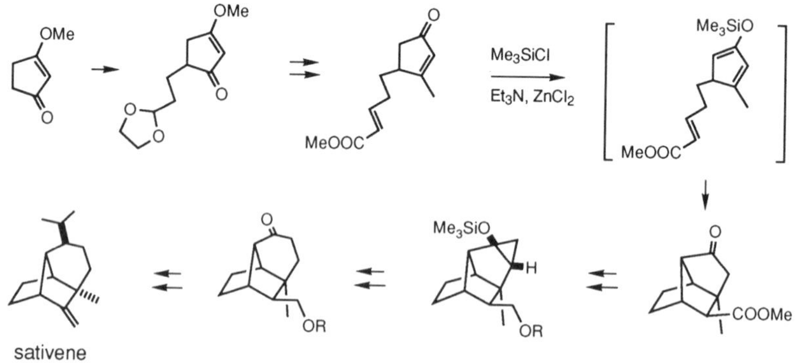

sativene

An analogous Diels–Alder reaction coupled with an oxy-Cope rearrangement was included in a scheme for synthesis of gascardic acid [Berubé, 1989]. This work has yet to be completed.

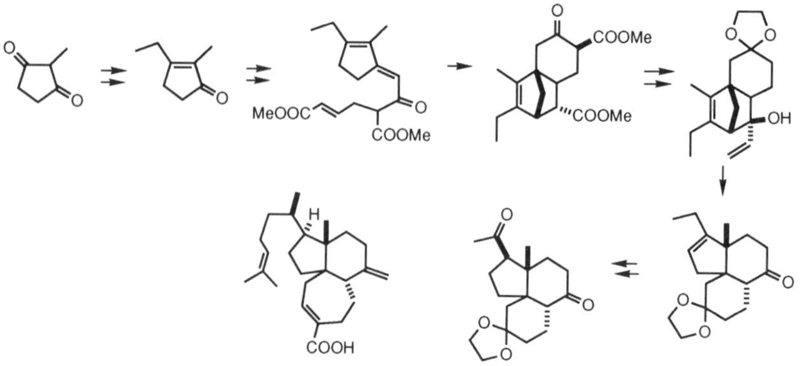

gascardic acid

6.7.4. Bond Formation at C-3, C-4 or at C-4, C-5

Kinetic alkylation of mono-*O*-alkyl derivatives of cyclic 1,3-diketones occurs at the α'-position, and on hydrolysis of the enol ethers, 4-substituted 1,3-cycloalkanediones are produced. It is also possible to effect a second alkylation prior to the hydrolysis in order to acquire the 4,4-disubstituted compounds. This protocol has been employed on numerous occasions, including a synthesis of 14-deoxyisoamijiol [Majetich, 1991b] which featured closure of the central ring of the target molecule by a Lewis acid-promoted intramolecular Michael reaction with an allylsilane as donor. The silicon substituent directed the regiochemistry.

14-deoxyisoamijiol

On the other hand, two consecutive alkylation reactions of a 3-alkoxy-2-cyclopentenone, under thermodynamic and kinetic control, respectively, were the crucial steps en route to methylenomycin-A [Koreeda, 1981]. The careful selection of the alkylating agents enabled transformation of the intrant residues into an exocyclic methylene group and a carboxylic acid without problem.

methylenomycin-A

Such alkylation reactions can be combined into one-pot operation by using a bifunctional alkylating agent [Koreeda, 1983], as shown in the construction of a diquinane precursor of coriolin.

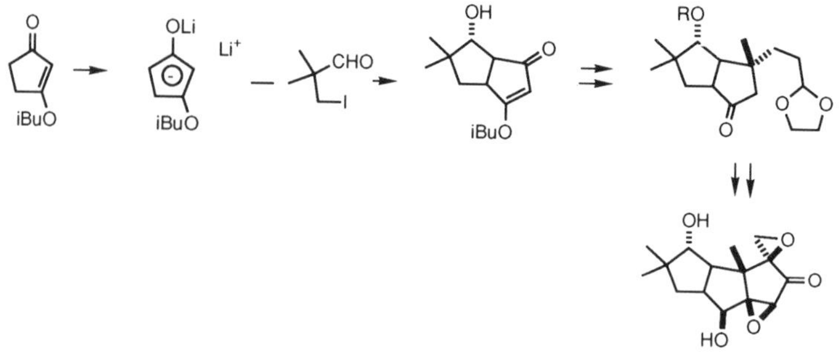

coriolin

Twofold kinetic alkylation of a similar enone initiated the preparation of pentalenolactone [Parsons, 1980]. Formation of the diquinane framework was performed by an intramolecular Claisen condensation.

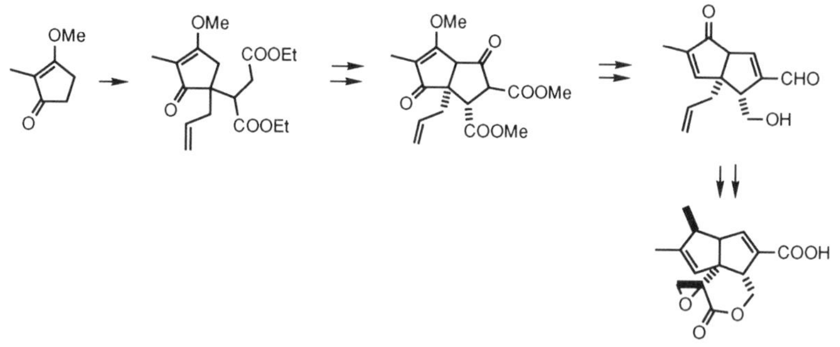

pentalenolactone

Concerted bond formation at C-4 and C-5 of an unsaturated analog of 1,3-cyclopentanedione is achievable by cycloaddition processes, for example the Diels–Alder reaction. Synthetic utility includes the preparation of coronofacic acid [H.J. Liu, 1984] and intermediates of prostaglandins [Brugidou, 1979].

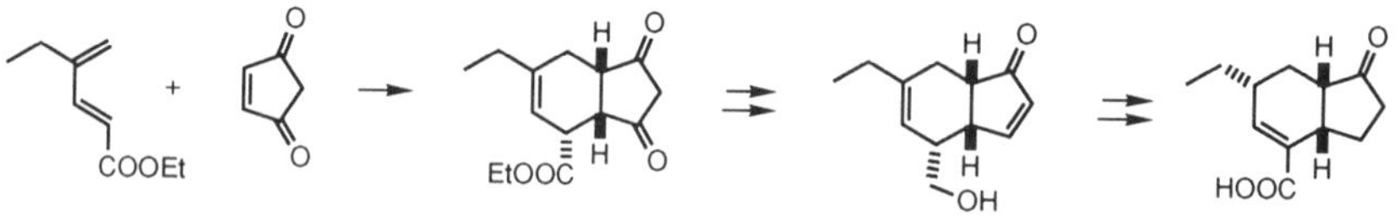

coronofacic acid

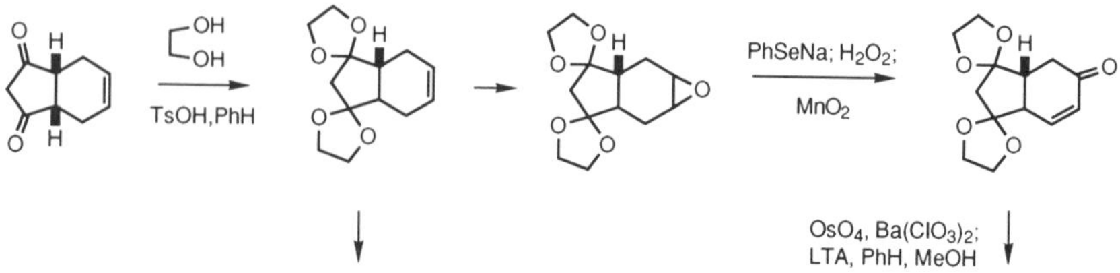

6.7.5. Bond Formation at C-2, C-3, C-4 or at C-2, C-3, C-5

The elaboration of the angular triquinane skeleton of silphiperfolene [Curran, 1987b] by a tandem free radical cyclization method is very efficient. Treatment of the properly constructed 3,4,4-trisubstituted 2-cyclopentenone ethyleneacetal with tin hydride reagent led to the desired product. The substrate was of course derived from 1,3-cyclopentanedione via the monoenol ether. It should be noted that protection of the ketone group rectified the steric course which otherwise led predominantly to the methyl epimer.

silphiperfolene

Rearrangement of a cyclization product of a 1,3-cyclopentanedione derivative afforded a ketone with the skeleton of copacamphor and ylangocamphor [Kasturi, 1988]. Accordingly, the ketone was submitted to oxidative degradation of the aromatic ring for the affixation of the isopropyl group.

copacamphor   ylangocamphor

6.7.6. Bond Formation Involving C-2 and/or C-4

Oudenone is a 2-alkylidene-1,3-cyclopentanedione which has succumbed to synthesis [Ohno, 1971]. Thus aldol condensation of 2-acetyl-1,3-cyclopentanedione with furfural gave an enone, hydrogenation of which led to oudenone and another product.

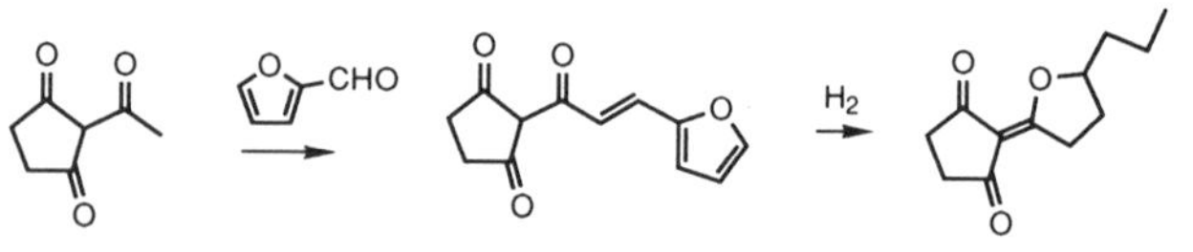

oudenone

A synthesis of (−)-bertyadionol [A.B. Smith, 1986] advocated the use of a cyclopentenone which is fused to a 1,3-dioxane unit as a latent cyclopentenealdehyde. The bicyclic heterocycle was prepared from 1,3-cyclopentanedione.

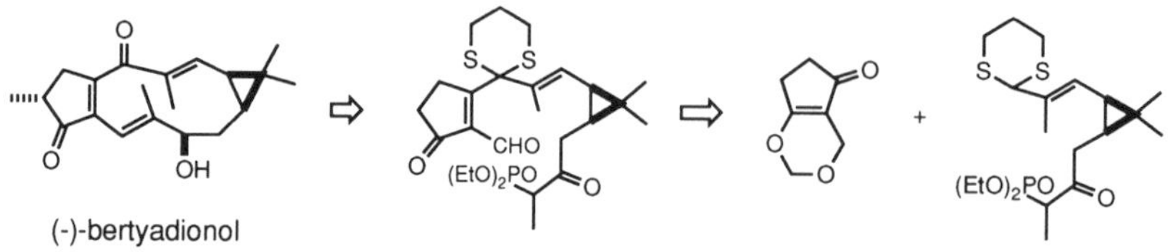

(-)-bertyadionol

Kjellmanianone is accessible by kinetic carbomethoxylation and oxygenation of 3-methoxy-2-cyclopentenone [Boschelli, 1981].

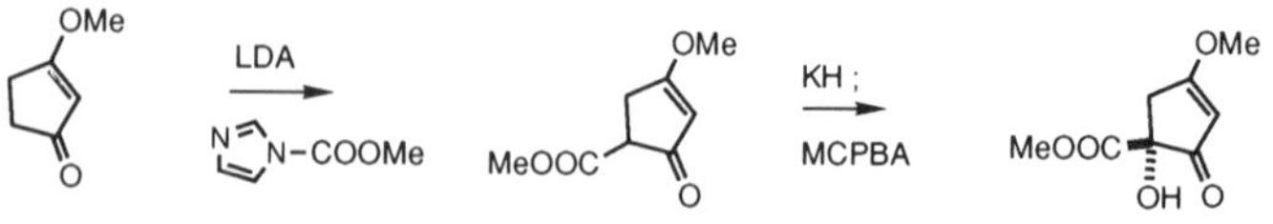

kjellmanianone

An acrylic ester substituted with a 1,3-cyclopentanedione nucleus at the α-position has been employed as the dienophile in a synthetic route to verrucarol [Trost, 1982a]. Excellent stereocontrol of reactions was observed, and the incursion of an intramolecular ene reaction involving one of the cyclopentanedione carbonyls proved to be beneficial to the endeavor.

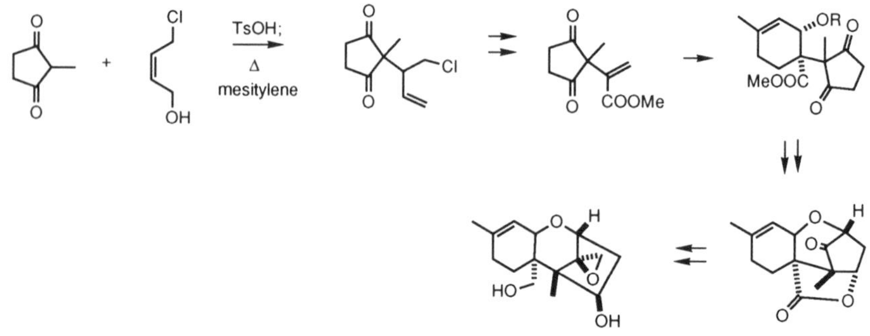

verrucarol

7

SIX-MEMBERED RING SYNTHESIS

7.1. CARBOXYLIC ACIDS AND ESTERS

By far the most popular building blocks that belong to the cyclohexane/enecarboxylic family are the tetrahydrophthalic acids, especially the Δ^4 isomers. Both *cis*- and *trans*-isomers of this tetrahydrophthalic acid and the anhydride and readily available from the Diels–Alder reaction. The utility of these substances in the synthesis of chiral compounds has been greatly increased in recent years by the discovery that the *cis*-dimethyl ester undergoes selective hydrolysis on treatment with pig liver esterase to give the half-ester in 98% yield and 96% optical purity [S. Kobayashi, 1984]. The absolute configuration of the center bearing the carboxyl group is (*R*), and that carrying ester is (*S*). This desymmetrization is very beneficial to synthetic manipulations.

Previous to the development of enzymatic desymmetrization, the racemic half-ester was produced on alcoholysis of the anhydride. It has been used as the A-ring precursor for 7,11-dideoxydaunomycine [Yadav, 1981]. 1,5-Naphthalenediol was the CD-ring component.

7,11-dideoxydaunomycinone

A kindred lactone useful for the elaboration of the AB-ring synthon of 6-deoxydaunomycinone [Penco, 1984] was derived from the dimethyl ester of the *trans*-diacid.

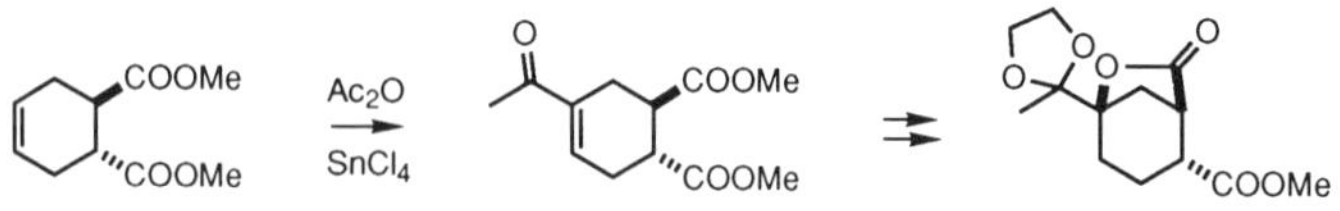

Recently a synthesis of anticapsin [Baldwin, 1993] from the monomethyl ester has been accomplished.

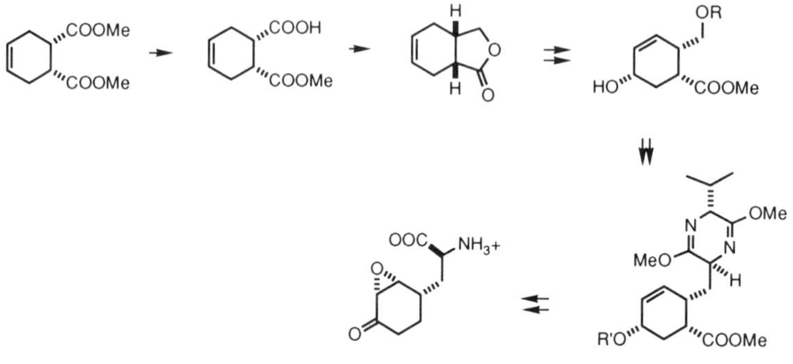

anticapsin

The half-ester undergoes halolactonization very readily. Highly stereoselective reactions can be performed on the bridged lactones. A methylated ester lactone so derived has been converted into geijerone and γ-elemene [Wakamatsu, 1987].

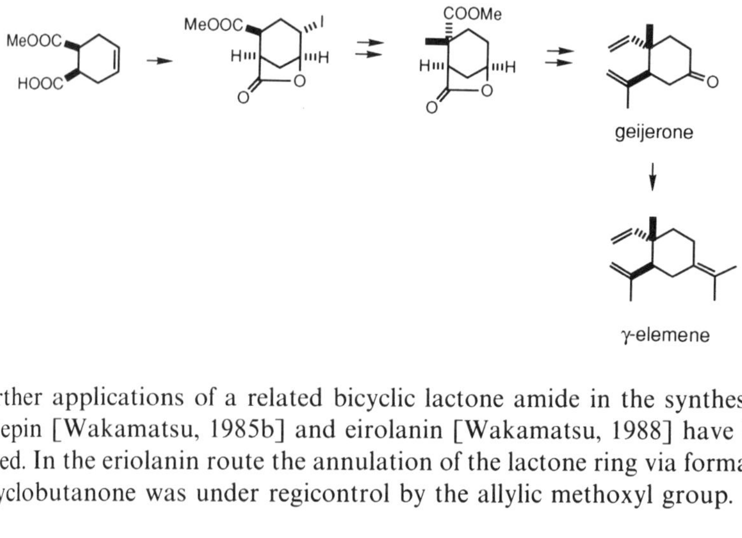

geijerone

γ-elemene

Further applications of a related bicyclic lactone amide in the synthesis of vernolepin [Wakamatsu, 1985b] and eirolanin [Wakamatsu, 1988] have been reported. In the eriolanin route the annulation of the lactone ring via formation of a cyclobutanone was under regicontrol by the allylic methoxyl group.

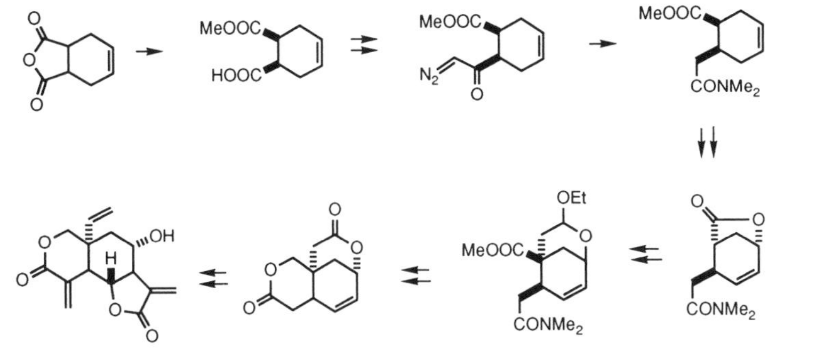

vernolepin

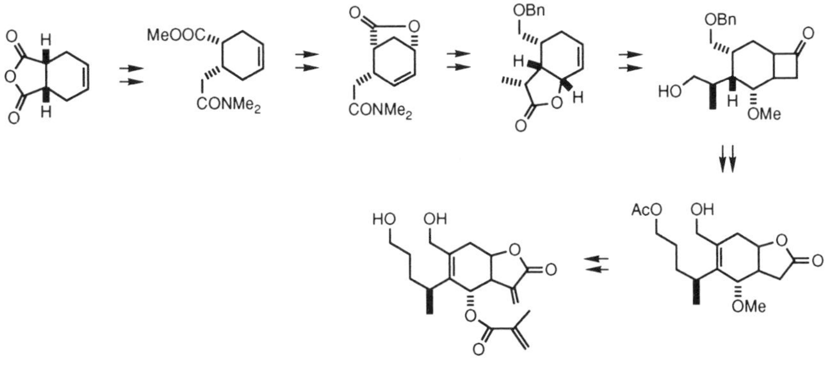

eriolanin

For synthesis of coccinine, montanine, and pancracine [Ishizaki, 1991] one of the carboxyl groups of the cyclohexenedicarboxylic acid was converted into an aryl ketone and the other degraded to give an amino function. The two *cis* substituents directed the *cis*-dihydroxylation of the double bond from the opposite face of the six-membered ring.

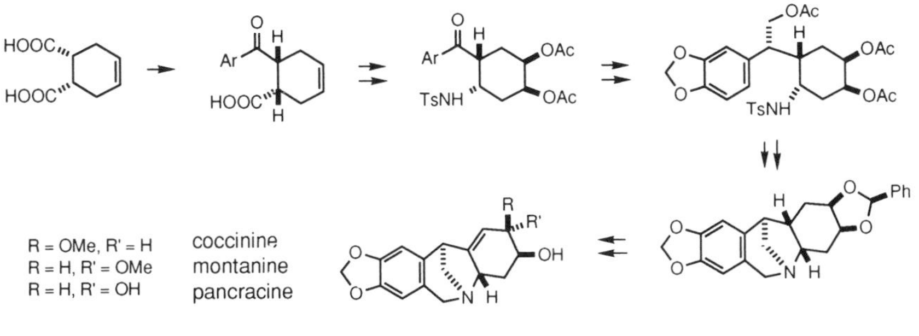

R = OMe, R' = H	coccinine
R = H, R' = OMe	montanine
R = H, R' = OH	pancracine

Chiral esters of the C_2-symmetrical *trans*-diacid are readily obtained by the Diels–Alder reaction. A synthetic application of such a substance is in the elaboration of (+)-papuamine [Barrett, 1994].

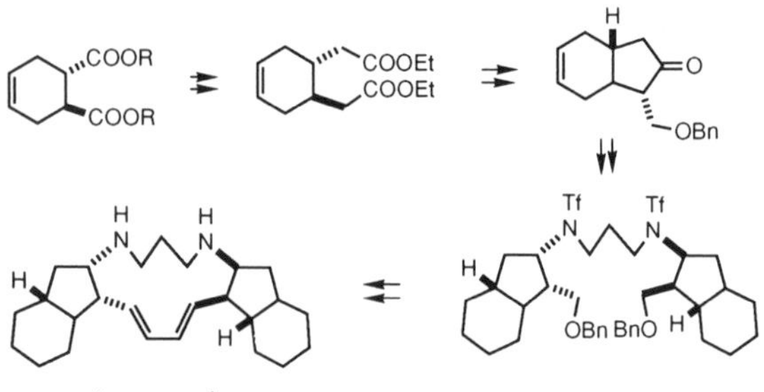

ent- papuamine

The *trans*-diester with (+)-menthyl groups underwent a formal Michael–Dieckmann cyclization tandem to give a bicyclic enone diester which contained all the skeletal carbon atoms of bilobalide [Corey, 1988]. With a C_2-symmetry the *trans*-diester was free from all the stereochemical problems pertaining to framework construction.

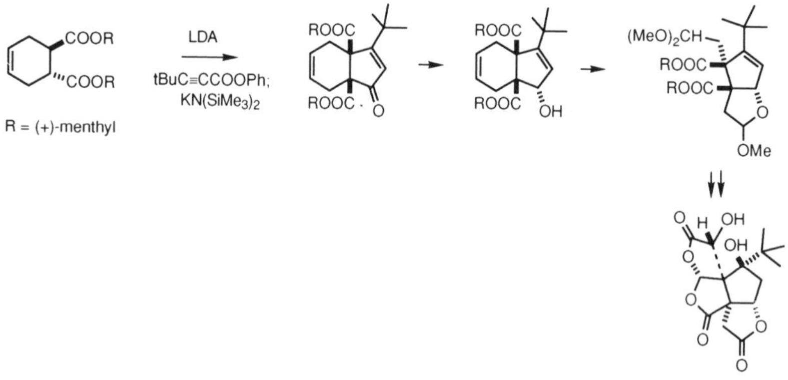

bilobalide

A slight variation of this strategy of annulation is by a tandem alkylation–Dieckmann cyclization, as shown in a synthesis of vetiselinene [Garratt, 1986]. The need for one of the original ester groups vanished after the ring was formed, and it was extirpated readily because of its relationship with the ketone group.

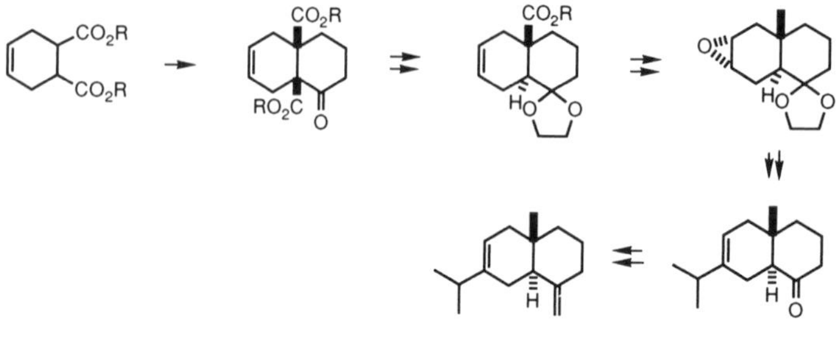

vetiselinene

The chiral half-ester obtained from pig liver esterase-mediated hydrolysis of the *cis*-dimethyl ester has found numerous synthetic applications, e.g., to fortamine [S. Kobayashi, 1990a].

fortamine

Since the remaining ester was to be replaced by the methylamino group the free acid was converted into a *t*-butyl ester which is more resistant to saponification. Although such a circuitous scheme must be instituted while progressing towards the target molecule of the correct absolute configuration, the value of microbial desymmetrization is undiminished.

Another use of the same chiral building block is its conversion into an A-ring synthon for vitamin-D_3 metabolites [S. Kobayashi, 1990b].

The clinical significance of the immunosuppressant antibiotic FK-506 has spawned many synthetic endeavors. A method for accessing a chiral segment of FK-506 extending from C-24 to C-34 [Kocienski, 1990] made use of the now familiar chiral half-ester. After removal of the carboxyl group the ester was hydrolyzed to the (R)-3-cyclohexenecarboxylic acid, which enabled a stereoselective introduction of the oxygen functionality to the double bond. On reduction to the aldehyde stage, a suitable handle for extension of the carbon chain was established.

The epoxides (diastereomers) of the *cis*-dimethyl cyclohexenedicarboxylate could be separated via transformation of the *trans*-epoxide into a lactone [Kuhn, 1989] by pig liver esterase. This lactone is of apparent synthetic utility.

A chiral *β*-lactam which is a known precursor of thienamycin has been acquired [Kaga, 1988] by degradation of the half-ester, i.e., converting the free carboxylic acid into an amino group, and formation of the lactam ring. The cyclohexene was further modified into a cyclopentanone unit and thence by Baeyer–Villiger reaction to establish the two carbon chains. When the lactone was opened, the α-chain was epimerized.

thienamycin

R = SiMe₂tBu

A *γ*-lactone is readily available from the half-ester by selective reduction of the acid function (via the acid chloride). The diacid obtained from cleavage of the double bond is pseudosymmetrical, therefore it gave only one cyclopentanone on cyclization. The oxadiquinane skeleton then induced a reduction of the ketone group from the *exo* face, leading to a compound in which two of the stereocenters were the same as those in brefeldin-A [Gais, 1984a, c]. The third stereocenter was epimeric to brefeldin-A and was readily reestablished by equilibration of its carbonyl pendant.

brefeldin-A

The oxadiquinanedione was found to be an equally viable building block for the prostaglandins [Gais, 1984b].

Related to the above work is an approach to pentalenolactone which has proceeded to a diquinane stage according to a report [Plavac, 1979]. Here the racemic *meso*-diol dimethyl ether was employed.

The enantiomeric lactone is obtainable from enzymatic oxidation of the racemic cyclohexene-4,5-dimethanol [Jakovic, 1982]. Opening of the lactone with thiophenoxide ion, followed by homologation of the carboxylic acid chain and Pummerer rearrangement led to a key intermediate of sweroside glucone methyl ether [T. Ikeda, 1984].

sweroside

It should be noted that the saturated *meso*-diol can be converted into a chiral lactone by horse liver alcohol dehydrogenase in a biphasic system [Matos, 1986]; this practice overcame product inhibition of the oxidation and facilitated separation.

The (+)-*trans*-cyclohexene-4,5-dimethanol is readily desymmetrized owing to its C_2-symmetry. In the following a scheme for the synthesis of vitamin-D_3 [Lythgoe, 1977] is presented.

vitamin-D_3

The *meso*-diacetate undergoes selective hydrolysis by porcine pancreatic lipase, which permits homologation at the free hydroxyl group to a nitrile. A chiral *cis*-octahydroisoquinoline has been acquired and its conversion into (+)-meroquinene hydrochloride [Danieli, 1991] signaled the clearance of a route to quinine. A chiral bicyclic δ-lactone derived from the monoacetate was used in a synthesis of the tetracyclic indole alkaloid (−)-antirhine [Danieli, 1994].

(+)-meroquinene
hydrochloride

Hydrolysis of the *meso*-diacetate with pro-(*R*) pig liver esterase provided a monoacetate with which the free alcohol group can be Homolgated to an isochrome-3-one and thence to (−)-alloyohimbane [Riva, 1987] by its condensation with tryptamine and another simple reaction.

(-)-alloyohimbane

The dimethyltetrahydrophthalic anhydride derived from a Diels–Alder reaction of (E,E)-2,4-hexadiene and maleic anhydride was found to be a useful substrate in building a large portion of the vitamin-E side chain [Berubé, 1984]. While the relative configurations of the secondary methyl groups of the anhydride correspond to those of the target molecule, one methylene group was missing. Fortunately this missing methylene could be inserted via dibromocarbene addition to the double bond with solvolytic ring opening and a reduction maneuver at a later stage when the six-membered ring was cleaved.

A few other symmetrical cyclohexane/enecarboxylic acids have found utility in synthesis. For example, 4-methoxycyclohexanecarboxylic acid was converted into the α,β-saturated congener and thence to an isocyanate. A formal [4 + 1]-cycloaddition of the unsaturated isocyanate with an isonitrile led to a hydroindolone product. N-alkylation and Pictet–Spengler cyclization then gave rise to a progenitor of erysotrine [Rigby, 1991].

erysotrine

7.2. CYCLOHEXANONE DERIVATIVES

Symmetrical cyclohexanones have enjoyed much popularity in synthetic adventures. Because of the enormous literature, a small selection is presented. These cyclohexanones are classified according to their substitution patterns.

We consider 1,2-cyclohexanedione and 1,3-cyclohexanedione as symmetrical molecules despite their occurrence mainly as enols. In a sempervirine synthesis [Potts, 1968] involving reaction of a β-carbolinium salt with 1,2-cyclohexanedione, the latter compound apparently behaved as a diketone in the two condensation steps.

sempervirine bromide

7.2.1. 4-Substituted Cyclohexanones

With one or two simple substituents at C-4, these cyclohexanones embody a prochiral carbonyl group. However, reactions such as monoalkylation and Wittig reaction to give trisubstituted olefins necessarily destroy the symmetry. It is of interest to see a return to the *meso* state after such a change by certain reactions [Mander, 1973].

The potentially valuable building blocks for synthesis, (R)- and (S)-4-hydroxy-2-cyclohexenones, have been obtained from 1,4-cyclohexanedione [Carreño, 1990]. The desymmetrization started from sulfinylation of the monoacetal with menthyl p-tolyl sulfone. Remarkably, diastereoisomeric alcohols were produced by reduction of the keto sulfoxide with diisobutylaluminum hydride in the presence or absence of zinc chloride. Acetal hydrolysis was accompanied by elimination of p-toluenesulfenic acid.

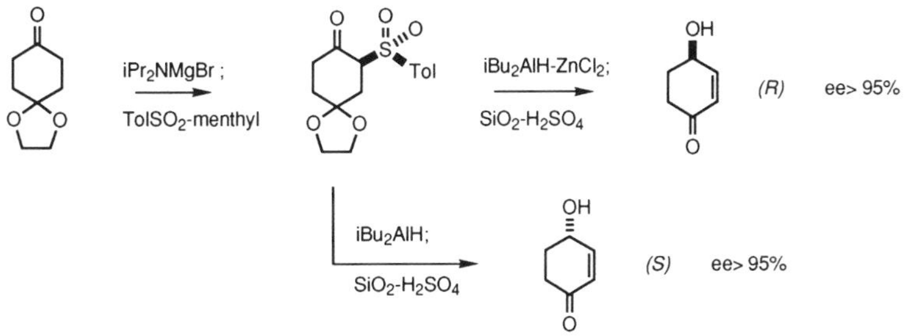

4-Hydroxycyclohexanone was the A-ring synthon in an approach to vitamin-D_3 [Inhoffen, 1958]. Its union by an aldol reaction with an unsaturated aldehyde was followed by a Wittig reaction and uv-induced isomerization of the triene.

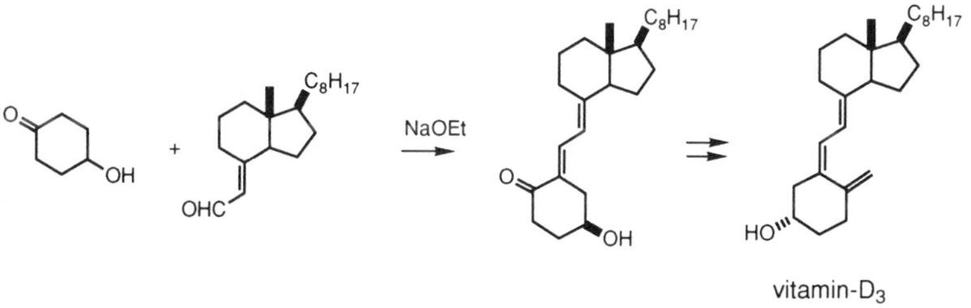

vitamin-D_3

7.2.1.1. Annulation

The elaboration of glycozoline [Chakraborty, 1969] from 4-methylcyclohexanone is rather straightforward, involving Fischer indolization and dehydrogenation.

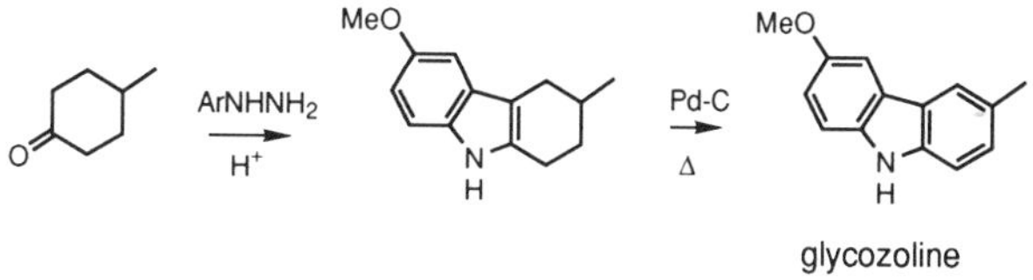

glycozoline

In identifying methyl 4-oxocyclohexanecarboxylate as a synthon for the B-ring of β-eudesmol [Carlson, 1972] the synthetic process implies methylation at the ketone group, annulation at the same point and the adjacent carbon atom, together with conversion of the ester function into an isopropylol side chain. In practice it was found expedient to complete the annulation first, leaving a bicyclic enone to which the angular methyl group was added with the aid of lithium dimethylcuprate.

β-eudesmol

Annulation using an enamine has become a standard procedure. For a synthesis of cadinene dihydrochloride [Piers, 1975] the pyrrolidinoenamine of 4-hydroxycyclohexanone was used.

cadinene
dihydrochloride

From a retrosynthetic analysis of juncusol, the suitability of 1,4-cyclohexanedione as precursor can be proclaimed, correlation being hinged on two annulations [Boger, 1982]. However, the synthetic problem was rendered more complicated by the different substitution patterns of the two aromatic rings, and furthermore, these two rings are on the same side of the cyclohexane ring as defined by the axis passing through C-1 and C-4.

Fortunately, the latter aspect is actually not a serious impediment to the synthesis. When 1,4-cyclohexanedione was replaced by 4-benzyloxycyclohexanone as its synthetic equivalent, the reactivity dilemma was resolved. Thus a controlled annulation and transformation of the product into a β-tetralone laid the foundation for the second phase of carbocycle construction via an α-pyrone. A Diels–Alder reaction constituted the aromatic system directly.

juncusol

It is quite apparent that 4-isopropenylcyclohexanone should serve well in a synthesis of nootkatone [Pesaro, 1968]. A cross-conjugated cyclohexadienone was the first bicyclic intermediate which underwent reaction with lithium dimethylcuprate. The presence of a carbomethoxy group at an α-position of the ketone facilitated the process and also helped rectification of the newly established stereocenter, the incorrect configuration having arisen from axial approach of the reagent. Thus, dehydrogenation and borohydride reduction (with hydride approach from the same face of the molecule as the methyl group) accomplished the task.

nootkatone

1,4-Cyclohexanedione can be protected as monoacetals with several diols. As long as the reaction conditions used to differentiate the functional groups do not undermine the stability of the acetal, chemo- and regioselectivity can be achieved. Since huperzine-A can obviously be disconnected to a 6-oxo-5,6,7,8-tetrahydroquinolin-2-one, a 1,4-cyclohexanedione monoacetal should be applicable to its synthesis [Kozikowski, 1991], using the ketone to initiate annexation of the pyridone ring. A particularly favorable attribute of the monoprotected diketone to this synthetic target is that the establishment of the aromatic moiety ensured subsequent acylation and alkylation to occur at the appointed benzylic position.

huperzine-A

Conversion of a monoacetal of 1,4-cyclohexanedione into the tricyclic intermediate of trihydroxydecipiadiene [Dauben, 1984] began with an organometallic reaction. An intramolecular [2+2]-photocycloaddition was achieved after regeneration of the ketone group and introduction of a conjugated double bond. This photocycloaddition furnished a tricyclic ketone in which the four-membered ring was bridged two *peri*-positions of a *cis*-decalin.

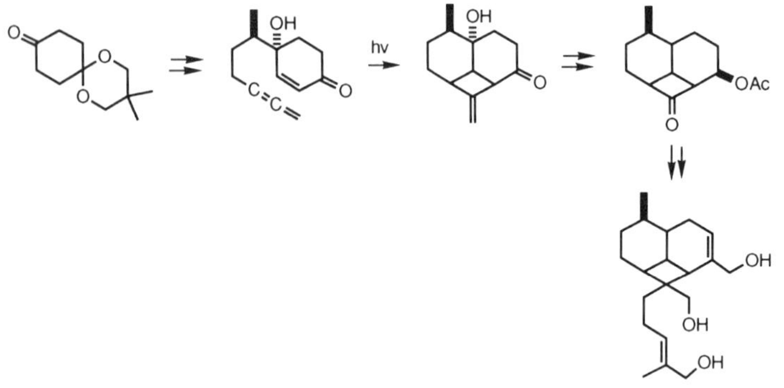

trihydroxydecipiadiene

An even more impressive annulation method pertains to the evolvement of the tetracarbocyclic skeleton of stemodin [Germanas, 1991] from a monoacetal of 1,4-cyclohexanedione. The key step was a cobalt-mediated [2+2+2]-cycloaddition which effectively formed C—C bonds between the A- and B-, B- and C-rings, and the C-6/C-7 bond. It is also remarkable that one diastereomeric form which specified the relative stereochemistry of C-9 and C-10 was favored. This stereochemistry corresponds to stemodin.

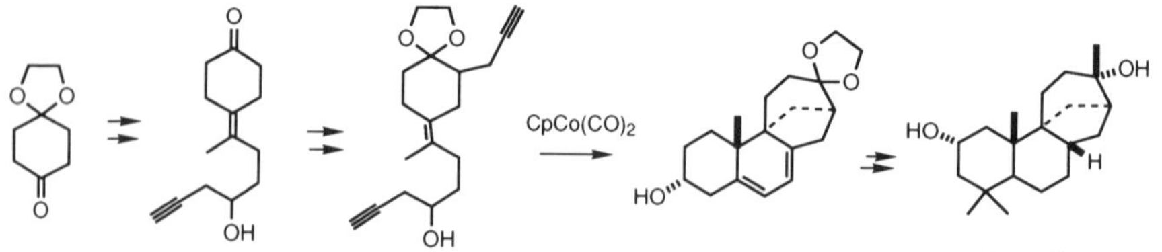

stemodin

A totally different approach to stemodin [Toyota, 1993] was based on intramolecular Diels–Alder reaction and Pd-catalyzed cyclization.

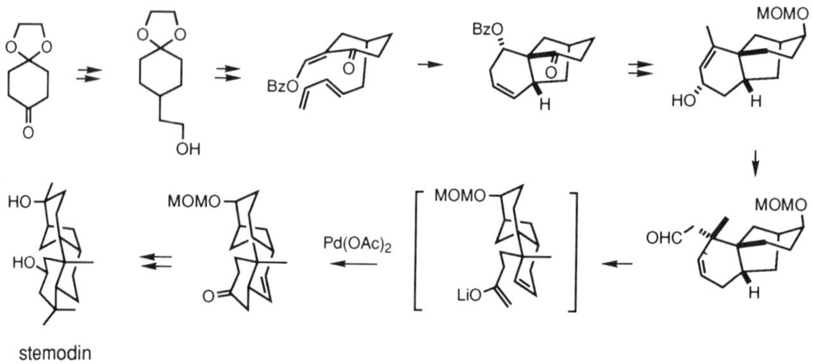

stemodin

The Dötz reaction has become an important procedure for the construction of 1,4-dioxygenated benzene derivatives, and it permits a rapid and convergent assembly of anthracyclines. In one approach [Wulff, 1984] keyed to B-ring closure, a monoacetal of 1,4-cyclohexanedione was used to prepare the unsaturated Fischer carbene complex.

A 4,4-disubstituted cyclohexanone was considered as an expedient precursor of elwesine [Sanchez, 1980]. This identification is rather more obvious after a Pictet–Spengler transform and de-N-alkylation. In the synthetic process an intramolecular Michael reaction may be planned in view of the *cis*-fusion of the hydroindole nucleus.

elwesine

Heptelidic acid has been prepared from 4-isopropylcyclohexanone [Danishefsky, 1988]. Annulation of the seven-membered ring was initiated by desymmetrization to afford the cross-conjugated enone ester, a measure instrumental to the attachment of a dioxygenated isobutenyl chain, while providing the lactone carbonyl. It is significant that the structure of the target molecule is amenable to assembly from a locally symmetrical isobutenyl derivative; this possibility helped circumvent stereochemical problems pertaining to lactonization. The approach also benefited from 1,2-asymmetric induction by the isopropyl group during the Michael reaction with the alkenylcuprate reagent.

heptelidic acid

A neat synthesis of cystodytin-A [Ciufolini, 1991] was based on a pyridine annulation from 2-alkylideneketones such that a 2,6-bisarylidenecyclohexanone would form an 8-arylidene-5,6,7,8-tetrahydroquinoline, and a ketone group would be generated by splitting the remaining arylidene group. The apparent waste of the second equivalent of the araldehyde was compensated by smooth access to a symmetrical product from a readily available cyclohexanone. The alternative would have been to involve protection of an α-methylene group of the same ketone; consequently, selective oxidation to generate a ketone group after the formation of the hydroquinoline would also be required.

diplamine   cystodytin-A

It is known that many syntheses have run into dead ends. A truly disappointing result is the failure of transposing a hydroxyl group in an approach to vernolepin [Marshall, 1979].

The synthetic plan was formulated on the observation that the two carbon chains at C-5 and C-7, which constitute the lactone rings, are the same.

Furthermore, relocation of the C-8 oxygen function to C-9 in various precursors engenders the possibility to introduce the angular vinyl group and its geminal methylene by alkylation with regio- and stereocontrol.

vernolepin

Desymmetrization of a 1,4-cyclohexanedione monoacetal by the introduction of a conjugated double bond and reaction of the unsaturated ketone with an arylithium reagent created a substance which contains elements for three of the four rings of 3-demethoxyerythratidinone [Danishefsky, 1987b]. Interestingly, the fourth ring was created by an S_H2' reaction.

3-demethoxyerythratidinone

An approach to the related alkaloid erysotrine [Ishibashi, 1985, 1988] employed the ethyleneacetal to condense with a phenethylamine derivative. The formation of the hydroindole system was followed by a tandem Pictet–Spengler-type cyclization. Thus intramolecular interception of a Pummerer rearrangement intermediate by an enamide produced the N-acylinium species.

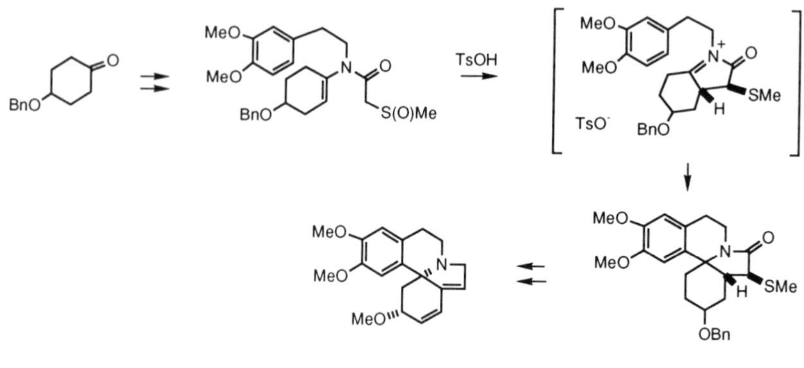

erysotrine

A synthesis of O-methyljoubertiamine [Strauss, 1978] started from arylithium reaction of the same substituted cyclohexanone. However, this route, which featured a Claisen rearrangement as the key step, suffered from great difficulties in functionalization of the allylic position.

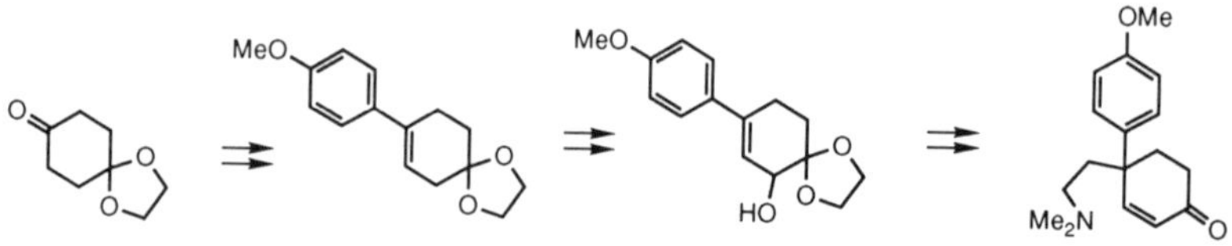

O- methyljoubertiamine

A free radical cyclization was featured in a route to mesembranol [Ishibashi, 1991]. Using the same cyclohexanone an N-dichloroacetyl derivative of a 2-cyclohexenylamine was made. Its treatment with tributyltin hydride gave rise to a bicyclic lactam of O-benzyloxymesembranol.

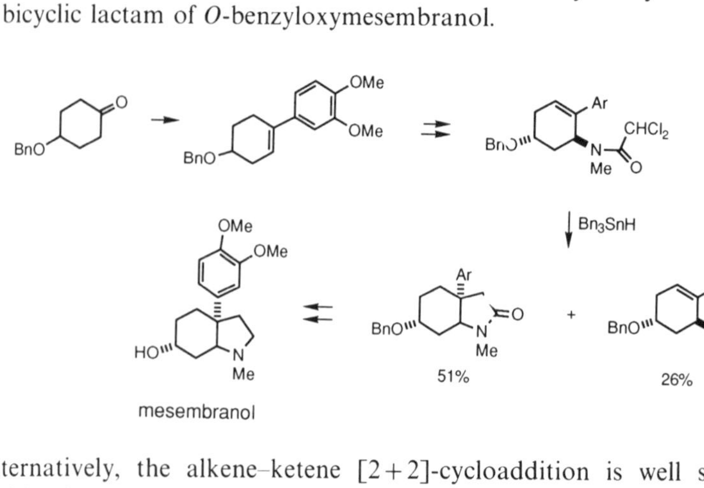

51%   26%

mesembranol

Alternatively, the alkene–ketene [2+2]-cycloaddition is well suited for assembly of mesembranol and O-methyljoubertiamine [Jeffs, 1982] on account of the excellent regioselectivity from reactions of styrenes. The required styrene was obtained readily from 4-hydroxycyclohexanone. For ease of generation and interception, dichloroketene was used in the cycloaddition. After dechlorination, insertion of the nitrogen atom to the four-membered ring was effected.

mesembranol

Perhaps a synthesis of erysotrine [Mondon, 1970] from a 4-oxygenated cyclohexanone presented another persuasive case regarding the role of symmetry. Thus problems of regioselectivity for reactions of an unsymmetrical compound (e.g., the 3-oxygenated cyclohexanone) and pitfalls inherent to an aldol system can be avoided. Despite the incorrect location of the methoxy group which entailed transposition, this substituent was crucial for the establishment of the conjugate diene system.

erysotrine

Spiroannulation of 4-benzyloxycyclohexanone constituted an alternative method for the synthesis of the key tricyclic intermediate of gibberellic acid

[Corey, 1979b]. Note that the relative stereochemistry of the two-carbon chain introduced to the cyclopentenone moiety of the spirocycle is not crucial because the ring closure (C-ring of gibberellic acid) would be subject to self-regulation to produce a compound with a *cis*-perhydroindan segment.

gibberellic acid

An apparent precursor of the acoradienes and cedrene has been obtained [Ziegler, 1987] from spirocyclization by means of a free radical reaction. The substrate was prepared from the ethyleneacetal of 1,4-cyclohexanedione.

α-acoradiene   β-acoradiene

Conventional approaches to pentalenene-type sesquterpenes entail annulation of a cyclopentane and the construction of the third ring. In light of this popular pursuit, the synthesis of pentalenolactone-E methyl ester [Taber, 1985] from 4,4-dimethylcyclohexanone is refreshing. In this method spiroannulation to form a pyran ring was accomplished; after contraction of the ketone and conversion of the acid into a diazoketo ester, treatment with Rh(II) acetate rendered the diquinane nucleus. Due to local symmetry of the pyran ring which invariably juxtaposed a C—H bond for insertion into the carbenoid species, this process created three stereocenters as those required for completion of the synthesis. The diquinane is perforce *cis*-fused.

pentalenolactone-E
methyl ester

7.2.1.2. Ring Expansion and Annulation

The strategy of solvolytic rearrangement resulting in ring contraction for a synthesis of bulnesol [Marshall, 1969] indicates the preparation of a bicyclo[4.3.1]decane derivative as a prerequisite. The acquisition of the seven-membered ring precursor of this intermediate by ring expansion of a 4-aryloxymethylcyclohexanone has certain merit. It supplied an activating ester group to help attachment of a side chain for the elaboration of the bridged ring, although several steps were needed to convert it into a methyl group. This methyl group was the chief regiocontroller in the rearrangement step, by stabilizing the incipient carbocation.

bulnesol

An interesting route to dactylol [Paquette, 1985] started from a ring expansion maneuver on 4,4-dimethylcyclohexanone to give a conjugated cycloheptenone. The synthesis continued by cyclopropanation and annulation to give an angular 3:7:5-fused tricycle which was further elaborated to an epoxide. Previously the Lewis acid-catalyzed isomerization of the epoxide to dactylol had been established.

dactylol

The dolastane-type diterpenes possess a 5:7:6 fused ring system. In (14S)-dolasta-1(15),7,9-trien-14-ol the hydroxyl-bearing carbon atom in the seven-membered ring is in a 1,4-relationship with the trigonally hybridized angular atom, so a 1,4-cycloheptanedione for synthesis of the ring system is indicated. Of course, ring expansion of a 1,4-cyclohexanedione monoacetal would be a viable pathway [Piers, 1986].

(14S)- dolasta-1(15),7,9-
trien-14-ol

The bicyclo[4.3.1] decane system of isoclovene is anchored at C-2 and C-9 by an unsaturated linkage. This structural feature suggests a bridged bicyclic intermediate in which the two carbon atoms adjacent to the same bridgehead are oxygenated; in turn such an intermediate is disconnectable by an aldol or Claisen transform to generate a monocyclic precursor. Since the option of forming a six-membered ring is by far the more efficient, the precursor should be a cycloheptanone. A 4,4-disubstituted cyclohexanone served as the starting material [Ahmad, 1989]. Ring expansion and aldol condensation delivered the bicyclic compound.

isoclovene

7.2.1.3. Ring Cleavage

The synthetic utility of cyclic compounds accrues on their service as parts of carbon chains. There are certain advantages in such tactics: availability of materials with proper distribution of functional groups and other substituents, and the much better stereocontrol in cyclic systems, at least in the common rings.

A retro-Dieckmann reaction of the β-keto ester obtained from ring expansion of 4,4-dimethylcyclohexanone was an important step for creating the central segment of moenocinol [Böttger, 1983]. The different reactivities at the α-positions of the two ester groups generated by ring cleavage of the sulfenylated ketone enabled the systematic chain elongation at the two termini.

moenocinol

A Reimer–Tiemann reaction product of 4-ethylphenol gives a 4,4-disubstituted cyclohexanone on hydrogenation. This ketone proved to be useful for the synthesis of eburnamonine [M.F. Bartlett, 1960] as oxidative ring cleavage led to a diacid containing a protected aldehyde group. On hydrolysis and condensation with tryptamine the pentacyclic skeleton of the alkaloid was obtained.

eburnamonine

A variation of this synthetic route consists of controlled fragmentation of the Haller–Bauer type with a 1,2-cyclohexanedione monthioacetal. For the synthesis of eburnamenine, quebrachamine, and tabersonine [Takano, 1976a, 1980a, b], ethyl 4-oxocyclohexanecarboxylate was submitted to the necessary transformations. The oxidation levels of two of the carbon chains were interchanged, and accordingly the subsequent reactions must respond to such changes.

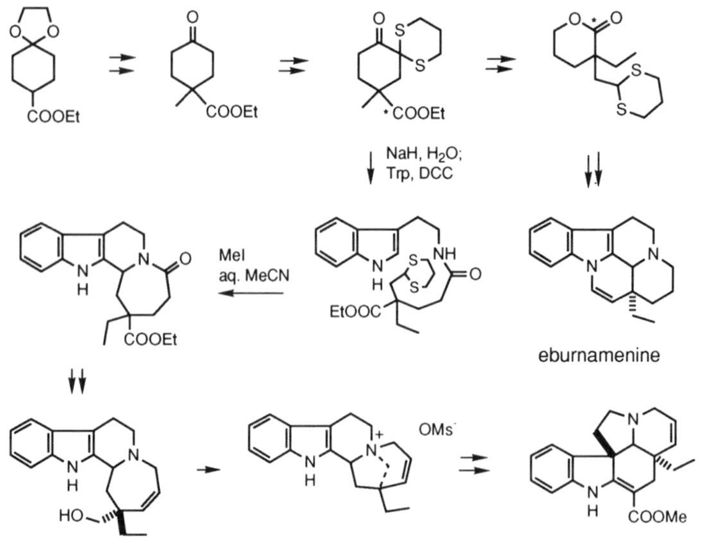

eburnamenine

tabersonine

The cleavage of the C—C bond connecting the ketone of 4,4-ethylenedioxycyclohexanone and one of its α-methylene groups to give an ω-formyl ester led the way toward norsecurinine [Heathcock, 1987]. When the chain was coupled to a proline derivative it became incorporated into the carbon fragment extending from the nitrogen atom to the lactone along the diene chromophore. Intramolecular Michael and aldol reactions served to erect the azabicyclo[3.2.1.]octane segment of the complete skeleton.

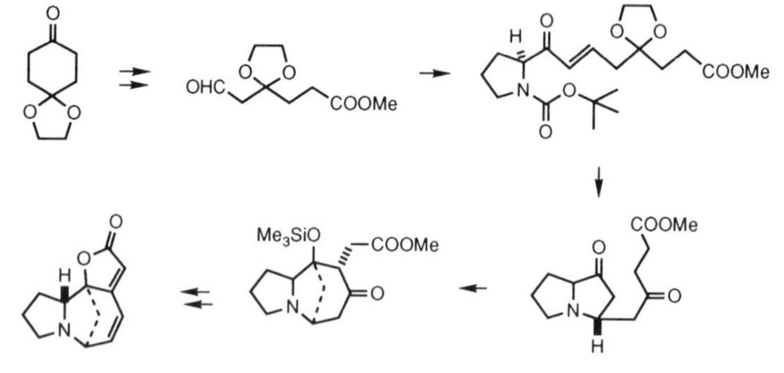

norsecurinine

Alkylidenation of 4-methylcyclohexanone, hydrogenation, and ring cleavage were the first three steps of a synthesis of muscone [Nair, 1964]. The properly substituted and protected diester was induced to undergo an acyloin condensation to form the macrocyle. Note that the ring cleavage step also served to transpose the carbonyl group, generating a β-methylated ketone.

muscone

Alteration of ring size via a cleavage and re-formation maneuver was also featured in the elaboration of the diene component for a Diels–Alder assembly of a common intermediate for illudol and fomannosin [Semmelhack, 1982]. In this case, the ring was contracted from six- to five-membered. The requirement was such that 4,4-dimethylcyclohexanone was the substrate. Naturally this modification involved desymmetrization.

illudol   fomannosin

Enantioselective enolsilylation of 4-methyl-4-*p*-tolylcyclohexanone followed by ring cleavage gave rise to a chiral diacid, enabling the preparation of (+)-α-cuparenone [T. Honda, 1993].

(+)-α-cuparenone

In a synthesis of pentalenene [Pattenden, 1987] from acid-catalyzed transannular cyclization, the required 5:8-fused ring system was acquired from an intramolecular deMayo reaction involving an enol derivative of a 1,3-cyclohexanedione derivative. The diketone was most readily prepared by a Claisen condensation, and in turn, the γ-keto ester from 4,4-dimethylcyclohexanone via alkylation, ring cleavage, and chain extension.

pentalenene

In synthesis of (−)-anisatin [Niwa, 1990] the β-lactone unit was crafted from a locally symmetrical cyclohexanone acetal. It seems expedient to build up such a spirocycle and then degrade it to release two geminal carbon chains because neighboring group participation provided a means to differentiate them. Since the two chains are of the same length, its cyclic precursor must be symmetrical, and in terms of easy preparation and degradation the cyclohexanone is the best. Furthermore, the direct bishydroxymethylation would not be superior due to overabundance of oxygen functionalities and the attendant complications which should be avoided. Admittedly these oxygen functions must be unveiled but it would be at a later stage of the synthesis.

(-)-anisatin

The use of 4,4-ethylenedioxycyclohexanecarbonitrile as a building block for hirsutic acid-C [Trost, 1979b] constituted a novel design. Actually the cyclohexane ring acted as a template on which two five-membered rings were scaffolded, which also fixed four stereocenters in the process. The most difficultly controlled quaternary carbon atom bearing the carboxyl group was thus established. When the bridged tricyclic intermediate in which the formerly desaturated six-membered ring underwent cleavage of the double bond, two aldehyde pendants thereby generated destined to be the two methyl groups of the target.

hirsutic acid-C

7.2.2. 2,6-Disubstituted and 3,5-Disubstituted Cyclohexanones

From 2,6-disubstituted and 3,5-disubstituted cyclohexanones a number of syntheses have been initiated. 2,6-Dimethylcyclohexanone was used in the elaboration of several terpenes because the octalone produced by Robinson annulation has the methylation pattern corresponding to the eudesmanes, and it possesses a carbonyl group at a position corresponding to the origin of a branched three-carbon chain of such sesquiterpenes. The configuration of the secondary methyl group can be changed because it is attached to the γ-position of an enone system. Shift of the double bond toward the methyl-bearing carbon fulfills other synthetic needs.

By manipulating the enone system of the octalone, it has been possible to transform it into geosmin [Marshall, 1968] and cybullol [Ayer, 1976a]. It should be emphasized that in the bicyclic structure the two methyl groups are *trans*.

geosmin

cybullol

A somewhat more complex series of transformations led to frullanolide [Still, 1977]. A Wharton rearrangement furnished the allylic alcohol in which the hydroxyl group was situated at an angular position of a *cis*-octalin. A [3.3]sigmatropic rearrangement of derived ester delivered the required carbon chain to the correct location stereoselectively. An alternative route to the same sesquiterpene [Semmelhack, 1981] involved intramolecular reaction of a π-allylnickel species with an aldehyde originated from the same octalone.

frullanolide

frullanolide

The established photochemical rearrangement pattern of cross-conjugated cyclohexadienones was essential to mapping synthetic routes for the vetispirane sesquiterpenes [Caine, 1976]. The useful dienones were obtained from the same octalone and its dehydro derivative (i.e., the extended dienone). The crucial aspect was that in the spirocyclic product the (latent) ketone branch (according to bisection of the five-membered ring by the plane containing the spirocyclic center) should be *cis* to the secondary methyl group.

α-vetispirene

The original plan of the reaction sequence was described in a synthesis of β-vetivone [Marshall, 1970].

β-vetivone

The hydrindanone analog was used to elaborate pyroangolensolide [Y. Fukuyama, 1973].

pyroangolensolide

An apparent intermediate for the synthesis of eudesmanolides is a *trans*-decalin annexed to a furan [Nemoto, 1987]. This interesting compound was acquired by the [3.3]sigmatropic rearrangement of a ketenacetal derived from a lactone in consort with one of the furan double bonds. The required lactone originated from 2,6-dimethylcyclohexanone via cyanoethylation, saponification, and reaction with a 2-furylmetallic reagent.

Pinguisone has been synthesized [Gambacorta, 1988] from 2,6-cyclohexanone based on a rather unusual strategy in which the hydrindan framework evolved from a bridged ring system via rearrangement. The advantage of this route is that the stereochemistry of the methyl groups on three contiguous carbon atoms was unfolded with excellent control; the conjugate addition to the bridged enone from the *exo* face established one pair and the rearrangement step necessarily resulted in a *cis*-fusion.

pinguisone

Of interest is also a synthesis of ascochlorin [Safaryn, 1986] from 2,6-dimethylcyclohexanone because of the stereochemical disposition of the three stereocenters in the six-membered ring. In fact, once the tertiary aldehyde group was introduced from the face opposite to the secondary methyl group the most stable relative stereochemistry of the molecule became that of the natural product. Chain elongation and coupling to the aromatic moiety were routine tasks.

ascochlorin

An unusual yet concise synthesis of frontalin [Utaka, 1983] featured α,α'-dibromination of 2,6-dimethylcyclohexanone and base-catalyzed oxygenation.

frontalin

A delightful employment of dimethyl 2-oxocyclohexane-1,3-dicarboxylate is in the construction of the azabicyclo[3.3.1]nonane moiety of atisine [Ihara, 1990a]. This compound underwent a double Mannich reaction to give a symmetrical bicyclic heterocycle, the desymmetrization of which began after homologation at the ketone and reduction of the esters. While one of the chains was eventually totally deoxygenated, the other chain was fashioned into an unsaturated ester which then participated in a reflexive Michael cyclization to assemble the three remaining carbocycles in one operation.

atisine

There has been growing interest in the application of chiral acetals as auxiliaries in organic synthesis [Alexakis, 1990], and of great popularity are those deriving from the C_2-symmetric 2,4-pentanediol. Recently in an elegant synthesis of (−)-lardolure [Kaino, 1990] the stereocontrolled transformation of such a 1,3-dioxane spiroannulated with *cis*-3,5-dimethylcyclohexanone to an all *cis*-skipped tetramethylated carbon ribbon was described. The process involved elimination and intramolecular alkylation to form a new C—C bond with inversion of configuration at the carbinol center, and cleavage of the intercyclic bond of the bicyclic hemiacetal. Of special significance is that, due to the symmetry properties of the spiroacetal, the elimination step did not give rise to an isomeric mixture.

(-)-lardolure

A chiral carbon chain with syndiotactic methyl substituents has been convened [Matsukama, 1991] by treatment of (3R,5R)-dimethylcyclohexanone with boron trifluoride etherate. Apparently a self-condensation of the ketone preceded acetalization and fragmentation to afford a cyclohexene.

Because of its σ-symmetry *cis*-3,5-dimethylcyclohexanone undergoes Baeyer–Villiger reaction to give only one unsymmetrical lactone. This desymmetrization was a crucial step by which the terminal pyran ring of monensin was fashioned [T. Fukuyama, 1979]. The lactone was converted into a Wittig reagent which was coupled with a lactol containing the BC-ring portion.

The Baeyer–Villiger desymmetrization of *meso*-cycloalkanones has been rendered more powerful by the possibility of an enantioselective process with enzymatic catalysis [Taschner, 1988].

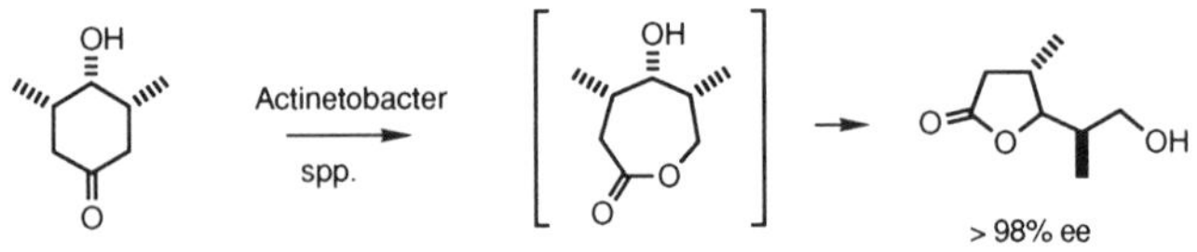

cis-3,5-Dioxycyclohexanones are available from all-*cis* 1,3,5-cyclohexanetriol. A synthetic use of this substance in the preparation of 6-substituted 4-hydroxy-δ-valerolactone has been demonstrated [Prasad, 1984a]. Note that this lactone has the same stereochemistry as that of compatin.

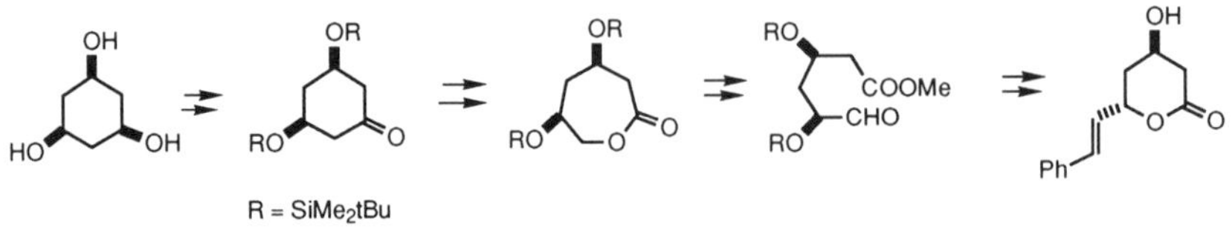

R = SiMe₂tBu

Retrosynthetic analysis of the internal acetal moiety of tirandamycin-A reveals a carbon chain that is correlated to a symmetrical *cis*-3,5-dimethyl-2,4,6-trioxycyclohexanone via a Baeyer–Villiger transform [Boeckman, 1986]. Further consideration on the basis of its stereochemistry indicates the possibility of using *cis*-2,6-dimethyl-1,4-cyclohexanedione to elaborate the required substrate.

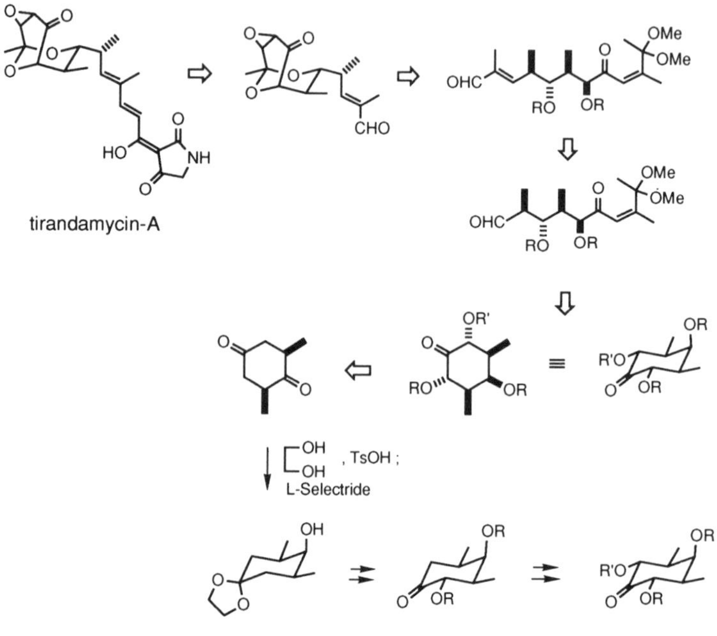

tirandamycin-A

The conversion of a bridged, symmetrical 2,2,3,5-tetrasubstituted cyclohexanone to α-santalene [Monti, 1978] by ring contraction, alkylation, and reduction maneuvers is noteworthy.

α-santalene

7.2.3. 1,3-Cyclohexanediones

1,3-Cyclohexanediones are important building blocks for synthesis. We classify here their utility according to the sites of bond formation/cleavage.

7.2.3.1. Bond Formation at C-1, C-2, or C-1, C-2, C-3

The synthesis of α-cyperone from dihydrocarvone via the Robinson annulation gave very poor results due to unfavorable stereoelectronic effects. Improvement has been possible by using 5-isopropenyl-2-methyl-1,3-cyclohexanedione [Agami, 1993]. However, the octalindione intermediate was racemic, thus optical resolution and removal of the isolated ketone group were extra steps to accomplish.

α-cyperone

Photoinduced bond formation at C-1 and C-2 of the original dione which had been converted into an alkene was a key step toward an elaboration of α-acoradiene [Oppolzer, 1983]. However, in unreveling the spirocyclic system one of the bonds was severed by reduction.

α-acoradiene

The spirocyclic alkaloid histrionicotoxin is a 3-aminocyclohexanol. Consequently, a 1,3-cyclohexanedione provides an excellent template for its synthesis. Aimed at developing reliable methods for the construction of the spirocyclic skeleton [Overman, 1975; Godleski, 1981], a number of studies was based on the simple compound.

For obvious tactical expediency the synthesis of histrionicotoxin [T. Fukuyama, 1975] enlisted 2-(3-butenyl)-1,3-cyclohexanedione, and new C—C and N—C bonds were formed at C-1 to create the spirocyclic heterocycle. The latter operation was an acid-catalyzed intramolecular Michael reaction with an amide as the donor.

histrionicotoxin

With a totally different strategy of spiroannulation [Winkler, 1989] 1,3-cyclohexanedione was employed to form a photocycloaddition substrate. The photocycloadduct was induced to fragment after proper adjustment of functional groups to yield a precursor of histrionicotoxin. The synthesis was rendered asymmetric by incorporating (+)-glutamic acid into the scheme.

Another component of the dendrobatid frog secretion is gephyrotoxin, a synthesis of which from 1,3-cyclohexanedione and *N*-benzylsuccinimide [Fujimoto, 1980] is particularly relevant to the theme of this monograph, as both building blocks are symmetrical. An examination of the two α-carbon atoms shows there are two two-carbon subunits attached to them in a *cis* relationship. The subunits are terminated by an alcohol group and at the cyclohexane moiety, respectively; simplification of the synthetic route is possible. Accordingly, *cis*-2,5-bis(hydroxyethyl)pyrrolidine could be used to carry out the synthesis, and desymmetrization (monoprotection of an alcohol group) was performed on it before condensation with 1,3-cyclohexanedione. Ring formation involving the α-carbon of the resulting enone and the side chain of the pyrrolidine could then be contemplated.

gephyrotoxin

A common method for the preparation of 3-substituted enones is via organometallic reaction of the monoethers of the enolized 1,3-diones. Reduction of the enol ethers with a complex metal hydride gives the enones. Of course these steps are desymmetrizing. In a synthesis of dihydromaritidine [Keck, 1982] a 3-aryl-2-cyclohexenone was further converted into an *N*-acylnitroso compound which underwent intramolecular ene reaction with a nearby cyclohexene ring to give a hydroindole derivative.

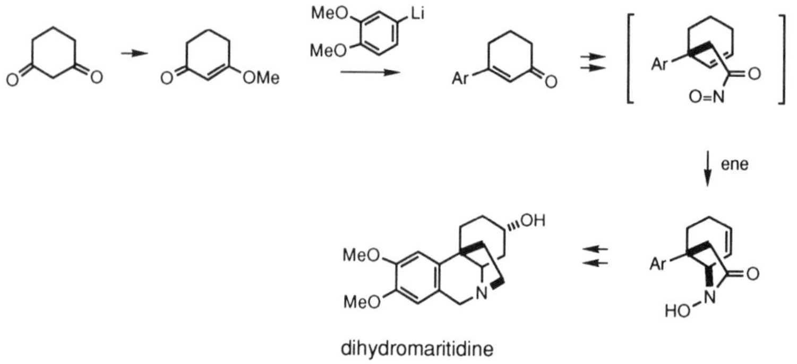

dihydromaritidine

2-Cyanoethyl-5-methyl-2-cyclohexenone, derived from the 1,3-dione, served in two synthetic routes for lycopodine [Heathcock, 1982b]. The enone permitted the stereoselective attachment of a side chain at the *β*-position for the construction of the second carbocycle. The heterocycle was formed by a Schiff condensation.

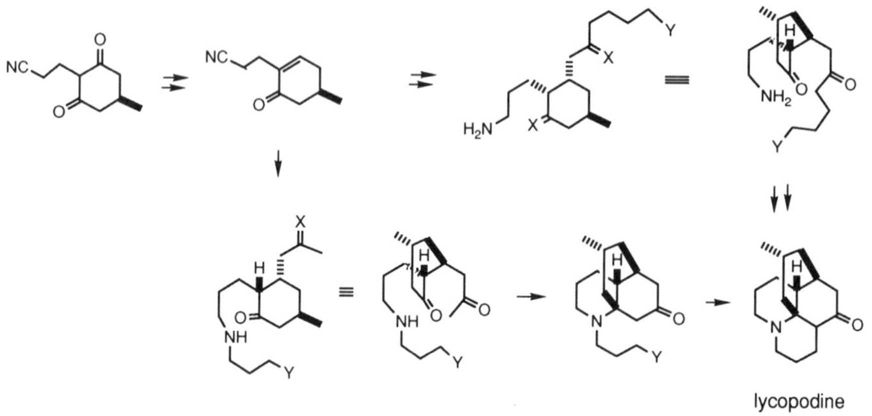

lycopodine

In a model study for the synthesis of aspidosperma and strychnos alkaloids a 3a-(*o*-nitrophenyl)perhydroindol-4-one was obtained [Solé, 1991] via sequential arylation and allylation of 1,3-cyclohexanedione at C-2, followed by degradation of the allyl group to an aminoethyl chain, and cyclization.

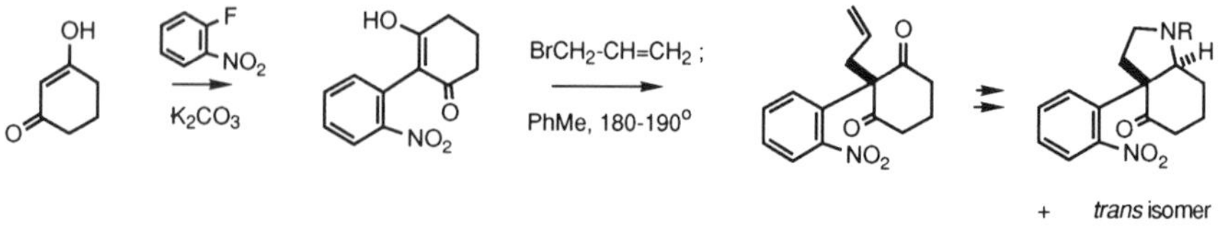

+   *trans* isomer

An approach to sporogen-AO-1 [K. Mori, 1988c] propounded 2,3-dimethyl-4-hydroxycyclohexanone as an intermediate. The *trans* stereochemistry of the vicinal hydroxyl and methyl groups defined its genesis from a Δ^3-cyclohexenone by way of, for example, hydroboration of a derivative. Thus acetalization of 2,3-dimethyl-2-cyclohexenone would furnish such a substrate, and the enone is available from 2-methyl-1,3-cyclohexanedione.

sporogen-AO-1

2,3,4-Trimethyl-2-cyclohexenone was required for a synthesis of ishwarone [Cory, 1979]. This compound was acquired from 2-methyl-1,3-cyclohexanedione via kinetic methylation of the enol ether and reaction of the product with methyllithium. The cyclohexene ring was introduced by a stepwise procedure involving conjugate alkylation, activation of the side chain, and cyclization. The steric effect exerted by the methyl group at C-4 of the cyclohexnone on the attack of cuprate reagent during the conjugate addition determined the *cis* relationship of the *vic*-dimethyl substituents in the product.

The remarkably efficient formation of the five-membered ring, leading to ishwarone, by a intramolecular carbene insertion into an angular methyl group must be due to the conformational preference that resulted in placing the two centers in proximity. The cause of this conformational bias in a *cis*-decalone is the secondary methyl group.

ishwarone

Reaction of the ethyl enol ether of ethyl 3,5-dioxocyclohexanecarboxylate, which is available from 3,5-dihydroxybenzoic acid [van Tamelen, 1956], with methyllithium, followed by acid workup, led to a cyclohexenone bearing an isopropylol side chain. A synthesis of eremophilone [Ficini, 1977] was realized by an annulation maneuver which was initiated by conjugate addition to the enone.

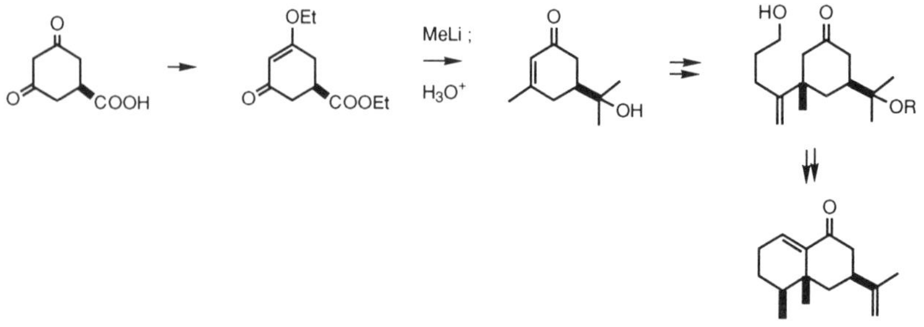

eremophilone

Michael reaction of the acetoxymethyl enone derived from the same enol ether with a sulfonylphthalide served to construct three of the condensed rings of γ-citromycinone [Hauser, 1984].

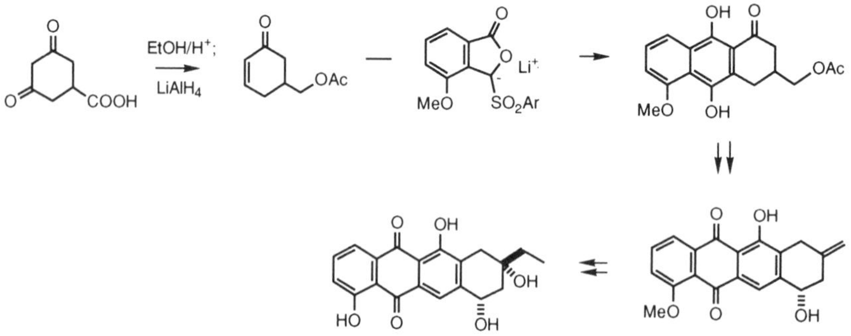

γ-citromycinone

The enol acetate of ethyl 3,5-dioxocyclohexanecarboxylate was used to build an AB-ring synthon via photocycloaddition with 1,1-diethoxyethene in a route to daunomycinone and adriamycinone [Boeckman, 1983a]. The adduct eliminated acetic acid and then underwent electrocyclic opening. In the presence of juglone methyl ether the reactive diene was intercepted to afford two tetracyclic products. The major compound proved to be the desired regioisomer.

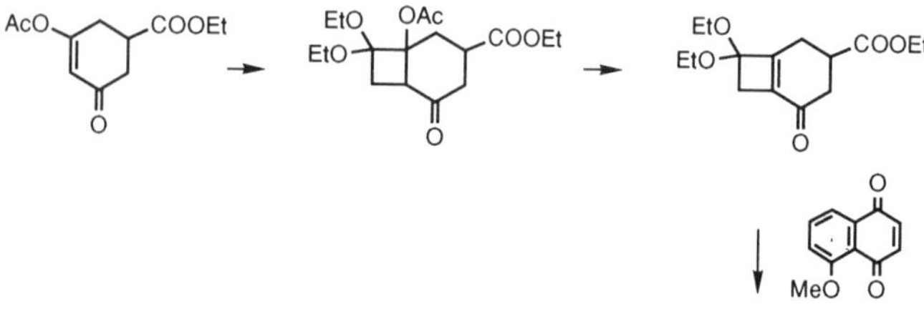

adriamycinone   (major)

An analogous strategy was delineated for the preparation of aklavinone [Boeckman, 1982]. In this case the starting material was 1,3-cyclohexanedione.

An *ar*-annulation approach to rabelomycin has been considered [Kraus, 1991]. The ring system of a model was assembled by the Michael reaction between 5-methyl-1,3-cyclohexanedione and a 2-acetyl-1,4-naphthoquinone. Another Michael reaction, now an intramolecular version furnished a product with the anthracene nucleus.

R = OH   rabelomycin

A synthetic route leading to lycoramine [Yee, 1979] entailed the preparation of a 3-substituted 2-aryloxy-2-cyclohexenone which underwent photocyclization to generate a fused dihydrobenzofuran system. An ester group was provided in the aromatic ring such that it appeared at the *peri* position in the photocyclized product, the purpose being participation in the azepine ring formation.

The ketone function in the photocyclization substrate is a necessity. But because in lycoramine the secondary hydroxyl group is situated at one carbon atom remove, a transposition maneuver is required in the latter part of the synthesis.

lycoramine

Oxygen function transposition was also required before completing a synthesis of crinine [Whitlock, 1967]. Both this and the above syntheses are concerned with method development and the advantages seem to outweigh the additional steps for the functional group adjustments.

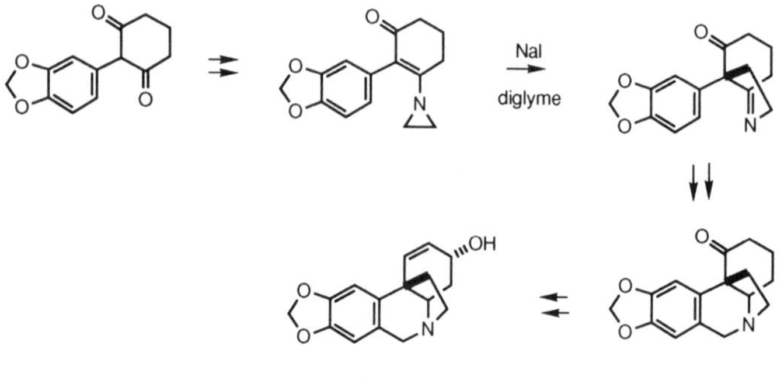

crinine

A building block for the DE-ring of certain pentacyclic triterpenes has been elaborated from the symmetrical 2,5,5-trimethyl-1,3-cyclohexanedione [Barltrop, 1962] upon desymmetrization and annulation. Its union with bicyclic synthons would lead to C-seco derivatives of the triterpenes.

Actually an enol lactone derived from the same compound has been converted into olean-11,13(18)-diene [Corey, 1963].

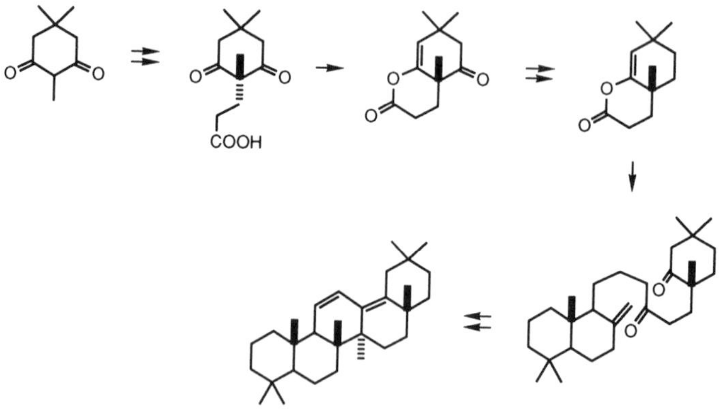

olean-11,13(18)-diene

A different method for preparation of the DE-ring synthon is demonstrated in a synthesis of δ-amyrin [van Tamelen, 1972] which constituted a formal synthesis of β-amyrin. The work essayed the polyene cyclization initiated by ionization of an epoxide.

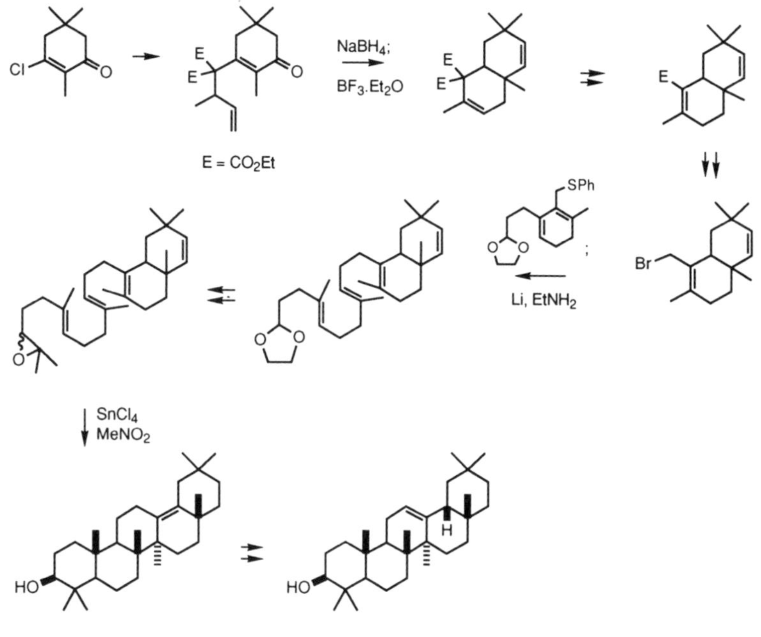

E = CO₂Et

δ-amyrin   β-amyrin

Evodone is a latent 1,3-cyclohexanedione in which one of the carbonyl groups is locked up in a furan ring. A relatively simple approach to evodone [Srikrishna, 1988] is outlined below.

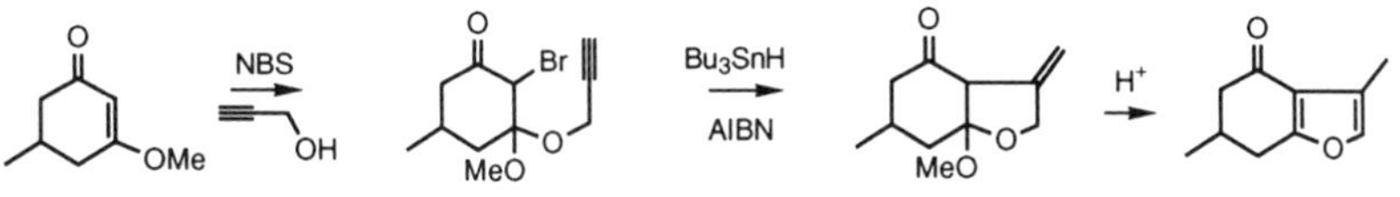

evodone

Synthesis of bisabolangelone [Riss, 1986] starting from 5-methyl-1,3-cyclohexanedione is very logical. Interestingly, desymmetrization of the dione occurred with the preparation of 5-methyl-2-cyclohexenone, which was then used to effect a 1,3-dipolar cycloaddition.

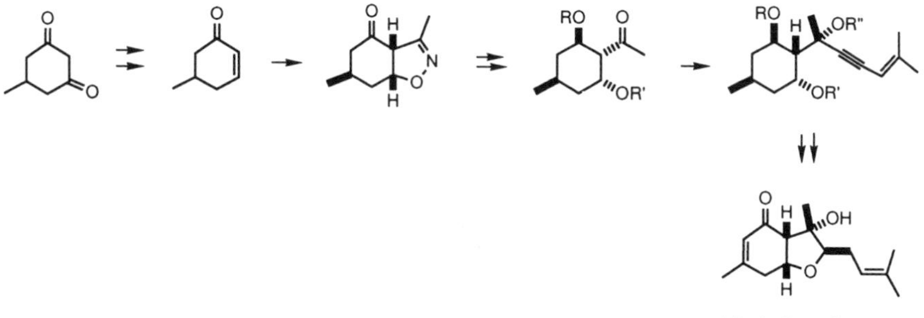

bisabolangelone

In the synthesis of an *Eumorphia prostata* furanosesquiterpene [Bohlmann, 1981] the transformation undergone by 5-methyl-1,3-cyclohexanedione is quite different from any other shown above. The six-membered ring was fashioned between C-2 and C-4. Homologation at C-1 to bear the 3-furoyl group was indicated.

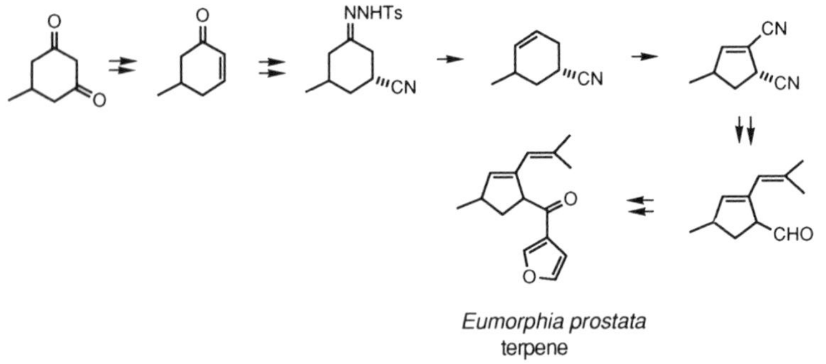

Eumorphia prostata
terpene

Perhaps it should be mentioned that fused furans are readily formed from 1,3-cyclohexanediones by anchoring a two-carbon unit at one of the carbonyl oxygen atoms and at C-2. Among the most convenient methods are those using a sulfone reagent [Padwa, 1990] and Rh-catalyzed decomposition of 2-diazo-1,3-cyclohexanedione in the presence of substituted acetylenes. The latter method permitted a facile synthesis of isoeuparin [Pirrung, 1994b], pongamol and lanceolatin-B [Pirrung, 1994a].

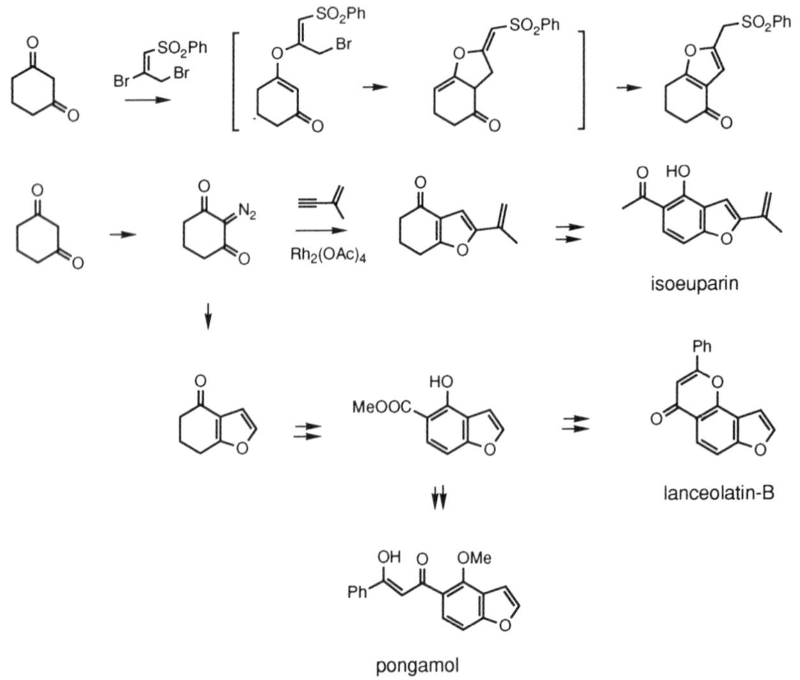

isoeuparin

lanceolatin-B

pongamol

Of related interest is the access to dihydropyran derivatives by a tandem Knoevenagel/hetero-Diels–Alder reaction sequence [Koser, 1993].

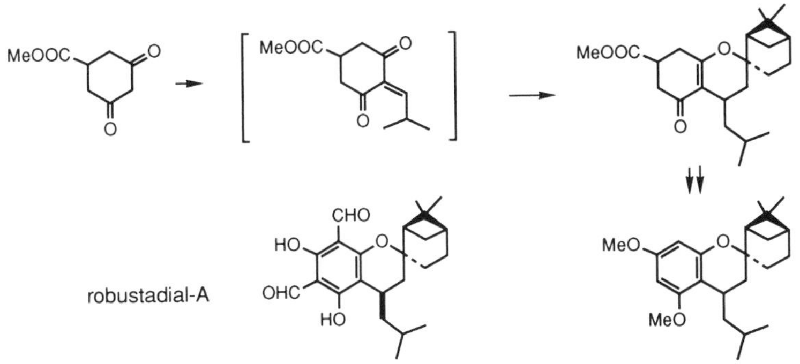

robustadial-A

7.2.3.2. Bond Formation at C-1, C-2, C-3, or C-1, C-2, C-3, C-4 or C-1, C-3, C-4

It is abundantly clear that 1,3-cyclohexanedione is a convenient source of 3,3-disubstituted cyclohexanones. Such a building block has served in the elaboration of axamide-4, axisonitrile-4, and axisothiocyanate-4 [Chenera, 1992].

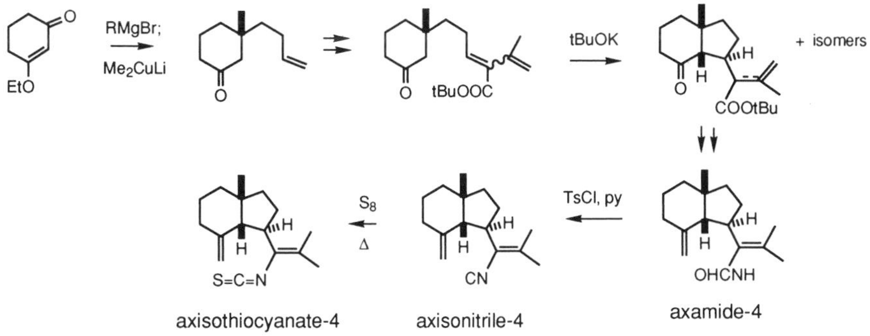

axisothiocyanate-4   axisonitrile-4   axamide-4

While the above synthesis involved a Michael reaction in the close of the second carbocycle, intramolecular alkylation was delineated in the schemes for synthesis of β-eudesmol [Vite, 1988a] and valeranone [Vite, 1988b].

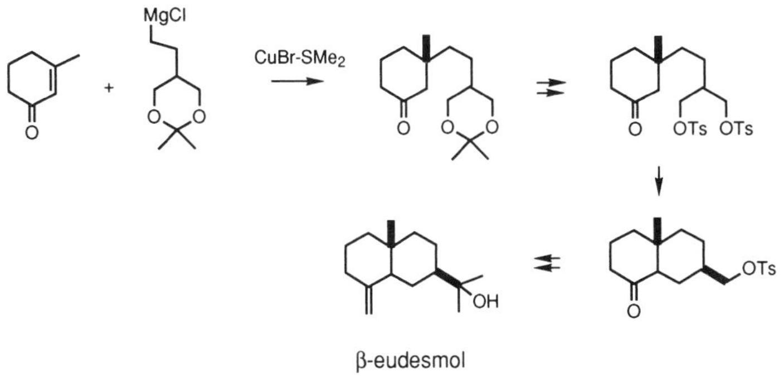

β-eudesmol

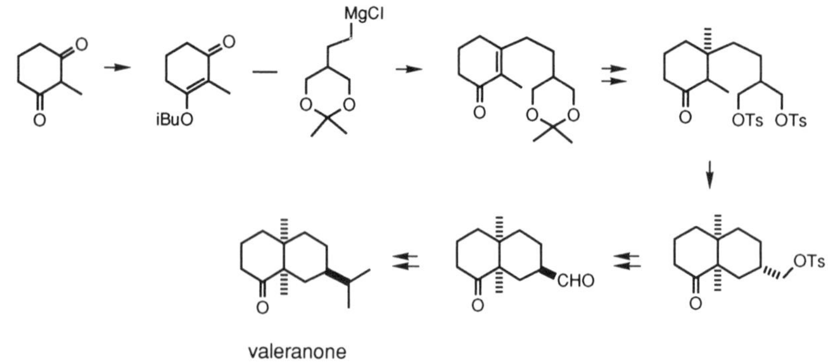

valeranone

Yet another variation concerned the synthesis of β-eudesmol [Kawamata, 1988] in which the key step was an intramolecular aldolization.

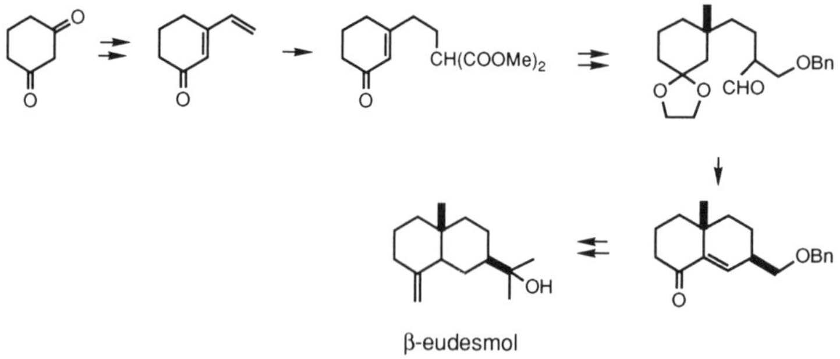

β-eudesmol

An intramolecular acylation served to establish the tricyclic skeleton of ligularone and furanoeremophilan-14,6α-olide [Tada, 1980].

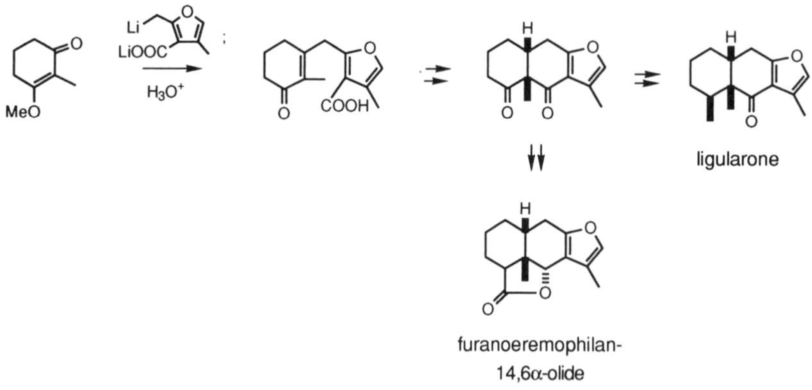

ligularone

furanoeremophilan-
14,6α-olide

An early success in the synthesis of the steroid skeleton is that of methyl 3-ketoetionate [Wilds, 1950, 1953]. In this work 1,3-cyclohexanedione was designated as the building block for the C-ring.

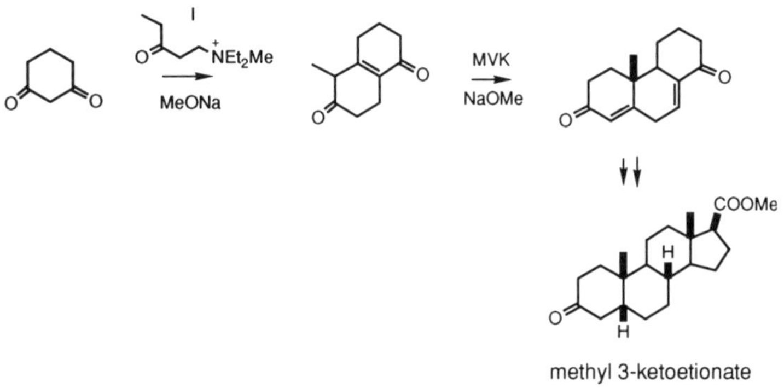

methyl 3-ketoetionate

The formation of 4-oxotetrahydrocarbazoles by photocyclization of 3-arylamino-2-cyclohexenones was the basis of a method for synthesizing aspidofractinine [Dufour, 1989]. The carbazole derivative was properly functionalized such that alkylation at the benzylic position gave rise to a product whose cyclization into an enamide not only established another ring, it also enabled a facile introduction of structural elements or the elaboration of the remaining subunits.

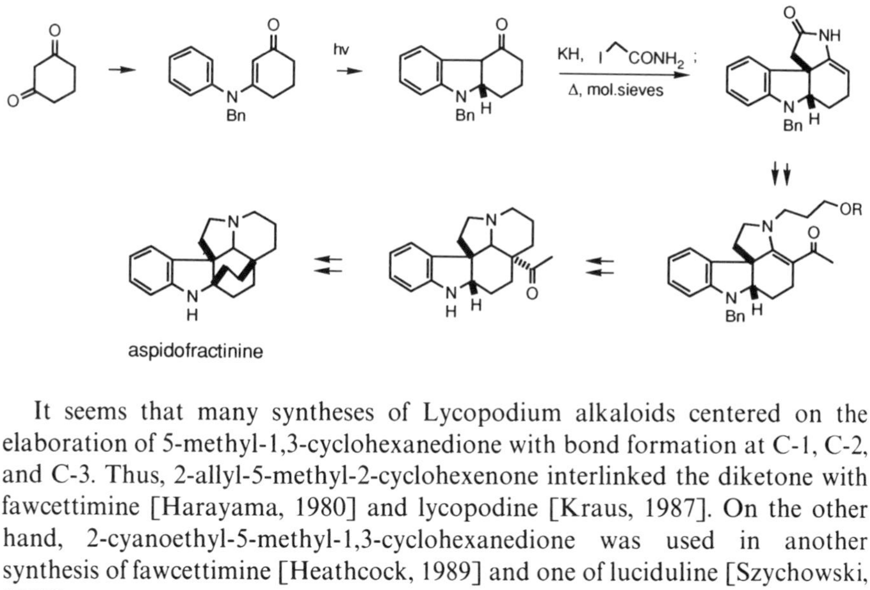

aspidofractinine

It seems that many syntheses of Lycopodium alkaloids centered on the elaboration of 5-methyl-1,3-cyclohexanedione with bond formation at C-1, C-2, and C-3. Thus, 2-allyl-5-methyl-2-cyclohexenone interlinked the diketone with fawcettimine [Harayama, 1980] and lycopodine [Kraus, 1987]. On the other hand, 2-cyanoethyl-5-methyl-1,3-cyclohexanedione was used in another synthesis of fawcettimine [Heathcock, 1989] and one of luciduline [Szychowski, 1979].

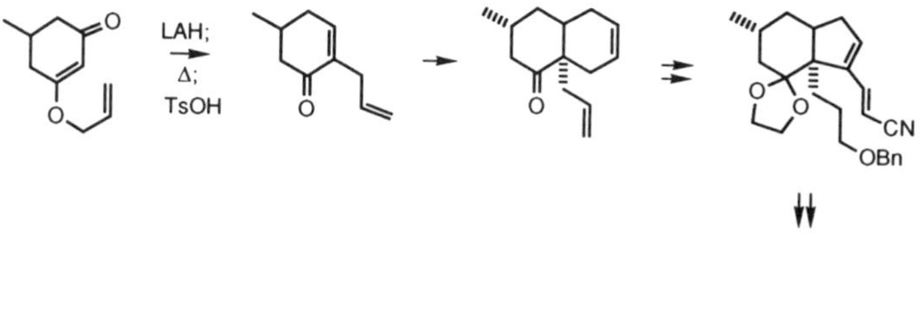

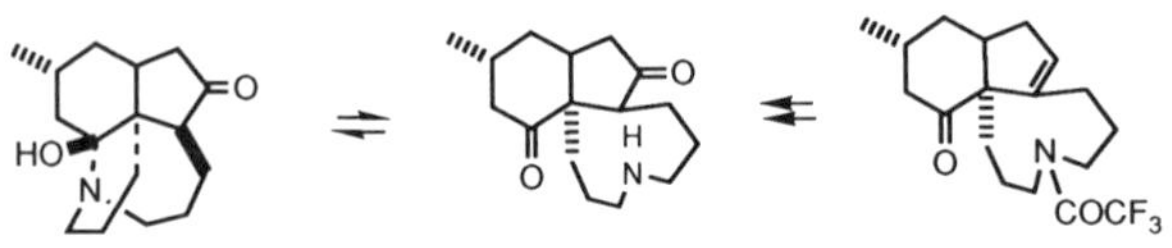

fawcettimine

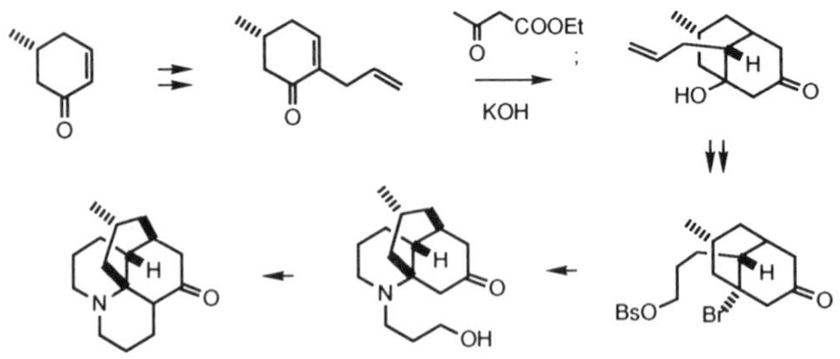

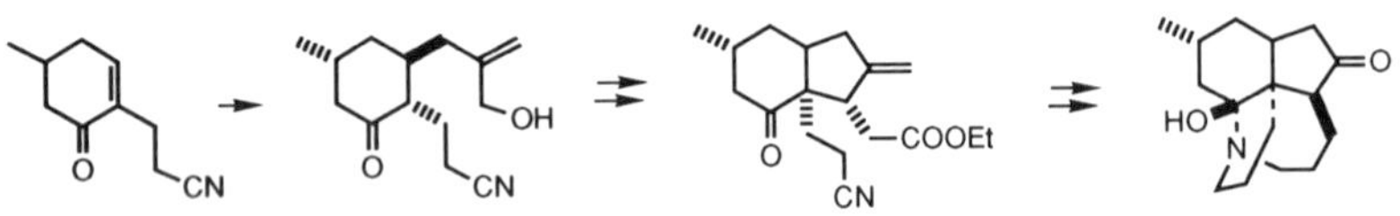

fawcettimine

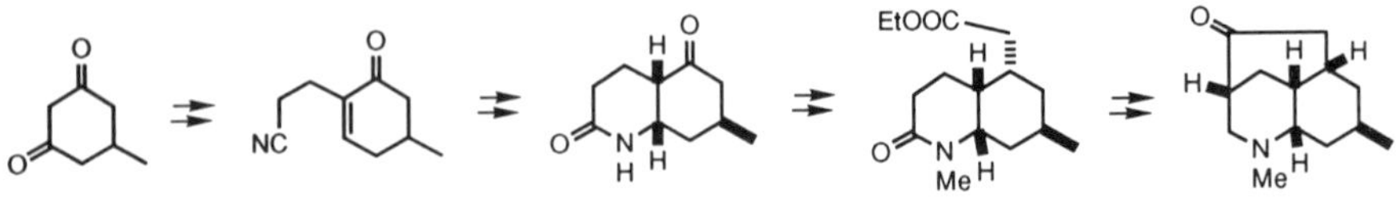

luciduline

The dioxohydroquinoline obtained by acid-catalyzed cyclization of 2-cyanoethyl-5-methyl-1,3-cyclohexanedione was the building block for 12-epilycopodine [Wiesner, 1968]. The establishment of a new C—C bond at C-1 of the original diketone by an intramolecular [2+2]-photocycloaddition caused the stereochemical problem that prevented the attainment of lycopodine.

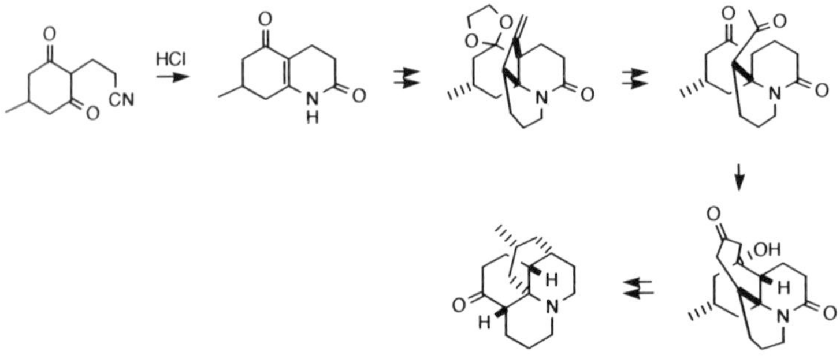

12-epilycopodine

The evolvement of ibogamine [Nagata, 1968] from a dioxocyclohexane-carboxylic acid involved formation of two C—C bonds and two N—C bonds. All the oxygen functions of the cyclohexane derivative were replaced; the crucial steps were an intramolecular nitrene addition to a double bond and the subsequent acylative ring opening to form the tryptamide. The seven-membered ring was closed by an intramolecular alkylation at the α-carbon of the indole nucleus.

ibogamine

Verticillene has been constructed [Begley, 1990] from 1,3-cyclohexanedione by systematic attachment of alkyl groups, culminating in the formation of the 12-membered ring by McMurry coupling. The major product of this reaction is a rearranged olefin whose conjugate diene moiety could undergo 1,4-reduction.

verticillene   (major)

The elaboration of the epimeric alcohol of 3β-hydroxy-7β-kemp-8-en-6-one [Paquette, 1992] from 2-methyl-1,3-cyclohexanedione involved kinetic methylation of the monoenol ether, organometallic reaction, and annulation to construct an octalone intermediate. The latter compound underwent a Pd-catalyzed trimethylenemethane cycloaddition to establish the five-membered ring. Completion of the cyclic framework was by an intramolecular aldol reaction.

3β-hydroxy-7β-
kemp-8-en-6-one

Another synthesis consisting of bond formation at C-1, 2, 3, and 4 is that of morphine [Toth, 1987]. In this route 2-allyl-1,3-cyclohexanedione was converted into 2-allyl-3-benzenesulfonyl-2-cyclohexenone which was hydroxylated at C-6. After linkage of the sulfonyl compound with an aromatic ring in the form of an ether, two rings were formed by treatment with *n*-butyllithium. The existing functional groups of the tetracyclic sulfone were adequate for modification and piperidine ring closure to yield a precursor of morphine.

morphine

A synthesis of lycopodine [Stork, 1968] that started from a 5-(*m*-methoxybenzyl)-1,3-cyclohexanedione proceeded via *trans*-3-benzyl-5-methyl-cyclohexanone but suffered from the asymmetry of the latter compound, as annulation to obtain a hydroquinolone derivative was not regioselective. However, this scheme propounded an efficient assembly of the bridged ring system by an intramolecular Friedel–Crafts alkylation with an acyliminium ion.

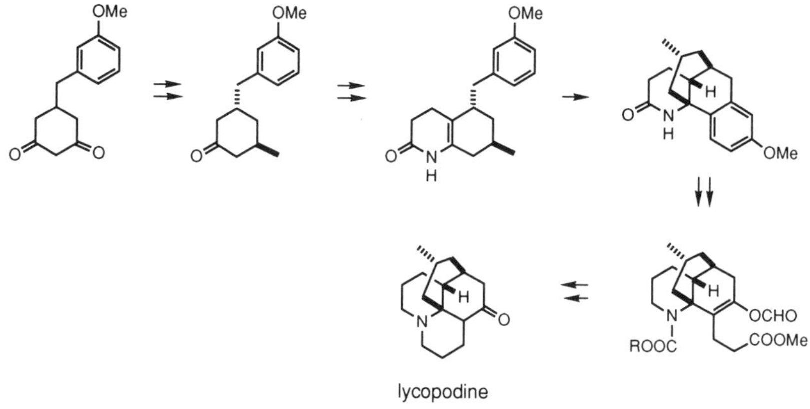

lycopodine

A much simpler compound that yielded to synthesis from 1,3-cyclohexanedione is β-cyperone [Gammill, 1976]. In this work the C-1 ketone was converted into an alkene bearing an isopropyl group and C-4 became disubstituted; one of the side chains was condensed with enone carbonyl to complete the bicyclic skeleton.

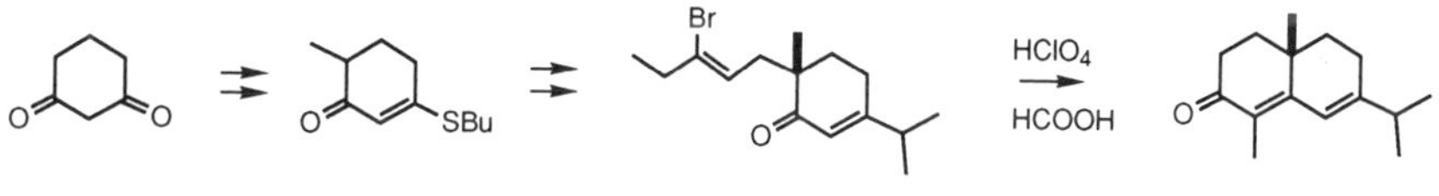

β-cyperone

7.2.3.3. Bond Formation at C-1, C-3

Relatively few syntheses are known which resulted from bond formation at the two carbonyl sites only. This section contains an example.

An intramolecular ketene–alkene [2 + 2]-cycloaddition was the basis of forming a 1,2-annulated bicyclo[3.3.1]nonane skeleton [Funk, 1988]. For the conversion of such a compound to clovene the cyclobutanone ring was expanded and the necessary functional group adjustment then made.

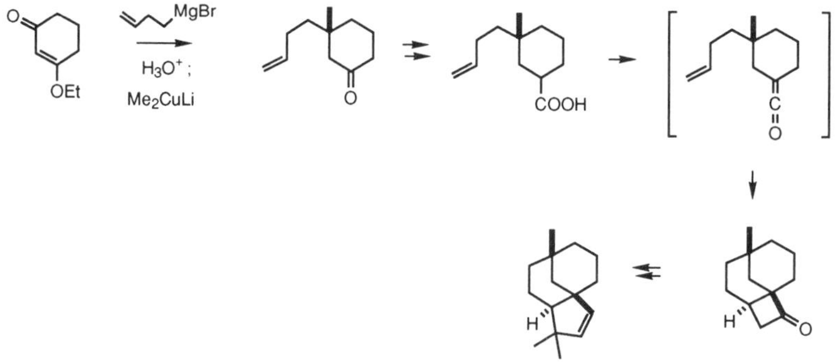

clovene

The oxahydrindan portion of milbemycin has been constructed [Fujiwara, 1992] from 1,3-cyclohexanedione. All the ring carbon atoms except for C-2 have been modified, with C—C bond formation at C-1, C-3, and C-4, and oxygenation at C-4, C-5, and C-6.

milbemycin
(bottom portion)

7.2.3.4. Bond Formation at C-1, C-4 or C-1, C-4, C-6

A simple synthesis of (E)-1-chloro-2,4-ochtodien-6-ol [Zegarski, 1985] from dimedone comprised deoxygenation, allylic bromination. Wittig reaction, and conversion of the chloride into a hydroxyl group.

1-chloro-2E,4-ochtodien-
6-ol

The following is representative of the many synthetic uses of 3-methyl-2-cyclohexenone, which can be conveniently prepared from 1,3-cyclohexanedione. Conversion of the enone into 1,3-dimethyl-1,3-cyclohexanediene is a routine procedure, and the Diels–Alder reaction of the latter compound with acrolein furnished a bicyclo[2.2.2]octenal amenable to transformation into norpatchoulenol [Teisseire, 1974].

A conceptually similar synthesis of seychellene [Jung, 1978] started from transformation of 2-methyl-1,3-cyclohexanedione into 1,2-dimethyl-3-trimethylsiloxy-1,3-cyclohexanedione, which was used to condense with methyl vinyl ketone to construct the bicyclic intermediate.

norpatchoulenol

An intramolecular Diels–Alder reaction was the key step in the construction of 9-pupukeanone [Piers, 1982], and thence, 9-isocyanopupukeanane. The required substrate was derived from 4,6-dimethyl-1,3-cyclohexanedione via alkylation and several well-known operations.

9-isocyano-   9-pupukeanone
pupukeanane

7.2.3.5. C-1/C-2 Bond Scission

As β-dicarbonyl compounds are readily cleaved by retro-Claisen reaction, a bifunctional carbon chain results when 1,3-cyclohexanedione is subjected to proper reaction conditions. Consequently the synthetic utility of 1,3-cyclohexanediones is magnified.

Ring cleavage occurred upon bissulfenylation of such compounds. The product derived from methyl 3,5-dioxocyclohexanecarboxylate has been processed into desethylquebrachamine [Takano, 1980c].

An intramolecular $C \rightarrow O$ transacylation of a hydroxypentynyl 1,3-cyclohexanedione results in ring expansion [Schore, 1987]. The reaction became uncurbed with the triple bond ligating to a dicobalt hexacarbonyl residue.

δ-Alkynyl carbonyl compounds are now very readily available from 1,3-cyclohexanediones by their conversion into 2,3-epoxycyclohexanones via the cyclohexenones, with implementation of the Eschenmoser–Tanabe fragmentation. Among the many applications of this reaction sequence are syntheses of *exo*-brevicomin [Kocienski, 1976] and dendrobatid indolizidine-207A [Holmes, 1991].

exo-brevicomin

dendrobatid
indolizidine-207A

For accessing optically active cecropia juvenile hormones JH-I and JH-II [K. Mori, 1988a] from 2-ethyl-2-methyl-1,3-cyclohexanedione, the latter compound was reduced by yeast ((2S,3S)isomer 99% ee). On Baeyer–Villiger reaction a lactone containing two adjacent oxygen-bearing chirality centers was produced. As these centers relate to the configurations of the oxirane carbon atoms in the target molecules by retention at the quaternary site and inversion at the tertiary carbon, activation of the latter to induce intramolecular displacement at an appropriate stage of the synthesis was in order. Thus the desymmetrization maneuver together with oxidative cleavage of the six-membered ring constituted an operational solution to this and associated synthetic problems.

R = Et   JH-I
R = Me   JH-II

Enol derivatives of 1,3-cyclohexanediones readily react with dihalocarbenes and the adducts are susceptible to dehalogenation with ring expansion. A retrosynthetic analysis showed that the Prelog–Djerassi lactone is amenable to construction from a tetrasubstituted cycloheptanone in the proper enol form, and the latter enol derivative would be derivable from 4,6-dimethyl-1,3-cyclohexanedione [Stork, 1979]. It should be noted that the stable *meso* mode of the dione in which the two methyl groups are *cis*-diequatorial has the required configurations, and the desymmetrized enol derivative had no adverse effect on the subsequent transformations.

Prelog-Djerassi
lactone

A significant method for the construction of eight-membered ring compounds is via [2+2]-photocycloaddition of enol derivatives of 1,3-cyclohexanediones, because upon liberating the hydroxyl group the cycloadducts are prone to fragmentation which relieves ring strain of the condensed cyclobutane moiety. Approaches to fusicoccin-H [Grayson, 1984] and epiprecapnelladiene [A.M. Birch, 1980] have been delineated. The latter represents an intramolecular version of deMayo reaction.

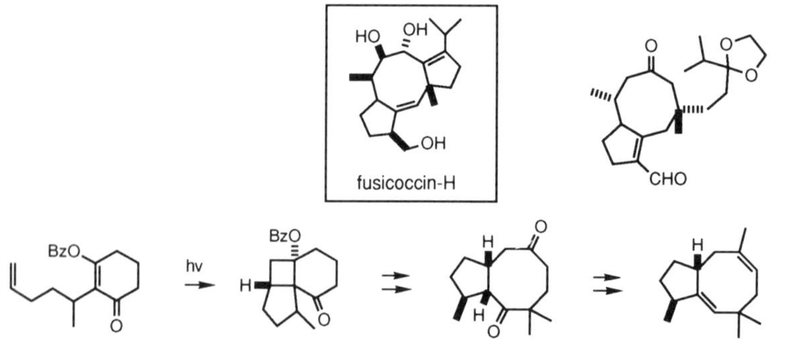

fusicoccin-H

epiprecapnelladiene

When dimedone (in its enol form) was submitted to photoreaction with 2-methyl-2-cyclopentenol the adducts underwent retro-aldol fission directly. The minor regioisomer has been used in a synthesis of hirsutene [Disnayaka, 1985].

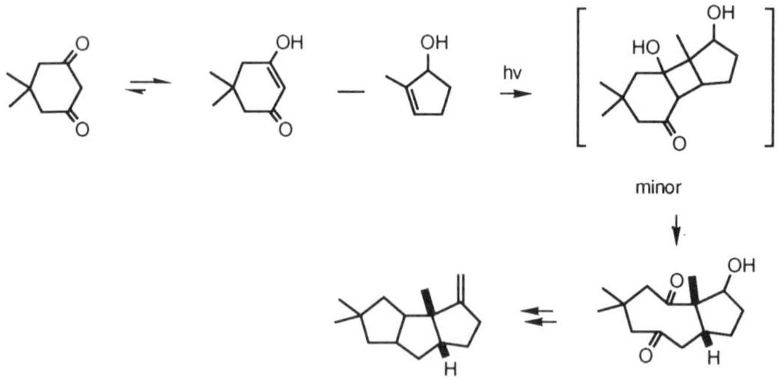

minor

hirsutene

A route to pentalenene [Pattenden, 1984] involved a slightly modified method. Cleavage of the C-1/C-2 bond of the original cyclohexanedione was delayed after the ketone group was converted into a tertiary alcohol. The fragmentation still proceeded well, and the modification served to differentiate the two carbonyl groups which would have been generated by a direct retro-aldol process.

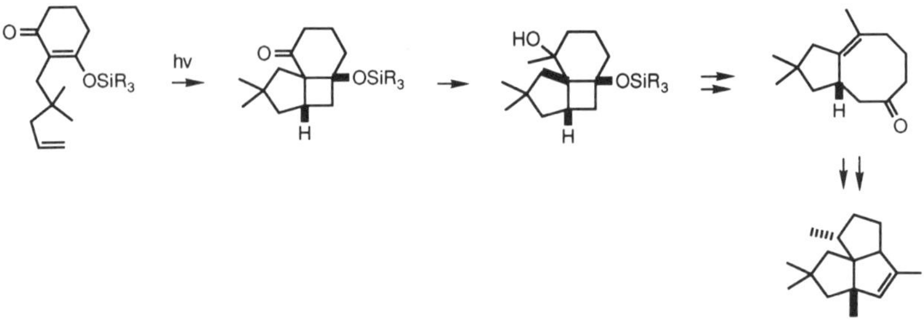

pentalenene

Quite relevant to the above examples is a synthesis of isocomene [Pirrung, 1981] which involved intramolecular photocycloaddition of a side chain to a cyclohexenone, itself obtained from a 1,3-cyclohexanedione derivative after kinetic methylation. The C-1/C-2 bond was not severed, but it was induced to migrate, giving rise to the angular triquinane.

isocomene

An effective transformation of 2,2-dialkyl-1,3-cyclohexanediones into 3-alkyl-2-cyclohexenones with insertion of a carbon atom involves treatment with dimethyl lithiomethylphosphonate [Y. Yamamoto, 1990]. The addition of the organolithium reagent to the diketone caused a retro-aldol fission which was followed by a Horner–Wadsworth–Emmons reaction. This method of cyclohexenone synthesis is illustrated in a synthesis of α-acoradiene.

α-acoradiene

In the role of a building block for quadrone [Burke, 1981, 1982], dimedone must undergo extensive structural changes. First, a spirocycle was created from one of the carbonyl sites, next the six-membered ring was severed to give a 4,4-disubstituted 2-cyclopentenone in which the aldehyde chain participated in an intramolecular Michael reaction as the donor. In subsequent aldol condensation its role was reversed to an acceptor.

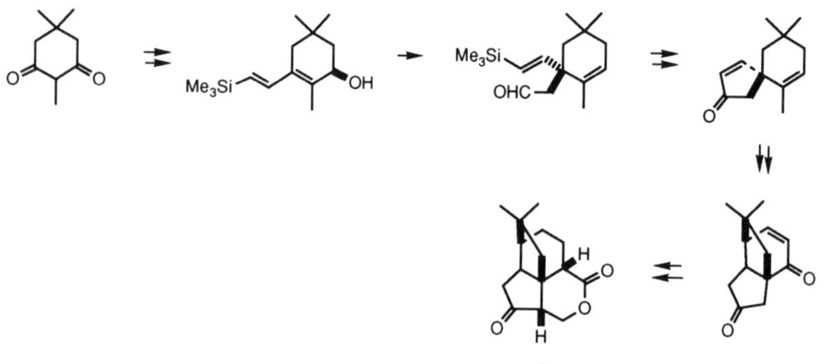

quadrone

A special case pertaining to alkylation at C-1 and C-3 with subsequent cleavage of the C-1/C-2 bond of a 1,3-cyclohexanone system is the ring expansion of a bicyclo[2.2.2]octane-2,6-dione via an anionic oxy-Cope rearrangement, in connection with a synthetic endeavor toward cerorubenic acid-III [Paquette, 1988].

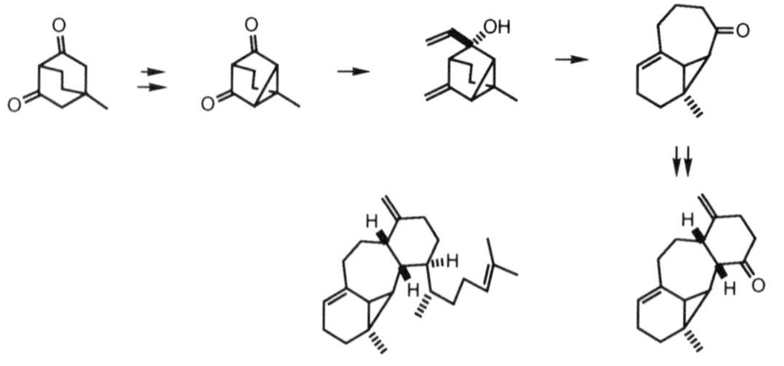

cerorubenic acid-III

Oxidative cleavage of a bicyclic dione after selective enzymatic reduction laid the foundation for a gradual accumulation of structural features that eventually led to (E)-endo-bergamoten-12-oic acid in both the α- and the β-form [K. Mori, 1994b].

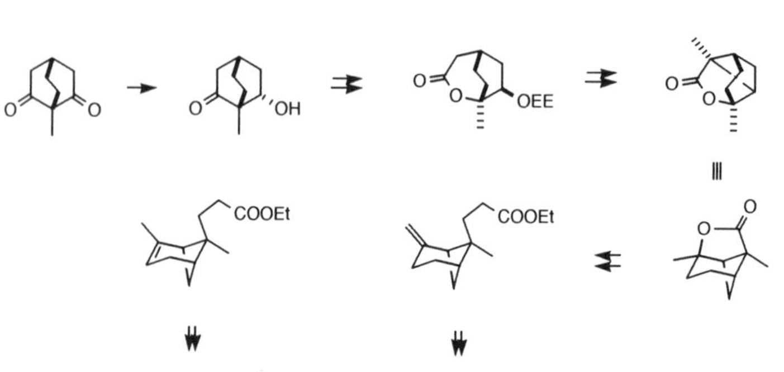

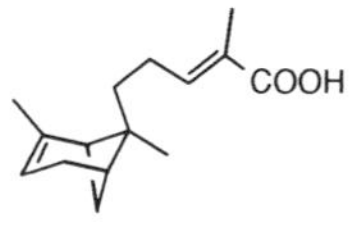

α-(E)-endo-bergamoten-   β-(E)-endo-bergamoten-
12-oic acid   12-oic acid

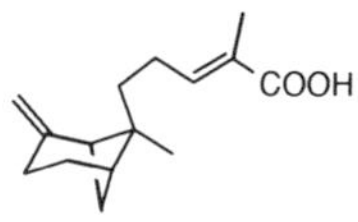

After trimethylation of a symmetrical bicyclo[2.2.2]octenedione the photochemical oxadi-π-methane rearrangement transformed it into another tricyclic compound embodying a functionalized diquinane. Further elaboration of the rearrangement product led to (−)-coriolin [Demuth, 1986].

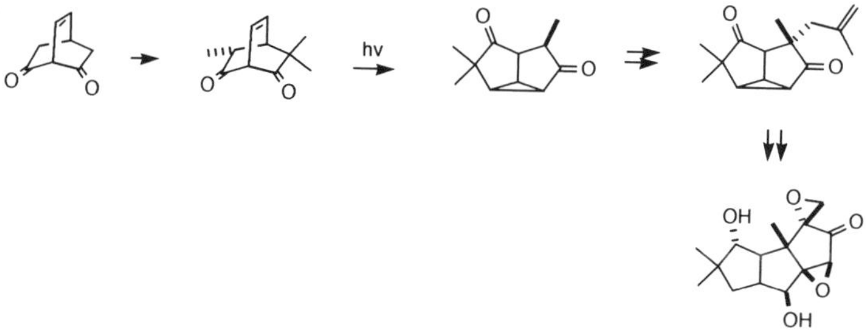

(-)-coriolin

7.2.3.6. Bond Formation at C-2, C-3 or C-2, C-4 or C-2, C-3, C-4 or C-2, C-3, C-6 or C-2, C-4, C-5, C-6

The 1,2,3,5-tetrasubstituted benzene ring synthon for maytansine is easily correlated with a 5-substituted 1,3-cyclohexanedione, one of the substituents in the benzene derivatives being a methoxy group. In the event, two preparative methods for such synthons have been developed [Corey, 1977; Foy, 1977].

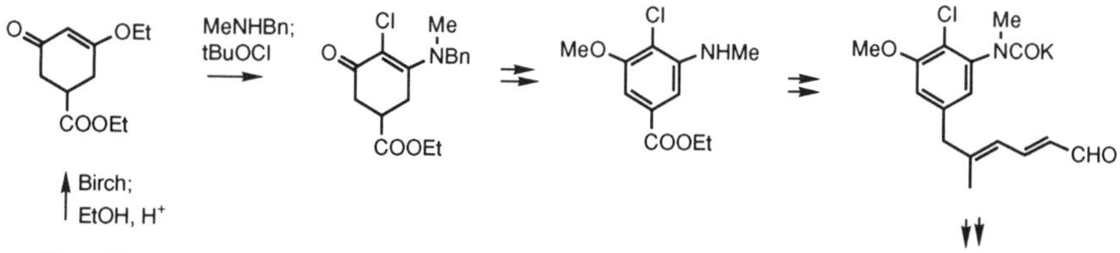

gallic acid

maytansine

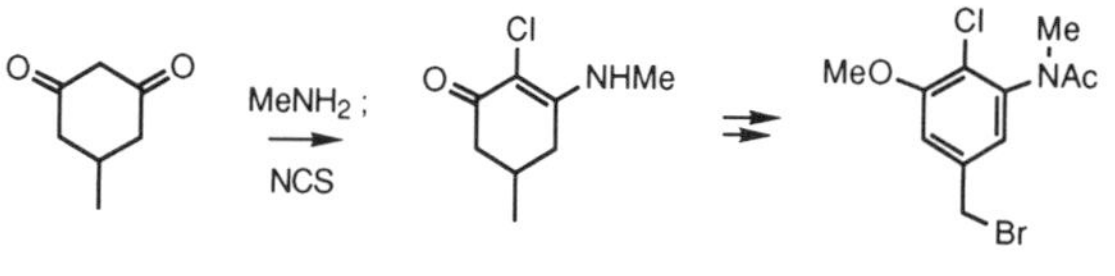

The assembly of complex molecules involving bond information at C-2 and C-4 of a symmetrical 1,3-cyclohexanedione is exemplified by a synthesis of vernolepin [Kieczykowski, 1978]. Even in the desymmetrized form of the monomethyl ether, annexation of the pyran ring onto the existing carbocycle did not cause any regiochemical problems. In fact the synthetic inquiry into vernolepin is less formidable in the regiochemical aspects because a related target, vernomenin, is a lactone isomer, and the local C_2-symmetry about C-7 and extending to both C-6 and C-8 is beneficial to both planning and execution of a synthesis. As the intramolecular alkylation that formed the pyran ring determined the structure of the enone, which unfolded immediately thereafter, the construction of the γ-lactone could be implemented accordingly.

vernolepin

A similar method was employed in erecting curzerenone [Miyashita, 1981]. The presence of a fused furan in the cyclohexanedione precursor was advantageously monoprotected in a latent form of the required structural subunit. Such protection (other than with a simple enol ether) is also quite crucial to specify the alkylative attachment of the three-carbon side chain.

curzerenone

It was more expedient to effect a reflexive Michael reaction on an enol silyl ether of a 1,3-cyclohexanedione to construct a bicyclo[2.2.2]octanone intermediate for a synthesis of eriolanin [M.R. Roberts, 1981]. Subsequent fragmentation unraveled a 4,5-disubstituted 2-cyclohexenone.

eriolanin

A synthesis of sanadaol [Nagaoka, 1987] was realized in the overall process of bond formation at C-2 and C-4 of 1,3-cyclohexanedione in a cycloheptane unit, with final homologation of the C-1 ketone to an unsaturated aldehyde. Interestingly, the initial bond formation was at C-1 and C-4, but the C—C bond joining C-1 was a temporary device to help cyclization at C-2.

sanadaol

Diazotization of 1,3-diones followed by the Wolff rearrangement gives rise to β-keto esters in which one set of the original bonding partners is exchanged. For cyclic diketones ring contraction is a consequence. Such a transformation on 5,5-dimethyl-1,3-cyclohexanedione entails molecular desymmetrization of considerable significance, as demonstrated in a synthesis of velleral [Froborg, 1978] from the cyclpentanonecarboxylic ester.

velleral

A different way of desymmetrizing the cyclic diketone led to a cyclopentenealdehyde [Bergman, 1986] which has found even wider application as a building block of certain terpenes. Approaches to hirsutene (Hudlicky, 1980] and pentalenene [Hudlicky, 1987] are only two of many examples.

hirsutene

Bond formation at C-2, C-3, and C-4 of 5-isopropyl-1,3-cyclohexanedione is witnessed in a synthesis of hibiscone-C [Koft, 1982]. This sesquiterpene poses an interesting synthetic problem because of its *peri*-fused furan ring. An excellent solution involved an intramolecular [2+2]-photocycloaddition to form a cyclobutene unit, the oxidative cleavage of which led to the precursor of the furan.

hibiscone-C

A *β*-himachalene synthesis [Piers, 1979] was devised on the basis of homo-Cope rearrangement to form a 1,4-cycloheptadiene annexed to a cyclohexanone. The new ring contained three methyl groups in the desired positions and two double bonds of very different substitution patterns, which warranted selective hydrogenation of one of them, as required by the target molecule. The remaining unsaturation and the methyl group in the six-membered ring were easily fashioned from the ketone group.

β-himachalene

Conversion of 1,3-cyclohexanedione into an aromatic ring by the Alder–Rickert reaction fulfilled the need for establishing the skeleton of mycophenolic acid [Patterson, 1993].

mycophenolic acid

7.2.3.7. *Bond Formation at C-3, C-4*

Many syntheses have been accomplished via bond formation of C-3 and C-4 of 1,3-cyclohexanedione and derivatives. Thus, the initial stage of a method for securing trisporol-B [Miyaura, 1986] involved twofold kinetic alkylation of a 1,3-cyclohexanedione derivative. After serving as the activator the exposed ketone group was attacked by a lithium acetylide reagent. Hydrolysis delivered a cyclohexenone with proper substituents for a relatively straightforward synthesis of the target compound.

trisporol-B

Dimedone proved to be a useful precursor of β-damascenone [Torii, 1979]. The desymmetrization by enol ether formation permitted a regioselective alkylation to append the four-carbon side chain.

dimedone

β-damascenone

It is also easy to identify the cyclohexenone portion of nootkatone as derivable from 5-methyl-1,3-cyclohexanedione [Heathcock, unpublished]. Accordingly, double alkylation at the α-position of the monoethyl enol ether under kinetically controlled conditions, with iodomethane and isoprenyl bromide, led to the desired intermediate; the stereochemistry of the newly created quaternary carbon center was established by 1,2-asymmetric induction of the methyl group in the adjacent carbon atom. Reaction of the ketone group with vinyllithium and exposure of the tertiary alcohol to acid gave a conjugate dienone which had been previously converted to nootkatone [Dastur, 1974].

(major)

nootkatone

A silicon-containing analog of the nootkatone precursor has been acquired [Majetich, 1985]. Its conversion into nootkatone was more direct, the ring closure being mediated by a Lewis acid-catalyzed reaction accompanied by desilylation.

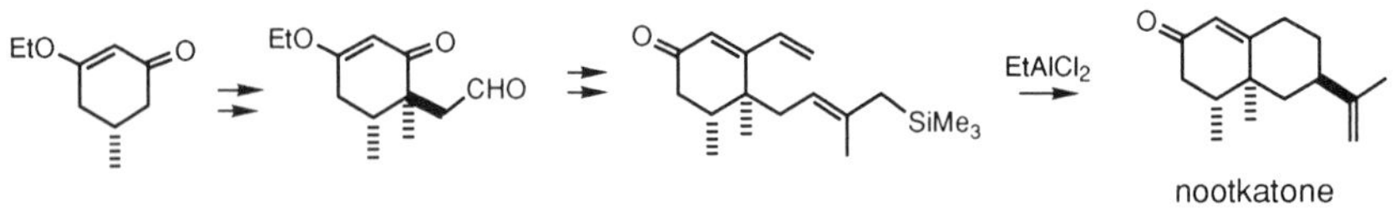

nootkatone

Similar strategies were formulated for the synthesis of perforenone [Majetich, 1987] and neolemnane [Majetich, 1991]. In the latter scheme the formation of an eight-membered ring was initiated by reaction of the allylsilane residue with fluoride ion.

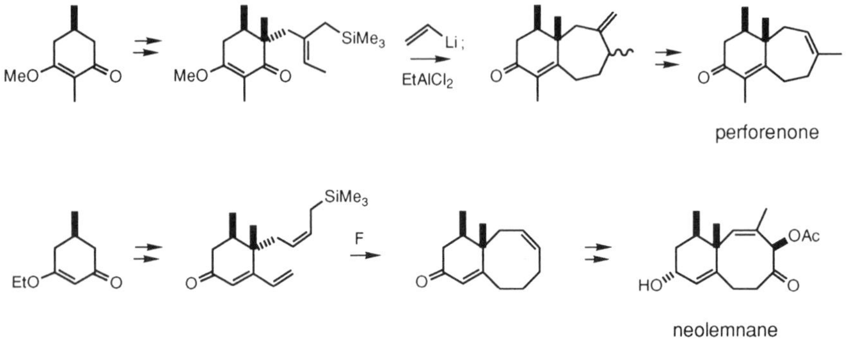

perforenone

neolemnane

A recent route for cortisone [Horiguchi, 1989] started from elaboration of 1,3-cyclohexanedione to a 3,4-disubstituted 2-cyclohexenone which became the C-ring. The five-membered ring was created via conjugate alkenylation and an ene reaction.

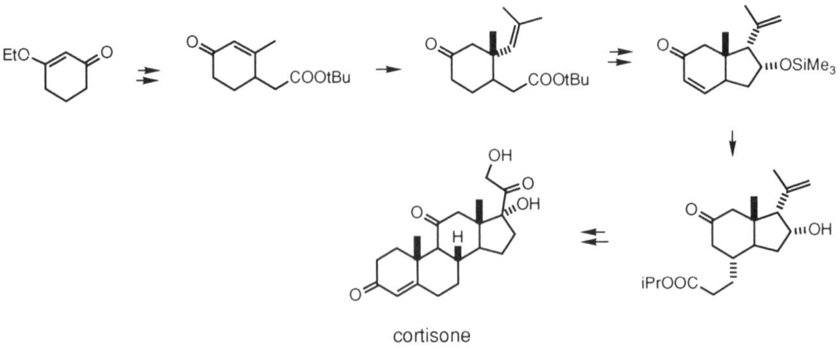

cortisone

A synthetic intermediate for vernolepin could be assembled [Torii, 1977] via kinetic dialkylation at C-4 of a 1,3-cyclohexanedione derivative and an intramolecular Michael reaction of the derived enone to complete the oxabicyclic system.

Many terpenes contain a *gem*-dimethylcyclohexane subunit. Synthetic approaches to these substances from 2,2-dimethyl-1,3-cyclohexanedione have become attractive after the discovery of a facile differentiation of the two ketone groups by asymmetric reduction with baker's yeast. (3S)-Hydroxy-2,2-dimethylcyclohexanone thus available has been exploited in syntheses of (−)-karahana ether [K. Mori, 1985a], dihyroactinidiolide [K. Mori, 1986b], and (2S,5R)(6E,8E)-2,5-epoxy-6,8-megastigmadiene [K. Mori, 1986d], by relatively simple transformations.

(6E,8E)-2(S),5(R)-epoxy-   dihydroactinidiolide
6,8-megastigmadiene

Using the Diels–Alder reaction of a diene derived from the optically active hydroxycyclohexanone as the key step several biologically active sesquiterpenes and diterpenes have been prepared. These include (+)-baiyunol [K. Mori, 1987], the complement inhibitor (−)-K 76 [K. Mori, 1988b], (−)-warburganol [K. Mori, 1989b], (−)-pereniporin-A and -B [K. Mori, 1989c], (+)-3-hydroxytanshinone [Haiza, 1990], and both enantiomers of polygodial [K. Mori, 1986e].

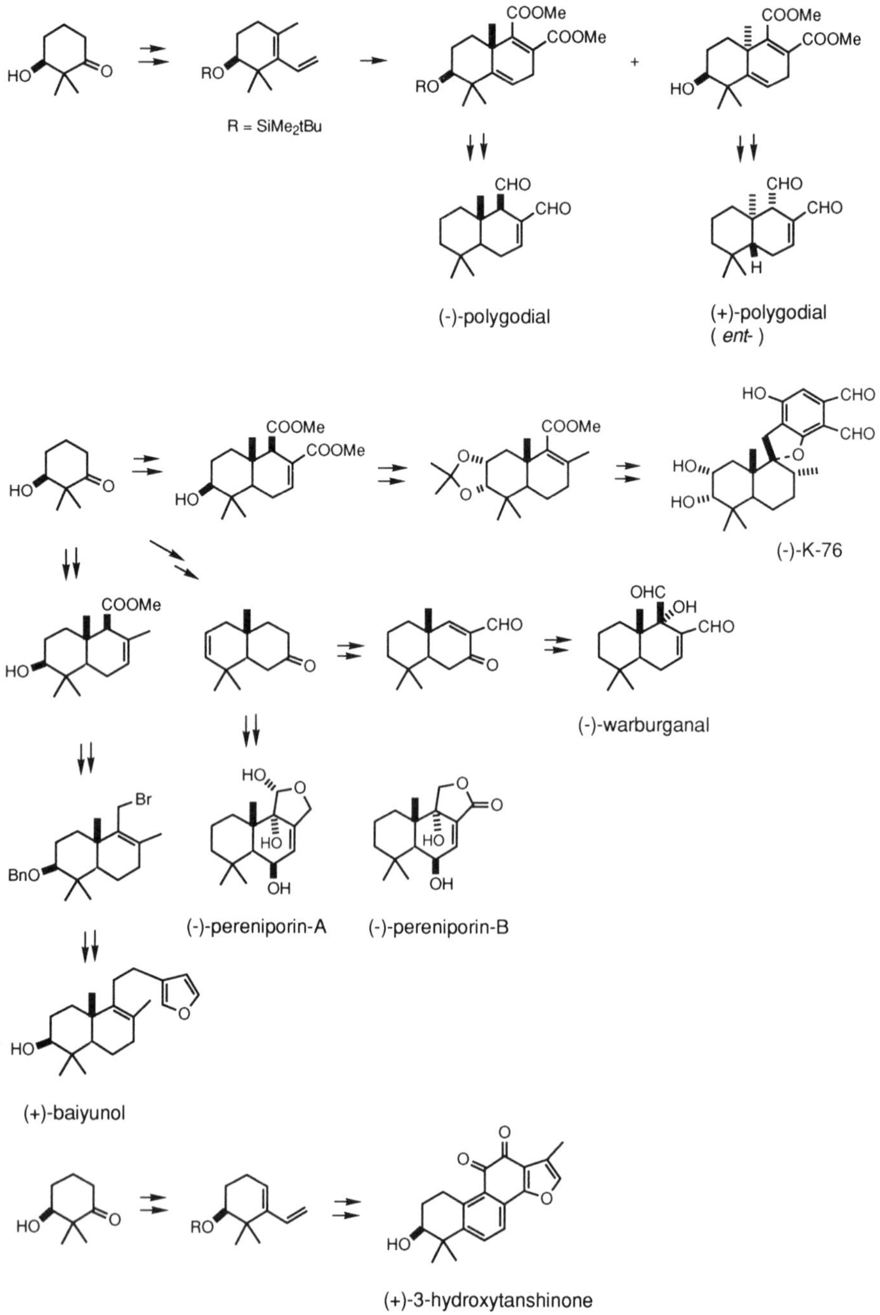

R = SiMe₂tBu

(-)-polygodial

(+)-polygodial
(ent-)

(-)-K-76

(-)-warburganal

(-)-pereniporin-A   (-)-pereniporin-B

(+)-baiyunol

(+)-3-hydroxytanshinone

Other applications include *O*-methylpisiferic acid [K. Mori, 1986a] and glycinoeclipin-A [K. Mori, 1989d; Murai, 1989]. Only one C—C bond was formed with C-4 in the latter synthesis.

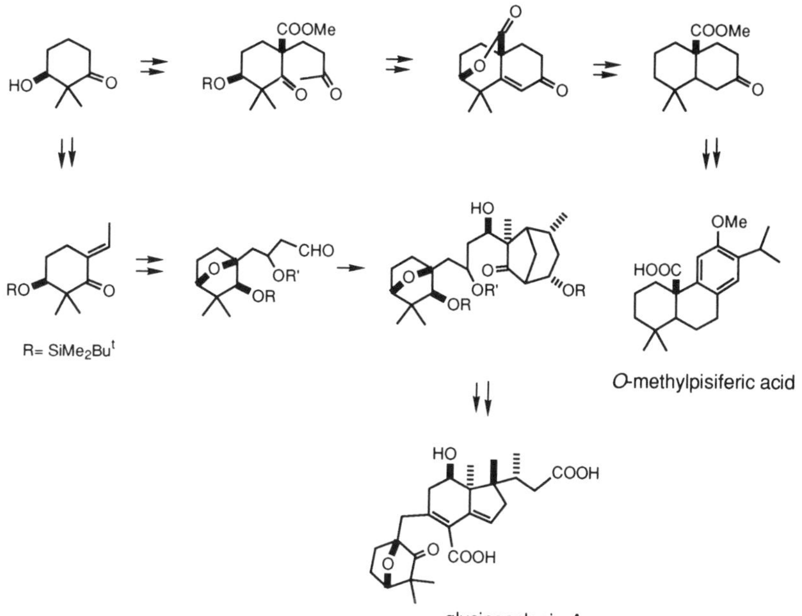

O-methylpisiferic acid

glycinoeclepin-A

In a seychellene synthesis [Fukamiya, 1973] 2-methyl-1,3-cyclohexanedione was fashioned into a 2,4-cyclohexadienone in which a side chain double bond participated in an intramolecular Diels–Alder reaction to form the tricyclic skeleton.

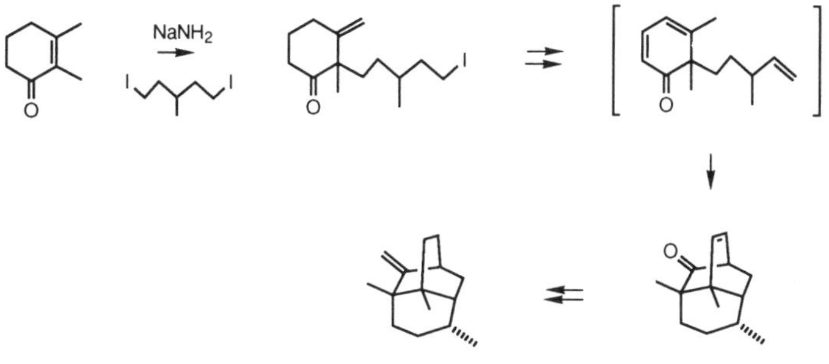

seychellene

There is a more obscure correlation between 1,3-cyclohexanedione and *O*-methylpallidinine with respect to synthetic operations. It is now evident that designation of C-3 of the diketone as the quaternary benzylic site has certain advantages, as the aromatic ring can be introduced via an organometallic reaction, and functionalization at C-4 is attainable with a 3-cyclohexenone acetal [McMurry, 1984].

O-methylpallidinine

The elaboration of the American cockroach sex excitant, periplanone-B [Still, 1979] is quite an accomplishment. The method was based on ring expansion by an anionic oxy-Cope rearrangement, the substrate of which was secured via kinetic alkylation of a 5-alkoxymethyl-2-cyclohexenone and reaction of the enone with vinylmagnesium bromide. As indicated previously, the substituted cyclohexenone was derived from the diketone.

periplanone-B

7.2.3.8. Bond Formation at C-4, or C-4 and C-6

One of the most efficient and elegant syntheses of β-vetivone proceeded by twofold kinetic alkylation of the monoethyl enol ether of 5-methyl-1,3-cyclohexanedione [Stork, 1973]. As these alkylation steps are subjected to 1,2-asymmetric induction, the electrophilic chain always orients itself toward the face opposite to the secondary methyl group. Consequently, the spirocyclic product has the same relative configuration as β-vetivone. Note that the first alkylation involved the more reactive allylic chloride.

A modification of the method which involved activation of the nucleophilic site permitted the use of a weaker base for the alkylation [Eilerman, 1981]. And for the cyclization step the enolate species was generated from a decarbethoxylation process.

An alternative pathway to β-vetivone and may other related sesquiterpenes [Dauben, 1977] employed the α'-formyl derivative of the diketone monoenol ether. The formyl group served as an activator in the reaction with (1-carbethoxycyclopropyl)triphenylphosphonium bromide. In one operation a homo-Michael reaction occurred which was followed by an intramolecular Wittig reaction involving the formyl group.

The Diels–Alder/fragmentation sequence can be used to control the spirocyclic center of synthetic targets such as solavetivone [Murai, 1981b]. Note that the Diels–Alder reaction formed C—C bonds with both C-1 and C-4, but after the fragmentation only the bond at C-4 was retained. Thus, depending on the configurational demand a particular substituent at C-4 would have to be attached in the Diels–Alder reaction or in the previous alkylation step.

Megaphone has a cyclohexenone moiety in which three additional substituents are presentat C-4 and C-6. One of the synthetic routes developed for this substance [Zoretic, 1983] called for elaboration of 1,3-cyclohexanedione as shown.

megaphone

Segmentation of the six-membered ring of 2,2,5,5-tetramethyl-1,3-cyclo-hexanedione is easily achieved by oxidation of the dianion. This bicyclo[3.1.0]hexanedione is still a *meso-* compound, but desymmetrization occurs on ring cleavage. *cis*-Chrysanthemic acid is readily made from the ring-opened product. [Krief, 1993].

cis-chrysanthemic
acid

7.3. WIELAND–MIESCHER KETONE AND ANALOGS

Wieland–Miescher (WM) ketone is the Robinson annulation product of 2-methyl-1,3-cyclohexanedione with methyl vinyl ketone. Its extensive applications to natural product synthesis are due to its substitution pattern being amenable to further elaboration, particularly with respect to the convenient differentiation of the ketone groups. Furthermore, the optically active ketone can be obtained via microbial reduction [Prelog, 1956], resolution of a derived hemiphthalate [Newkome, 1972], and most conveniently by asymmetric cyclization of 2-methyl-2-(3-oxobutyl)-1,3-cyclohexanedione. An effective catalyst for the cyclization is proline [Eder, 1971; Hajos, 1974; Gutzwiller, 1977] ((+)-WM ketone with (S)-proline). The *meso*-2,2-disubstituted 1,3-cyclohexanedione is usually formed by alkylation of 2-methyl-1,3-cyclohexanedione.

Many different ways to modify the bicyclic ketone as required by various synthetic targets are evident. It should be emphasized that most of the synthetic targets are unsymmetrical.

7.3.1. Retention of the Bicyclic Structure

For a synthesis of (+)-pallescensin-A [A.B. Smith, 1984] the route involved selective acetalization of the isolated ketone group, thiomethylation, reductive methylation, Wolff–Kishner reduction, and deacetalization to afford the trimethyldecalone intermediate. The furan ring was then constructed via allylation and cleavage of the double bond.

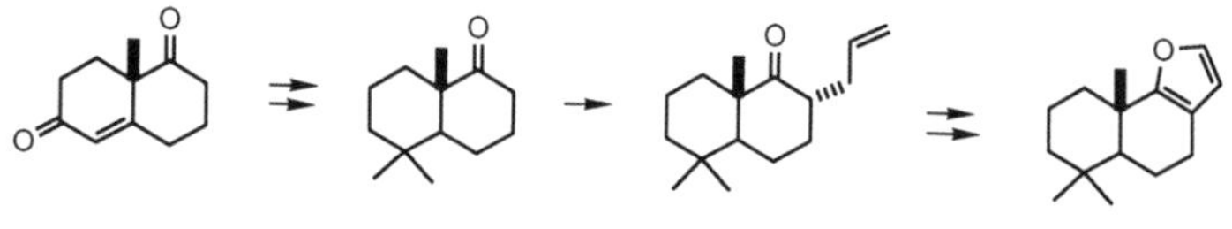

(+)-pallescensin-A

It is obvious from a structural analysis that WM ketone is suitable for a synthesis of (+)-dysideapalaunic acid [Hagiwara, 1988]. The ketone groups provided the necessary activation for appending the side chains.

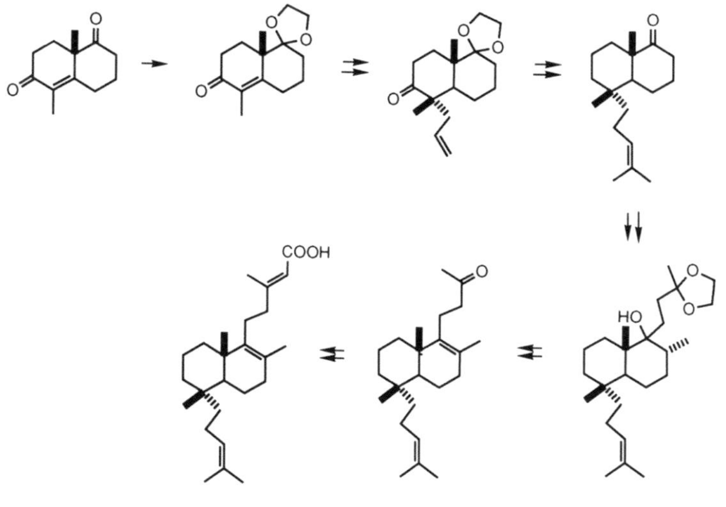

(+)-dysideapalaunic
acid

Similarly, zonarol and isozonarol have been synthesized [Welch, 1978] by proper alkylation and redox manipulations of WM ketone.

isozonarol   zonarol

Muzigadial is a rearranged drimane sesquiterpene which exhibits potent antifeedant activity against insects such as the African armyworm. Its synthesis is therefore of significance in terms of gaining the substance for crop protection. The racemate has been prepared [Bosch, 1986] from Wieland–Miescher ketone.

muzigadial

A simple correlation in structural features exists between WM ketone and costal. In fact, the synthesis of costal [H.J. Liu, 1985] from 4-methylated WM ketone was even simpler, as it involved removal of the conjugated ketone, transposition of the double bond, and attachment of the branched side chain. The latter operation depended on the subangular ketone group and was accomplished via conversion into the enone, Wharton rearrangement, and subsequent redox manipulations to afford an octalone.

costal

It is instructive to compare the synthesis of costal with a route to (−)-vetiselinenol [Teisseire, 1980] in terms of functional group utilization. In the vetiselinenol synthesis the enone carbonyl served as a base for the appendage of the three-carbon side chain, while the missing exocyclic methylene group was introduced via double bond migration, hydroboration, oxidation, and Wittig reaction. The saturated ketone group of WM ketone was not required, and it was removed at an early stage of the synthetic process.

(-)-vetiselinenol

An approach to valeranone [Banerjee, 1973] consisted of ketone reduction and protection of the secondary alcohol, conjugate addition of a methyl group, and subsequent introduction of the isopropyl side chain, taking advantage of the activation property of the ketone. The subangular carbonyl group was regenerated at a late stage of the synthesis.

valeranone

In a synthesis of lindestrene [Minato, 1968] the conjugated ketone of WM ketone served as the pedestal for the furan ring, whereas the isolated ketone group was converted into a double bond.

lindestrene

As reductive alkylation of the enone system of a partially reduced WM ketone is under steric control, the compound is useful for synthesis of avarol [Sarma, 1982].

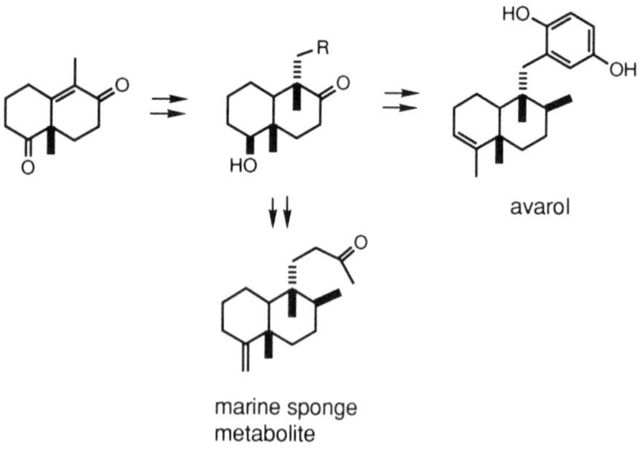

avarol

marine sponge
metabolite

A slightly different tactic was employed in the introduction of the analogous side chain of annonene [S. Takahashi, 1979]. A Claisen rearrangement established both an acetaldehyde subunit and an exocyclic methylene group in one step. The aldehyde group was further extended to incorporate the furan residue.

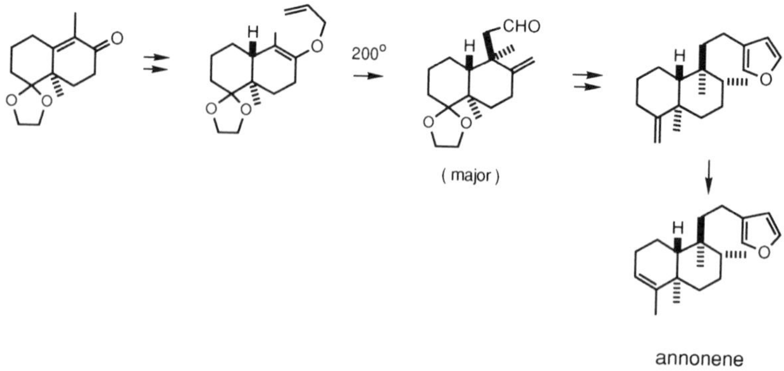

(major)

annonene

The use of Wieland–Miescher ketone in a synthesis of ajugarin-IV [Kende, 1982] must implement the necessary oxygenation at the β-position of the conjugated ketone.

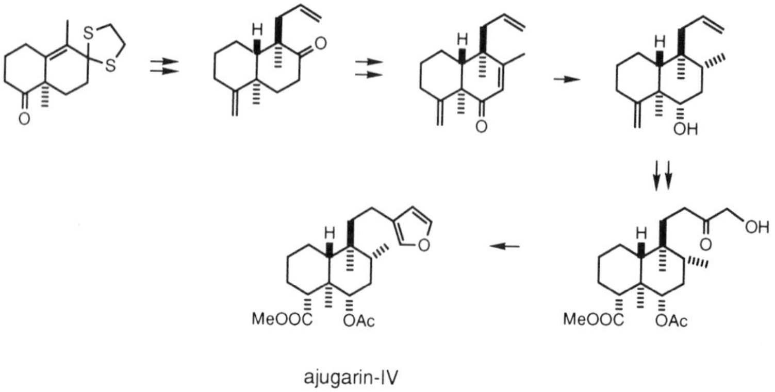

ajugarin-IV

Many terpenes contain an oxygenated side chain geminal to a methyl group. A method for installing such a functional group is by reductive carboxylation, as shown in a synthesis of the antifungal metabolite, LL-Z1271α [Welch, 1977].

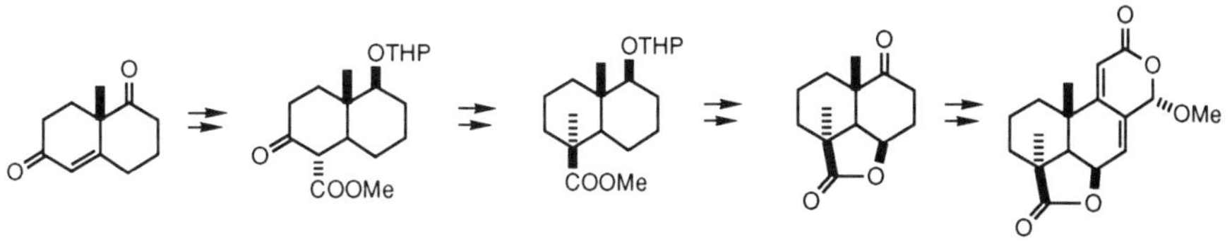

LL-Z1271α

An analog of Wieland–Miescher ketone is the 4-carbomethoxy derivative which can be readily prepared by Robinson annulation of 2-methyl-1,3-cyclohexanedione with Nazarov keto ester. Methylation of the bicyclic diketo ester also furnished a product in which the two tertiary methyl groups are in a *trans* relationship, a substitution pattern required in the elaboration of isoiresin (diacetate) [Pelletier, 1966, 1968b], andrographolide lactone [Pelletier, 1968a], marrubiin [Mangoni, 1972], and 3β-hydroxynagilactone-F [Reuvers, 1986].

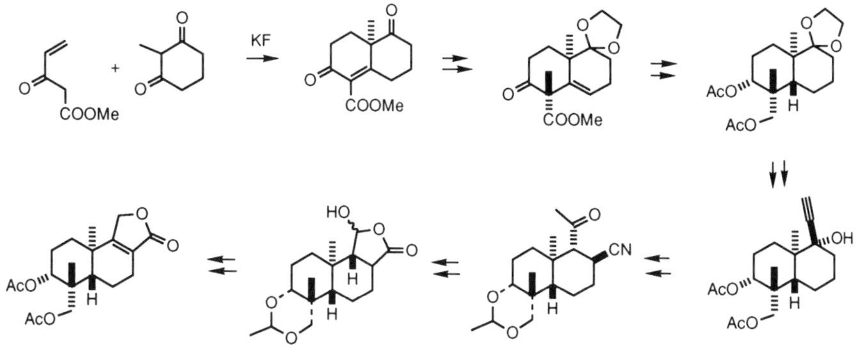

isoiresin diacetate

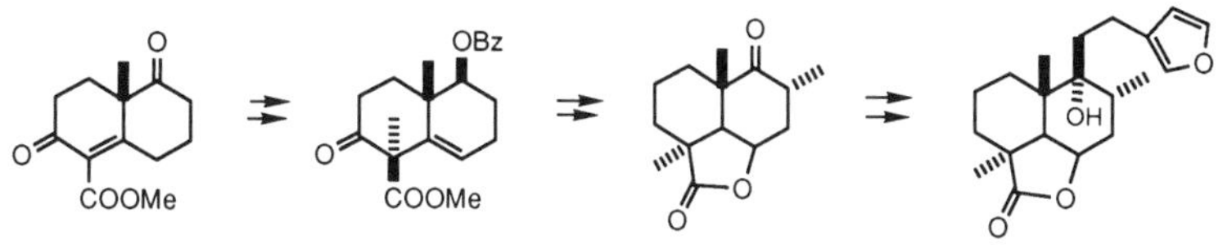

marrubiin

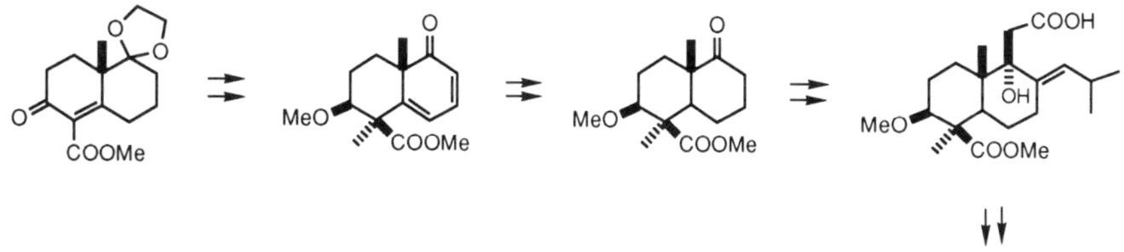

3β-hydroxynagilactone-F

It appears more appropriate to use 1,4-dimethoxy-2-butanone in the Robinson annulation to prepare an analog of Wieland–Miescher ketone for the synthesis of colorata-4(13),8-dienolide [de Groot, 1982]. The methoxy group of the bicyclic diketone provided a handle for the attachment of the exocyclic methylene group.

colorata-4(13),8-dienolide

Another analog of WM ketone is the 1-methyl derivative, obtained by annulation with 3-penten-2-one. Due to chelation effects, the favorable transition state for the Michael reaction step determines the *cis-vic*-dimethyl pattern of the annulation product. It is expedient to use this octalindione in the synthesis of valencene, valerianol, eremophilene, and eremoligenol [Coates, 1970b].

eremophilene   eremoligenol   valerianol   valencene

7.3.2. Cyclomutation

It has been recognized that the hydrazulene ring system is more readily assembled by rearrangement of the hydronaphthalene skeleton than cyclization procedures, especially when substituents of defined stereochemistry are present. In other words, the simultaneous ring expansion/ring contraction (cyclomutation) of decalin derivatives, which are very easily obtained, is often the preferred method for gaining access to the hydrazulenes.

Several sesquiterpenes have been synthesized based on this strategy, including α-bulnesene, bulnesol [Kato, 1970a; Heathcock, 1971] and kessane (Kato, 1970b].

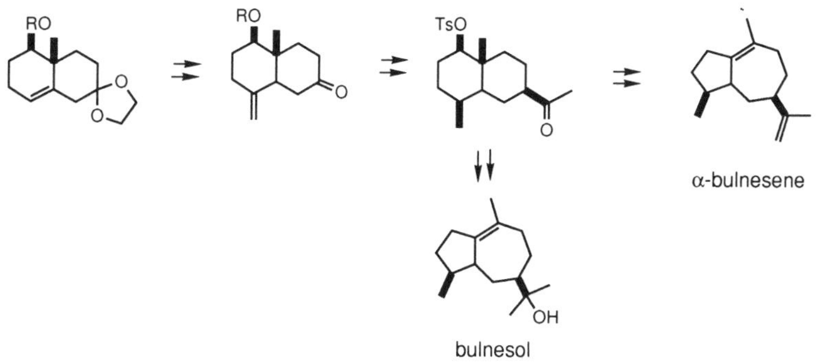

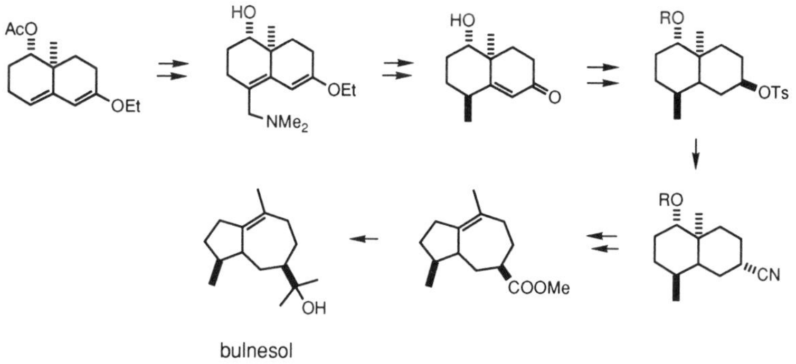

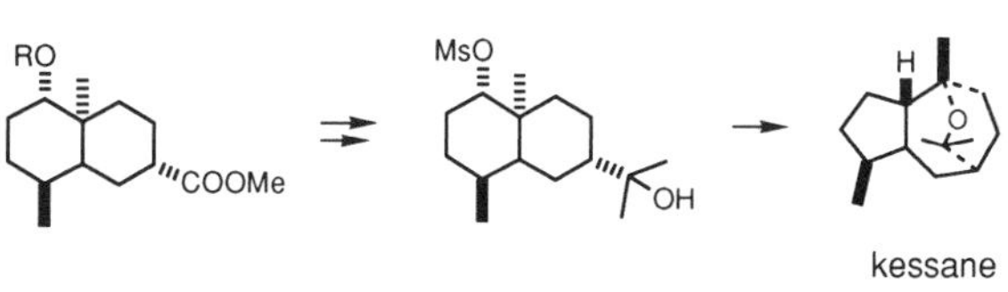

bulnesol

α-bulnesene

bulnesol

kessane

Noteworthy is the inconsequential stereochemistry of the angular methyl group in the decalin precursor in the bulnesol synthesis. Of course the products are tetrasubstituted olefins.

The base-induced pinaol rearrangement of α-glycol monotosylate or mesylate is very simple. When the glycol monosulfonate is placed at the angular and subangular positions of a decalin ring its arrangement product is a hydrazulenone. Thus, such an intermediate derived from Wieland–Miescher ketone has been converted into confertin [Heathcock, 1982a].

confertin

Intramolecular alkylation occurs when WM ketone is submitted to reduction with lithium in liquid ammonia. The major product of this reaction is a *trans*-fused dimethylperhydroindandione. The utility of this compound was recognized in a synthesis of butyrospermol [Reusch, 1977; Kolaczkowski, 1985], as the CD-ring portion of the terpene molecule features such structural elements in a latent form.

butyrospermol

Under a different set of reaction conditions the tricyclic ketol can undergo ring opening to give a spirocyclic diketone. The spirocyclic diketone derived from 4-methyl-WM ketone is a superb precursor of β-vetivone [Subrahamanian, 1978].

β-vetivone

7.3.3. Ring Cleavage

An alternative access to the hydrazulene skeleton from Wieland–Miescher ketone is exemplified by a route to globulol [Marshall, 1974]. After the introduction of an equatorial methyl group at the γ-carbon of the enone and conversion of the two carbonyl groups into an allylic alcohol and a mesylate, fragmentation of the bicyclic system was induced via hydroboration and treatment of the resulting borane with base. Since the two double bonds of the hydroxy-1,6-cyclodecadiene are in close proximity, ionization of the allylic alcohol induced participation by the trisubstituted double bond in a stereodefined manner to give a product with four consecutive stereocenters corresponding to those present in globulol. The double bond was then used to graft the cyclopropane ring.

globulol

A 3-methylene-1,5-cyclodecadiene was the key intermediate for synthesis of α- and β-longipinenes [Miyashita, 1974]. This compound, acquired from Wieland–Miescher ketone, was cyclized photochemically. Ring expansion was performed afterwards.

longipinenes

Approach to A-secoeudesmane sesquiterpenes is facilitated by choosing Wieland–Miescher ketone as the starting material. Thus, the two stereocenters of γ-elemene can be established in a decalin intermediate prior to ring cleavage to unveil the two unsaturated side chains [Kato, 1979].

γ-elemene

Because of the presence of a subangular ketone group, Wieland–Miescher ketone is structurally fit to be elaborated into ivangulin [Grieco, 1977a] and eriolanin [Grieco, 1978]. The conjugated ketone group served as a pivot in the construction of the lactone moiety.

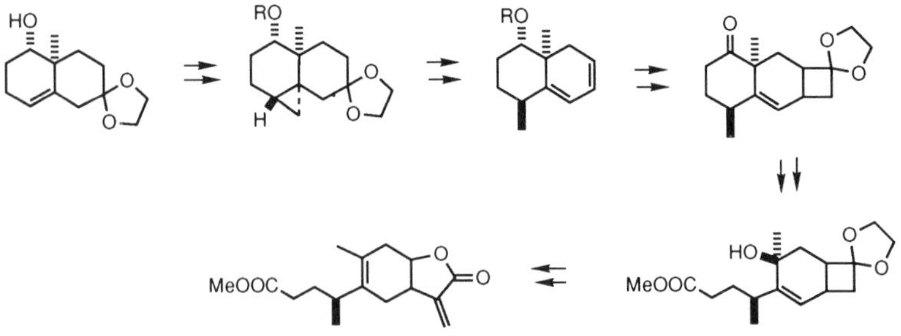

ivangulin

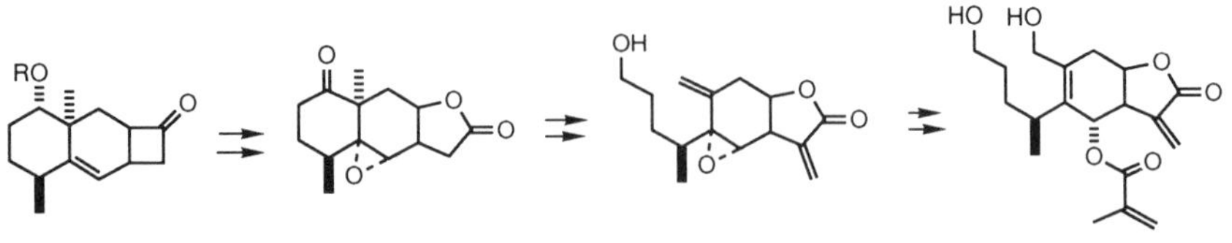

eriolanin

The $(S)(+)$-Wieland–Miescher ketone has also been degraded to provide an A-ring synthon for taxol [Golinski, 1993].

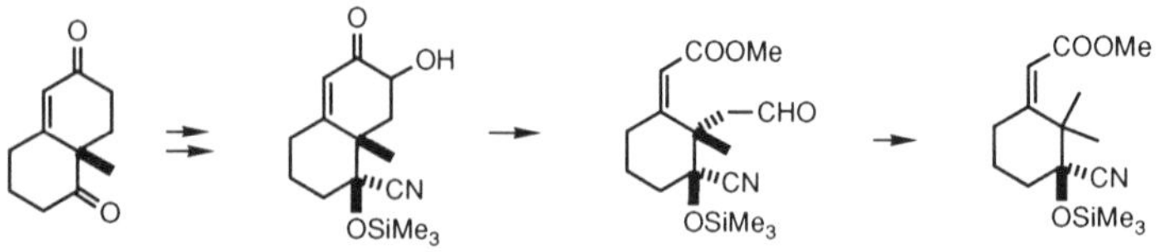

The formation of two and three rings by one intramolecular reaction of substrates derived from a seco WM ketone was each a key step in the synthesis of 5α-Δ^{16}-pregnen-3β-ol-20-one [Kametani, 1980] and $(+)$-atisirene [Ihara, 1986b]. The ring formation was effected thermally in one case, and by base treatment in the other. In other words, they are Diels–Alder and reflexive Michael reactions.

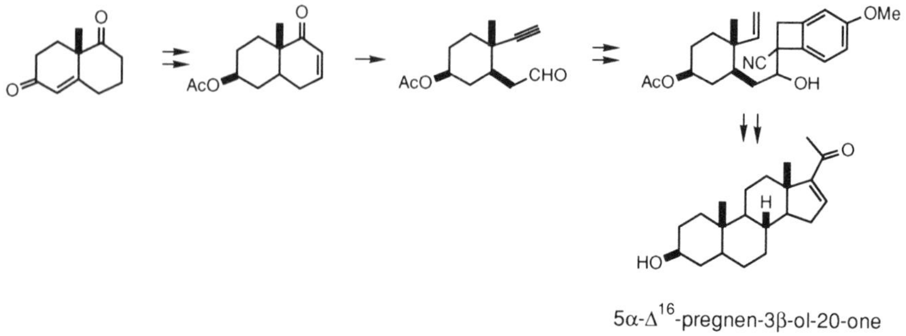

5α-Δ^{16}-pregnen-3β-ol-20-one

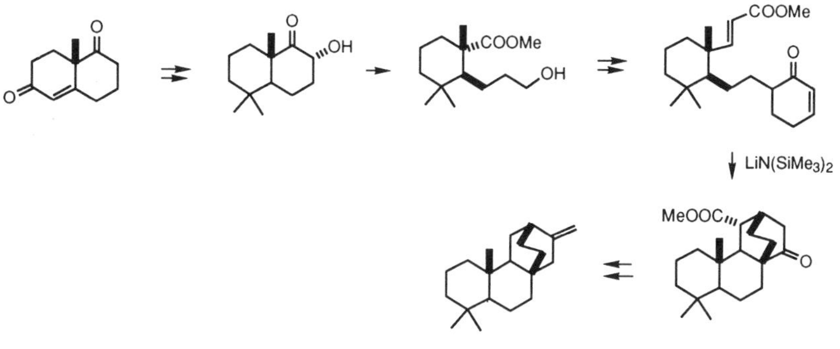

(+)-atisirene

A variation of the synthetic route to a fully equipped benzocyclobutenone, exploited for direct transformation into a tetracyclic precursor of (+)-chenodeoxycholic acid [Kametani, 1982a], is by a Grob fragmentation of a benzocyclobutenylcarbinol.

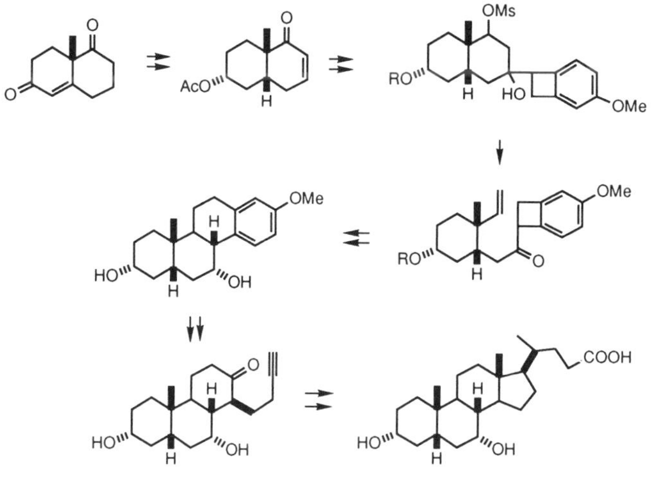

(+)-chenodeoxycholic acid

The chiral 4-methyl analog of Wieland–Miescher ketone was found suitable for the synthesis of (+)-methyl trisporate-B [S. Takahashi, 1988]. Comparison of the two structures suggests a process involving selective cleavage of the C—C bond linking the saturated ketone and its methylene carbon. An additional unsaturation must be introduced and chain lengthening is indicated.

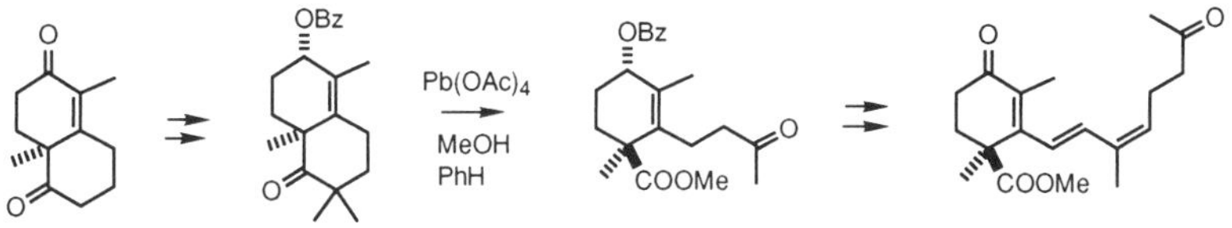

(+)-methyl trisporate-B

Cleavage of both rings of WM ketone was featured in a synthesis of zoapatanol [Kane, 1981]. The B-ring furnished the cyclic carbon atoms of the oxepane ring and the A-ring carbon atoms constituted part of the long side chain. The important aspect is that two stereocenters were inherited from the partially reduced WM ketone.

zoapatanol

7.3.4. Ring Annexation

The formulation of synthetic pathways to natural products possessing bridged ring systems from Wieland–Miescher ketone is perhaps more challenging. However, excellent work in this context has emerged.

The skeleton of seychellene can be considered as comprising a 2,7-bridged bicyclo[2.2.2]octane or a 1,7-bridged decalin. A synthesis based on the elaboration of a decalone [Piers, 1971] called for the preparation of an intermediate from Wieland–Miescher ketone by temporary masking of the saturated carbonyl function, angular methylation, and conversion of the β-decalone into a secondary methyl group while appending a tosyloxymethyl chain to the subangular site. Three stereocenters were created from the enone system and one of them was associated with an electrophilic residue which was to be employed in an intramolecular alkylation.

seychellene

A landmark synthesis of longifolene [Corey, 1964] featured an intramolecular Michael reaction of a homooctalindione which was derived from Wieland–Miescher ketone by ring expansion maneuver. This work is conceptually elegant as stereocontrol was self-regulated; only the *cis* bicyclic structure of the equilibrium mixture can undergo cyclization.

In terms of practicality, the reversible Michael reaction and the alternative pathways for retro-Michael reactions make the synthesis less than totally satisfactory. An improvement in the construction of a tricyclic skeleton of longifolene is by an intramolecular alkylation route [McMurry, 1972], the substrate of which was also secured from Wieland–Miescher ketone. This method predicates the formation of a *cis*-decalone for the cyclization followed by ring expansion.

The alkylation approach proved successful in the synthesis of sativene [McMurry, 1968]. The required intermediate is a *cis-α*-decalone having a tosyloxy group *cis* to the angular substituents, and in the adjacent carbon atom, an isopropyl chain in a *trans* relationship. This substitution pattern was conveniently established via hydroboration of a trisubstituted double bond; the borane approach was subject to steric control by the angular methine hydrogen.

Cyclobutane formation by intramolecular alkylation is made more difficult by the ring strain. However, an entry into the skeleton of α-copaene and α-ylangene [Heathcock, 1966] by such a process from a *cis-β*-decalone has been accomplished. The decalone was obtained from Wieland–Miescher ketone.

α-ylangene

Besides sectioning of the decalin system, annulation serves to augment the number of rings and the number of skeletal carbon atoms of Wieland–Miescher ketone. Many di- and triterpenes whose AB-ring portion is structurally akin to the drimane sesquiterpenes are profitably constructed from Wieland–Miescher ketone or its analogs.

The quassin cyclic array has an AB-ring segment that is also oxygenated at C-1 and C-7 (besides C-2), corresponding to those in Wieland–Miescher ketone. This pattern apparently influenced the development of a synthetic route involving a Diels–Alder cycloaddition [Grieco, 1980a; Vidari, 1984]. The C-7 oxygen was identified as an activator of the dienophile (Δ^8-double bond) and accordingly the double bond of Wieland–Miescher ketone must be moved, indirectly and stepwise, to the α',β'-position.

quassin

The method has also been extended to the synthesis of (−)-chaparrinone [Grieco, 1993] and (+)-picrasin [M. Kim, 1988]. Outlined below is another route to (+)-picrasin-B [M. Kim, 1988; Kawada, 1989] which consists of a different mode of C-ring formation.

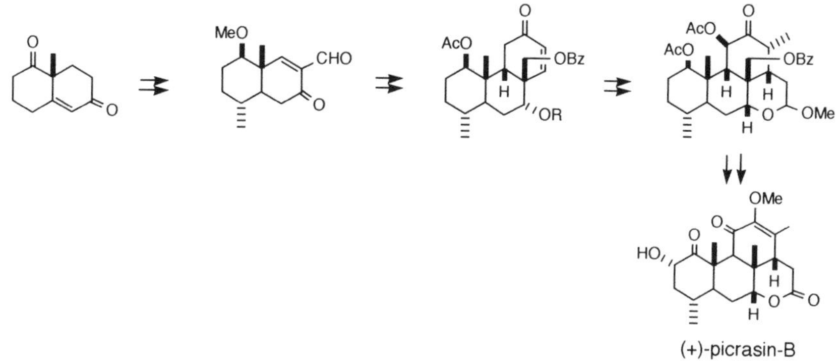

(+)-picrasin-B

Interestingly, in a synthesis of (+)-halenaquinol [N. Harada, 1988] a similar enone chromophone was created from the saturated ketone (with transposition) and used in another Diels–Alder reaction. Prior to this operation the original enone of Wieland–Miescher ketone was modified and extended into a dioxane ring which would become part of the oxofuran functionality. A slightly simpler compound, (+)-xestoquinone, has also been prepared [N. Harada, 1990].

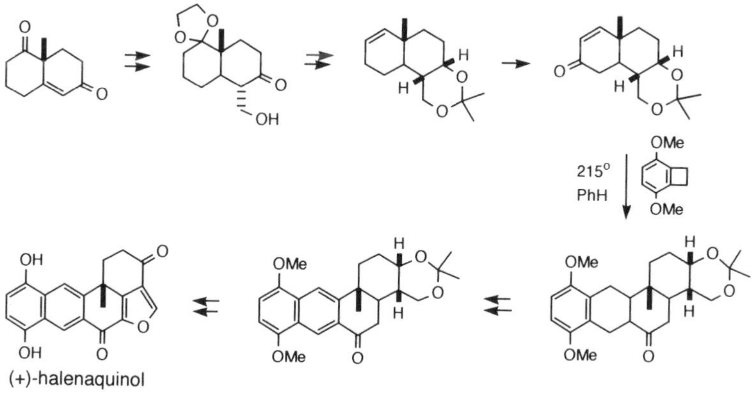

(+)-halenaquinol

In a yet to be completed synthesis of isoarborinol [Arseniyadis, 1991] a scheme outlined Wieland–Miescher ketone and its lower homolog as building blocks; Wieland–Miescher ketone was fashioned into a diene, and the lower homolog into a dienophile.

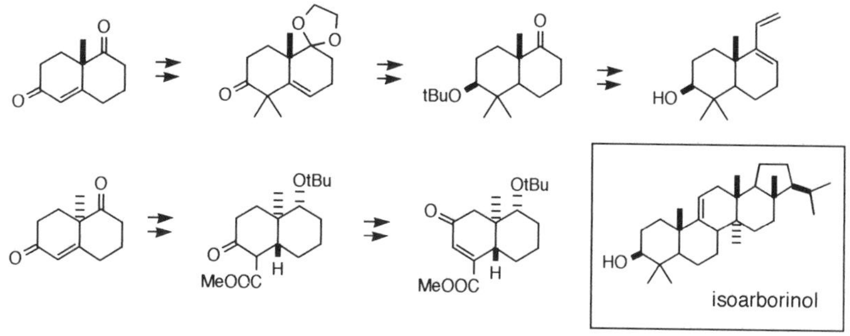

isoarborinol

Fundamentally, an approach to triptonide [Garver, 1982] was based on annulation of WM ketone and a series of oxidation reactions.

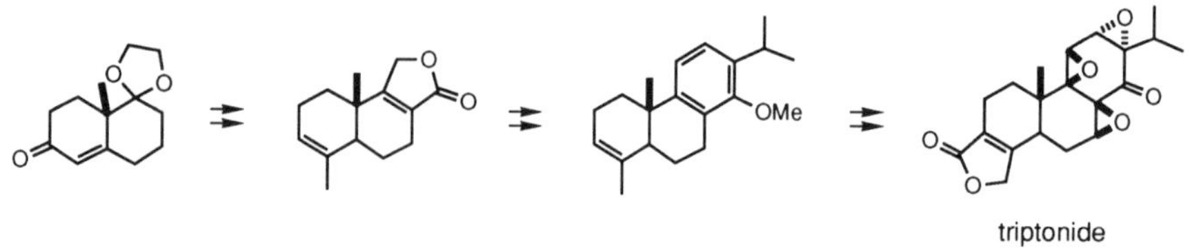

triptonide

Amarolide has been synthesized [Miyagi, 1984; Hirota, 1987] by a route that identified a methylated WM ketone as the BC-ring synthon.

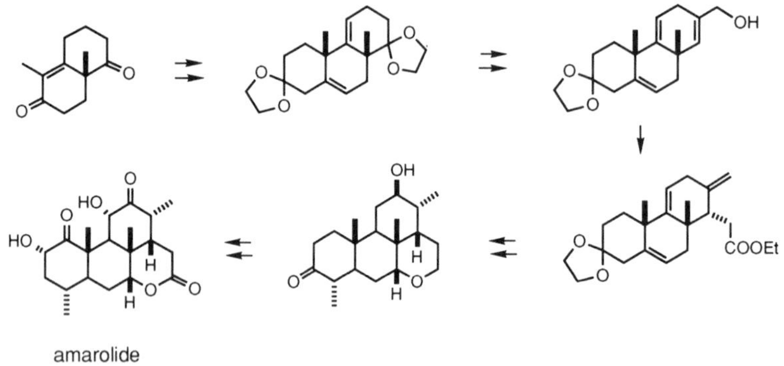

amarolide

For a synthesis of progesterone [Stork, 1967b, 1974] Wieland–Miescher ketone was used as a CD-ring synthon. Ring contraction was performed after annexation of the AB-ring portion. On the other hand, extension of a two-carbon unit from the saturated ketone, coupling with 2-methyl-1,3-cyclohexanedione and cyclization led to 5α-androst-8(14)-ene-3,17-dione [Ruppert, 1973].

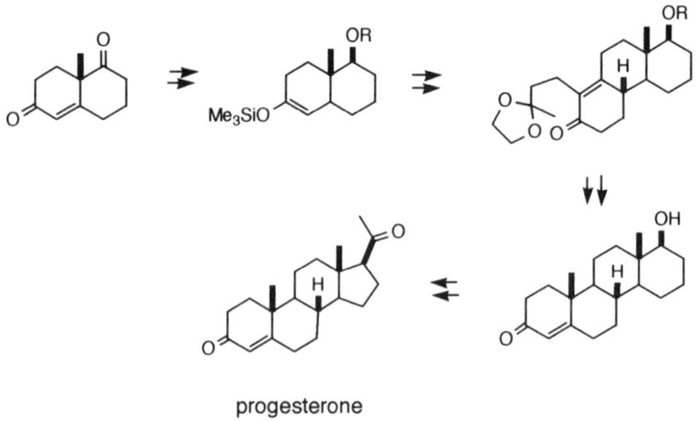

progesterone

The identification of Wieland–Miescher ketone as a structural unit in germanicol is not at all difficult. But the less obvious mapping of the bicyclic diketone with the BC-ring segment of the triterpene [Ireland, 1970] has a great advantage in that both ketone groups can be utilized to their fullest extent. Compare this to their use only as an activator for methylation, e.g., in mapping with the AB-ring.

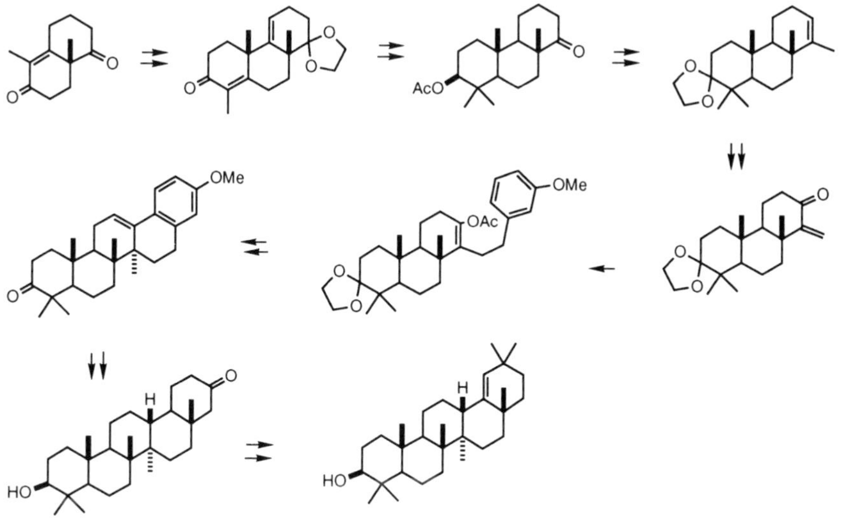

germanicol

The application of a Wieland–Miescher ketone analog in an elaboration of shionone [Ireland, 1975a, b] also shows masterly analysis of a topographic problem in synthesis. Thus the DC-ring of shionone finds a structural correspondence to the ketone.

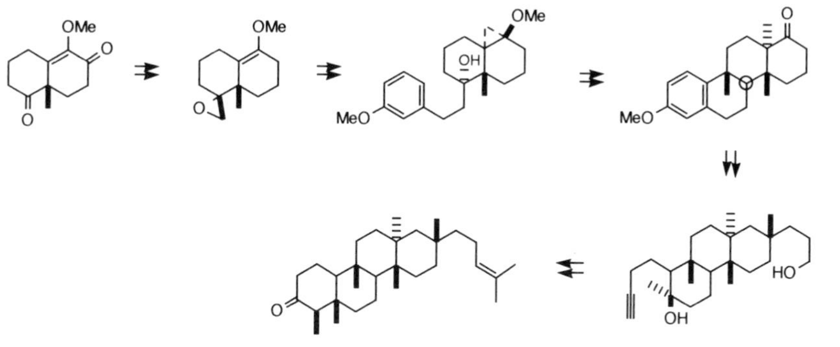

shionone

4-Methylated Wieland–Miescher ketone proved to be an excellent starting material for aphidicolin [McMurry, 1979; Trost, 1979b]. The 4α-hydroxymethyl group was introduced by reductive alkylation with formaldehyde, which also established the *trans* ring juncture.

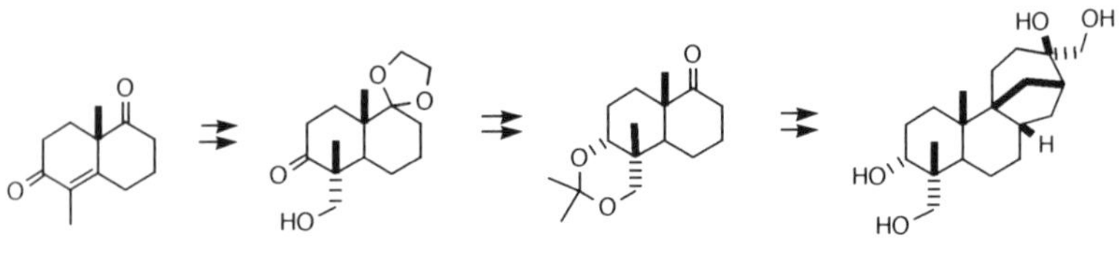

aphidicolin

The synthetic accomplishment of stemodin and maritimol [Piers, 1985a] was attended by a convergent evolvement of a keto ester from a mixture of two photocycloadducts. The intermediates are β-diketones which can undergo reversible Claisen cyclization–ring opening. The equilibration process is a case of mechanistic symmetrization–desymmetrization.

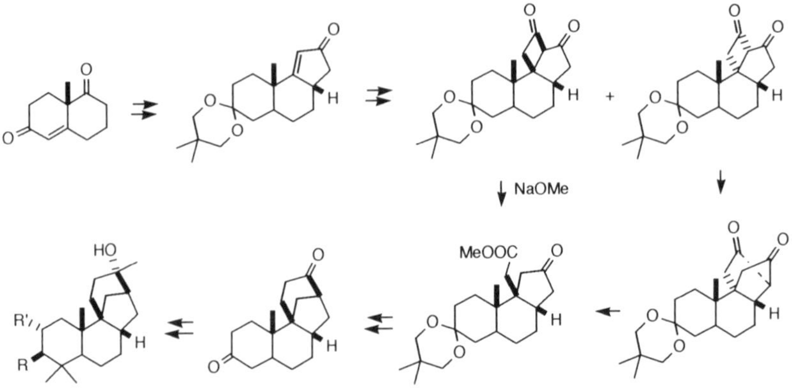

stemodin (R=H, R'=OH)
maritimol (R=OH, R'=H)

In a retrosynthetic perspective, removal of the C-ring of dolabradiene and Wittig transform at the vinylidene group would suggest the appropriateness of 4-methylated Wieland–Miescher ketone in the synthesis of the diterpene [Y. Kitahara, 1964], annulation onto the enone being so well precedented. Furthermore, a Robinson-type annulation would leave a ketone group in a position destined to bear a methyl group and a vinyl group, and this feature conduced to achieving the transformation.

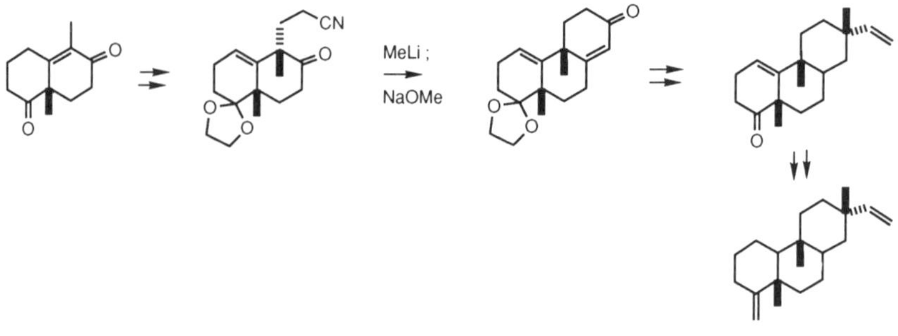

dolabradiene

A synthesis of methyl vinhaticoate [Spencer, 1971] from Wieland–Miescher ketone requires a routine substitution at C-4, a Robinson-type annulation from the saturated ketone (and the α-carbon), methylation at C-14, and annexation of a furan ring. To accomplish the last two tasks an isomer of the initially obtained tricyclic enone was elaborated; *vic*-difunctionalization of the conjugated double bond by addition of a methyl group and formylation solved the regiochemical problems.

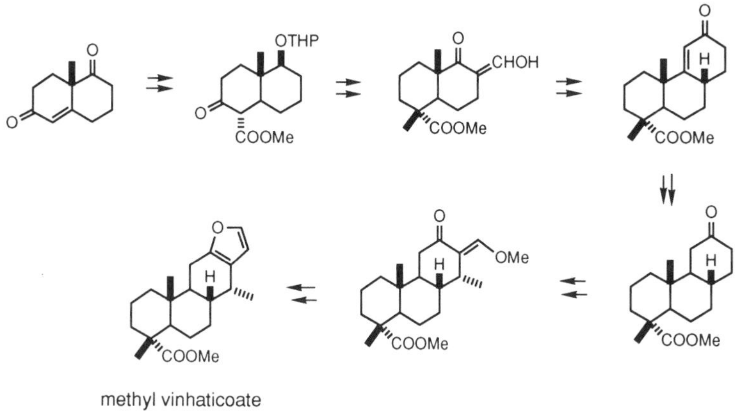

methyl vinhaticoate

Furannulation of a protected WM ketone at the enone site furnished a precursor of paspalicine [A. Ali, 1989; Guile, 1991]. A related compound is (−)-paspaline, which has also been synthesized [A.B. Smith, 1985].

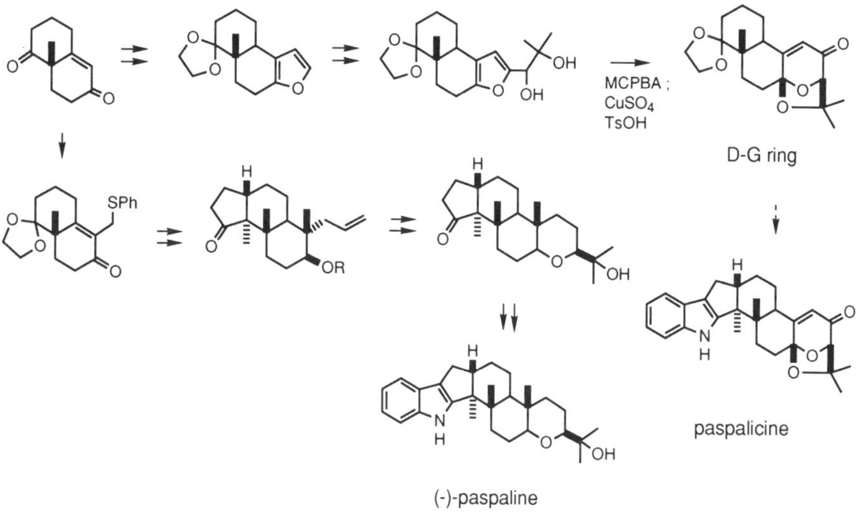

D-G ring

(-)-paspaline

paspalicine

An annulation which also involved skeletal rearrangement has been reported. The reaction lends itself to modification for synthesis of the taxane diterpenes

[Hayakawa, 1986]. The substrate for this tandem [2+2]-cycloaddition and [3.3]sigmatropic rearrangement was derived from Wieland–Miescher ketone. This result might have been unexpected; it is suspected that the researchers were aiming at construction of another ring system via a direct intramolecular [4+2]-cycloaddition.

taxusin

7.4. 1,4-CYCLOHEXANDIONES AND *p*-BENZOQUINONES

Despite the commonness and extensive literature concerning the application of *p*-benzoquinone and its symmetrical congeners, it is discussed along with 1,4-cyclohexanedione. Necessarily these examples are quite selective.

First, the S_2-symmetry of 2,2,5,5-tetramethyl-1,4-cyclohexanedione in connection with its suitability for a synthesis of *cis*-chrysanthemic acid [d'Angelo, 1983; Buisson, 1984] is considered. It is immediately apparent that, despite its desymmetrization by reduction of one of its carbonyl groups, there would be no adverse consequences to the plan involving Baeyer–Villiger reaction of the remaining ketone.

cis-chrysanthemic acid

From the synthetic viewpoint polyoxygenated cyclohexanes are logically approached from p-benzoquinone and appropriate derivatives. Many routes have been developed on this basis. Thus, in a synthesis of chaloxone [Fex, 1981] p-benzoquinone was converted into a C_2-symmetric dibromodihydroxycyclohexene on which desymmetrization reactions were performed.

chaloxone

In cases where a cyclic olefin is to be preserved the synthesis usually proceeds by proper protection in the form of a Diels–Alder adduct. For example. the 1:1-adduct of p-benzoquinone and a 9-alkoxyanthracene permitted the stereoselective reduction of the two ketone groups and dihydroxylation of the exposed double bond. Conduritol-A was released in the final step [Knapp, 1983].

conduritol-A

The 6,6-dimethylfulvene adduct of p-benzoquinone was used in a synthesis of phyllostine [Ichihara, 1976]. The purpose was to realize a hydroxymethylation on the enedione system. Note the symmetrical adduct was essential to the operation.

phyllostine

A reaction sequence leading to terrein [Klunder, 1981] involved ring contraction of the p-benzoquinone–cyclopentadiene adduct. After practically all the crucial functionalities were set in the correct positions a cyclopentene derivative was liberated by flash vacuum pyrolysis of the tricarbocyclic compound.

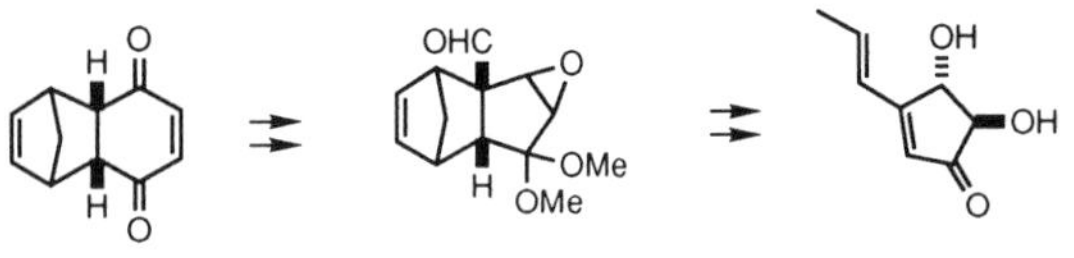

terrein

An interesting synthetic approach to pyrenolide-B [Asaoka, 1985] involved insertion of an oxa-alkylene segment between C-2 and C-3 of *p*-benzoquinone. The process was initiated on the cyclopentadiene adduct in which only one side of the molecule was exposed.

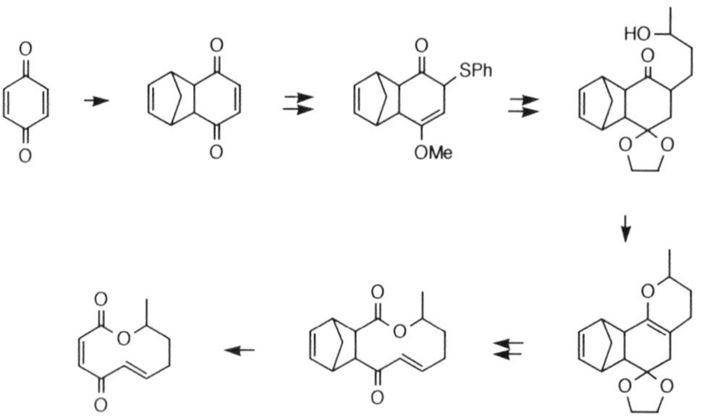

pyrenolide-B

p-Benzoquinone undergoes Diels–Alder reaction with certain styrenes. It is of interest that *p*-divinylbenzene formed a 1:2 adduct with *p*-benzoquinone, which has special chiroptical properties, the adduct being a member of the helicenes [L. Liu, 1990].

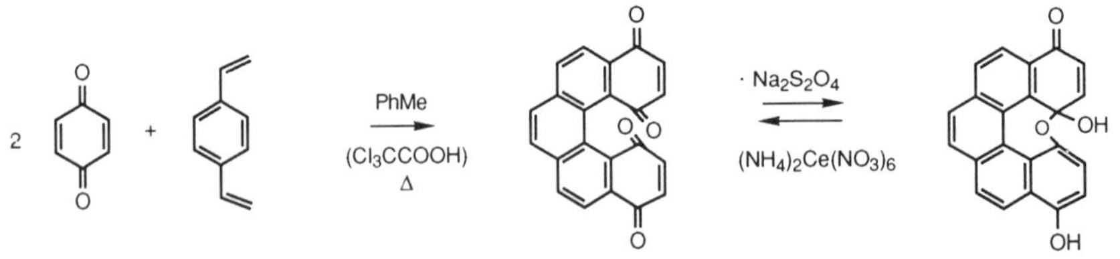

A very simple route to occidol [T.-L. Ho, 1973] exploited its pseudosymmetry in the sense that removal of the isopropylol group reduces the sesquiterpene to a symmetrical hydrocarbon. Thus the synthesis is simplified by employing a symmetrical precursor provided that a handle is retained for the attachment of the missing side chain. The symmetrical 1,4-dihydronaphthalene fulfilled this need.

occidol

Di-O-methylation of the tetrahydronaphthoquinone led to a dihydro-naphthalene. The latter compound, on isomerization of the double bond and C-formylation at the β-position of the resulting styrene, was transformed into a useful progenitor of the AB-ring segment of daunomycinone [Rama Rao, 1984]. It has been demonstrated that further reactions afforded an α-ketol which was converted into 4-demethoxydaunomycinone.

Ag$_2$CO$_3$-celite

R (-)

The hexalin nucleus of compactin is amenable to assembly from p-benzoquinone via its adduct with butadiene [Hsu, 1983]. Chemical reduction of the enedione double bond followed by microbial conversion of the dione by *Aureobasidium pullulans* led to (−)-diol accompanied by two other recyclable isomers. The (−)-diol has a C_2-symmetry and therefore any regiochemical problems associated with subsequent functionalization of the double bond were avoided.

(+)-compactin

Manipulation of a *cis*-octalindiol [Girotra, 1982, 1984] which culminated in compactin is noteworthy in view of the divergent routes developed from the two enantiomers of an enone. The σ-symmetric diol was obtained from lithium aluminum hydride reduction of the corresponding diketone.

(+)-compactin

The C_2-symmetric octalin diol has found an application in the synthesis of oppositol and prepinnaterpene [Fuzukawa, 1987]. Again it must be emphasized that the symmetry element facilitated the elaboration of the unsymmetrical five-membered ring. Subsequently, the participation of one of the hydroxy groups in lactone formation permitted modification at the other quarter of the molecule.

R = Me   oppositol

R = $\bigwedge\!\!\!\diagup\!\!=\!\!\big\langle$ prepinnaterpene

A novel approach to daunomycinone [Tamariz, 1984] involved the Diels–Alder reaction of p-benzoquinone and 2,3,5,6-tetramethylene-7-oxabicyclo[2.2.1]heptane. Reduction of the adduct to the hydroxy enone started a desymmetrizing maneuver by which the D-ring was evolved. The important consequence of obtaining the dihydronaphthalene, in which the styrenic double is closer to the methoxy group, as major product was that its condensation with methyl vinyl ketone furnished a 9:1 mixture of regiosomers in favor of the correctly substituted precursor of daunomycinone.

(7 : 3)

(9:1 regioisomers)

Two major contributions to organic synthesis pertain to the completion of yohimbine [van Tamelen, 1958] and reserpine [Woodward, 1958]. Interestingly, both syntheses were initiated by a Diels–Alder reaction of p-benzoquinone. In the former work, the adduct with butadiene was semihydrogenated (chemically with zinc/acetic acid), and homologated at one ketone by the Darzens reaction. The homologation process also epimerized the ring juncture. The corresponding acid which had the most stable configurations at three stereocenters was coupled to tryptamine to give an amide containing all (actually with one extra) skeletal carbon atoms of yohimbine. The octalone system in this amide was then cleaved at the double bond to release two aldehyde chains, one of which participated in a Pictet–Spengler cyclization immediately. The resulting product differs from yohimbine by being epimeric at C-3 and having an extra carbon unit in the lactol ring, which should be methyl ester. Degradation of the unwanted carbon unit did not present any problem, and in the fused lactol the ring juncture stereochemistry already coincided with the target molecule. In any event, the ester pendant of yohimbine is equatorial, so the correct configuration of C-16 would no doubt be reached. As for inversion of the configuration at C-3 the dehydro derivative was prepared and its hydrogenation led to the 3α-isomer.

yohimbine

reserpine

With five contiguous stereocenters in one cyclohexane ring, reserpine qualified as a particularly formidable synthetic target when its structure was elucidated. In the mid-1950s its successful synthesis, largely an exercise in stereochemistry, was probably beyond the capability of all but a few. How rapid progress has been!

This first synthesis of reserpine with exquisite stereocontrol employed the Diels–Alder reaction of *p*-benzoquinone with methyl 2,4-pentadienoate to established the CE-ring juncture of the alkaloid. Reduction and lactonization paved the way for dioxygenating the double bond which was at the *β,γ*-position of the carbonyl group, by enlisting the free hydroxyl as an internal director. An intricate yet elegant series of reactions was implemented to transform the enedioxy system into a transposed enone which was cleaved to give the projected aldehydo carboxylic acid intermediate. (The reader is urged to compare this enone generation process with the manner in which the CD-ring intermediate of cholesterol was made [Woodward, 1952]. Pay special attention to the oxygenation pattern of the precursors and note the effective use of the zinc dust reduction. The cholesterol synthesis also started with a Diels–Alder reaction, but since the benzoquinone dienophile was unsymmetrical this work is not discussed.)

The service of an ethyl congener of the Diels–Alder adduct in an approach to aklavinone [T. Li, 1981] should be noted. The enone derived from the adduct participated in an efficient annulation to afford the tetracyclic skeleton of the target compound.

aklavinone

Several synthetic approaches to ibogamine suffered from lack of control at the carbon atom bearing the ethyl chain. This perplexing problem was addressed in one solution [Sallay, 1967] by noting that all methine hydrogen atoms of the cyclohexane ring are *cis*, and three of them may originate from a Diels–Alder adduct of 1,3-hexadiene and *p*-benzoquinone. Comparison of ibogamine with this adduct indicates a requirement for reduction (double bonds and carbonyl groups), nitrogen atom insertion, and the addition of a methylene link, in addition to the attachment of the indole nucleus to the span of carbon atoms which was part of the quinone. In fact, only the conjugated double bond of the

Diels–Alder adduct did not serve a purpose in the synthesis. The less hindered ketone was protected and retained till the last stage, when it was used to achieve a Fischer indolization; the other ketone was converted into an oxime and its Beckmann rearrangement caused the required ring expansion/nitrogen atom insertion; and the isolated double bond was a conduit for functionalization and homologation at the carbon atom *para* to the nitrogen substituent so that the azabicyclo[2.2.2]octane skeleton could be erected. In essence, *p*-benzoquinone played an important role in this synthesis, and it was completely desymmetrized in the first step.

ibogamine

The identification of *p*-benzoquinone as the foundation of the C-ring in a cortisone synthesis [Sarett, 1952] was an important intellectual achievement, in view of the fact that functionalization at C-11 of the steroid skeleton is a difficult event. The preincorporation of an oxygen atom in that position greatly facilitated the synthetic approach. The construction of the B-ring by a Diels–Alder reaction also smoothed the pathway because in the tricyclic stage a selective manipulation on one of the two hydroxyl groups became feasible. Consequently, a hydroxy ketone was produced and elaboration of the D-ring could proceed.

cortisone

A long-standing problem in steroid synthesis (e.g., of estrone), based on the Diels–Alder reaction that unites the AB- and D-ring portions, had been associated with the orientation as well as the stereochemistry at the CD-ring juncture. A breakthrough [Dickinson, 1972] consisted of boron trifluoride catalysis. Furthermore, by using the symmetrical 2,6-dimethyl-*p*-benzoquinone the intermediates were more easily handled. Firstly the question of chemoselectivity for the cycloaddition step vanished; secondly the two carbonyl groups experienced different degrees of steric shielding so that the removal of the less hindered carbonyl was a relatively simple task; and thirdly the carbonyl group (at C-15) provided the opportunity for epimerization of the CD-ring juncture which was essential. Accordingly, a short synthesis of estrone methyl ether was developed.

(major)

estrone methyl ether

There is a slightly less obvious correlation between kempene-2 and the dimethyl quinone, however, a route to the tetracyclic diterpene from this dienophile has been delineated [Dauben, 1991]. It should be noted that once the *trans*-hydroxyoctalone was established, the two other equatorial carbon substituents in the saturated ring, a methyl group and a hydroxymethyl residue for later ring formation, were appended without any stereochemical issue. Tactically useful was the evolvement of the five-membered ring together with part of the cycloheptene constituent from a Diels–Alder adduct; the Diels–Alder reaction represented an excellent if not unique method for controlling the last two stereocenters of kempene-2.

kempene-2

Other work concerning skeletal construction by Diels–Alder reaction of 2,6-dimethyl-*p*-benzoquinone includes a potential intermediate of a forskolin [Bold, 1987]. Oxygenation of C-6 and C-9 and the presence of methyl substituents at C-8 and C-10 of forskolin point enticingly to the quinone and a 1-alkoxy-1,3-pentadiene as the proper addends.

(major regiomer)
(10 : 1.5)

forskolin

Structurally related to forskolin is erigerol. This compound has yielded to synthesis [Kienzle, 1988] by a method involving a Diels–Alder reaction of 3-ethoxypropenylidenecyclopropane with 2,6-dimethyl-*p*-benzoquinone. The cyclopropane ring of the adduct was the progenitor of the *gem*-dimethyl group.

erigerol

A rather unusual way of modifying 2,6-dimethyl-*p*-quinone resulted in a symmetrical spirocyclic ketone. The latter compound has been used, in the form of the morpholinoenamine, to condense with methyl 3,3-diemthycyclopropenoate, which resulted in a precursor of illudine-M [Franck-Neumann, 1989].

E = COOMe

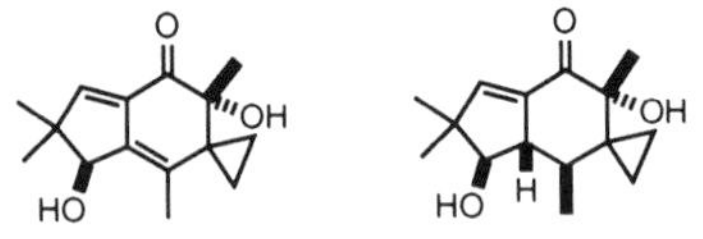

illudine-M

Bridged ring systems derived from *p*-benzoquinones and cyclic dienes are of synthetic value, as witnessed in the construction of trichodermol [Still, 1980b]. This synthesis featured a ring contraction and fragmentation, and its stereocontrol was implicitly laid in the inchoate stage of the Diels–Alder reaction. For example, the complete shielding of one face of a cyclopentenone unit in an intermediate was responsible for its conversion into an *exo*-1,3-diol.

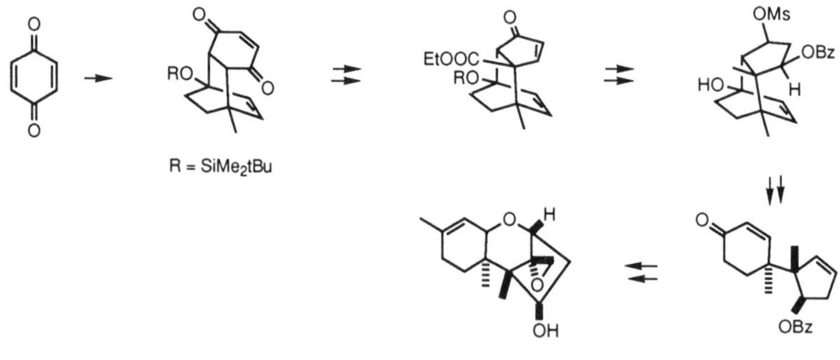

R = SiMe$_2$tBu

trichodermol

An excellent synthesis of quassin [Stojanac, 1979, 1991] from 2,6-dimethyl-*p*-benzoquinone is characterized by the intricate assembly of the hydrophenanthrene skeleton. Even after an identification of the origin of the C-ring from the quinone and a four-carbon unit from a diene, it is still not easy to unravel the plan which is shown.

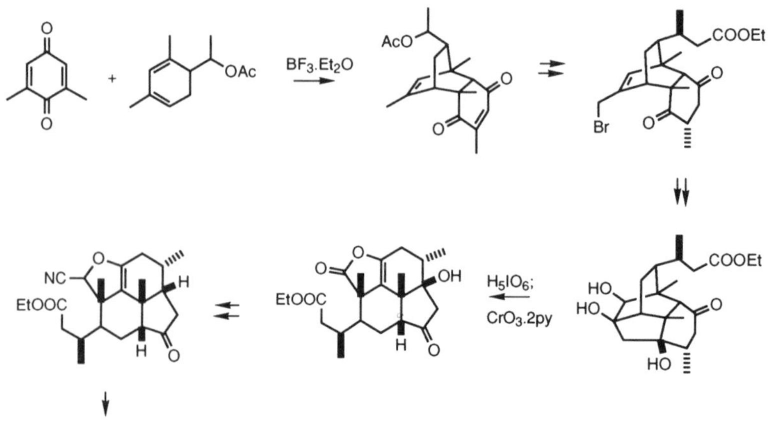

quassin

Retrosynthetically, a Baeyer–Villiger transform indicates a hydroacenaphthenone as a latent BCD-block. The relative stereochemistry at C-7 and C-10 (terpene numbering) is such that a gainful association of C-1 and the prolactonic carbonyl group is evident, as it solves the problem of stereocontrol at the two sites. For ready release of the hydroacenaphthenone its precursor should be equipped with a double bond or an α-glycol subunit. At this point the molecular framework is already suggestive of a Diels–Alder approach. When the methine hydrogen at the C/D-ring junction of quassin is replaced by a hydroxyl group the synthetic operation becomes realizable because a benzoquinone, specifically, 2,6-dimethyl-p-benzoquinone, is readily identified as the proper dienophile.

A synthesis of averufin [O'Malley, 1985] is of particular pertinence here. The synthesis consisted of bisannulation of 2,6-dichloro-p-benzoquinone to afford a symmetrical tetrahydroxyanthraquinone. Because of the relative reluctance of the *peri*-hydroxylated quinones to undergo alkylation in order to preserve the hydrogen bond, three of the hydroxyl groups of the anthraquinone could be selectively protected, leaving the fourth one to direct C—C bond formation at its *ortho* position. This reactivity desymmetrization due to subtle structural factors is very useful for synthetic purposes.

R = SiMe₃

R = H   averufin
R = OH   nidurufin

7.5. QUINONE ACETALS, QUINONEMETHIDES, AND QUINODIMETHANES

The monoacetals of *p*-benzoquinone are quite readily available. For example, reaction of hydroquinone monoalkyl ethers with thallium trinitrate in the presence of methanol gives rise to 4,4-dimethoxy-2,5-cyclohexadione, regardless of the original alkoxy group [McKillop, 1976]. Since these compounds possess two differentiated ketone functions, they show great promise in synthesis, and their applications are likely to expand in the future. Two variations on a synthesis of cherylline [Hart, 1978] are outlined.

cherylline

More significantly, *p*-benzoquinones can form cyanhydrin trimethylsilyl ethers with the more hindered carbonyl group. This chemoselectivity permitted a smooth access to an antibacterial metabolite of marine sponges, 4-carbethoxymethyl-4-hydroxy-2,6-dibromo-2,5-cyclohexadienone [Evans, 1977b].

While the synthetic potential of such quinone acetals has yet to be fully explored, it should be noted that fused anisole and benzonitrile derivatives are readily obtained from them [Evans, 1977a] by a cyclization process involving bond formation at the *meta* position.

In the cherylline syntheses above the reaction intermediates were protected quinonemethides. The generation and interception of quinonemethides as a central strategy for access to certain aromatic compounds have been under scrutiny. For example, in an approach to podophyllotoxin [Kende, 1977] the tetralin system was created by such a process.

podophyllotoxin   picropodophyllone

Silver oxide oxidation is a good method for the generation of p-quinonemethides [Angle, 1989a]. Suitable intramolecular interceptors include aromatic rings, allylsilanes, and β-keto esters [Angle, 1989b]. p-Quinonemethide formation in the presence of styrenes gives rise to indan derivatives [Angle, 1990]. In this last case there is desymmetrization of the phenol ring.

It is highly relevant to mention the advantage of passage through a symmetrical 4,4-disubstituted 2,5-cyclohexadienone during synthesis of crinine [S.F. Martin, 1987] in view of the 1,3-relationship of the hydroxyl group and the amino nitrogen atom. Correlation of the alkaloid with the proper cyclohexadienone precursor is easily made though oxidation and a Michael transform.

crinine

Comparison of the method with an oxidative coupling of the two aromatic rings to generate a cyclohexadienone, as shown in a route to oxocrinine [Kupchan, 1978], should be made.

oxocrinine

o-Quinodimethanes have been exploited as synthetic intermediates in a large body of work. Particularly significant are steroid syntheses based on trapping of such species by intramolecular Diels–Alder reactions. With respect to symmetrical *o*-quinodimethanes, they are considered to be the diene components in the synthesis of occidol [T.-L. Ho, 1972] and cordiachrome-B [Watabe, 1987]. In the latter process, epimerization of tricyclic product at the ring juncture was a serious problem.

cordiachrome-B

7.6. BENZENE DERIVATIVES

7.6.1. Substitutions

Symmetrical benzene derivatives are sources of many other symmetrical and unsymmetrical substances in which the original structure subunits are preserved or modified. In some cases the aromatic ring is transformed into a cyclohexane, cyclohexene, cyclohexadiene, or seco state. To adopt such a way of gaining access to a certain compound usually has a reason, most likely it points to the availability of the precursor.

Phenol undergoes an extensive change in structure when reacted with sodium hypochlorite. The intervention of a Favorskii rearrangement is indicated. As the constitution of the dihydroxycyclopentenecarboxylic acid product is easily correlated with the prostaglandins, it is appealing to develop a practicable method to convert the acid into a system of the biologically important substances [Gill, 1979].

As the synthesis of grifolin is concerned, a most direct and convergent route would be that involving geranylation of orcinol [S. Yamada, 1978].

grifolin

In principle, a synthesis of oxogambirtannine [Merlini, 1967] consisted of a coupling between tryptamine and 2,6-dicarboxyphenylacetic acid and then an esterification. No other manipulation was needed. This simple and direct way took full advantage of the hidden molecular symmetry, which greatly simplified the synthetic pathway.

oxogambirtannine

Employment of 2-methoxyisophthalic acid in a synthesis of naucleficine [Naito, 1986] via the enamide photocyclization methodology meant that the methoxy group was sacrificed.

n/aucleficine   58%   19%

For constructing maturone, the possibility of using a 5-methyl-2-naphthol, in which C-3 was further functionalized, became quite appealing, as an o-naphthoquinone and thence a 2-oxy-1,4-naphthoquinone may be obtained by oxidation, and the furan ring may be formed with the participation of the C-3 functionality [Ghera, 1986]. Consequently, a 2-bromomethyl-3-methylbenzyl phenyl sulfone was prepared from methyl 2,6-dimethylbenzoate. Its condensation with a butyrolactone led to a tetralin which already possessed elements of the furan ring. Aromatization and oxidation according to plan indeed gave a precursor of maturone. In retrospect, employment of the symmetrical benzoic ester to form the annulating component was a wise choice.

maturone

Another case featuring a symmetrical benzene derivative to start a synthesis dealt with chuangxinmycin [Kozikowski, 1982]. In this substance the benzene moiety of the indole nucleus is trisubstituted, but it hardly suggests a symmetrical precursor. Instead, an obvious intermediate would be a 1,2,3-trisubstituted benzene with three different pedigrees, perhaps one of them a nitro group. However, it is also obvious that such a compound would be difficult to acquire by way of nitration, for example. It is thus very gratifying that the diethylacetal of the symmetrical 2,6-dinitrophylacetaldehyde underwent displacement on treatment with methyl thioacetate in the presence of lithium hydroxide. At this stage the indole ring was formed and the missing substituent introduced to effect the closure of the thiopyran ring.

chuangxinmycin

Annulation onto a symmetrical benzene ring is necessarily desymmetrizing. Thus this very common practice during elaboration of many polycyclic aromatic substances, notably the polyketide natural products, embodies such a change in structural characteristics. However, some simple examples are those involving cyclization of *p*-disubstituted benzenes.

Several routes to ferruginol have adopted this strategy but with different preparations of the cyclization substrates. A tertiary alcohol obtained from the *p*-anisylacetylide addition to 2,2,6-trimethylcyclohexanone followed by hydrogenation [F.E. King, 1957] was treated with polyphosphoric acid to give the tricyclic intermediate. Formation of two rings in one step from a monocyclic substrate was also investigated [Fetizon, 1960]. Alternatively, a reaction of *p*-methoxyphenylacetyl chloride with 1,3,3-trimethylcyclohexene in the presence of aluminum chloride led to the tricyclic ketone directly [Wolinsky, 1972]. The last method appears to be superior in view of the possibility of regulating the ring juncture stereochemistry.

ferruginol

ferruginol

The strategy of desymmetrizing annulation of an aromatic ring, leading to only one polycyclic product, has been applied to an assembly of (−)-yohimbone [Kametani, 1982b] and yohimbinone [Miyata 1983]. This approach is convergent and concise.

(-)-yohimbone

Eseroline, a precursor of physostigmine, has been acquired from a route starting from hydroquinone dimethyl ether [Harley-Mason, 1954]. The bipyrrolidine moiety was annexed to be the benzene ring after a proper side chain was introduced to an *o*-position of the aromatic ether. The promptness of this synthesis is related to the symmetry of the aromatic ether.

eseroline

Certain compounds embody symmetrical aromatic units, therefore the most convenient way to synthesize them involves elaboration of the proper aromatic compounds. Examples are hypolepin-B [Y. Hayashi, 1972] and pterosin-E [Nambudiry, 1974].

hypolepin-B

pterosin-E

In an abandoned route to olivin [Hatch, 1978] the tricyclic framework was actually prepared quite readily. 3,5-Dimethoxybenzyl chloride was converted into a glutaronitrile. Then using an intramolecular Friedel–Crafts acylation, the glutaronitrile was totally desymmetrized with respect to the aromatic ring and the other portion of the molecule.

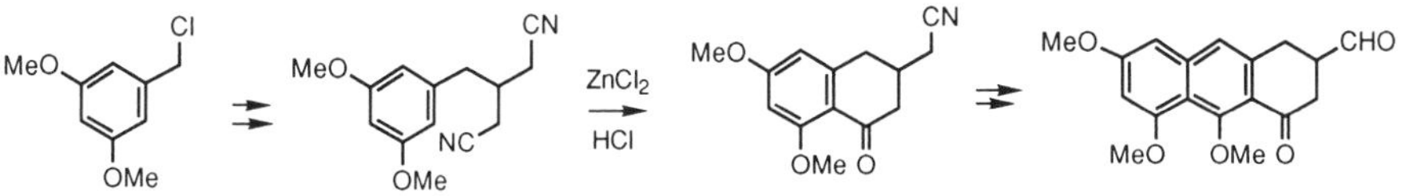

Tri-*O*-methylolivin was synthesized [Dodd, 1984] by different approach—conversion of an aromatic ring into a bisacetal of a 5-substituted 1,3-cyclohexanedione. Annulative formation of the naphthol section was then performed by condensation with methyl orsellinate dimethyl ester.

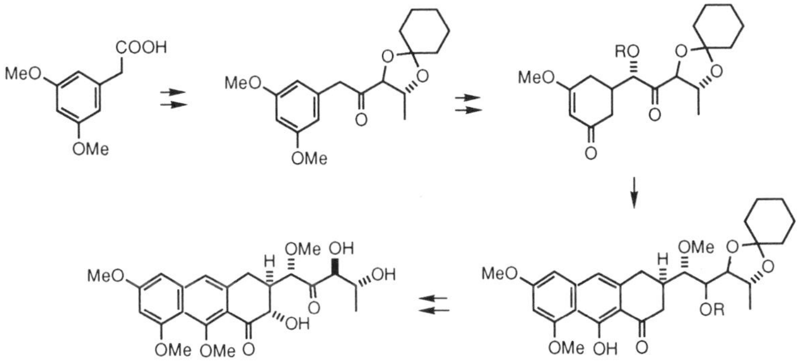

tri-*O*-methylolivin

Several synthetic approaches to curvularin are known. Generally, macrocyclization is delayed to the last step and the two popular and conventional ways to achieve it are lactonization and intramolecular acylation. Implementation of the desymmetrizing operations took place earlier in lactonization [Wasserman, 1981], whereas the aromatic portion may maintain a locally symmetrical state until the Friedel–Crafts cyclization [P.M. Baker, 1967; T. Takahashi, 1980a].

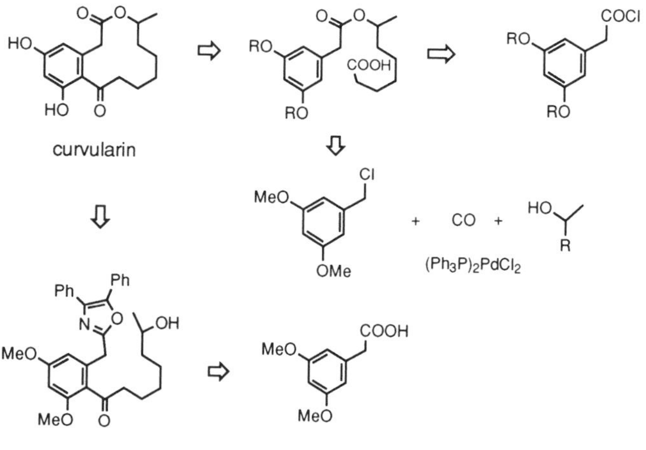

curvularin

A very early desymmetrization of the aromatic ring was implemented in a synthesis of zearalenone [T. Takahahi, 1980b]. Methyl 3,5-dimethoxybenzoate was converted into 3,5-dimethoxy-2-iodobenzyl phenyl sulfide, which was used to build an ester containing all the carbon and oxygen residues of the target compound.

zearalenone

Δ^1-Tetrahydrocannabinol is conveniently accessible from reactions of olivetol with various monoterpenes such as verbenol [Mechoulam, 1967], (+)-chrysanthenol [Razdan, 1975], and p-mentha-2,8-dien-1-ol [Petrzilka, 1967]. These syntheses invariably involved C-alkylation of olivetol to give a menthenyl cation intermediate which underwent O—C bond formation.

Δ^1-tetrahydrocannabinol

verbenol   (+)-chrysanthenol   p-mentha-2,8-dien-1-ol

Perezone is a sesquiterpene in which the quinone ring is unsymmetrically substituted. However, it has been synthesized from orcinol dimethyl ether via alkylation [Cortes, 1985]. The final step consisted of oxidation, and because oxygenation occurred at a p-position of an existing methoxy group, this desymmetrizing reaction led to only one product, despite the presence of two methoxy substituents. The consequence of such desymmetrization on the system is worthy of attention.

perezone

(Of some relevance to this desymmetrization leading to a trioxygenated benzene ring is a synthesis of dehydroneotenone [Weber-Schilling, 1969] in which the key step was deoxygenation condensation. The assembly from a benzopyranone and 4,5-dimethoxy-*o*-benzoquinone was via a Michael reaction, elimination, and tautomerization sequence. Of course the desymmetrization of the *o*-quinone occurred in the coupling step.)

dehydroneotenone

The Pechmann reaction of a symmetrical phenol is a desymmetrizing process and is valuable for the synthesis of naturally occurring coumarins. Interestingly, mammea-B/BB could be obtained from a trisubstituted coumarin prepared by further acylation and alkylation in a regiocontrolled manner [Crombie, 1985].

mammea B/BB

In a synthesis of aflatoxin-B$_1$ [Büchi, 1967] the Pechmann reaction product from phloroglucinol and ethyl acetoacetate was selectively methylated at the 7-hydroxyl group, an extremely favorable feature which was made possible by the preferential formation of the monoanion. A furo[2,3-*b*]benzofuran was then assembled via a β-acyl lactone rearrangement step, and the free hydroxyl group (actually regenerated from a temporarily masked species) was employed to construct the missing unit. It is interesting to note that this unit is also part of a coumarin system, however, the Pechmann reaction must be conducted carefully so as to avoid rupture of the terminal ring of the furobenzofuran unit and unwanted side reactions.

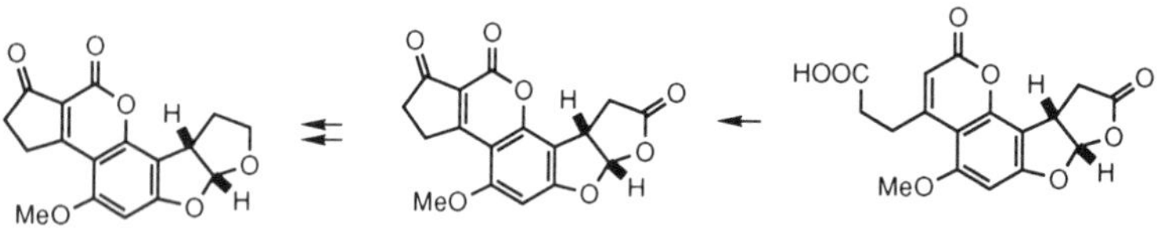

aflatoxin-B$_1$

The benzocyclobutene natural products, scillascillin and muscomsin, have been assembled from a 1-cyanobenzocyclobutene and phloroglucinol by two different ways. In both approaches the phoroglucinol moiety became unsymmetrical only after the formation of the pyranone ring. Thus scillascillin was derived from a phenone [Rawal, 1983], whereas muscomosin was obtained by an intramolecular Friedel–Crafts acylation followed by release of the hydroxyl groups from their protected form [T. Honda, 1992].

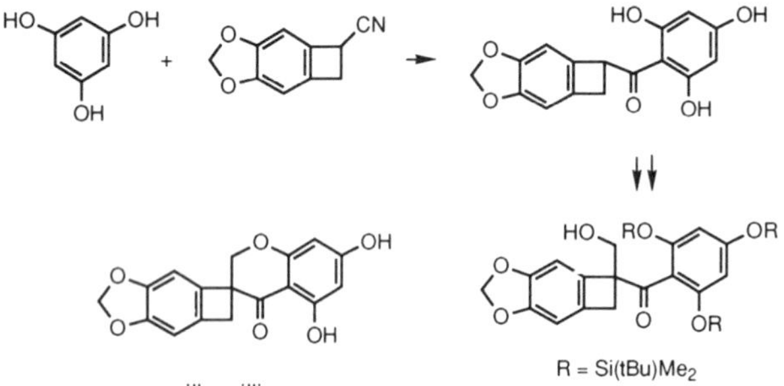

scillascillin

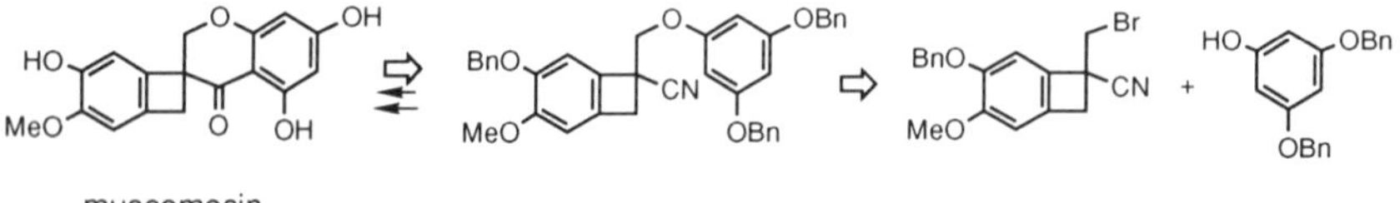

R = Si(tBu)Me$_2$

muscomosin

The adequacy of a coumaran-3-one to the synthesis of rocaglamide is readily recognizable. How to elaborate the heavily substituted cyclopentane unit is another matter to consider. Indeed, 2-(p-anisyl)-4,6-dimethoxycoumaran-3-one was prepared from phloroglucinol and processed into rocaglamide [Davey, 1991].

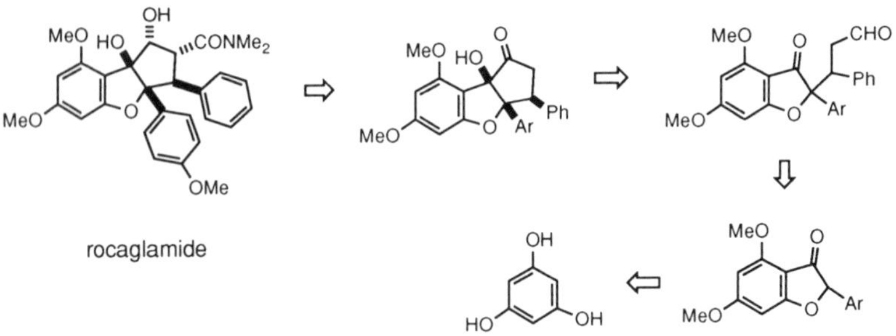

rocaglamide

The clear separation of a coumaran-3-one from a spiroannulated cyclohexenone in griseofulvin offers a strong allurement to synthetic design based on the combination of the two proper building blocks. While the candidate for the aromatic moiety is unmistakable, its double Michael reaction with a propenyl alkynyl ketone [Stork, 1964] is surprisingly effective and concise. The selective formation of griseofulvin in contrast to epigriseofulvin was attributed to kinetic control of the cyclization in which the transition state featured a better overlap of the donor enolate and the enone acceptor.

griseofulvin

A more recent synthesis of (+)-griseofulvin [Pirrung, 1991] in which a 2,2-disubstituted coumaran-3-one was formed by an intramolecular carbenoid interception[2.3]sigmatropic rearrangement tandem offers an interesting comparison with respect to methodology development. The substrate for this latter synthesis was also acquired from the symmetrical phloroglucinol dimethyl ether.

(+)-griseofulvin

A "cyclic exchange" process was witnessed in a synthesis of acronycine [Beck, 1968]. The formation of the acridone was accompanied by the cleavage of the tetrahydroquinone. The first tetrahydroquinoline intermediate was obtained from an intramolecular alkylation of a symmetrical precursor.

acronycine

Isorobustin has been synthesized from phloroglucinol monomethyl ether [Barton, 1990]. The *C*-acetylation started to desymmetrize the substitution pattern of the benzene ring.

isorobustin

A pathway to ficine [Anjaneyulu, 1969] from phloroglucinol trimethyl ether involved introduction of the pyrrolidine ring via a Friedel–Crafts reaction which produced an imine. Construction of the flavone ring system was rather routine.

ficine

Condensation of the dimethyl ether of phloroglucinol with *o*-nitrobenzoyl chloride gave rise to a symmetrical benzophenone from which a 2,2-dimethylchromene could be formed readily [Adams, 1981]. The use of this intermediate for synthesis of acronycine required a formal displacement of the oxygen function at C-5 of the chromene, but apparently this direction was difficult to control and the linear isomer was also produced.

acronycine

The reaction of phloroglucinol with isatin in the presence of alkali also served to desymmetrize the former compound. An acridine derivative was formed and it served as an intermediate for synthesis of acronycine [Hlubecek, 1970].

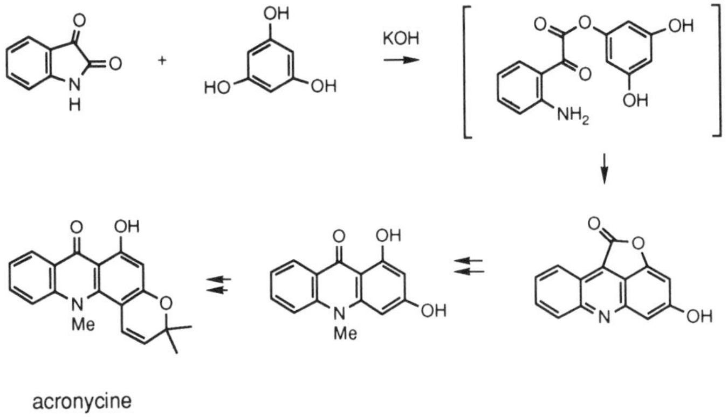

acronycine

Barakol is a tautomer of (2-acetyacetyl-3,5-dihydroxy)benzyl methyl ketone. Consequently, the symmetrical 3,5-dihydroxybenzyl methyl ketone was identified as a useful synthetic precursor [Bycroft, 1970].

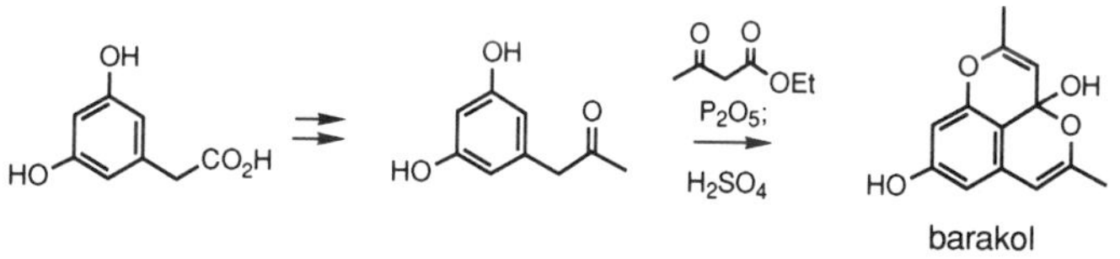

barakol

An intramolecular C-formylation delivered pulvilloric acid [Bullimore, 1966] from a symmetrically substituted benzoic acid. The synthetic pathway traversed a 3,5-dimethoxybenzyl ketone.

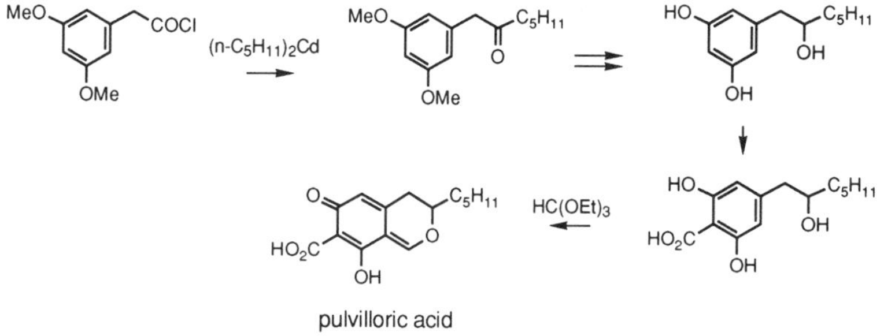

pulvilloric acid

A slightly more complex molecule, sclerotiorin, has been similarly assembled [Chong, 1969]. An approach to emodin and physcion [Hassall, 1970] is particularly significant in the context of this book. The two building blocks were symmetrical, but the Friedel–Crafts reaction led to desymmetrization of the donor moiety. A 9-cyano-10-hydroxyanthracene was then established, on oxidation it extruded the cyanide ion to afford the anthraquinone.

emodin   physcion

Chirality transfer from center to axis was featured in a synthesis of (−)-ancistrocladine [Bringmann, 1986]. The chiral tetrahydroisoquinoline and the naphthalene portions were linked by an ester bond and a Heck-type coupling was effected intramolecularly. The tetrahydroisoquinoline unit was readily prepared from 3,5-dimethoxybenzyl acetone.

(-)-ancistrocladine

Several known synthetic approaches to colchicine take advantage of the local symmetry about the trimethoxylated benzene ring. The C—C bond connecting it to the tropolone ether is strategic, and its formation does not cause regiochemical problems. The biogenetically patterned scheme, which avoided de novo formation of the tropolone ring, employed either purpurogallin as its source [A.I. Scott, 1965] or a source prepared from solvoytic fragmentation of an adduct of the substituted cyclopentadiene and dichloroketene [Kato, 1974]. The oxidative coupling of a symmetrical pyrogallol derivative resulted in desacetamidocolchiceine. Thus desymmetrization occurred in the cyclization step. Note that steric effects dictated bond formation with a less hindered site of the tropolone ring.

FeCl$_3$, H$_2$SO$_4$
CHCl$_3$, EtOH, H$_2$O

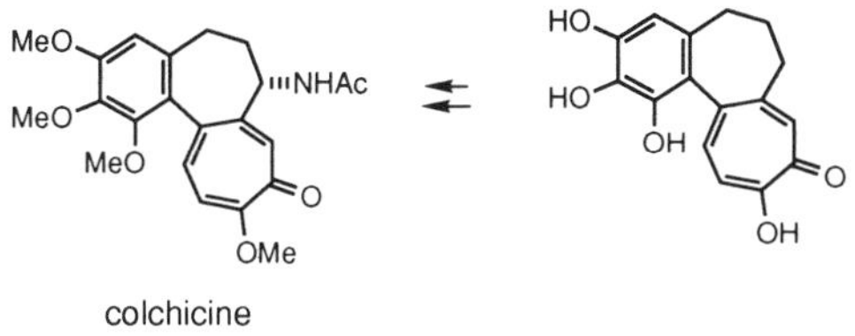

colchicine

In another route to colchicine [Evans, 1981] the B-ring was closed in conjunction with the evolement of the C-ring by a ring expansion rearrangement. Besides mechanistic intricacy the cyclization was efficient and it enjoyed the symmetrical consequences of having the aromatic molecule as its precursor.

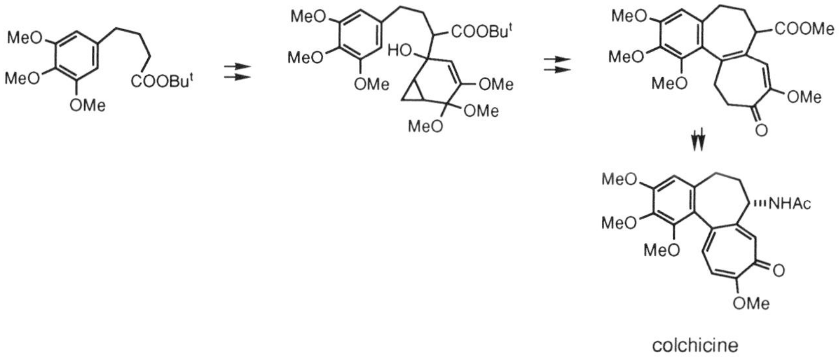

colchicine

Alternatively, methyl 3-(3,4,5-trimethoxyphenyl)propanoate was chloromethylated and then converted into a Wittig reagent which was used to react with ethyl 2-oxo-2-(2-vinylcyclopropyl)acetate [Wenkert, 1989]. Cope rearrangement of the Wittig reaction product and Dieckmann condensation led to the tricyclic skeleton of colchicine. In this case the desymmetrization step occurred much earlier.

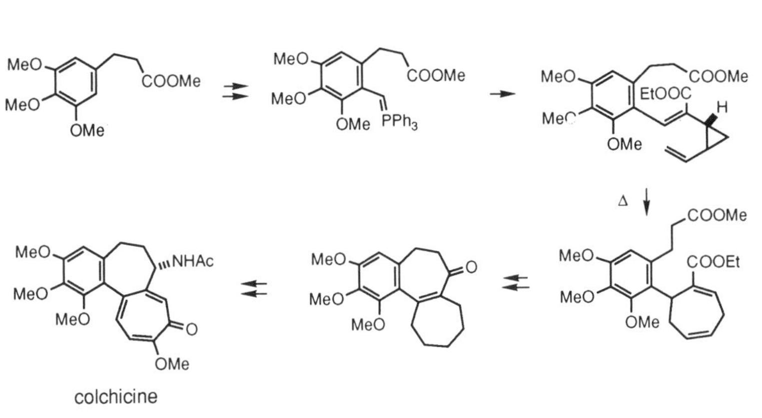

colchicine

A synthesis of staganacin [Kende, 1976] involved 3,4,5-trimethoxybenzyl bromide as alkylating agent to assemble a diarylbutane which was submitted to aryl–aryl coupling.

steganacin

The A-ring of imenine is fully substituted, and its synthesis from 3,4,5-trimethoxybenzaldehyde is a distinct possibility, in view of the existence of many methods for the construction of the isoquinoline nucleus and functionalization of C-1 of isoquinolines. Furthermore, the tetracyclic system typified by imenine is attainable from a 1-benzylisoquinoline. The fact that a synthesis of imenine [Cava, 1973] was accomplished in a straightforward manner was partially a consequence of the symmetry properties of the various monocyclic precursors.

imenine

Intramolecular alkylation occurs readily when an appropriate electrophilic chain is present in a phenol. A diazoketone fulfills the criterion, and the intrinsic directing effect of the phenolic hydroxyl group suggests the formation of spirocycles from the *p*-substituted phenols. For a synthesis of solavetivone [Iwata, 1981] a spiroannulated symmetrical cyclohexadienone was prepared, desymmetrization was first executed on the cyclopentanone moiety which enabled stereoselective reduction of one of the two double bonds. The successful realization of this plan affirms the notion that symmetry considerations are of considerable benefit in solving synthetic problems.

solavetivone

It is instructive to compare this with the reaction strategy for initiating the desymmetrization operation by a Birch reduction [Murai, 1981a].

β-vetivone   solavetione

α-Cedrene synthesis from a spirocyclic intermediate [Corey, 1969a; Crandall, 1969] entailed a convenient access of the latter compound. An intramolecular alkylation of a 4-substituted phenol met the essential requirement.

α-cedrene

The spirocyclic sesquiterpene anhydro-β-rotunol has been synthesized from 2,6-dimethyl-4-hydroxybenzyl alcohol [J.N. Max, 1987].

anhydro-β-rotunol

The symmetrical 2,4,6-trimethylphenol can be urged to undergo allylation at C-4. The cyclohexadienone thus obtained was identified as a building block for erythronolide-B [Corey, 1978] in view of its ready transformation into a "slightly" unsymmetrical hexasubstituted cyclohexanone. The asymmetry induced by the presence of a carbon chain at one of the α-carbon atoms of the ketone was crucial to the introduction of an oxygen functionality to C-6 (erythronolide numbering) and the unveilment of the macrolactone carbonyl group by a Baeyer–Villiger reaction.

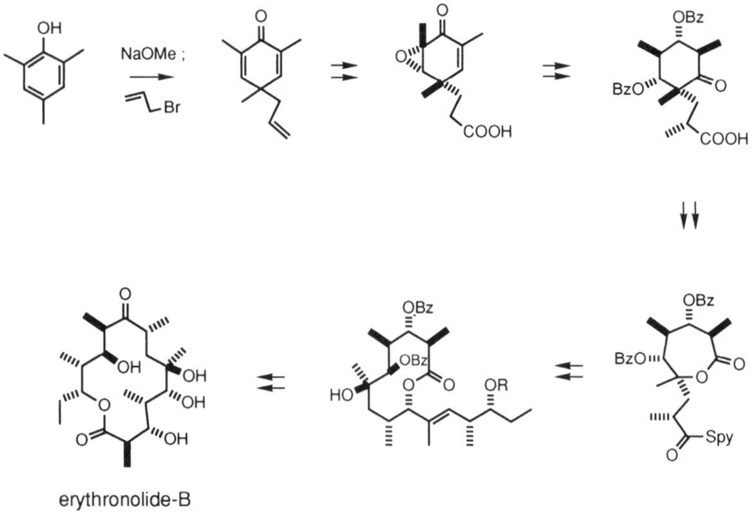

erythronolide-B

C-Alkylation of 2,6-dimethylphenol with simple electrophiles is exempted from regiochemical problems. In light of this consequence it is of interest to correlate seychellene with the alkylation product from 5-bromo-3-methyl-1,3-pentadiene by a Diels–Alder transform. In seychellene none of the cyclohexane moieties of the bicyclo[2.2.2]octane system has a 1,3-dimethyl pattern! However, a correlation is possible in a homotwistane or an *endo*-2,6-trimethylenated bicyclo[2.2.2]octane. A homotwistadienone has indeed proved to be a useful intermediate [Frater, 1974].

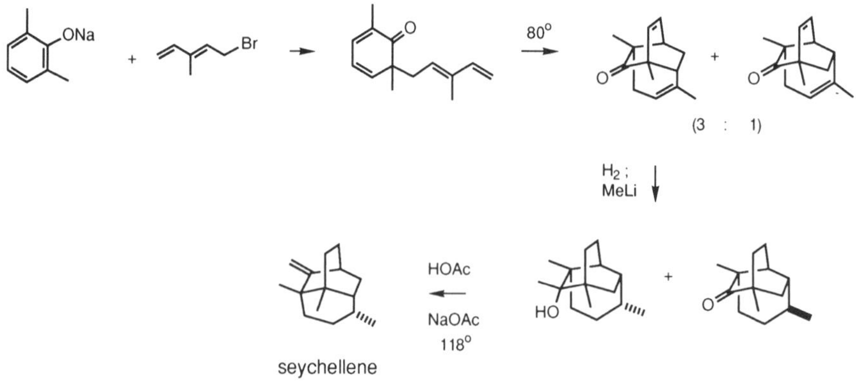

seychellene

An earlier synthesis of patchouli alcohol [Danishefsky, 1968] featured an intermolecular Diels–Alder reaction in which the diene was prepared from *o*-methylation of 2,6-dimethylphenol.

patchouli
alcohol

The unsaturated tricyclic ketone isomer was not applicable because one of its methyl groups was wrongly positioned.

For a study of the linear oxygen inversion barrier in aryl methyl ethers, 9,18-dimethoxydinaphtho[*a,j*]anthracene was synthesized [Gupta, 1991]. The twofold annexation of hydroquinone dimethyl ether to create this U-shaped molecule employed the same reaction sequence for each side, owing to the symmetry of the molecule. However, the initiation of the second series of reactions by mercuration was completely nonregioselective, also due to local symmetry with respect to both structure and reactivity, and the consequence was the generation of the Z-shaped isomer.

Although biphenomycin-B is an unsymmetrical molecule, its synthesis can take advantage of the fact that the aromatic segment is symmetrical. Of course it is crucial to be able to differentiate the two carbon chains of such a symmetrical intermediate. This strategy has been followed to completion in one case [Schmidt, 1991] and is being continued in another [Carlström, 1991].

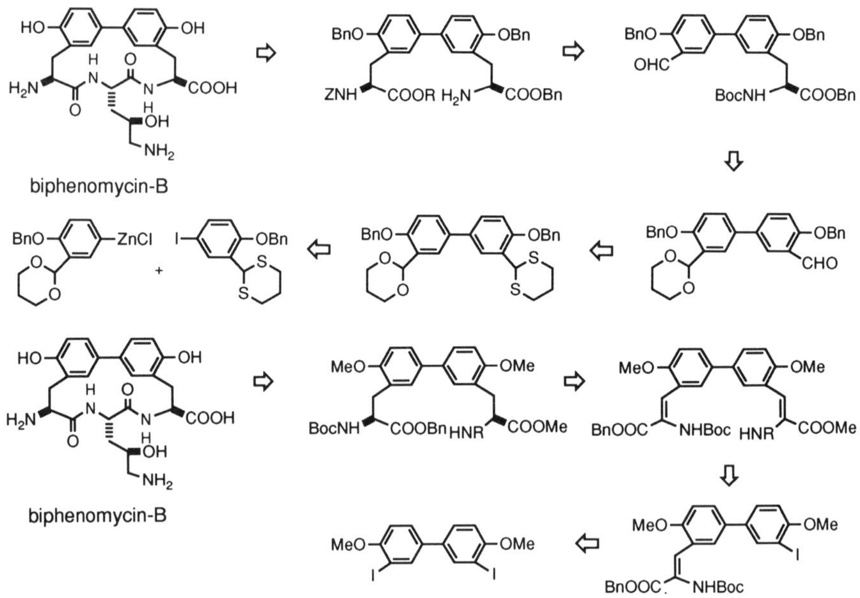

biphenomycin-B

biphenomycin-B

A method for the introduction of substituents into biphenyl is via the bis(tricarbonylchromium) complex [Yang, 1991] by reaction of the derived dianion with an electrophile, then a nucleophile, and followed by oxidative decomplexation. Apparently the product is rendered unsymmetrical.

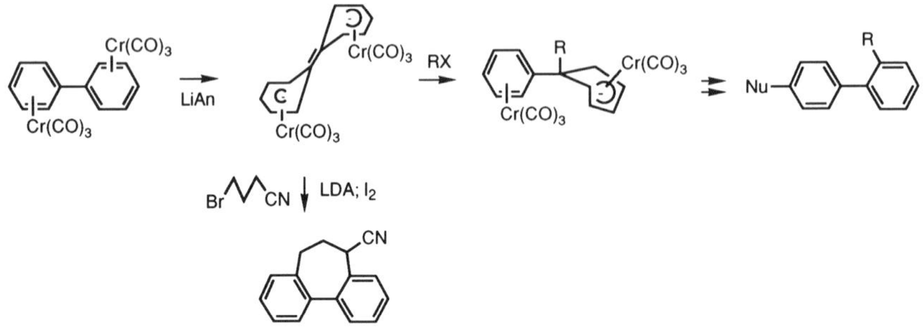

Erythrinadienone has been prepared by an intramolecular oxidative phenol coupling [Kametani, 1972] of a bisphenethylamine. The two identical aromatic rings were not symmetrically substituted, but the coupling and oxidation produced a symmetrical 4,4′-diphenoquinone. The molecular symmetry was destroyed upon N—C bond formation (Michael reaction).

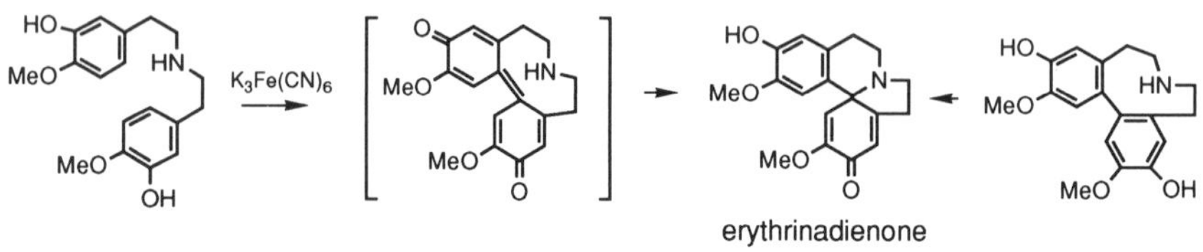

erythrinadienone

It may be relevant to mention that recognition of the biogenesis of usnic acid facilitated its chemical synthesis [Barton, 1956] via oxidative phenol coupling, although both the substrate and the product are not symmetrical.

usnic acid

Similarly, substances such as carpanone, which are derived from oxidative dimerization of shikimate metabolites, are expediently accessible in a biogenetically patterned synthesis [O.L. Chapman, 1971; M. Matsumoto, 1981].

carpanone

Oxidative coupling of carboxylic acid dianions is a convenient method for assembling precursors of lignans of the γ-butyrolactone and tetrahydrofuran types. An example is the synthesis of enterolactone [Belletire, 1986]. Desymmetrization, serving to differentiate the two carboxyl groups, was by methanolysis of the anhydride.

enterolactone

7.6.2. Dihydroxylation

Access to polyhydroxycyclohexanes and -cyclohexenes has been greatly facilitated by the discovery and populatization of microbial oxidation of benzene and its derivatives. For example, benzene gives *cis*-5,6-dihydroxy-1,3-cyclohexadiene by incubation with *Pseudomonas putida*, and this diol has been converted into pinitol [Ley, 1987]. Naturally, the conduritols are readily prepared from the same diol [Carless, 1989; Dumortier, 1992]. More recently quebrachitol has been synthesized [Carless, 1993].

conduritol-D   conduritol-A

(Some simple benzene derivatives such as bromobenzene are excellent substrates for the microbial oxidation. Although the products are unsymmetrical and their synthetic uses are outside the scope of our discussion, a synthetic correlation for an effective approach to pinitol [Hudlicky, 1990] may be presented.)

(+)-pinitol   (-)-pinitol

Of significance is the development of synthetic pathways for both enantiomers of conduritol-C [C.R. Johnson, 1991c] via acetonide formation, and acetylation with isopropenyl acetate in the presence of P-30 lipase. In this last step the (−)-isomer was obtained. The differentiation of the two allylic hydroxyl groups enabled selective configuration inversion of one or the other.

(-)-conduritol-C   (+)-conduritol-C

The dihydroxycyclohexadiene and its *trans*-isomer were available previously. Their preparation was also in connection with conduritol synthesis [Nakajima, 1957, 1959; Aleksejcyzk, 1985], and *chiro*-inositol 2,3,5-triphosphate [Carless, 1990].

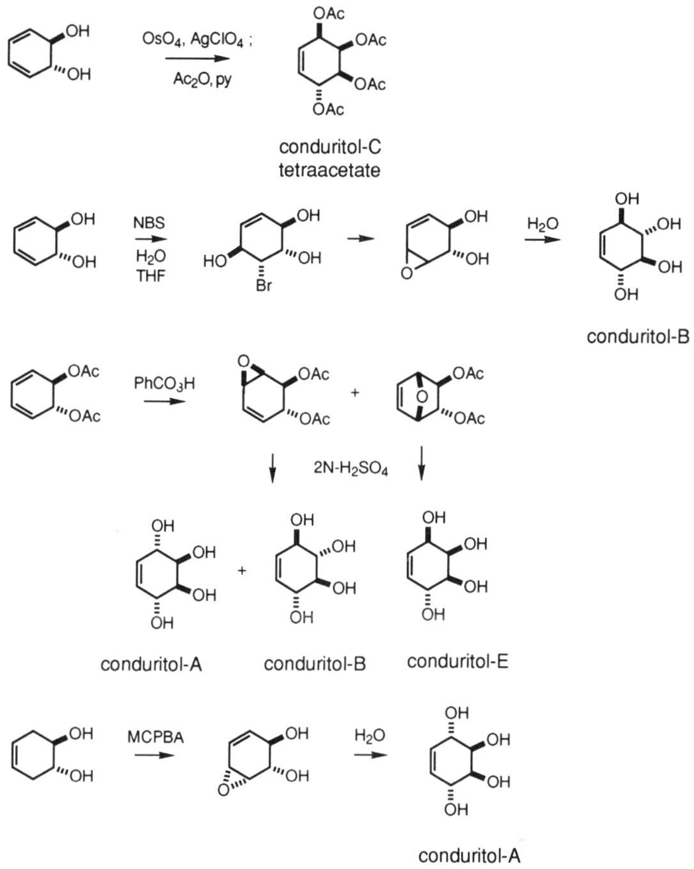
OsO4, AgClO4 ;
Ac2O, py
conduritol-C
tetraacetate
NBS
H2O
THF
H2O
conduritol-B
PhCO3H
2N-H2SO4
conduritol-A
conduritol-B
conduritol-E
MCPBA
H2O
conduritol-A

Lycoricidine represents a more complex structure which has been elaborated from *cis*-5,6-dihydroxy-1,3-cyclohexadiene [S.F. Martin, 1993b].

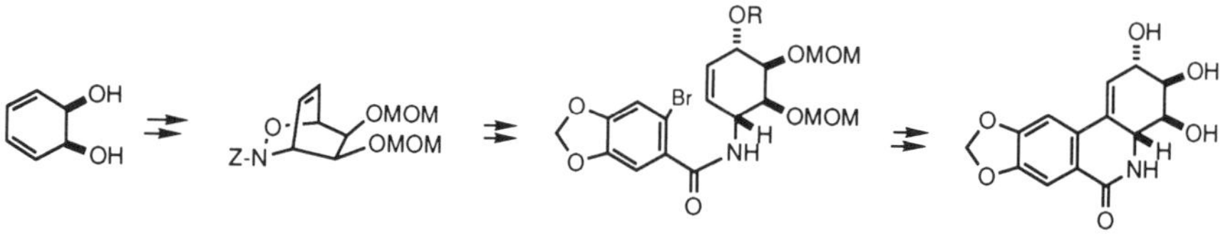

lycoricidine

7.6.3. Nuclear Reduction

The benzene ring is reducible catalytically or chemically. Because of the availability of a large number of benzene derivatives the preparation of hydroaromatic compounds may take advantage of this reservoir by applying reduction steps. For example, the general applicability of the Birch reduction to such substrates has influenced the formulation of many synthetic routes.

1,4-Cyclohexadiene undergoes epoxidation and the hydrolytic ring opening of the epoxide gives a C_2 diol. The diacetate has been converted into the ($-$)-diol monoacetate by incubation with pig liver esterase [Suemune, 1989], and then a potential intermediate of eldanolide.

Synthesis of the conduritols from 1,4-cyclohexadiene requires only conventional methods [Sütbeyaz, 1988; Secen, 1990]. Usually one of the double bonds is protected while the other is undergoing modification/functionalization.

Quite a number of symmetrically substituted benzenes have been used as starting materials for complex syntheses via the Birch reduction. Directed alkylation at the bisallylic carbon is made possible when the ether chain bears a good donor which can stabilize the metal associated with that site. For example, dimethylaminoethyl p-isopropylphenyl ether has been converted into 2-alkyl-4-isopropyl-2-cyclohexenones on Birch reduction followed by alkylation. Introduction of a 2-(4-methyl-3-penten-1-yl) substituent was involved in a plan to synthesize oplopanone [Köster, 1981], although 8-epioplopanone was likely to be the final product [Piers, 1986]. Torreyol was obtained in a similar manner [Franke, 1984].

oplopanone   8-epioplopanone

The synthesis of chlorismic acid and shikimic acid [Coblens, 1982] from 1,4-dihydrobenzoic acid involved desymmetrizing addition of bromine.

shikimic acid   chorismic acid

An outstanding application of 1,4-dihydrobenzoic acid is in an elaboration of reserpine [Pearlman, 1979]. In this approach *trans*-dihydroxylation with performic acid gave a compound in which the configurations of its three substituents agreed with those of the E-ring of reserpine. The introduction of the missing carbon chains for union with 6-methoxytryptamine was achieved by using the pro-C-18 hydroxyl to form an ether with a butenolide and by effecting an intramolecular [2 + 2]-photocycloaddition, with subsequent hydrolysis, Baeyer–Villiger reaction, and retro-aldol fragmentation. The intramolecular photocycloaddition was a crucial step to render the necessary stereocontrol of the *cis*-DE-ring juncture.

reserpine

Significant to the development of synthetic methodology is the intramolecular asymmetric Heck reaction as demonstrated in the elaboration of an intermediate of (+)-vernolepin [Kondo, 1993]. The prochiral substrate was acquired from methyl 1,4-dihydrobenzoate.

(+)-vernolepin

In the quest for vernolepin and vernomenin, many research groups conceived the same idea of exploiting a dihydrobenzoic acid or derivatives as the foundation of the carbocycle. In one synthesis [Zutterman, 1979] the dihydrobenzoic acid was methoxymethylated, reduced with lithium aluminum hydride, and esterified. With the ester side chain in the form of a diazomalonate, the copper-catalyzed decomposition led to a carbenoid, the intramolecular interception of which resulted in a tricyclic lactone. This desymmetrization created opportunities for further selective modifications of the molecule which culminated in the sesquiterpene.

5-Hydroxymethyl-5-vinyl-1,3-cyclohexanedione is a potential precursor of vernolepin [Scheffold, 1976]. It has been prepared from 3,5-dimethoxybenzoic acid in a manner similar to the above. An advanced intermediate of paniculide-B and -C has also been prepared [R. Baker, 1984].

vernolepin

paniculide-B

The value of the method for procuring 5-substituted 1,3-cyclohexanediones is further demonstrated by a synthesis of linderalactone and its isomers [Gopalan, 1980]. The method is complementary to the Claisen cyclization from keto esters (cf. curzerenone synthesis). This work is significant in that stereocontrol elements were not required.

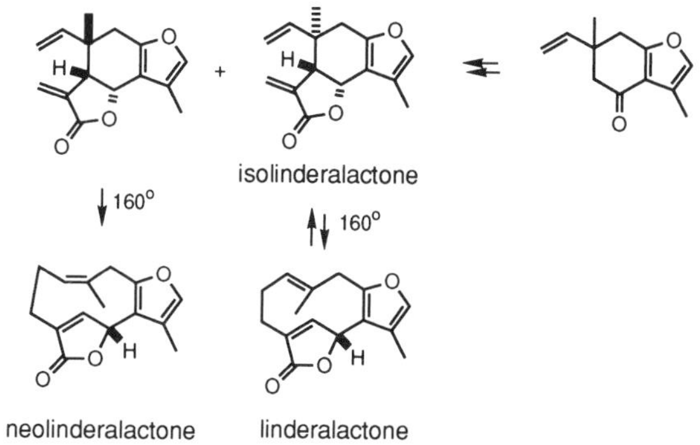

isolinderalactone

neolinderalactone   linderalactone

Another variation involves sequential alkylations of 3,5-dimethoxy-1,4-dihydrobenzoic acid in a synthesis of furoventalene [Castanedo, 1987]. The alkylation steps introduced an isohexenyl chain and a segment for furan ring formation. The final stage restored the aromatic ring with elimination of functionalities.

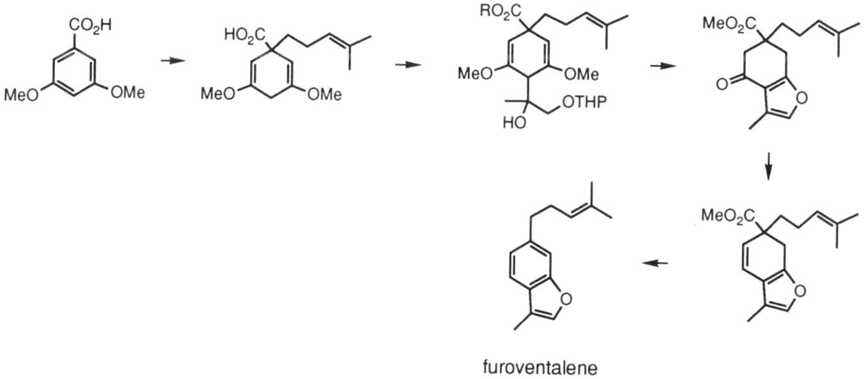

furoventalene

An interesting strategy for the synthesis of 15-deoxyeffusin [Kenny, 1986] consisted of formation of a tetralincarboxylic acid from 3,5-dimethoxybenzoic acid via Birch reduction, alkylation, rearomatization and other operations. The required methoxylation pattern of the tetralin is such that this method appears to be the most expedient. It should be noted that the late stage involving oxidative cleavage of the B-ring is biomimetic.

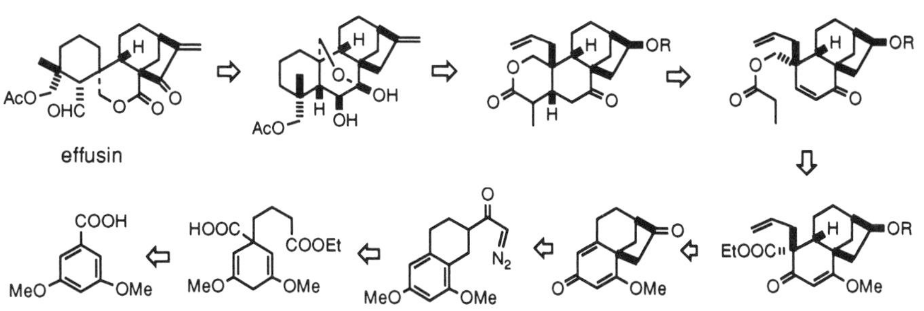

effusin

The Birch reduction has been coupled with cyclopropanation to constitute an alkylation method. Thus the formation of 4,4-disubstituted 2-cyclohexenones is delineated in the following schemes for the syntheses of trichodermin [Colvin, 1973], and 9-isocyanopupukeanane [Schiehser, 1980]. The differential reactivities of the two double bonds should be noted, and the manipulations must be consistent with such characteristics.

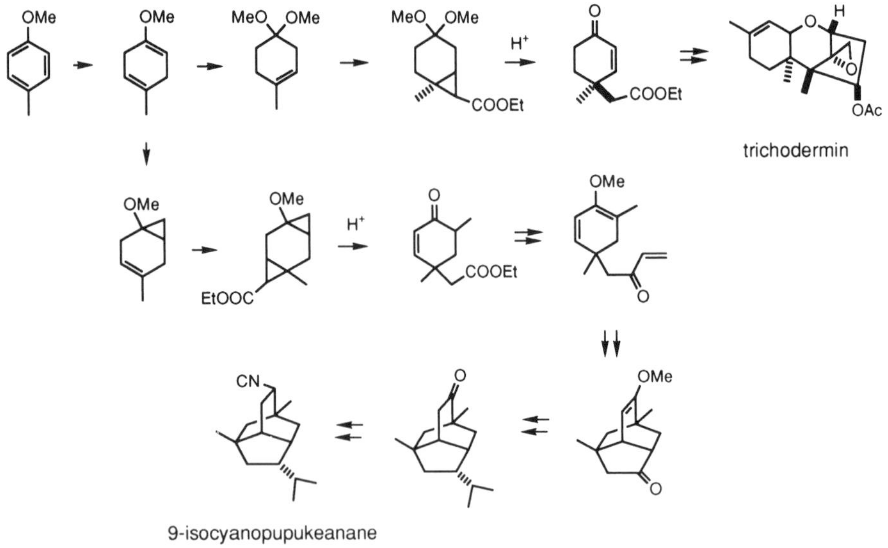

trichodermin

9-isocyanopupukeanane

The derivation of a synthon for the cyclohexene moiety of milbemycin-β_1 [Anthony, 1987] from 3,5-dihydroanisole also followed a similar strategy.

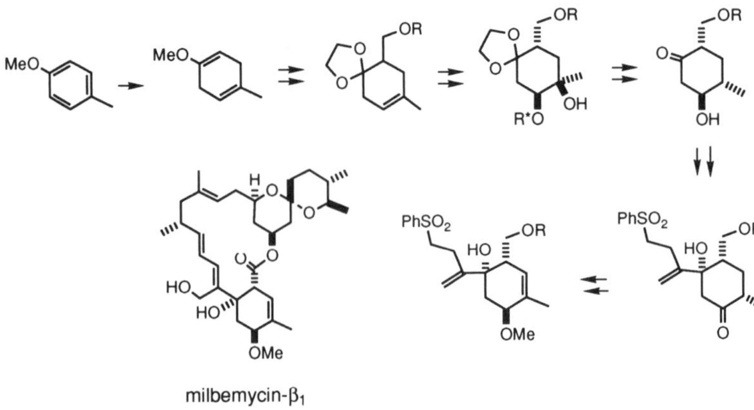

milbemycin-β_1

A tactically different manipulation of a dihydroanisole leading to vernolepin [Isobe, 1977] should be noted. After the conversion of the enol ether to a more stable ethyleneacetal and the oxidation of the alcohol to an aldehyde, an allyl chain, destined to become the vinyl residue at a ring juncture, was introduced. Functional group manipulations to give a cyclohexenone containing a malonic ester side chain ushered in the second stage of the synthesis. An intramolecular Michael reaction afforded the bicyclic framework which lacked only the γ-lactone, with respect to the complete skeleton of the sesquiterpene.

vernolepin

4-Methyl-3-cyclohexenone, obtained by mild hydrolysis of 1-methoxy-4-methyl-1,4-cyclohexadiene, underwent a Diels–Alder reaction of inverse electron demand with 3-carbomethoxy-α-pyrone in a highly regioselective manner. The *cis*-fused bicyclic ketone, generated from an in situ decarboxylation, has a homonuclear diene unit and an angular methyl group corresponding to occidentalol; accordingly a synthesis of the sesquiterpene was executed [Watt, 1972]. The intermediate proved suitable for elaboration of the copaenes and ylangenes [Corey, 1973b].

occidentalol

β-copaene

α-copaene

In an approach to tazettine [Abelman, 1990] the cyclohexene moiety was derived from 4-methoxybenzyl alcohol. The key step of this synthesis was an intramolecular Heck reaction.

tazettine

It should be noted that hydroquinone methyl ether can be reduced to 4-methoxy-3-cyclohexen-1-ol. Apparently this compound is an activated 4-hydroxycyclohexanone, and after protection of the hydroxy group an α-bromoketone is accessible. The preparation of such a bromoketone and its employment in a lycoramine synthesis [K.A. Parker, 1992] showed a good topological judgement of the target and the properties of intermediates. Thus in a 1,3-disubstituted cyclohexane the *cis*-diequatorial isomer is more stable, and the 4-alkoxy-2-aryloxy-cyclohexanone must exist in this form predominantly. Observe the correspondence in the relative stereochemistry of the carbinyl centers with those of lycoramine; the configuration of the quaternary benzylic carbon generated upon closure of the dihydrofuran ring and it was thus predetermined.

lycoramine

Dihydroanisole and its analogs undergo Diels–Alder reactions via in situ conjugation of the double bonds. The adducts, as such or after modifications, are fragmented owing to the presence of a 1,3-dioxygenated circuit. The overall process presents an expedient synthesis of 4-substituted 2-cyclohexenones by reductive alkylation of anisole. The synthetic examples illustrating its power include *threo*-juvabione [A.J. Birch, 1970], nootkatone [Dastur, 1974], and vetispirane sesquiterpenes such as hinesolone and solavetivone [Murai, 1981c].

threo- juvabione

The bicyclo[2.2.2]octenone derived from 3,4,5-trimethylanisole gave a trimethylhydrindanone on ring expansion and photorearrangement. The product, with distinct functionalities including the methyl groups in all-*cis* stereochemistry, proved to be an ideal precursor of pinguisone [Uyehara, 1985].

pinguisone

There is an alternative to the Diels–Alder method for indirect substitution at C-4 of a cyclohexenone. The Birch reduction products from anisoles are readily transformed into cyclohexadienyl cations stabilized by a tricarbonyliron group. These species are good electrophiles for enol derivatives, and the reactions occur at the *p*-position of the methoxy group. Consequently the cyclohexenone produced by this method bears a C-4 substituent and it is quite different from that produced by the Diels–Alder method, in terms of functional group distribution. The use of this approach to a synthesis of trichodermol [O'Brien, 1989] demonstrates the point.

trichodermol

Generally the cyclopropanation of dihydroanisole and congeners occurs at the electron-rich double bond. The adducts with dihalocarbene readily undergo ring expansion by an electrocyclic opening. Thus the decrease in symmetry by Birch reduction is inherited in products such as nezukone [A.J. Birch, 1968].

nezukone

p-Cymene is an inexpensive source of α-terpinene, itself converted into precarabrone [Bohlmann, 1982] via cyclopropanation with ethyl diazoacetate. A chain elongation process culminated in the sesquiterpine molecule.

α-terpinene

precarabrone

Dihydroaromatic substances can play an important role in synthesis due to their susceptibility to ring cleavage. In a synthesis of the cecropia juvenile hormone JH-I [Corey, 1968a], 1-methoxy-4-methyl-1,4-cyclohexadiene served admirably as a precursor for the terminal segment. Thus, the desired compound was secured by removal of the alcohol group from the hydroxy ester obtained from cleavage of the electron-rich double bond. There is no ambiguity regarding the geometry of the trisubstituted double bond and the purity of such a critical intermediate.

· cecropia JH-I

Sacrificial deletion of three carbon units from a benzene derivative could have unique qualification for the preparation of a 1,3-dicarbonyl compound. For example, Birch reduction of [12]metacyclophane followed by ozonolysis provided 1,3-cyclopentadecanedione [Y. Li, 1991] which is a precursor of muscone.

muscone

The access to a β-lactam containing an ester group at the α-position [Bringmann, 1991] is another example. Note that the lactamization step was subjected to asymmetric induction by the β-substituent, therefore the *trans*-isomer was formed preferentially.

The availability of *cis*-1,3,5-cyclohexanetriol from hydrogenation of phloroglucinol qualifies it as an excellent starting material for synthesis. In its synthetic applications involving desymmetrizing modifications a prime example is the synthesis of prostaglandin F$_2$ [Woodward, 1973].

A trioxahomoadamantane derivative was formed by reaction with glyoxalic acid. Deprotection at one site was accomplished by complex metal hydride reduction. This operation permitted desymmetrization aimed at establishing a lactol ring. Thus, by dehydration to generate a cyclic double bond, the solvolysis of the primary mesylate was attended by the double bond participation. The lactol ether unit is a masked aldehyde to be used in the homologation process.

The oxygen atoms of the lactol ether also has a *cis*-1,3-relationship with respect to the six-membered ring, identical to the F prostaglandins except that the prostaglandins are cyclopentanediols. Therefore, to complete the prostaglandin synthesis, the cyclohexane intermediate must undergo ring contraction.

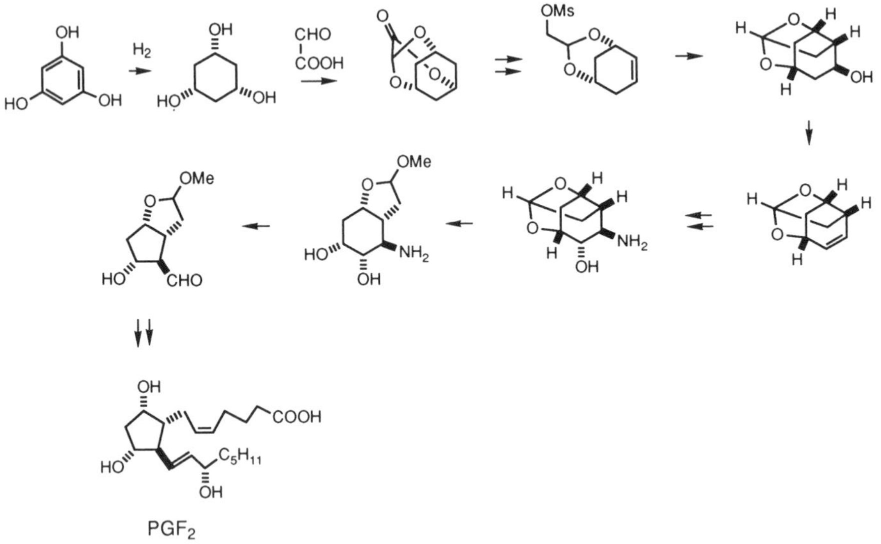

PGF₂

The ready protection of cis-1,3,5-cyclohexanetriol in the lactone acetal form facilitates the preparation of other useful building blocks such as cyclohexenediol derivatives and both enantiomers of 5-benzyloxymethyl-2-cyclohexenone [Carda, 1990]. The (−)-enone has been converted into (−)-mintlactone [Carda, 1991].

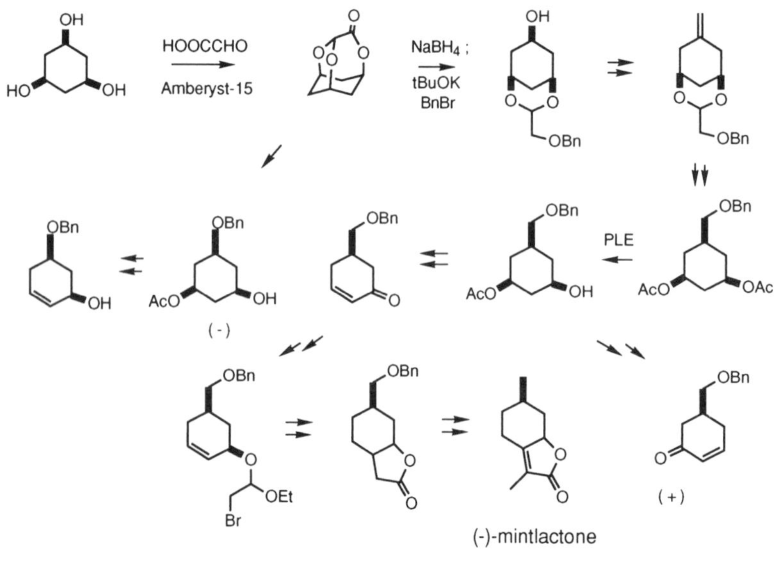

(-)-mintlactone

A 6-substituted 4-benzyloxy-δ-valerolactone has also been synthesized in a chiral form [Suemune, 1990] from *cis*-1,3,5-cyclohexanetriol. Thus the diacetate monobenzyl ether was selectively hydrolyzed with catalysis by pig liver esterase then oxidized to provide a cyclohexenone. The oxidation was accompanied by the elimination of acetic acid. Reduction of the carbonyl group, protection of the alcohol, and ring cleavage resulted in a lactone. Note that this compound has the same absolute configuration as the lactone moiety of (+)-compactin.

compactin

With two hydroxyl groups of the cyclohexanetriol protected, dehydration is a desymmetrizing reaction. Ozonolysis of such a compound would give rise to a *syn*-1,3-dioxygenated six-carbon chain [Prasad, 1984b]. The functional groups at the chain termini can also be differentiated as they are formed, by virtue of the electronic effects of the allylic oxygen atom [Schreiber, 1982]. This reaction sequence has practical value in the establishment of skipped polyol chains.

major

It is of interest to consider the structural relationship between the all-*cis* 5-hydroxycyclohexane-1,3-dicarboxylic acid and chaminic acid, with respect to the conversion of the former into the latter [Gensler, 1973]. Formation of the anhydride and its reaction with methylmagnesium chloride gave an unsymmetrical dihydroxy carboxylic acid. Exchange of the tertiary hydroxyl group into a chlorine atom was simple, and after esterification and oxidation of the cyclic alcohol, cyclization by an intramolecular alkylation could be effected. The remaining tasks were functional group adjustments.

chaminic acid

Hexahydrogallic acid can supply six carbon units to emetine [Burgstahler, 1960]. A synthesis featuring this use involved deletion of the central hydroxylated carbon atom, and amide bond formation of the side chain atom with a phenethylamine upon its detachment from the cyclohexane ring during Wolff rearrangement. It is interesting to note that a continuous three-carbon subunit including the ethyl side chain was inserted into the carboxyl group and the ring carbon of a hexahydrogallic acid derivative. This subunit originated from 1-diazopropane.

The amide formation was followed by the unveilment and cleavage of the triol. Because of the local symmetry of the 3-substituted glutaraldehyde, the Pictet–Spengler cyclization selected the chain so that a more stable product would be generated. Consequently, the resulting lactam contained two equatorial substituents.

emetine

5-Hydroxymethyl-2-cyclohexenone was obtained from 3,4,5-trimethoxybenzoic acid via reduction (and reductive demethoxylation). After protection of the hydroxyl group the enone underwent [2+2]-photocycloaddition with 1,1-diethoxyethene to provide a precursor of paniculide-B [A.B. Smith, 1983].

paniculide-B

7.7. POLYCYCLIC AROMATIC COMPOUNDS

Synthetic chemists generally use a polycyclic aromatic compounds to target substances which contain a similar aromatic moiety. Symmetrical naphthalene and anthracene derivatives are eligible starting materials.

The role of 1,4-naphthalenediol in the synthesis of radermachol [Joshi, 1994] is evident, and the intermediates are outlined here. It is interesting to note that five other approaches toward the tetracyclic diketone lacking the isopentenoyl chain were blocked either at the stage of closing the seven-membered ring, or in the preparation of a deMayo reaction intermediate.

radermachol

A potential intermediate for anthracycline synthesis has been conveniently assembled from 1,5-naphthalenediol diallyl ether [I. Takahashi, 1990] by way of Claisen rearrangement, chain elongation, and cyclization. The same symmetry element was maintained to this point, and desymmetrization occurred on the aromatization of one of the terminal rings. Because of the symmetry the aromatization did not encounter chemoselectivity problems.

The same naphthalenediol has also been used to elaborate nanaomycin-A [Kometani, 1983], after initial oxidation to the naphthoquinone.

nanaomycin-A

The monoallyl monomethyl ether of 1,5-naphthalenediol underwent Claisen rearrangement to initiate a synthesis of daunomycinone [K.A. Parker, 1980]. Linkage of the naphthalene derivative to an A-ring synthon relied on a Michael reaction.

daunomycinone

In a previous section, mention was made of using 1,5-naphthalenediol in a synthesis of 7,11-dideoxydaunomycinone [Yadav, 1981]. The two hydroxyl groups were separately derivatized into a methyl and a methoxymethyl group so that the latter functioned as a selective activator for o-lithiation. The arylithium thus generated reacted with a bridged lactone to afford a B-seco intermediate of the anthracycline aglycone.

In a route to shikonin [Terada, 1987; Braun, 1991] the dimethyl ether of 1,5-naphthalenediol was first converted into 1,4,5,8-tetramethoxynaphthalene. On account of symmetry this latter compound yielded only one formylation product, which was employed in the elaboration of the side chain.

(R)- shikonin

(S)- alkannin

The first total synthesis of morphine [Gates, 1950, 1956] started from 2,6-naphthalenediol. After protection of one of the hydroxyl groups, an orthoquinone was created as a means to introduce the catechol pattern present in the target. Interestingly, the originally protected hydroxyl was liberated and fashioned into another *o*-quinone to direct the construction of the hydrophenenanthrene skeleton by a Diels–Alder reaction; the nonbenzylic ketone group also participated in the formation of the azacycle at a later stage.

morphine

3,6-Dimethyl-2,7-naphthalenediol dimethyl ether proved to be an excellent starting material for a synthesis of parvifolin [Covarrubias, 1991]. A *β*-tetralone, formed by reduction, was subjected to enamine annulation and fragmentation to deliver the 6:8 fused ring system. It should be emphasized that the symmetrical nature of the naphthalene derivative made the synthesis less complicated.

parvifolin

The two central rings of anthracycline antibiotics are symmetrical by virtue of a rapid internal redox process. The predisposed existence of one particular form of the dihydroxynaphthoquinone moiety in these compounds does not exclude their attainment from a precursor possessing the other arrangement. Accordingly, many efficient routes to gain access to the aglycones of these antibiotics exploited such characteristics. These approaches include Diels–Alder assembly for daunomycinone [T.R. Kelly, 1980, 1984] and 2-hydroxyaklavinone [H. Tanaka, 1984]. Remarkably, either the formation of the A-ring or the D-ring may take precedence.

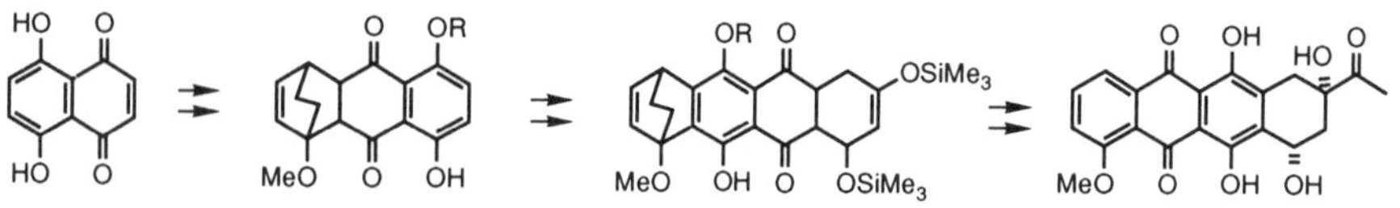

daunomycinone

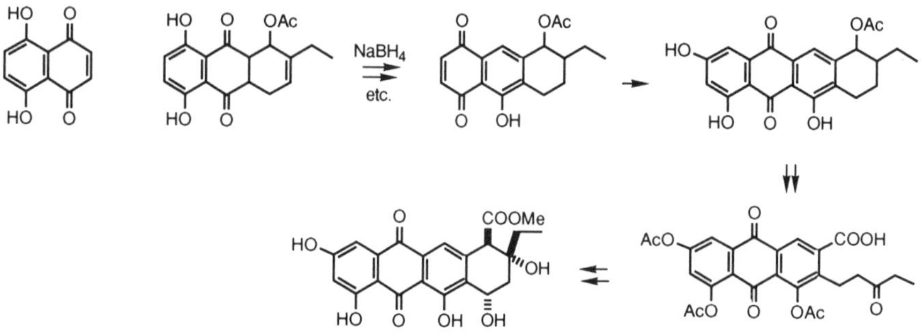

2-hydroxyaklavinone

An expedient approach to shikonin [Tanoue, 1987] is via a Friedel–Crafts alkylation of naphthazarin (5,8-dihydroxy-1,4-naphthoquinone).

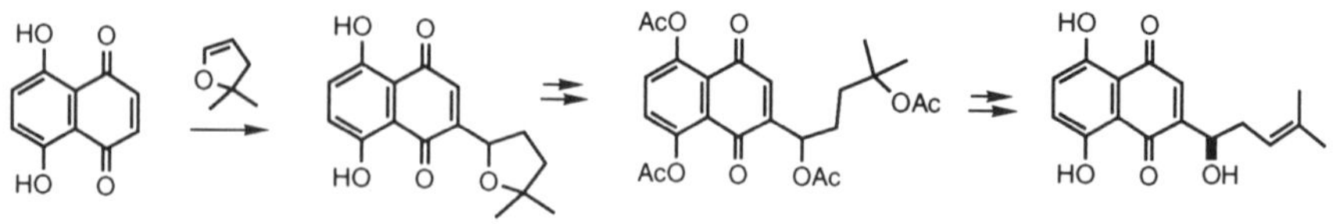

shikonin

Emphasis must be made on the fact that naphthazarin is readily desymmetrized by monoacylation, and such a compound was needed to provide regiocontrol to the Diels–Alder reaction with 1-methoxy-1,3-cyclohexadiene. During the second Diels–Alder reaction to form the A-ring an internal migration and redox exchange of the naphthoquinone system enabled the cycloaddition to take place at the terminal ring.

Symmetry considerations led to the pursuit of a synthesis of 7-conc-*O*-methylnogarol from 1,4,5,8-tetramethoxynaphthalene [Kawasaki, 1988]. Thus, linkage to a highly functionalized carbon chain was initiated in preparation for generation of the bridged sugar subunit.

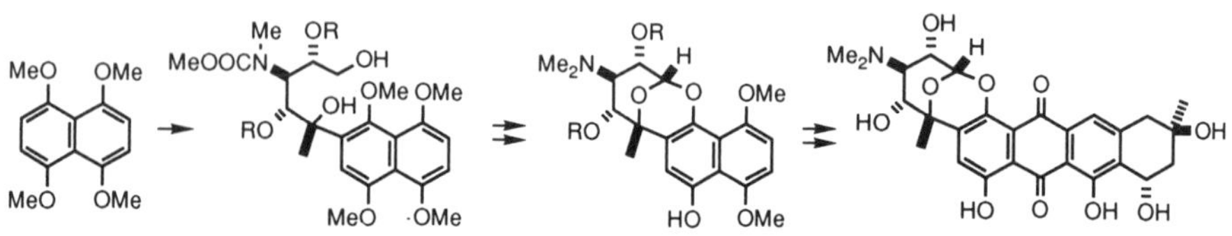

7-conc-*O*- methylnogarol

The expediency associated with a synthesis of circumanthracene [Broene, 1991] was the maintenance of a symmetrical intermediate which was acquired by a Wittig reaction of the symmetrical reagent derived from an anthracene derivative.

circumanthracene

The Claisen rearrangement of 1,4-dihydroxy-9,10-anthraquinone bis(2-chloroallyl)ether [Wong, 1979] is interesting, because only one furan ring was formed on further cyclization of the product. This desymmetrization should be exploitable.

Another efficient method for the preparation of daunomycinone [F. Suzuki, 1978] used anthrarufin (1,5-dihdroxy-9,10-anthraquinone). Protection as the monomethyl ether permitted selective alkylation and chain building at the *o*-position of the phenolic hydroxyl group to secure a carboxylic acid for closure of the A-ring. The hydroxylation which made amends to the missing oxygen function at C-6 (anthracycline numbering) was also rendered possible.

Anthrarufin also served as starting material for vineomycinone-B$_2$ methyl ester [T. Matsumoto, 1991; Tius, 1991]. The structural correlation in this case is more apparent, as it requires to append two different units to each of the *o*-positions flanking the hydroxyl groups.

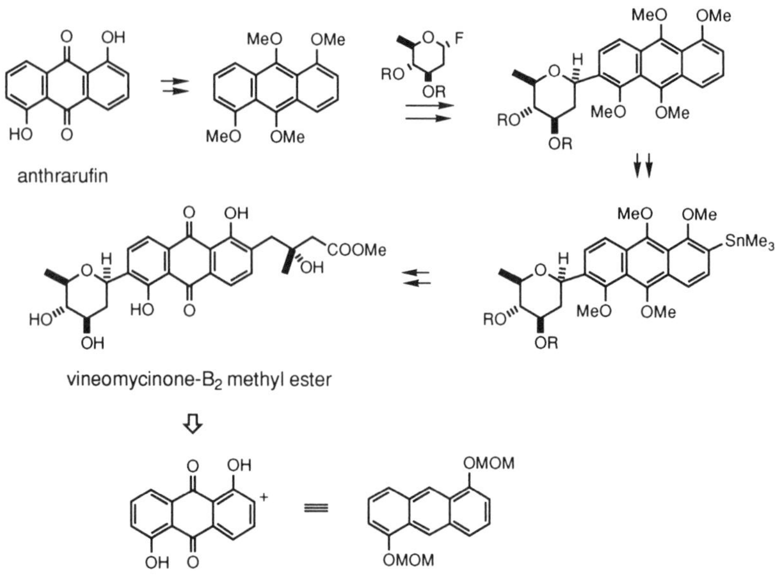

anthrarufin

vineomycinone-B$_2$ methyl ester

Chrysazin is 1,8-dihydroxy-9,10-anthraquinone, and a conspicuous candidate for the synthesis of aklavinone. Such an approach has been considered [Murphy, 1984].

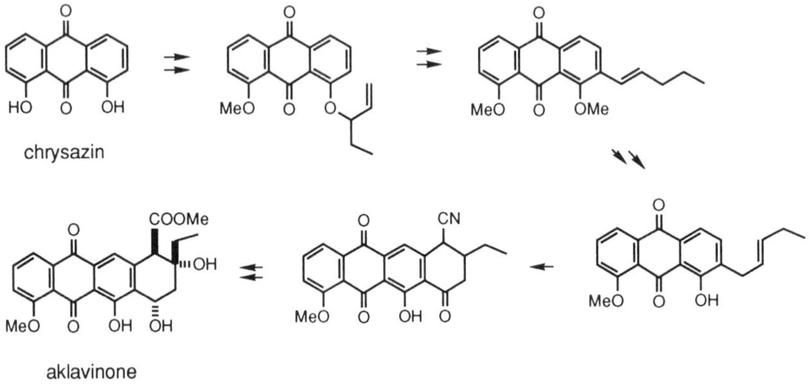

chrysazin

aklavinone

The aromatic portion of tylophorine is a symmetrically substituted phenanthrene. Accordingly it is expedient to elaborate this alkaloid from a 9,10-difunctionalized phenanthrene derivative using desymmetrization techniques. Application of the method featuring intramolecular cycloaddition of azide to alkene, thermal elimination of dinitrogen, rearrangement to an imine, and alkylation with a bystanding halide to form the indolizidenium species on 2,3,6,7-tetramethoxyphenanthrene-9,10-dimethanol led to a synthesis of tylophorine [W.H. Pearson, 1994].

tylophorine

8

NONCOMMON-SIZED RING SYNTHONS

8.1. THREE- AND FOUR-MEMBERED RING SYSTEMS

Except for a few classes of derivatives, the difficulties in the preparation and handling (due to volatility and instability) of cyclopropanes and cyclopropenes have restricted their synthetic applications. A few representatives relevant to the present context are shown below.

3,3-Dimethylcyclopropene undergoes addition with Grignard reagents. A particularly short synthesis of *cis*-chrysanthemic acid [Lehmkuhl, 1982; Nesmeyanova, 1982] involved carboxylation of the adduct with 2-methylpropenylmagnesium bromide, whereas reaction of the vinylmagnesium bromide adduct with acetone led to a tertiary alcohol which on exposure to perchloric acid was converted into santolina alcohol [Moiseenkov, 1982]. This addition-electrophilation sequence is also tenable in acetals of cyclopropenone [Nakamura, 1988], the organocuprate reagents seem to be the more general and effective nucleophiles.

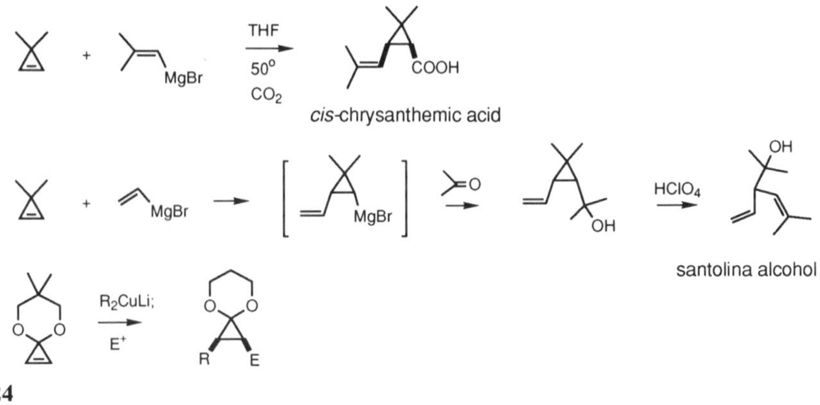

cis-chrysanthemic acid

santolina alcohol

The propyleneacetal of cyclopropenone is under thermal equilibrium at about 75°C with a species which can be formulated as a delocalized singlet vinylcarbene. Thermal [3 + 4]cycloaddition reactions with dienes have been observed and a remarkable application is a formal synthesis of colchicine [Boger, 1986b]. Interestingly, the cycloadduct of the pressure-promoted Diels–Alder reaction failed to decarboxylate, therefore this latter process is useless in terms of tropone preparation.

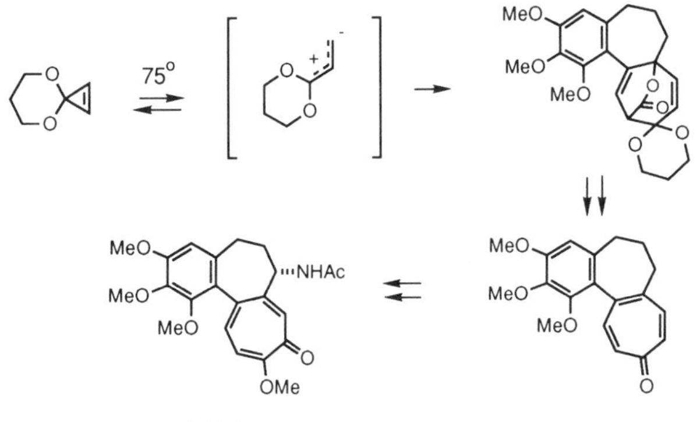

colchicine

Among the many synthetic uses of cyclopropane-1,1-dicarboxylic acid and its esters are the syntheses of morphine [Evans, 1982a] and talaromycin-B [Kocienski, 1984].

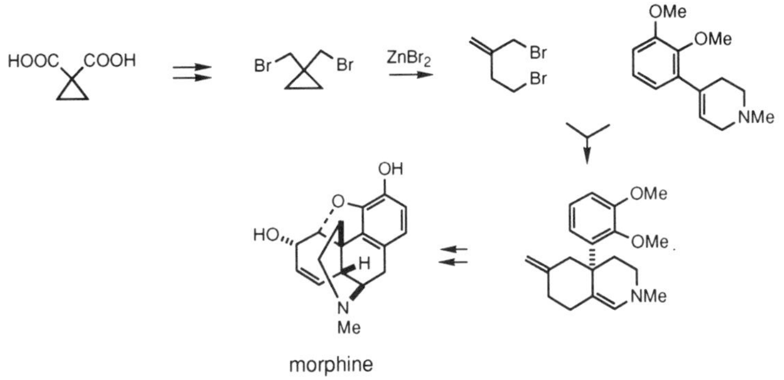

morphine

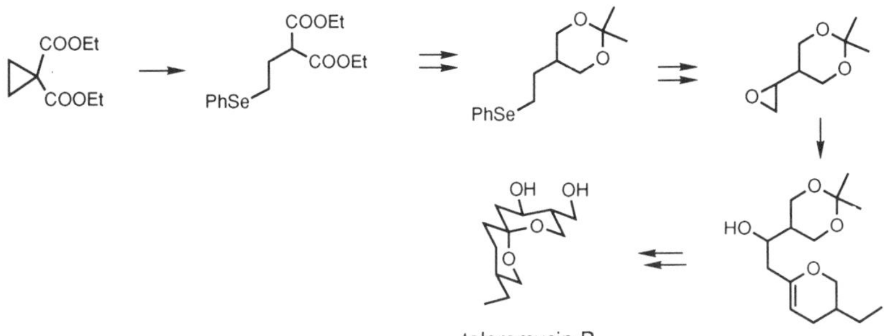

talaromycin-B

Enzymatic hydrolysis of a racemic, diastereomeric mixture of dimethyl 3,3-dimethylcyclopropane-1,2-dicarboxylate [M. Schneider, 1984] produced the (R)-monocarboxylic acid which is a precursor of (1R,3R)-trans-chrysanthemic acid.

Symmetrical cyclobutane derivatives possess a plane of symmetry either passing through C-1 and C-3 or between the C-1/C-2 bond. Desymmetrizing manipulations for synthetic purposes are illustrated in approaches to lineatin [K. Mori, 1983; Aljancic-Solaja, 1987] and eucannabinolide [Still, 1983b]. In accessing the latter compound two of the skeletal carbon atoms were incorporated into the ten-membered ring while the other two atoms became part of a side chain.

lineatin

eucannabinolide

Cyclobutanone appears only infrequently as a starting material. However, in a route to nicotine [Alberici, 1983] it was reacted with 3-pyridyllithium and the resulting alcohol was subjected to ring expansion by a Schmidt reaction.

myosmine   nicotine

Synthesis of chiral γ-butyrolactones such as (−)-methylenolactocin starting from symmetrical 3-substituted cyclobutanones is possible by taking advantage of enantioselective deprotonation with chiral amide bases and subsequent alkylation [T. Honda, 1994].

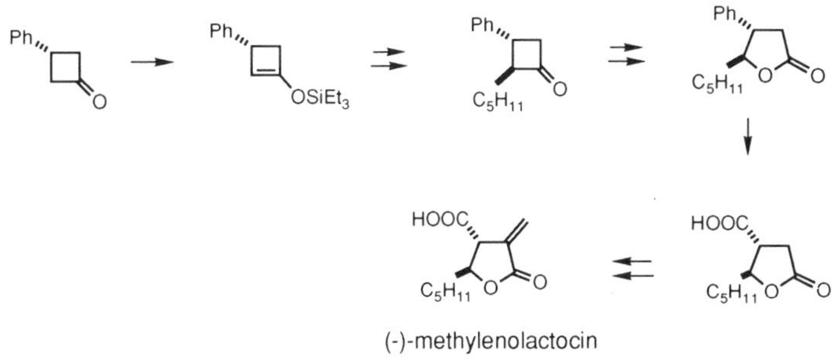

(-)-methylenolactocin

Asymmetric hydroboration of a bicyclic cyclobutene was a crucial step in an enantioselective synthesis of hirsutic acid-C [Greene, 1985]. Later steps included ring expansion and annulation.

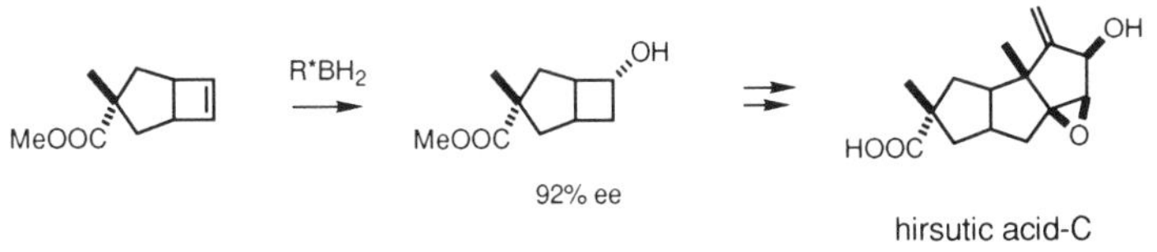

92% ee

hirsutic acid-C

1,2-Bis(trimethylsiloxy)cyclobutene is a readily available compound from acyloin condensation of succinic diesters. The Lewis acid-catalyzed condensation of this reactive enediol derivative with acetals constitutes a valuable chemical process because of the great synthetic potential of the products. The syntheses of showdomycin [Inoue, 1980] and pentalenene [Y.-J. Wu, 1991] are just two of the many attainable structural patterns.

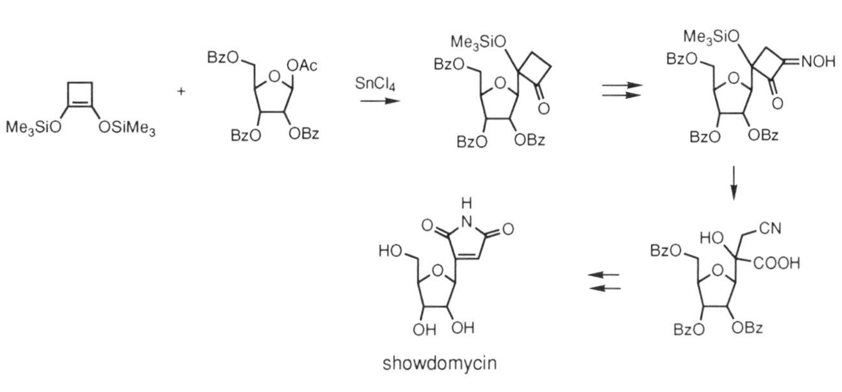

showdomycin

pentalenene

The photocycloaddition adducts derived from 1,2-bis(trimethylsiloxy)cyclo-butene are readily cleaved to give fused 1,4-cyclohexanediones. The process has been applied to a synthesis of 1-oxocostic acid [Van Hijfte, 1982].

1-oxocostic acid

Selective oxidation of *cis*-1,2-cyclobutanedimethanol by enzyme catalysis has cleared a route to chiral grandisol [J.B. Jones, 1982].

grandisol

The molecular symmetry of septicine is broken by the nitrogen atom, consequently its synthesis from   pyrrolidine unit with the proper (Z)-stilbene derivative would benefit from the absence of regiochemical problems in the union of these building blocks and possibly a simplified preparation of the latter species owing to its symmetry. Indeed a route has been developed from dimethyl squarate via 3,4-bis(3,4-dimethoxy-phenyl)-1,2-cyclobutenedione [Yerxa, 1994]. Since the two ketone groups in the diketone are equivalent, reaction with N-lithiopyrrole led to only one adduct.

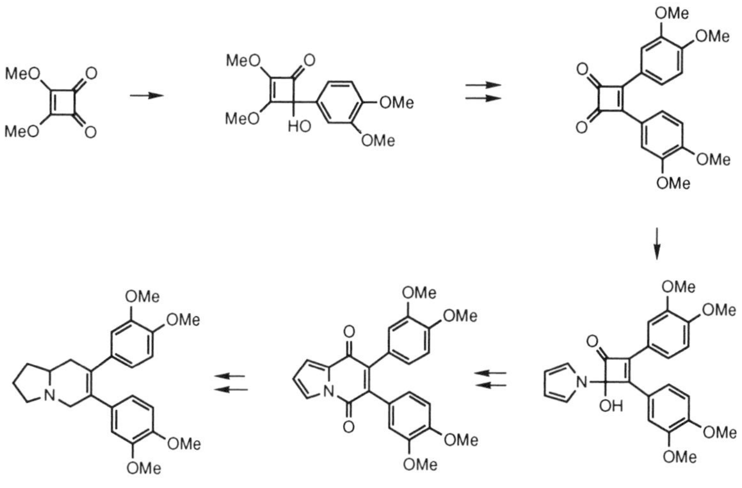

septicine

Other natural product syntheses taking advantage of the symmetry properties of squaric acid derivatives and skeletal changes via electrocyclic reactions include perezone [Enhsen, 1990] and lawsone [Heerding, 1991]. The syntheses of (E)-basidalin [Y. Yamamoto, 1994a] and (Z)-multicolanic esters [Y. Yamamoto, 1994b] demonstrated a new method for the access of substituted furanones.

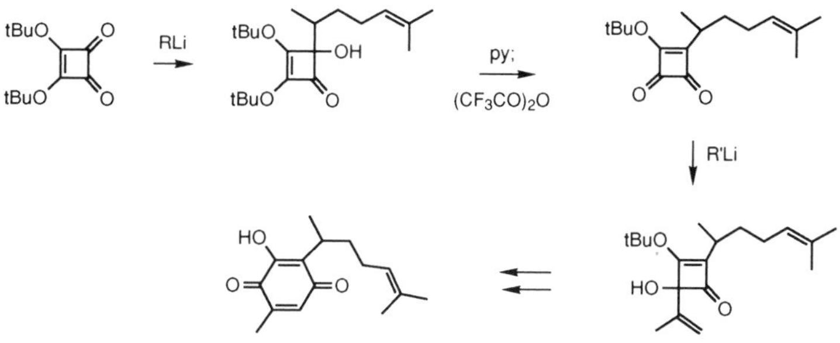

perezone

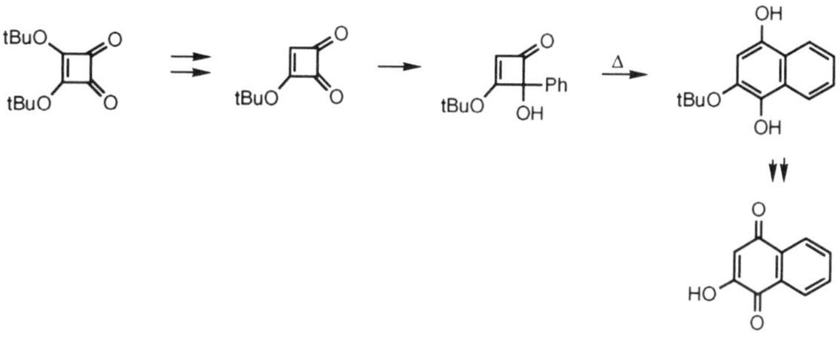

lawsone

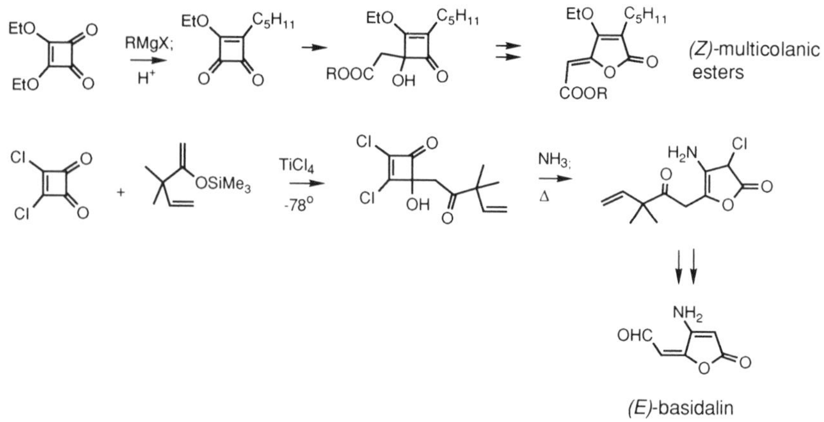

(E)-basidalin

It may be mentioned that bisethynylation of 3,4-dichloro-3-cyclobutene-1,2-dione initiated formation of cyclooligomeric enediynes which are potential cyclocarbon systems [Rubin, 1991]. Bisdecarbonylation products of such enediynes have been detected by laser desorption mass spectrometry.

8.2. SEVEN-MEMBERED AND LARGER RING SYSTEMS

An elegant synthesis of biotin [Confalone, 1980] started from cycloheptene. The key reactions included an intramolecular 1,3-dipolar cycloaddition which formed the tetrahydrothiophene ring.

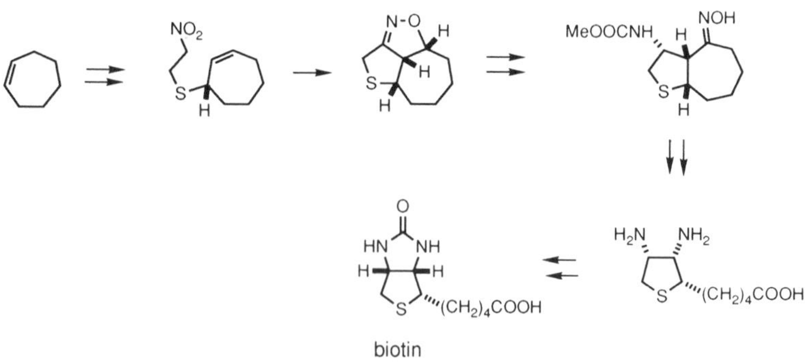

biotin

4-Cycloheptenone has been used as a template for the construction of the B-ring of vernolepin [Harding, 1981]. Although the synthetic work has yet to be completed, the design is excellent in terms of stereocontrol at three asymmetric centers.

vernolepin

The transformation of 4-cycloheptenone into an amino ester, followed by oxidative ring cleavage and cyclization involving the amino group and the nascent dialdehyde functionality, led to the known precursors of trachelanthamidine and isoretronecanol [Borch, 1977]. An alternative approach to the pyrrolizidine ring system is via Beckmann rearrangement of 4-cycloheptenoxime, reduction, and transannular reaction [Wilson, 1981].

trachelanthamidine   isoretronecanol

The synthesis of anatoxin-a has at least on two occasions taken advantage of molecular symmetry, although the target molecule is devoid of symmetry elements. The applicability of 4-cycloheptenone in the synthesis [Danheiser, 1985] was noted on considering the formation of the allylic C—N bond by an intramolecular trapping of the allyl cation (or S_N2 reaction). The allyl cation is available from electrocyclic opening of a fused dibromocyclopropane. The derivation of the substrate from 4-cycloheptenone is apparent.

anatoxin-a

Retrosynthetic analysis of hanegokedial readily reveals a bicyclic enone synthon. In principle this enone is also derivable from 4-cycloheptenone. Since the ketone group would cause complications along the synthetic sequence its modification was necessary [Taylor, 1982].

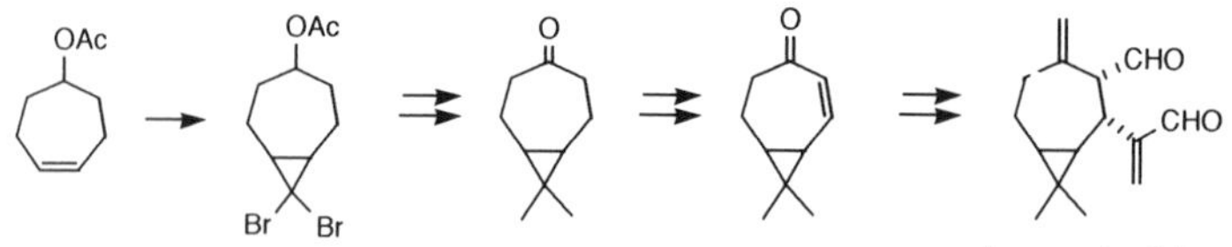

hanegokedial

The functionalized carbon chain which constitutes half of pyrenophorin has been prepared from the ethyleneacetal of 4,5-bis(trimethylsiloxy)-4-cycloheptenone [Wakamatsu, 1985b]. The methylation step served to desymmetrize the intermediate.

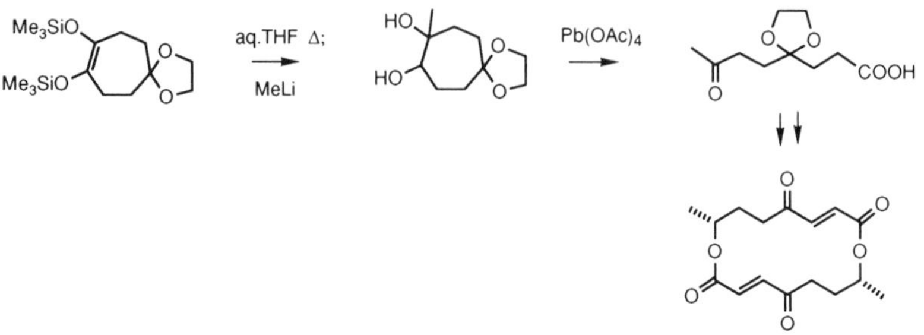

pyrenophorin

Synthetic routes to the Prelog–Djerassi lactone have been developed [Pearson, 1986, 1987, 1988]. Either 1,3-cycloheptadiene or cis-5,7-dimethyl-1,3-cycloheptadiene can be employed as starting material. For access to the chiral product, a lipase-catalyzed hydrolysis of the meso diacetoxycycloheptene achieved the desymmetrization.

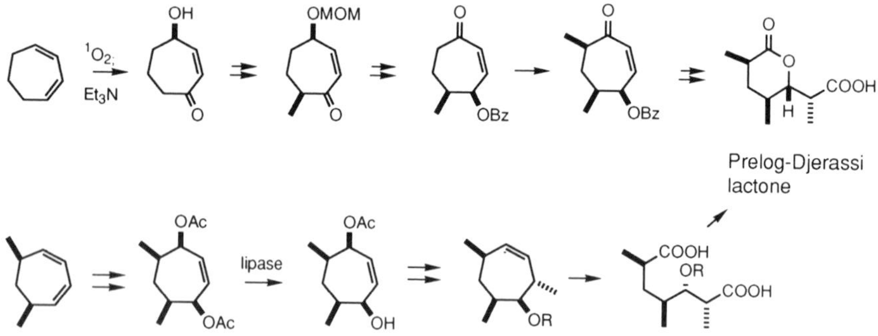

Prelog-Djerassi lactone

The use of a symmetrical cycloheptadiene as template to establish the stereochemical features of the Prelog–Djerassi lactone has its origin in the first synthesis [Masamune, 1975]. Interestingly, the cis-2,4-cycloheptadiene-1,6-dicarboxylic acid intermediate was acquired by degradation of the symmetrical bicyclo[4.2.1]nona-2,4,7-triene.

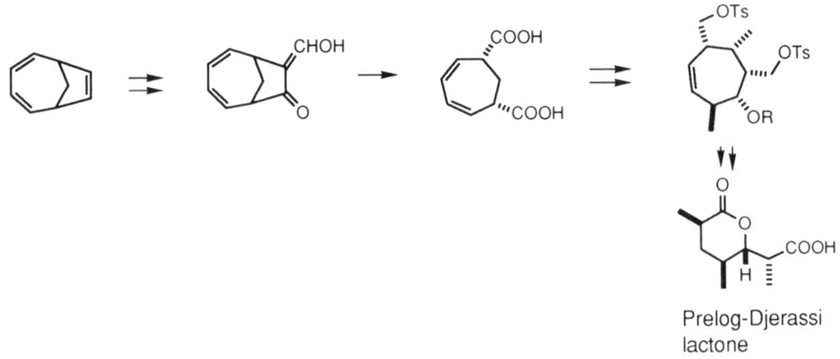

Prelog-Djerassi
lactone

As noted previously, two different methods for the elaboration of *O*-protected 3,5-cycloheptadienol into tropane bases have been reported [Iida, 1984b; Schink, 1991]. These routes involve 1,4-functionalization of the diene unit and in a later step an intramolecular displacement to form the heterocyclic system.

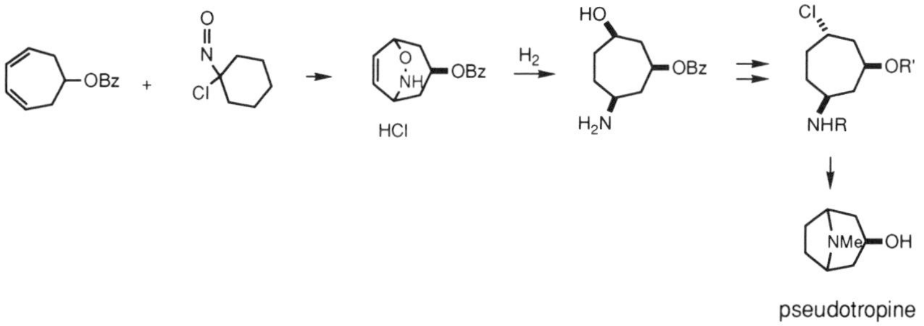

pseudotropine

Potential intermediates for skipped polyol systems can be made from 3,5-cycloheptenol. Lipase-catalyzed transesterification of the *meso*-diols provides chiral products [C.R. Johnson, 1991a]. Synthesis of four stereoisomers of 2,4-dideoxyhexose has been demonstrated [C.R. Johnson, 1991b].

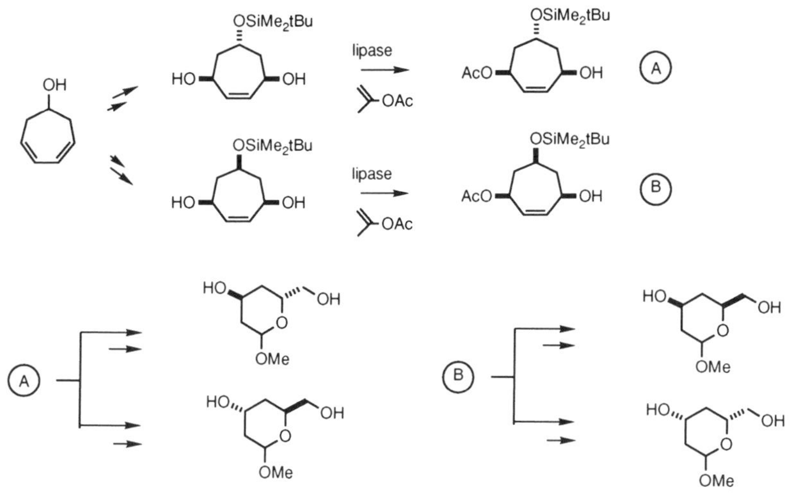

1,8-Addition to tropone desymmetrizes the cyclic ketone. Such reaction was employed in a synthesis of grosshemin [Rigby, 1987]. The acetic acid side chain thus introduced was fashioned into a diazoketone to effect an intramolecular cyclopropanation.

grosshemin

An outstanding synthetic application of cyclooctene is in prostaglandin-E$_1$ [Corey, 1968a, 1969b], serving as the α-chain and the foothold carbon atom of the cyclopentane by its transformation into 8-oxooctanenitrile and thence 9-nitro-8-nonenenitrile. The unsaturated nitro compound was to participate in either a Diels–Alder reaction or a Michael reaction.

PGE$_{1\alpha}$

1,3-Cyclooctadiene monoepoxide undergoes transannular cyclization on treatment with lithium diethylamide. The resulting bicyclic alcohol is very versatile material for synthesis of cyclopentanoids including allamcin, specionin, sarracenin, and coriolin. A convenient optical resolution of the alcohol has been achieved [Marotta, 1993].

A very interesting synthesis of the Corey lactone intermediate for the prostaglandins involved a ring expansion of 1,5-cyclooctadiene and an anionic oxy-Cope rearrangement [Paquette, 1980].

Corey lactone

It is essential to have reliable methods for the construction of *meso-β,β'*-dialkoxyoxepane units before attempts can be made to synthesize marine toxins such as ciguatoxin. One intriguing approach involves functionalization of 1,5-cyclooctadiene [E. Alvarez, 1990, 1991a]. The deviation from and resuscitation of symmetry in the process is noteworthy.

Synthetically, 1,5-cyclooctadiene is a convenient source of the central eight fragment of gossyplure, the female pink bollworm moth pheromone [Anderson, 1975]. By monoepoxidation, conversion into the α-ketol and subsequent cleavage to provide an aldehydo ester, systematic extension of the carbon chain at both ends could be orchestrated.

gossyplure

Semi-ozonolysis of 1,5-cyclooctadiene with oxidative workup led to (Z)-4-octenedioic acid, the monomethyl ester of which could be obtained by partial saponification and separated from the diester and the diacid by solvent extraction. This monoester underwent Kolbe coupling electrolytically with tenfold excess of pelargonic acid to furnish (Z)-4-pentadecenoic acid [Klünenberg, 1978] which was converted into disparlure.

disparlure

The monoepoxide of 1,5-cyclooctadiene is convertible into 4-cyclooctenone whose oxime undergoes Beckmann rearrangement. The symmetrical 1-aza-5-cyclononene obtained from one of the lactams has been elaborated into δ-coniceine [Wilson, 1979]. The alternation between symmetrical and unsymmetrical intermediates is quite interesting.

δ-coniceine

1,5-Cyclooctadiene forms a 1:1 adduct with dibromocarbene. The ready accessibility of (2Z,6Z)-cyclononadienone from this adduct by electrocyclic opening and conventional transformations has enabled the development of an isocaryophyllene synthesis [Kumar, 1976].

isocaryophyllene

It is of interest to note that three 1:2 adducts, each having the *cis–syn–cis*, *cis–anti–cis*, or *cis–anti–trans* stereochemistry, converged into one bicyclic allene after one treatment with one equivalent of methyllithium, whereas further debromination of the product afforded the *meso*-1,2,6,7-cyclodecatetraene [Dehmlow, 1990]. The *trans–syn–trans* adduct yielded a stereoisomeric bicyclic allene but eventually gave the same *meso* diallene. The final product from the *trans–anti–trans* adduct is the racemic diallene.

It may be considered that the heterocyclic systems of fawcettimine and serratinine are segmented azacyclononanes. A route to a tricyclic intermediate starting from cyclopentenone annexation to 1,5-cyclooctadiene [Mehta, 1991] exploited the remaining double bond for insertion of the nitrogen atom.

fawcettimine

An oxidation product of 1,5-cyclooctadiene is 1,5-cyclooctanediol. It served as starting material for a synthesis of pentalenene [Zhao, 1991]. An interesting feature of the process, after desymmetrization to give a cyclooctenone and the subsequent conjugate addition–electrophilation sequence, is the metal carbenoid insertion to form the five-membered ring.

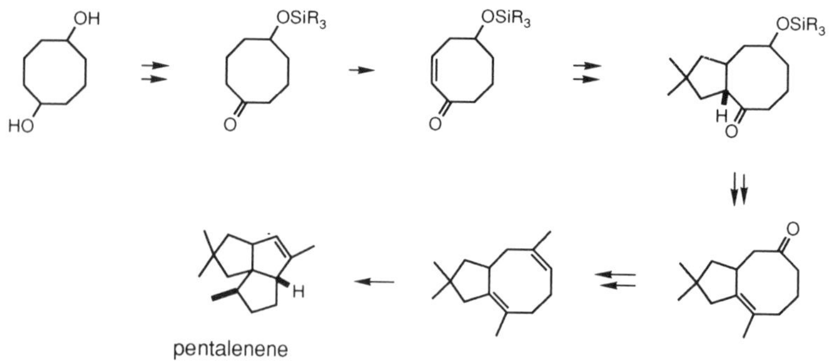

pentalenene

1,5-Dimethyl-1,5-cyclooctadiene is a cyclodimer of isoprene. By careful functionalization of one of its double bonds, such as hydroboration, and compartmentation of the ring by transannular reaction to address structural needs of several popular synthetic targets, novel routes to such molecules can be established. Among these routes are those leading to iridomyrmecin [Matthews, 1975] and pentalenene [Mehta, 1986a].

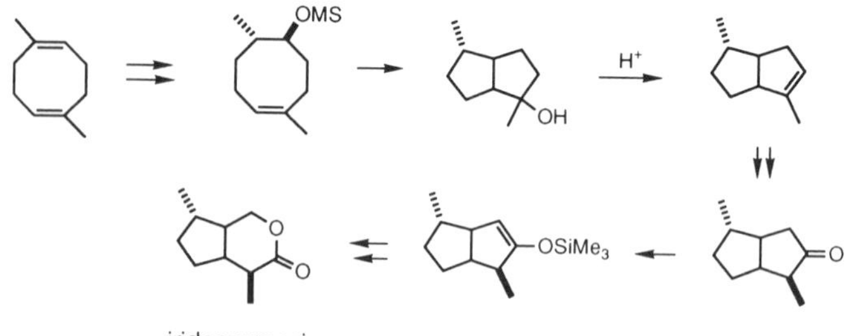

iridomyrmecin

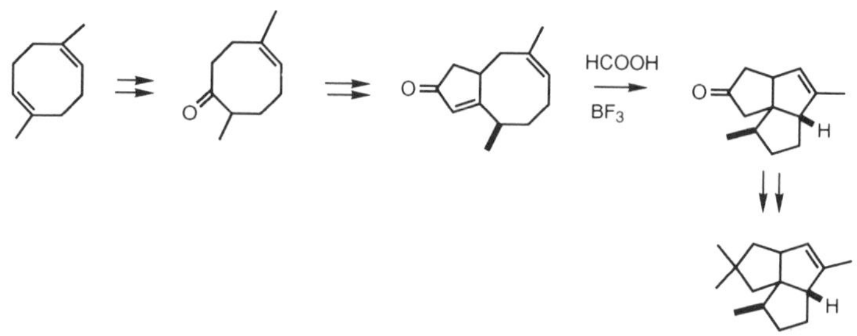

pentalenene

Olefin metathesis of 1,5-dimethyl-1,5-cyclooctadiene with isobutene gave 2,6-dimethyl-1,5-heptadiene. The latter compound was converted into hydroxycitronellal [Van Helden, 1977] upon hydration and hydroformylation.

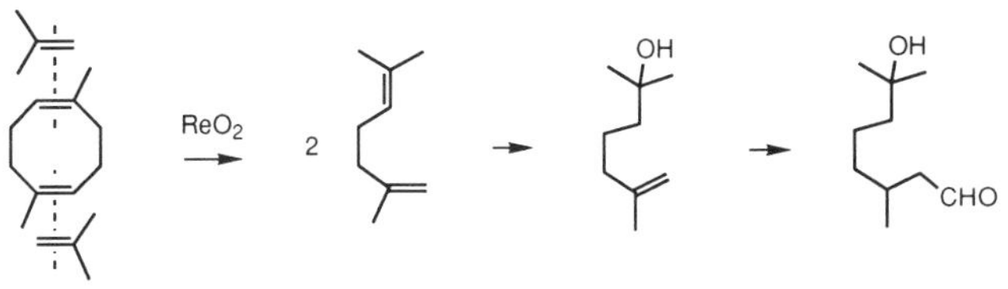

Although cyclooctanone is too simple and common a substance to be included in the present discourse, its one use in a synthesis of (+)-recifeiolide [Kaino, 1990] deserves special mention. An indirect alkylation was accomplished through derivatization into a chiral acetal and subsequent ring cleavage into an enol ether, which paved the way to cyclization. Note also an earlier synthesis [Gerlach, 1976] of the racemic compound from cyclooctanone.

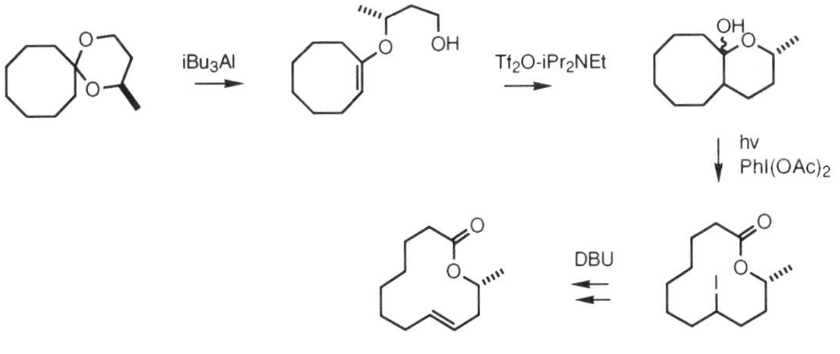

(+)-recifeiolide

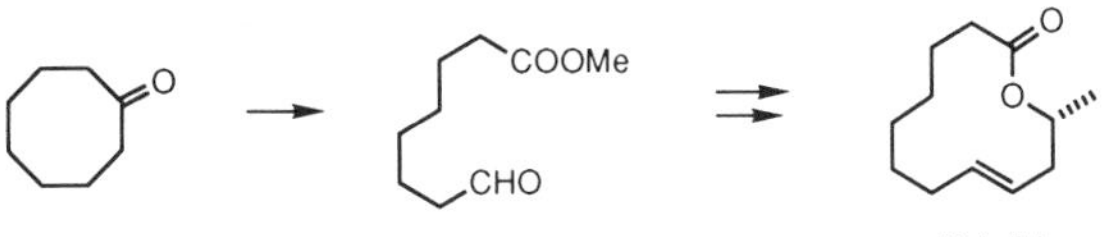

recifeiolide

The synthesis of anatoxin-*a*, an asymmetric alkaloid, from the symmetrical 5-methylaminocyclooctanone [Wiseman, 1986] is intriguing. α-Bromination of the ketone destroyed the symmetry.

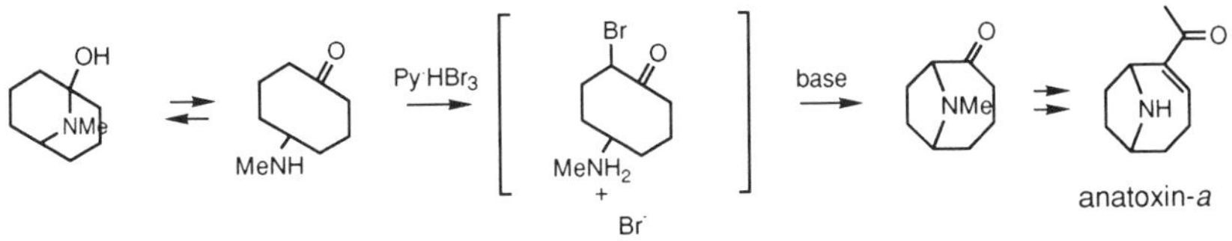

anatoxin-*a*

The use of bicyclo[7.1.0]deca-4,5-diene in a synthesis of isocaryophyllene [Bertrand, 1974] consisted of an initial [2+2]-cycloaddition with dimethyl ketene. Because of symmetry of the allene only one cycloadduct was obtained.

isocaryophyllene

Cyclononanone has been elaborated into recifeiolide [Schreiber, 1980] in a few steps. Also by way of oxidative ring cleavage it was readily converted into the queen substance of the honeybee [Trost, 1976].

recifeiolide

(1Z,6Z)-Cyclodecadiene is generated by isomerization of the unsymmetrical (1E,5Z)-isomer with sodium dispersed in alumina. The former compound was required in a synthesis of the fused oxane–oxepane systems [E. Alvarez, 1991b] which are structural units of certain marine toxins. The molecular symmetry simplifies the reaction patterns, permitting stereoselective formation of an 11-oxabicyclo[4.4.1]undecane intermediate by a transannular reaction.

Hydroboration of (1E,5E,9E)-cyclododecatriene leads to perhydroboraphenalene which is readily converted into the aza analog. The utility of the latter compound as precursor of the ladybug alkaloids (precoccinelline, hippodamine, hippocasine, myrrhine propyleine, and isopropyleine) is apparent [Mueller, 1984].

Fifteen-membered ring formation via expansion of the cyclododecatriene has been achieved by a reaction sequence of cleavage to the dialdehyde and a Horner–Wadsworth–Emmons reaction with 1,3-bis(dimethylphosphono)propanone [Büchi, 1979b]. The resulting ketone was readily converted into muscone.

muscone

The inexpensive (1*E*,5*E*,9*Z*)-cyclododecatriene deserves intense investigations into its synthetic potential. The (*E*)-double bond is more reactive toward epoxidizing agents and the monoepoxide can be isomerized to give an allylic alcohol which undergoes asymmetric epoxidation [Zarraga, 1989]. This chiral epoxy alcohol has been used to synthesize the *trans-syn-trans*-oxatricyclic systems [Zarraga, 1991] present in many microalga polyether toxins.

On monoozonolysis of (1*E*,5*E*,9*Z*)-cyclododecatriene, protection of one of the aldehyde groups, and methallylation, a mixture (differing in alkene geometry) was produced. The mixture underwent macrobicyclization by the Prins reaction [Schulte-Elte, 1979]. Double bond isomerization and hydrogenation of the product at high temperatures gave muscone.

A process of probable economical value is the preparation of phytone (diastereomers) [Pond, 1975], a component to be integrated into vitamin-E, via hydrogenation of 1,5,9-trimethylcyclododecatriene to leave only one double bond followed by oxidative cleavage generated a keto aldehyde. Selective homologation at the aldehyde terminus can be effected by reaction with dimethylketene, and the product was hydrogenated to provide phytone.

phytone

Selective ozonolysis of the (*E,E,E*)-triene gave an acyclic precursor of the cecropia juvenile hormone JH-III [Odinokov, 1985]. The advantages of this route consist not only of the industrial scale availability of the building block, but also the inheritance of the central (*E*)-double bond of the target molecule from the cyclic triene, which greatly simplifies the overall process.

Cyclododecanone is by far the most popular macrocyclic building block due to its availability. Many syntheses of muscone have started from cyclododecanone [D. Felix, 1971; Fehr, 1979; Tsuji, 1980], usually via transitory annulation of a five-membered ring and subsequent cleavage of the fusion bond.

muscone

Muscopyridine owes its asymmetry to a single methyl substituent. Synthetic considerations indicate that attachment of the methyl group at a late stage would greatly ease the effort. The *β*-position of the methyl group in a C-2 side chain of a pyridine nucleus is fortunate because the *α*-position can be oxygenated via rearrangement of a pyridine *N*-oxide, and methylation of a 2-(1-oxoalkyl)pyridine is conceivable. Thus a symmetrical aza analog of [10]metacyclophane is seen as a viable subtarget [Biemann, 1957].

The synthesis was carried out by annulation of cyclododecanone. Nitrogen atom insertion into a bicyclo[10.3.0]pentadecene followed by aromatization provided the key intermediate. The subsequent desymmetrizational introduction of the methyl group that required as many steps as the framework construction protocol is perhaps a weakness of this labor.

muscopyridine

The major source of cyclotridecanone is cyclododecanone by way of ring expansion; its attactiveness as a starting material is relatively low. However, an interesting approach to muscone [Nozaki, 1969] involved α,α'-alkylation with 2-chloromethyl-3-chloropropene to form a bridged ring system, hydrogenation, Norrish Type-I photocleavage, and degradation. Intermediates up to the photocleavage are symmetrical.

muscone

Since methods for achieving one-carbon expansion of ring systems are numerous, it is not surprising that conversions of cyclotetradecene and derivatives into muscone have been explored. Such work is usually performed in the context of methodology evaluation, for example, the silylmetallation of allenes [Morizawa, 1984] and the oxidative cleavage of 1,2-bis(trimethylsilyl)cyclopropanes [Ito, 1977].

muscone

The reaction scheme, left to right:

A 14-membered ring bearing two OSiMe₃ groups on a double bond, with reagents CH₂I₂ / Et₂Zn, converts to a 14-membered ring with a cyclopropane bearing two OSiMe₃ groups, with reagents FeCl₃ / HCl, converts to a 15-membered ring diketone, which converts to muscone.

9

SYMMETRICAL HETEROCYCLES AS BUILDING BLOCKS

It is often possible to discern certain fundamental heterocyclic subunits in complex molecules such as alkaloids. The synthesis of alkaloids is greatly facilitated by choosing heterocycles as building blocks because they are generally available and the heteroatoms offer handles for activation of carbon sites in their proximity. According to the scope of this monograph the following discussion is concerned mainly with synthetic applications of symmetrical heterocycles.

9.1. AZACYCLIC BUILDING BLOCKS

Heterocycles containing nitrogen atoms are readily prepared, particularly the 5- and 6-membered ring systems. They have found many uses in synthesis.

Aziridine is the smallest azacycle. Ring strain makes aziridines very susceptible to ring opening. Acceptor-substituted aziridines often behave as 1,3-dipolar reagents via electrocyclic tautomerization. Synthetic designs involving symmetrical aziridines are relatively rare, but indirect generation of an active aziridine equivalent for a cycloaddition approach to renierone [K.A. Parker, 1984] should be mentioned. The local molecular symmetry permitted an uncomplicated union of a symmetrical reactant with an unsymmetrical reactant.

renierone

In most applications featuring ring fission the heterocycle is invariably converted into unsymmetrical intermediates. Thus the C—N bond cleavage following *N-p*-methoxybenzylation of an arizidine made the access to corgoine and sendaverine [Otomasu, 1982a] quite efficient.

R=Me sendaverine
R=H   corgoine

Aziridine itself has been employed in a synthesis of necine bases [Kametani, 1984] and crinine [Whitlock, 1967]. This synthetic intermediate is still locally symmetric, therefore C—N bond cleavage is the desymmetrizing step. In the crinine synthesis this step also accomplished a cyclization to give a hydroindole from which only reduction and Pictet–Spengler reaction were required to complete the skeletal construction.

crinine

In a synthesis of dasycarpidone and epidasycarpidone [Dolby, 1968] acylation of *N*-methylaziridine initiated the preparation of the piperidone intermediate.

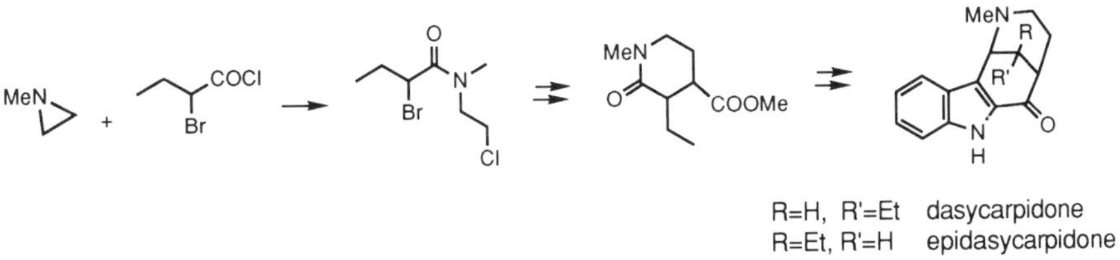

R=H, R'=Et   dasycarpidone
R=Et, R'=H   epidasycarpidone

Aziridines fused to a succinimide moiety undergo electrocyclic opening to generate dipolar 2,6-dioxopiperazines. In a synthetic design for quinocarcin [Garner, 1993] trapping of such a 1,3-dipole with an acrylamide was the key step. This highly diastereoselective desymmetrizing process is remarkable. In fact the desymmetrization occurred only when new bond formation started.

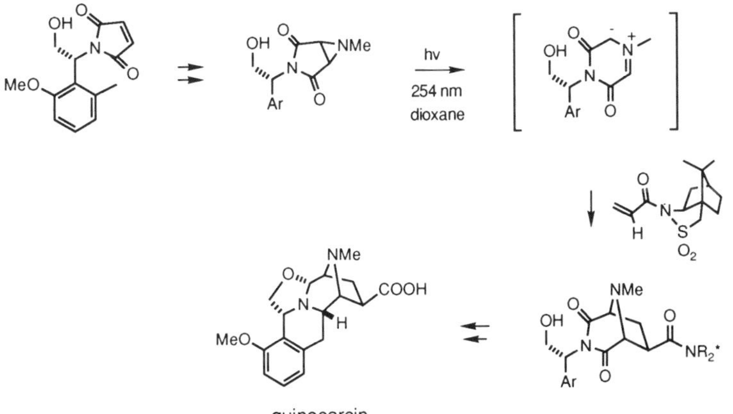

quinocarcin

Symmetrical 3-azetidinols are readily available from epichlorohydrin by reaction with primary amines. These substances can be transformed into 1-acyl-1-azabutadienes [Jung, 1991] via the 1-acyl-2-azetines.

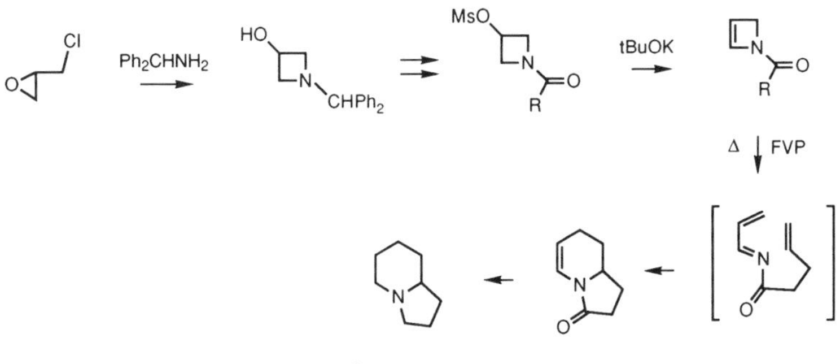

δ–coniceine

Pyrrole, pyrrolidine, and oxygenated derivatives are excellent building blocks for alkaloid synthesis. Any reaction that results in attachment of a carbon chain or a functional group to a ring atom (except the nitrogen) of these compounds perturbs the molecular symmetry.

The use of pyrrole to access necine bases such as isoretronecanol and trachelanthamidine [Brandänge, 1971] is a classical example of synthetic manipulation via α-functionalization. In this route an intramolecular Wittig reaction served to elaborate the bicyclic skeleton of the target compounds.

trachelanthamidine   isoretronecanol

The synthesis of elaeocarpine and isoelaeocarpine [T. Tanaka, 1970] also proceeded via a 2-substituted pyrrole. In a route to anatoxin-a [Bates, 1979] a trichloroacetyl group was also introduced at the α′-position of N-methylpyrrole to activate intramolecular cyclization. At that stage the molecule was already converted into a 5-substituted N-methylproline.

elaeocarpine   isoelaeocarpine

anatoxin-a

A viable synthetic route to epibatidine involves Heck-type coupling of the pyridine moiety to a symmetrical 7-azabicyclo[2.2.1]hept-2-ene [Clayton, 1993].

The most convenient access of the bicyclic substrate is via a Diels–Alder reaction of *N*-carbomethoxypyrrole. However, the requirement of an activated dienophile makes the preparation slightly lengthier.

epibatidine

Synthetic approaches to the ergot alkaloids from simple indole derivatives must contend with the problem of regioselective alkylation at C-4. One way to circumvent this difficulty is to form 4-substituted indoles de novo. From *N*-carbomethoxypyrrole such a process has been developed, thereby opening new avenues to chanoclavine-I [Natsume, 1981a], clavicipitic acid-I and -II [Muratake, 1983], (−)-α-cyclopiazonic acid [Muratake, 1985] and hapalindole-J and -M [Muratake, 1990].

clavicipitic acid-I   clavicipitic acid-II

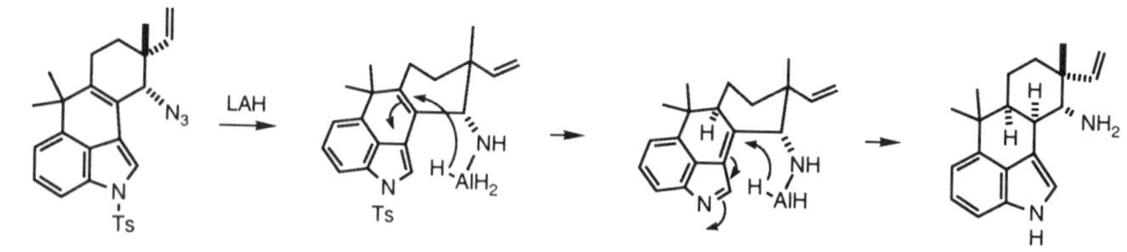

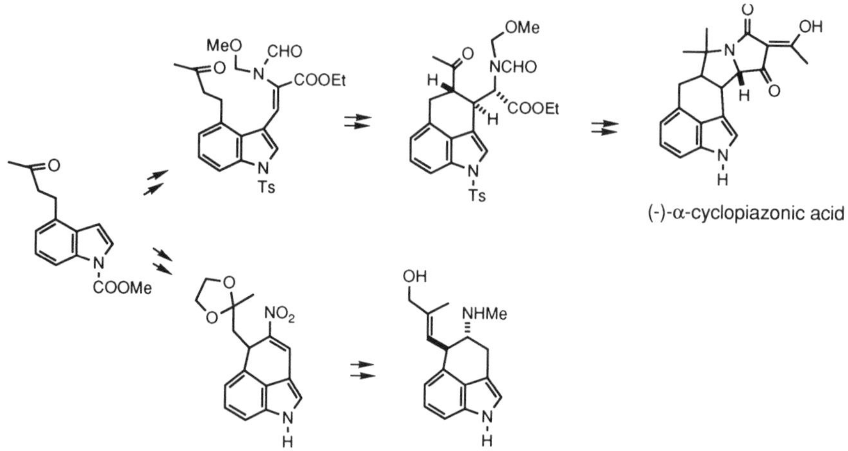

(-)-α-cyclopiazonic acid

chanoclavine-I

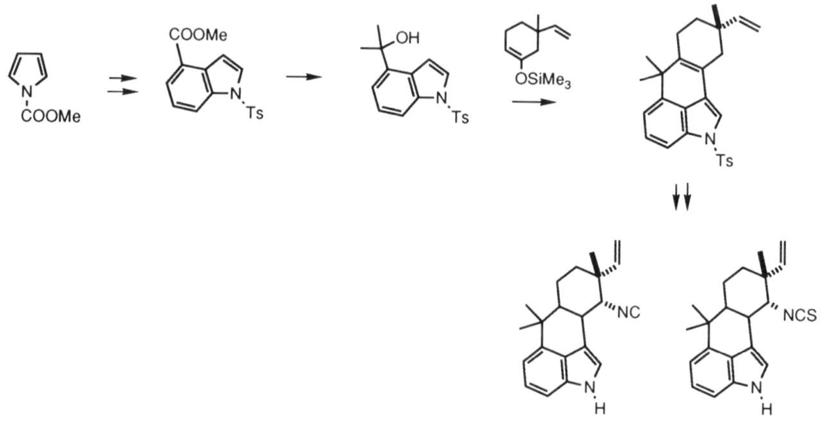

hapalindole-J.   hapalindole-M.

It is evident that the strategy of elaborating indole derivatives from pyrrole is all the more expedient in dealing with synthesis of the teleocidins [Muratake, 1987].

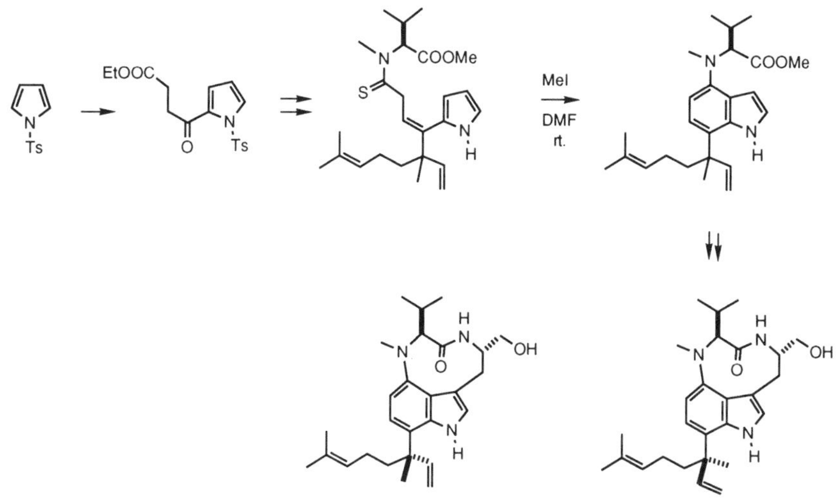

teleocidin-A2

lynbyatoxin-A
(teleocidin-A1)

The α-position of pyrrole is rendered electrophilic by conversion into an endoperoxide, as demonstrated in a synthesis of (−)-*cis*- and (−)-*trans*-trikentrin-A [Muratake, 1980]. Pyrrole itself must be first protected as the N-benzenesulfonamide to prevent oxidative degradation.

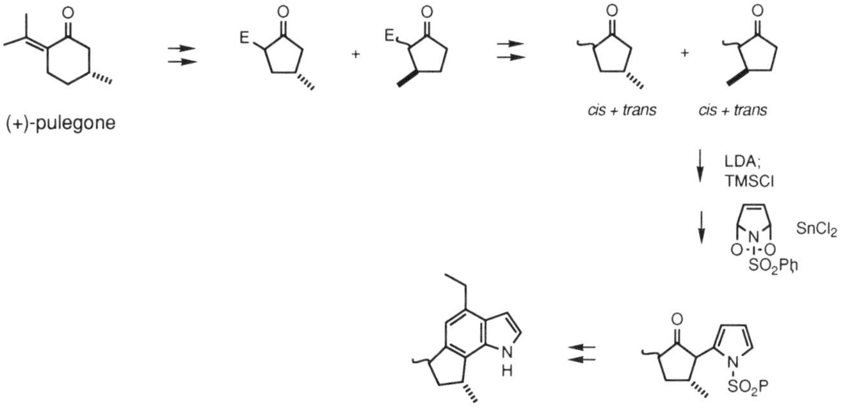

(+)-pulegone

cis + trans   cis + trans

(-)-*cis*-trikentrin-A
(-)-*trans*-trikentrin-A

Pyrrole undergoes oxidation with hydrogen peroxide to afford a mixture of 3- and 4-pyrrolin-2-ones. These compounds behave as electrophilic agents (2-oxopyrrolidin-5-ylium ion) toward phenol and indole [Bocchi, 1969].

The local symmetry of the heterocyclic portion of a *Reniera* metabolite renders possible its synthesis from dimethyl pyrrole-3,4-dicarboxylate. The benzoquinone moiety was created in the process [Frincke, 1982].

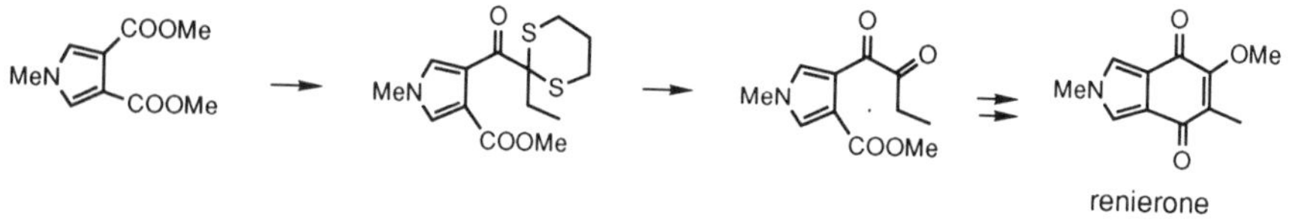

renierone

Apparently symmetry was a prominent factor which determined the synthetic route of hemin [Fischer, 1928, 1929]. The work was greatly simplified by condensation of two dipyrrylmethene subunits representing the AB- and CD-synthons because the latter is *meso*. No regiochemical problem arose. The situation for chlorophyll synthesis [Woodward, 1961] is different because an unsymmetrical porphyrin precursor was required.

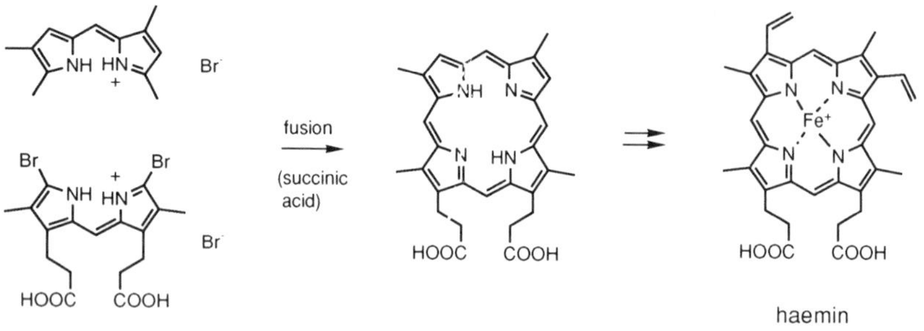

haemin

2,5-Dihydropyrrole derivatives can be alkylated at C-2. By this procedure a useful intermediate for an indolizidinediol isolated from *Rhizoctonia leguminicola* was obtained [Colegate, 1984]. A repeated alkylation at C-5 is possible, as demonstrated by an approach to monomorine-I [Macdonald, 1980].

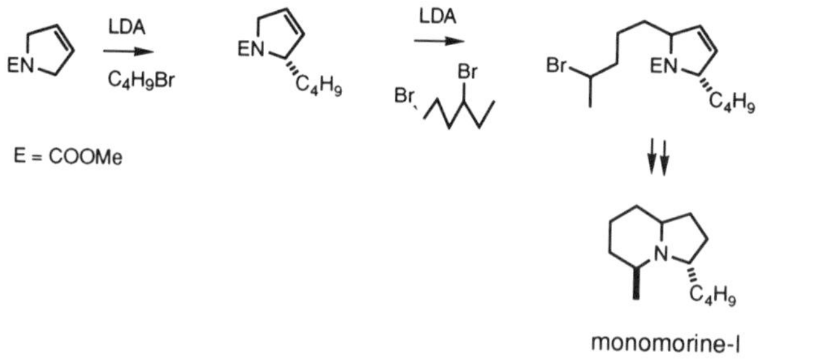

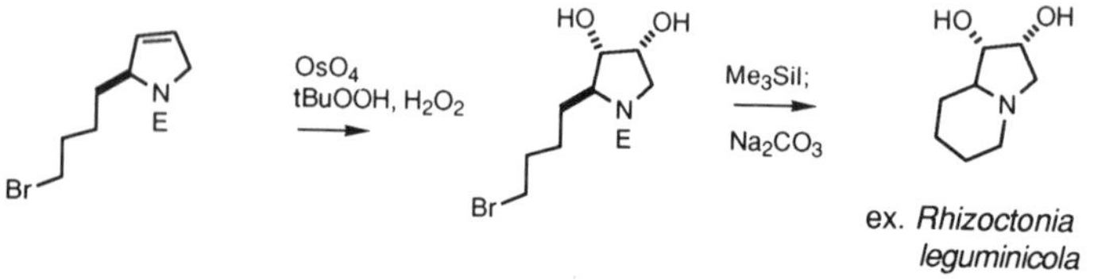

monomorine-I

ex. *Rhizoctonia leguminicola*

When the nitrogen atom of 2,5-dihydropyrrole is incorporated into a chiral formamidine, enantioselective alkylation is achievable [Meyers, 1987a]. With a chiral auxiliary derived from valine, the alkylation product was transformed into (+)-anisomycin (unnatural).

ent- anisomycin

A synthetic route to (*E*,*E*)-1,5-dienes consists of manipulating the symmetrical *N*-hydroxycyclobutanopyrrolidine through oxidation and 1,3-dipolar cycloaddition sequences to introduce two side chains at the α- and α′-positions, and cycloreversion [Tufariello, 1987b]. Both symmetrical and unsymmetrical products may be acquired. (Note the formation of 1,3-dienes by deamination of 3-pyrrolines [Lemal, 1966].)

N-Substituted pyrrolidines undergo dehydrogenation on treatment with mercuric acetate. The oxidation, in combination with an attack by proper nucleophiles, constitutes a useful method for structural modification. In this manner a synthesis of peripentadenine was completed [Lamberton, 1983].

peripentadenine

1-Pyrrolinium methylide, accessible from *N*-methylpyrrolidine *N*-oxide [Chastanet, 1985] or the trimer of 1-pyrroline [Terao, 1982], is a precursor of pyrrolizidines.

(four isomers)

1-Pyrroline N-oxide is available from N-ethylpyrrolidine via N-oxide formation, Cope elimination, and oxidation. It has found enormous utility in creating a 2-(β-hydroxyalkyl)pyrrolidine by reaction with an alkene followed by reduction of the N—O bond of the 1,3-dipolar adduct [Tufariello, 1979a]. Elaboration of supinidine [Tufariello, 1975], isoretronecanol [Iwashita, 1982], elaeocarpine and isoelaeocarpine [Tufariello, 1979b], elaeokanine [Tufariello, 1979c; Otomasu, 1982b], ipalbidine [Iida, 1985a], septicine [Iwashita, 1980], cocaine [Tufariello, 1979c], anatoxin-a [Tufariello, 1985], and a Nuphar alkaloid [Tufariello, 1987a] from the cycloadducts derived from the proper alkenes has been accomplished. Note the excellent regioselectivity of the cycloaddition due to frontier orbital control. In the approaches to cocaine and anatoxin-a the cycloaddition process was applied on two different occasions, when the second was an intramolecular version.

supinidine

elaeocarpine   isoelaeocarpine

elaeokanine-A   elaeokanine-C

elaeokanine-A   elaeokanine-C

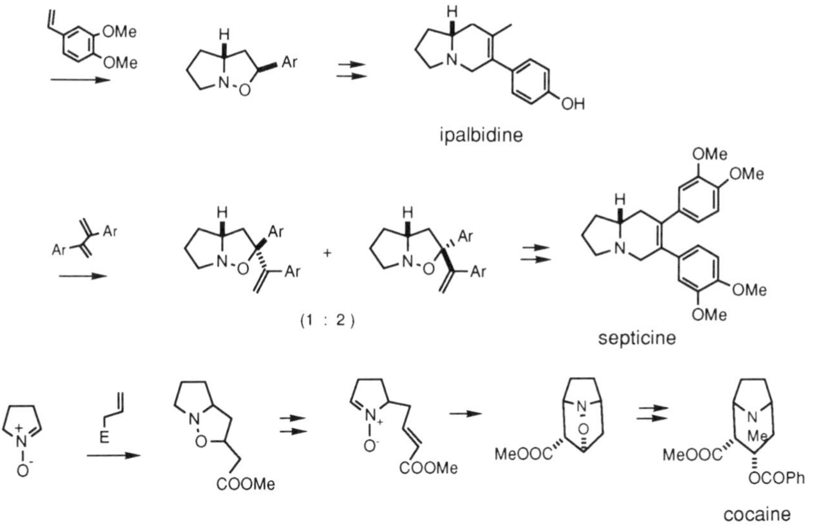

ipalbidine

(1 : 2)

septicine

cocaine

N-Alkoxycarbonylpyrrolidines are conveniently activated by electrochemical methoxylation at C-2. The products are masked N-acyliminium ions and they can undergo condensation in the presence of a Lewis acid with enol derivatives to provide precursors of hygrine [Shono, 1981], elaeokanine-C [Shono, 1984], or with a malonic ester to furnish an intermediate for the necine bases isoretronecanol and trachelanthamidine [Shono, 1984]. When the N-acyl group contains a nucleophile, a subsequent intramolecular reaction of the electrooxiation product is most profitable. An approach to isoretronecanol and trachelanthamidine [Blum, 1984] evolved.

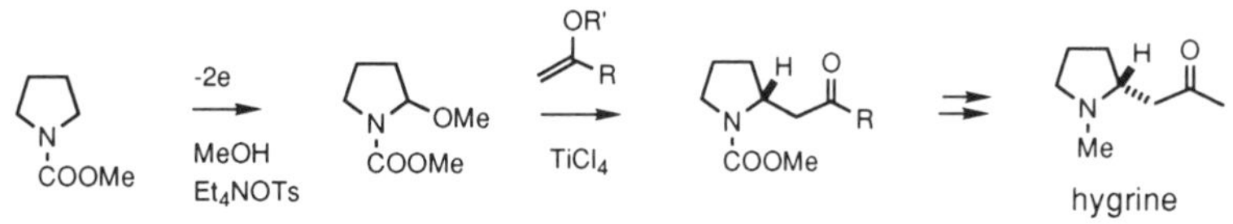

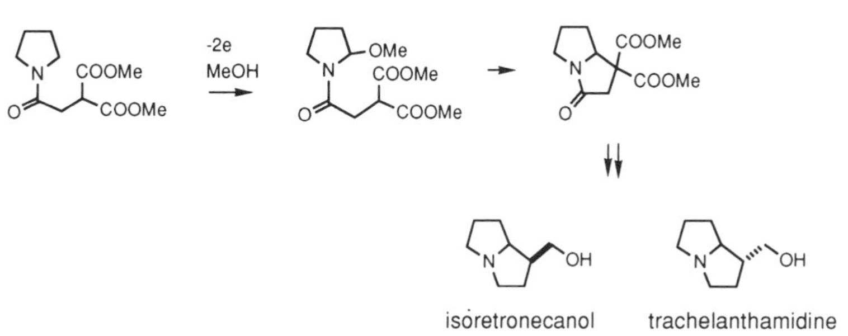

hygrine

isoretronecanol   trachelanthamidine

The methyl ester of oxalylacetylpyrrolidine is photoreactive, being transformed into a pyrrolidizine which can be used in a synthesis of isoretronecanol [Gramain, 1985]. Abstraction of the α-H by the (n,π*)-state carbonyl oxygen was followed by cyclization during the desymmetrization process.

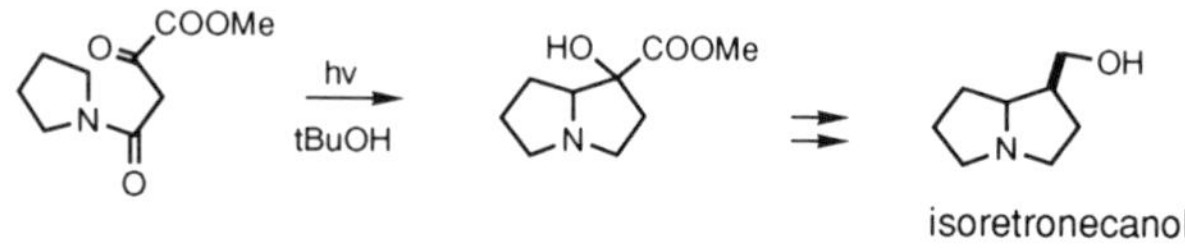

isoretronecanol

The double anodic methoxylation of N-carbomethoxypyrrolidine affords a substance in which both α-carbon atoms of the heterocycle become electrophilic. The skeleton of anatoxin-a was constructed [Shono, 1987] in a very concise manner from this compound.

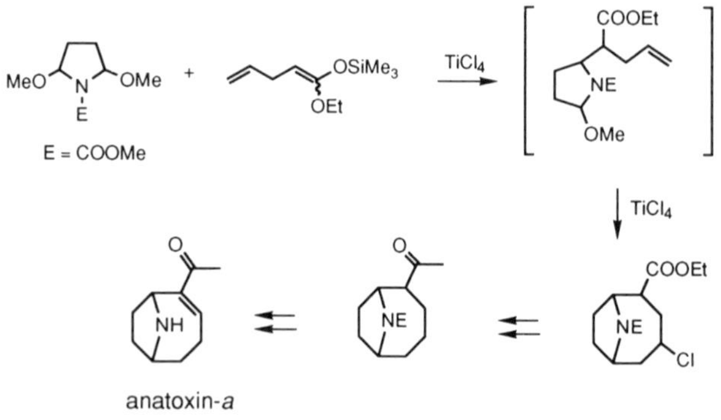

E = COOMe

anatoxin-a

N-Benzyl-2,5-dicyanopyrrolidine undergoes twofold alkylation with one or two different electrophiles. The products are readily transformed into 1,4-diketones [K. Takahashi, 1987]. Therefore the pyrrolidine is a valuable succinyl dianion synthon.

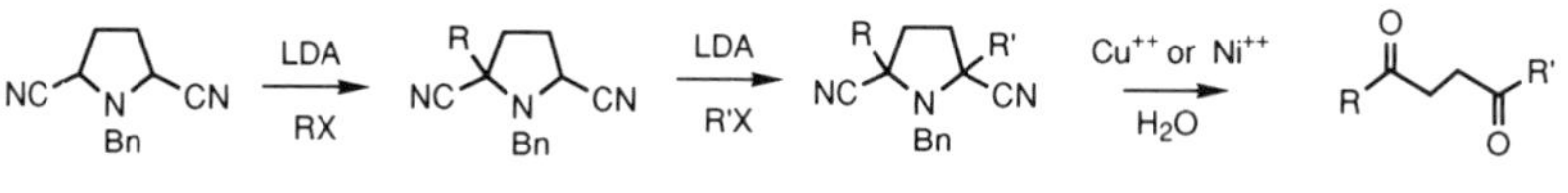

The pyrrolobenzazepine portion of cephalotaxine has been assembled from a maleimide [Tse, 1976]. Reaction with methylmagnesium iodide followed by dehydration provided a dienamide which was photochemically induced to form the benzazepine system.

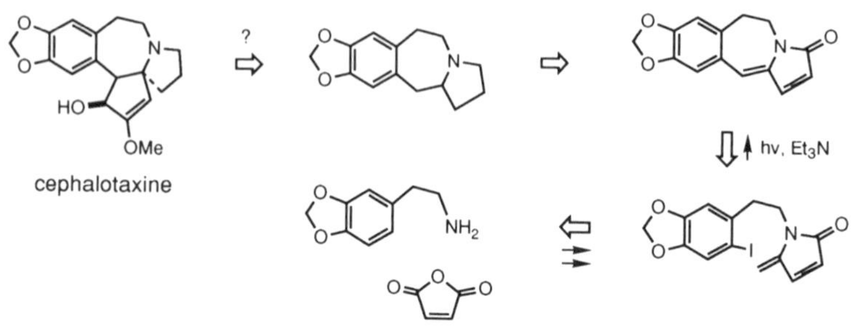

cephalotaxine

Direct desymmetrization of succinimide by Grignard reaction and reduction to furnish a γ-substituted butyrolactam constituted the initial stage of a synthetic pathway to anatoxin-*a* [Esch, 1987]. A synthesis of ($+$)-anisomycin [Wong, 1969] started from a Grignard reaction on (2R,3R)-N-benzyltartrimide.

(+)-anisomycin

While the Wittig reaction is quite inadequate on an imidic carbonyl group, the intramolecular version that forms five- and six-membered rings is acceptable in terms of yields. Examples involve syntheses of elaeokanine-C [Flitsch, 1987] and necine bases [Muchowski, 1980; Flitsch, 1983]. In the latter method the phosphorane was generated from a cyclopropyltriphenylphosphonium salt.

N-Substituted succinimides, reduced by alkali metal borohydrides at controlled pH to give the γ-hydroxy-γ-lactams [Hubert, 1975], are hence desymmetrized. Synthetically useful reactions, sometimes after their alcoholysis or mesylation, are those with nucleophiles. If activated in situ by N-pivalylation, 5-ethoxy-2-pyrrolodine reacts with Grignard reagents to give γ-keto aldehydes [Savoia, 1991]. However, the most popular applications of these cyclic carbinolamides and derivatives are related to alkaloid synthesis, for example, cuscohygrine from N-methylsuccinimide [Speckamp, 1983], isoretronecanol and trachelanthamidine from N-(2-bromoethyl)succinimide [Kraus, 1980], and chiral indolizidine alkaloids 167B and 209D from an N-(S)-α-phenylethylsuccinimide [Polniaszek, 1990].

cuscohygrine

The condensation of 5-acetoxypyrrolidone with a chiral organostannyl enol ether proceeded with good diastereoselectivity [Nagao, 1988], thereby a very short synthesis of (−)-trachelanthamidine was at hand.

Particularly efficient syntheses of izidine alkaloids are based on intramolecular trapping of the *N*-acyliminium species. A double bond in the *N*-substituent is sufficiently nucleophilic to attack the cation. In a route to elaeokanine-A [Taber, 1991] and elaeokanine-B [Wijnberg, 1981] the double bond was conjugated to an ester and an acetal, respectively. In the elaboration of necine bases [Nossin, 1979; Chamberlin, 1984] sulfur atoms of a ketene dithio acetal or phenylthioacetylene exerted regiocontrol.

Note that the functional group distribution of the indolizidinone thus produced is most expedient for the preparation of elaeokanine-A [Chamberlin, 1984].

elaeokanine-A

Mention should be made about the transformation of an adduct of maleimide and 10-mercaptoisoborneol into an unsymmetrical succinimide which underwent regioselective reduction. The subsequent of cyclization occurred only from the *exo* side to give a precursor of (+)-trachelanthamidine [Arai, 1991].

(+)-trachelanthamidine

The silicon atom of a vinylsilane side chain attaching to the nitrogen atom of these building blocks tends to facilitate such cyclizations which adopt a chairlike transition state [Overman, 1983; Hiemstra, 1985]. A substrate containing an allylstannane residue behaved similarly [Keck, 1985]. A synthesis of streptazolin [Flann, 1987] from a C_2-symmetric tartrimide derivative enlisted this process as the key step.

streptazolin

Of special significance is an enantiodivergent indolizidine synthesis [Heitz, 1989] from (+)-isoascorbic acid. Direct cyclization of an N-acyliminium ion to a compound of one optical series, whereas reduction and conversion of the lactam to the iminium ion before cyclization channeled it into the enantiomer.

A 5-oxopyrrolinium species has been generated from a phenyl thioether analog via transposition of the thio group to a carbenoid site [Kametani, 1986]. Collapse of the resulting zwitterion resulted in the pyrrolizidine system. A version of free radical cyclization leading to a precursor of isoretronecanol [Hart, 1984] has also been reported.

2-Aza-1,5-hexadienes in which the C=N moiety is part of an *N*-acyliminium ion are prone to undergo Cope rearrangement. This reaction is followed by recyclization involving the transposed alkenyl side chain if the formation of a relatively stable (tertiary or benzylic) carbocation is feasible. An exploitation of the method to synthesize necine bases has been accomplished [Hart, 1985; Ent, 1986].

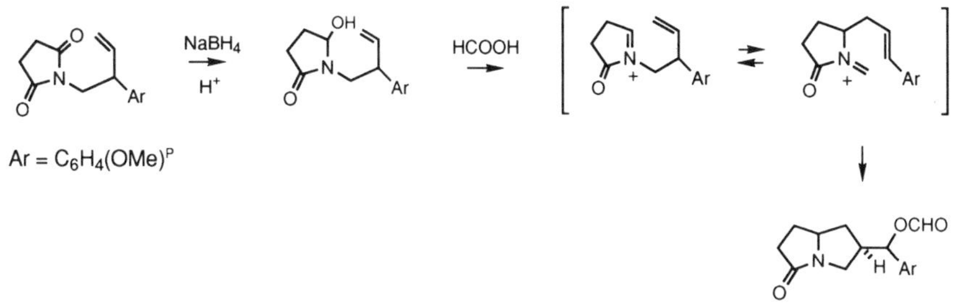

Ar = C$_6$H$_4$(OMe)^P

The combination of reductive desymmetrization of substituted amides and α-alkylation has also been employed in assembling even more complex natural products, for example, gephyrotoxin [Hart, 1983] and indolizomycin [G. Kim, 1990]. During the side chain attachment en route to indolizomycin the cyclopropane ring of a fused succinimide was the stereocontrol element.

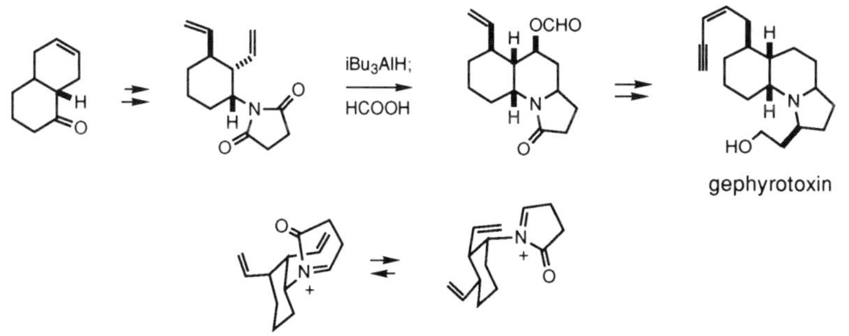

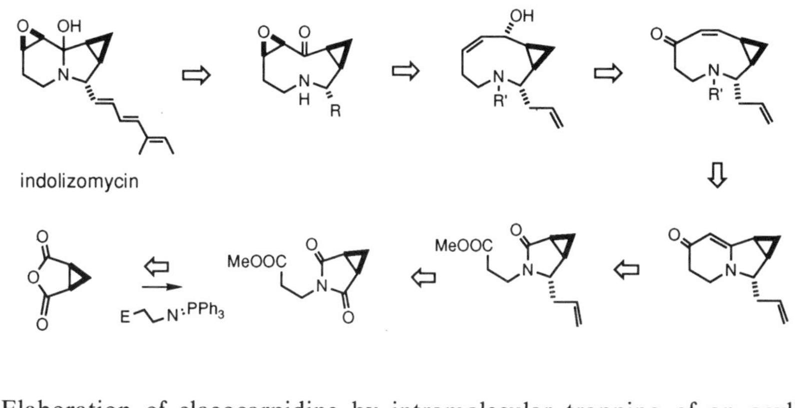

gephyrotoxin

indolizomycin

Elaboration of elaeocarpidine by intramolecular trapping of an acylium species is conceivable. Indeed a most expedient method is that by reduction of an imide [Harley-Mason, 1969].

elaeocarpidine

An obvious route to maleimycin consists of allylic hydroxylation of a symmetrical maleimide derivative [Singh, 1976].

maleimycin

(3S,4S)-Tartarimide proved to be useful for elaboration of swainsonine [Dener, 1988]. The C_2-symmetry allowed uncomplicated annexation of the six-membered ring, but it also required inversion of the oxygen functionality at the subangular carbon atom.

swainsonine

The symmetrical imide of pyrrolizidine affords the carbinolamide on reduction with LiAlH$_4$. Various 2-alkylpyrrolidines and pyrrolizidines are conceivably available by proper manipulation [K. Takahashi, 1989].

Tropinone and analogs are readily available by the Robinson–Schöpf method. An excellent application of such a bicycloheterocyclic compound in natural product synthesis consists of unraveling dimethyl betalamate, the progenitor of betanidin and indicaxanthin [Büchi, 1978]. By virtue of the bridged ring system various intermediates enjoyed much higher stability. Most of the intermediates are also symmetrical therefore encounters with isomeric mixtures were minimized.

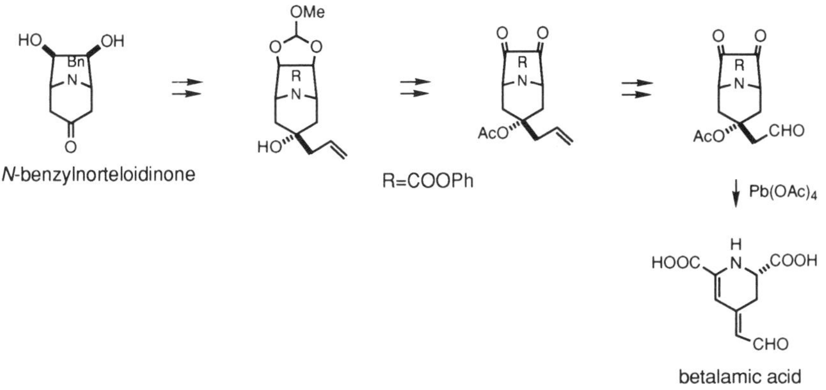

N-benzylnorteloidinone

R=COOPh

betalamic acid

A tropanone can be desymmetrized by regioselective enolization [Momose, 1990]. Oxidative cleavage of such an enol silyl ether paved the way to (+)-monomorine-I. On the other hand, N-chloronortropane undergoes rearrangement to the dehydropyrrolizidinium ion on exposure to a silver salt [Schell, 1980]. Hydride reduction returns it to a symmetrical base.

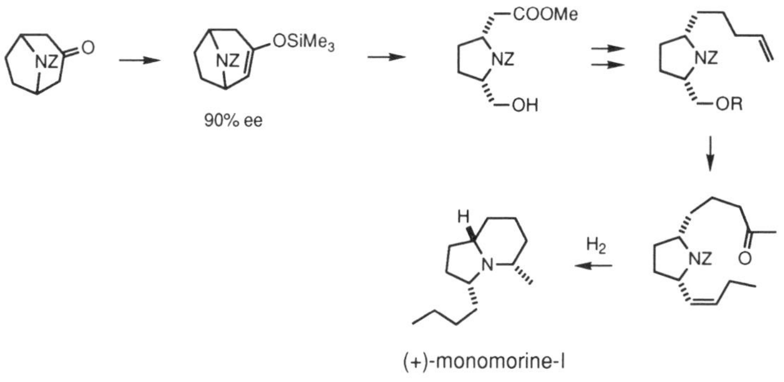

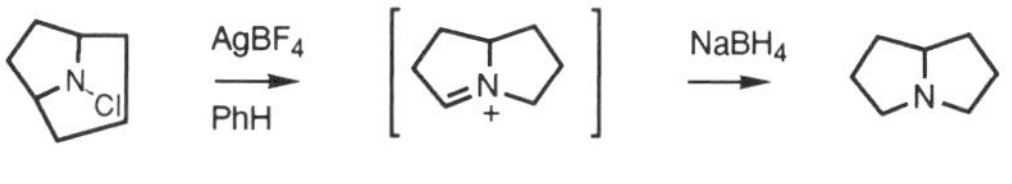

90% ee

(+)-monomorine-I

It is interesting to note the conversion of an unsymmetrical pyrrolidone to a pseudosymmetrical 1,8-dehydropyrrolizidine [Arata, 1979] via a Dieckmann-type cyclization and decarbonylation. In fact the penultimate precursor of the dehydropyrrolizidine is a symmetrical 5-azacyclooctanone.

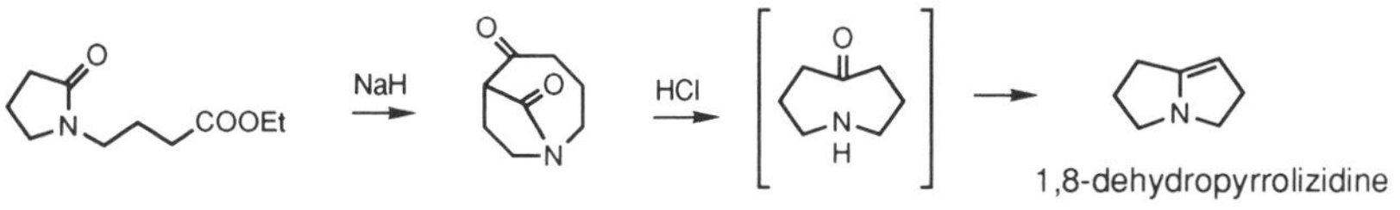

1,8-dehydropyrrolizidine

Biotin has a symmetrical imidazolidinone portion which can be exploited synthetically, for example, by a [2+2]-photocycloaddition pathway [Whitney, 1981]. The four-membered ring of the photoadduct was later expanded by heteroatom insertion (Baeyer–Villiger reaction and replacement of the cyclic oxygen atom with a sulfur atom).

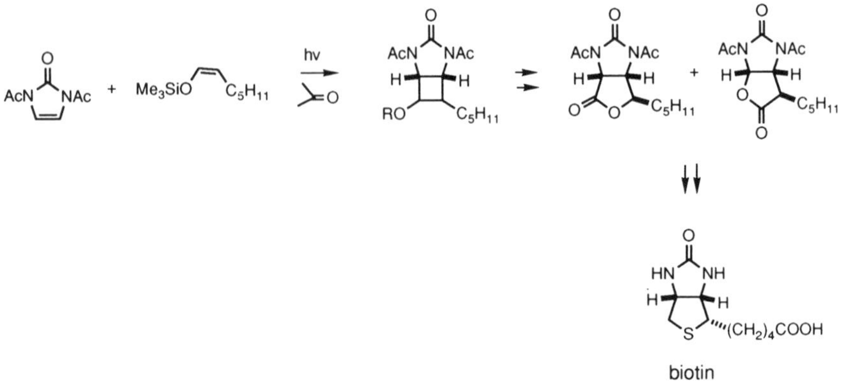

biotin

N-Protected *meso*-imidazolidin-2-one-4,5-dicarboxylic esters are readily converted into the ring skeleton of biotin. Introduction of the carboxylic acid side chain via alkylation of the proper sulfoxide has been demonstrated [Lavielle, 1978].

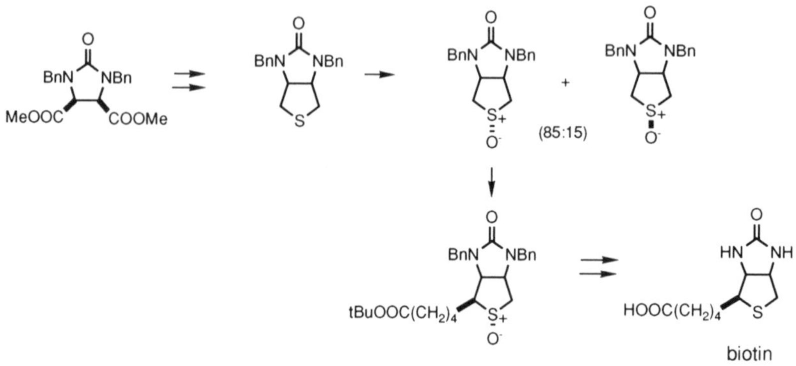

biotin

Pig liver esterase-catalyzed hydrolysis of the diacetate of *cis-N,N'*-dibenzylimidazolidin-2-one-4,5-dimethanol afforded a monoacetate which was readily transformed into (+)-biotin [Y.-F. Wang, 1984a].

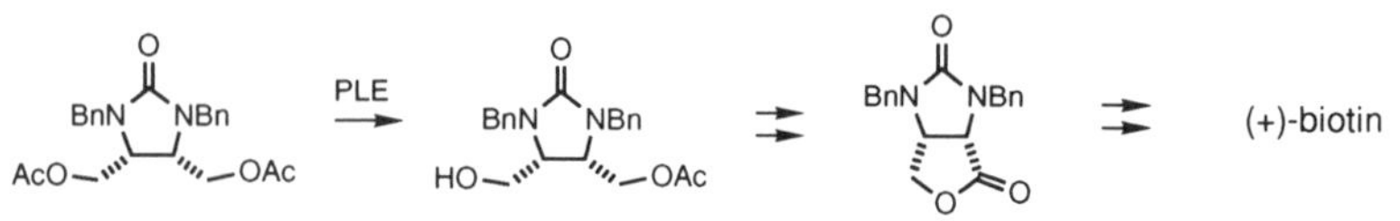

A number of alkaloids have been synthesized from pyridine or its simple derivatives via desymmetrizing manipulations. Among several methods the complex metal hydride reduction of pyridinium salts is frequently used because dihydro- or tetrahydropyridines are versatile synthetic intermediates. Interestingly, the reduction of a chiral pyridinium chloride derived from Zincke's salt and (R)-$(-)$-phenylglycinol gives a bicyclic oxazolidine which is susceptible to Grignard reactions to afford 1,2-dialkyl-1,2,3,6-tetrahydropyridines [Mehmandoust, 1989]. The properly substituted compounds have been converted into $(+)$-anatabine and $(+)$-coniine.

Zincke's salt

(+)-anatabine   (+)-coniine

N-Carbomethoxy-1,2-dihydropyridine is a suitable diene for the Diels–Alder reaction. For example, its cycloaddition with methyl 2-acetoxyacrylate provided an adduct which was further transformed into a substrate for a Cope rearrangement. The resulting *cis*-hydroisoquinoline represented a viable DE-ring synthon for reserpine [Wender, 1987].

reserpine

Instead of hydride reduction, an organometallic reaction can be used to introduce a carbon chain to C-2 of the pyridine ring accompaying the partial saturation. An interesting utility of these substituted dihydropyridines is the elaboration of solenopsin-A [Ogawa, 1984b] via a Diels–Alder reaction with methyl cyanodithioformate. The more complex cannabisativene structure has also been synthesized [Ogawa, 1984a].

Since N-acyl-1,2-dihydropyridines are susceptible to sensitized photo-oxygenation to give reactive endoperoxides, further functionalization at C-6 is possible. Thus, activation of the endoperoxides by stannous chloride permits attack by nucleophiles from C-6 at the same side as the existing C-2 side chain, due to S_N2 reaction, and the stereocontrol of the photooxygenation by the substituent at C-2 [Natsume, 1986]. The allylic alcoholic function that is generated in the process is another exploitable feature.

The method has proved quite valuable in the synthesis of piperidine alkaloids, including carpamic acid [Natsume, 1980b], azimic acid [Natsume, 1980b], pseudocarpamic acid [Natsume, 1980c], prosafrinine [Natsume, 1980c], prosophyllin [Natsume, 1981c], sedacryptine [Natsume, 1983], sedinine [Ogawa, 1985], cannabisativine [Ogawa, 1984a], and the indole alkaloid epiuleine [Natsume, 1980a]. Several other syntheses of certain natural products from unsymmetrically substituted pyridines based on the same theme have also been accomplished but they are outside the scope of our discussion.

sedacryptine

cannabisativine

carpamic acid

epiuleine

Another method of desymmetrization and 2,6-difunctionalization of a pyridine ring is illustrated by a synthesis of solenopsin-A and dihydropinidine [Comins, 1991b]. By means of *N*-acylation, C-2 of 4-chloropyridine is activated for Grignard reaction. It also permits a heteroatom-directed deprotonation at C-6 in the next step so that another substituent can be introduced by alkylation. With careful choice of reducing agents either the *cis*- or the *trans*-isomer of a 2,6-disubstituted piperidine is obtained.

	cis	*trans*
EtSi₃H	75%	15%
NaBH₃CN	12%	77%

dihydropinidine   solenopsin-A

An analogous operation on 4-methoxypyridine was developed to access indolizidine-209B [Comins, 1991c] and pumiliotoxin-C [Comins, 1991a]. With certain variations such as employing a chiral auxiliary in the *N*-acylation and introducing a triphenylsilyl group to C-2 it is possible to prepare chiral piperidines [Comins, 1992].

indolizidine-209B

pumiliotoxin-C

A trimethylstannyl group at C-4 of the pyridine blocks entry of nucleophile at that position, and the tactic was employed in a synthesis of epilupinine [Comins, 1986].

epilupinine

A σ-symmetric pyridinium ion was transformed into 18,19-dihydroantirhine [Chevolot, 1975] via the corresponding piperidine. A modified Polonovski reaction served to activate (and desymmetrize) the molecule for a Pictet–Spengler cyclization.

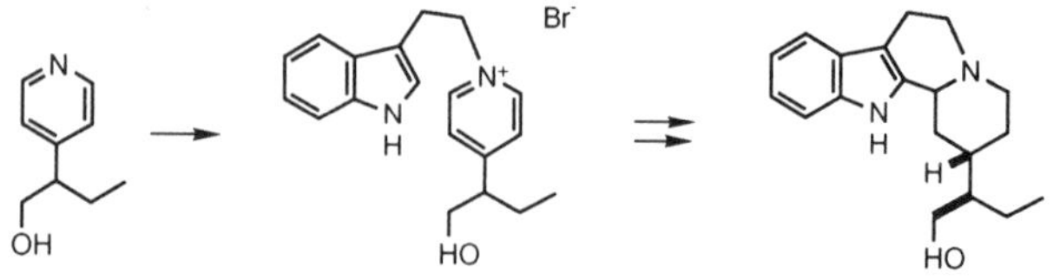

18,19-dihydroantirhine

Other 4-substituted pyridines that have been used in alkaloid synthesis include isonicotinaldehyde in the elaboration of olivacine [Besselièvre, 1976], N,N-diethylisonicotinamide and isonicotinic anhydride in the preparation of ellipticine [Watanabe, 1980; Kano, 1981], and isonicotinyl chloride in the assembly of nauclefine [Naito, 1979]. In each case the pyridine ring (the [c] bond) became part of a condensed ring system and desymmetrized.

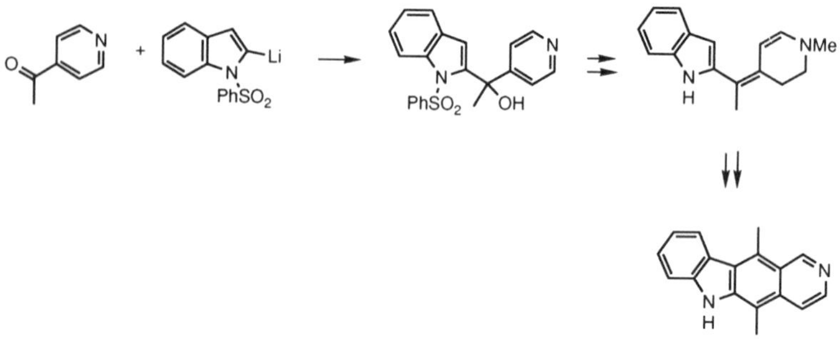

ellipticine

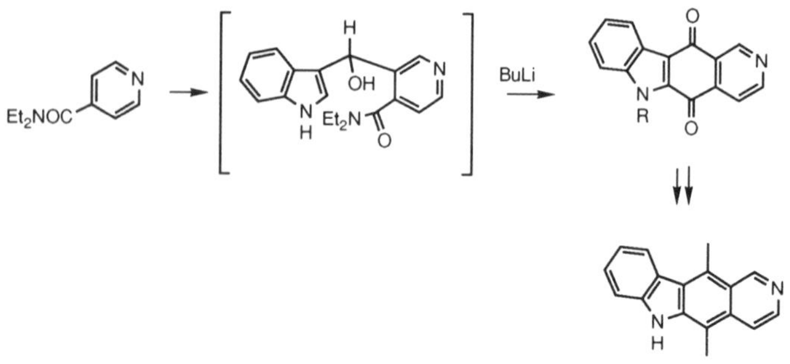

ellipticine

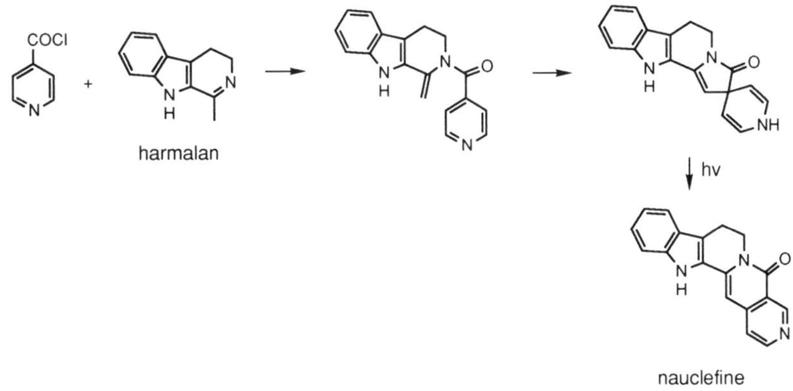
COCl
harmalan
hv
nauclefine

The use of 4-acetylpyridine as a building block [Besselièvre, 1975, 1976] may involve dearomatization, although the aromatic state must be restored. It is particularly interesting that in certain variations [Besselièvre, 1981; Naito, 1981] the heterocycle was severed completely in the intermediate stages. This strategy was inspired by biogenetic considerations.

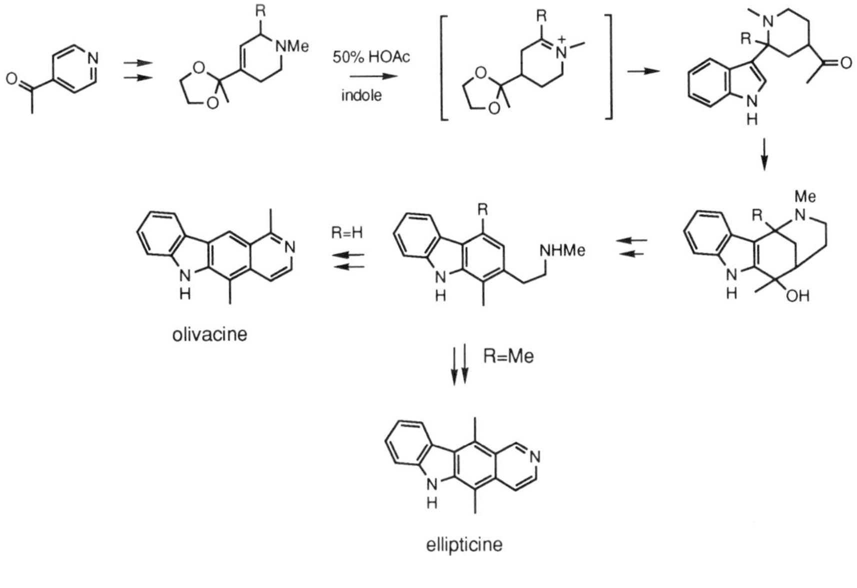
50% HOAc
indole
R=H
NHMe
R=Me
olivacine
ellipticine

Attachment of an indolyl group to C-2 of 4-acetylpyridine after quaternization, *N*-carbobenzyloxylation, and endoperoxidation constituted an important step toward synthesis of 3-epiuleine [Natsume, 1980a].

The most straightforward approach to muscopyridine is that involving nickel-catalyzed, twofold coupling of 2,6-dichloropyridine [Tamao, 1975]. Although a mixture of intermediates was produced, they converged to a single product.

muscopyridine

Lythranidine is almost C_2-symmetric, the symmetry being broken by methylation of one of the two phenolic hydroxyl groups. This structural feature has been exploited in its synthesis from the Wittig reagent derived from 2,6-lutidine [Fuji, 1980]. The symmetric properties permitted the ready isolation of an amine with the relative configuration at four stereogenic centers corresponding to that of the natural alkaloid. An earlier desymmetrization was subsumed in a preparation of solenopsin-A [Fuji, 1979].

lythranidine

solenopsin-A

More than half a century ago 2,6-lutidine was already employed in a synthesis of lobelanine [Scheuing, 1929] by a strategy of chain extension from both methyl groups. A synthesis of dehydrosolenopsin-C [MacConnell, 1971] involved monolithiation of 2,6-lutidine, subsequent alkylation, and reduction of the heterocycle. Both the acidity of the benzylic hydrogen atoms and equivalency of the methyl groups are important factors for the method to be applicable. Two individual chain extension processes converted s-collidine into a useful intermediate for the synthesis of myrrhine and hippodamine [Ayer, 1976b].

lobelanine

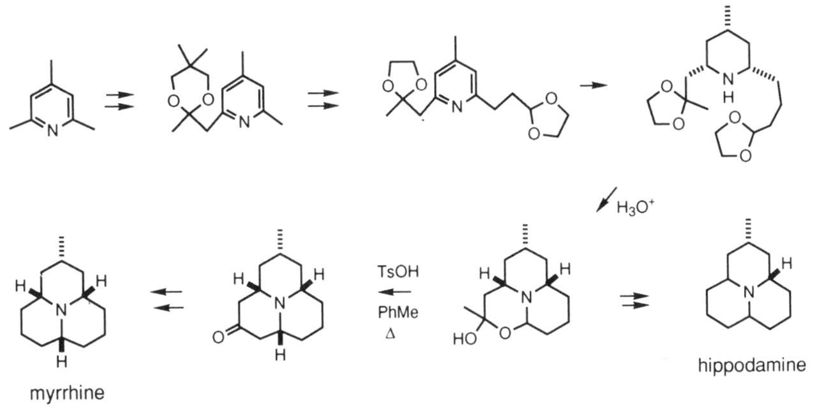

myrrhine

hippodamine

2,6-Lutidine is also a masked 2,6-heptanedione. Birch reduction of the heterocyclic system (and hydrolytic workup) unravels the chain. Construction of a 19-norsteroid [Danishefsky, 1974] was readily achieved by exploitation of this property and the alkylation potential of 2,6-lutidine.

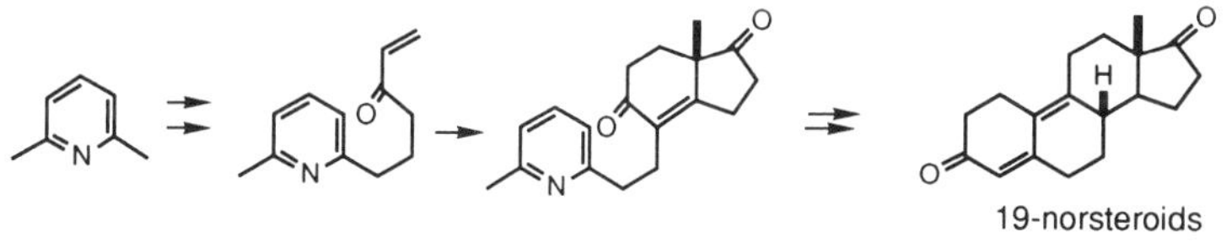

19-norsteroids

It is not surprising that 2,6-pyridinedicarboxylic acid is a viable starting material for the synthesis of berninamycinic acid [T.R. Kelly, 1984]. Once converted to two different amides nuclear functionalization was possible. The two subsequently formed thiazole rings were retained respectively as part of the molecular framework and until the final step. Actually the redundant heterocycle guarded the amide against hydrolysis, which must be performed on the other amide.

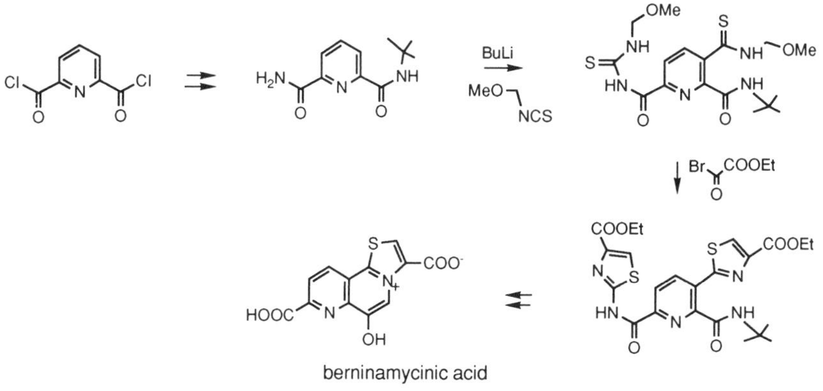

berninamycinic acid

In cytisine the saturated portion of the quinolizidine moiety is a
2,3,5-trisubstituted piperidine in which both C-3 and C-5 are one carbon long
and converge to the same nitrogen atom. A synthetic design [Bohlmann, 1956]
called for commencement from pyridine-3,5-dicarboxylic acid.

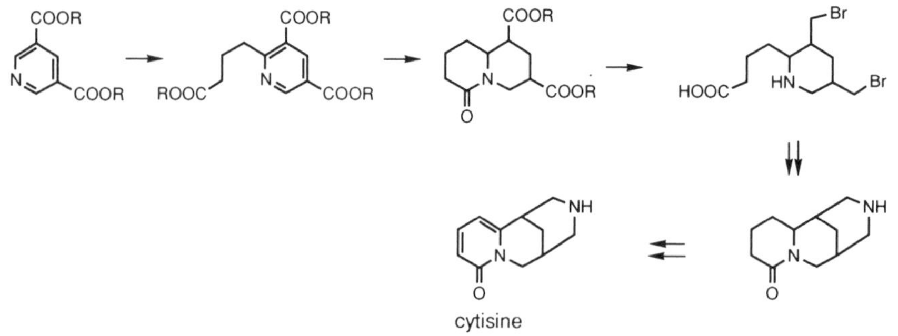

cytisine

2,2-Bipyridine is an excellent bidentate ligand for metal ions, but it has rarely
been used as a substrate for the construction of other interesting molecules. A
unique application of 6,6′-dimethyl-2,2′-bipyridine is its incorporation into the
skeleton of sexipyridine [Newkome, 1983].

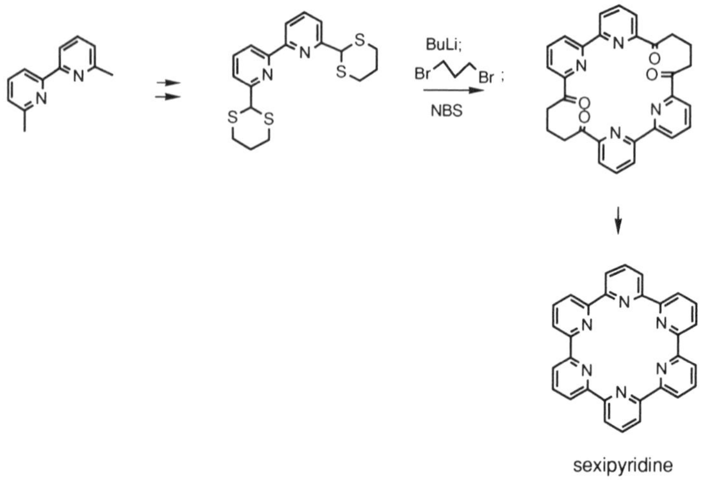

sexipyridine

An unusual synthesis of nicotine [Plaquevent, 1993] from 3,3′-bipyridine
involved monomethylation, reduction, ring contraction, and deformylation.

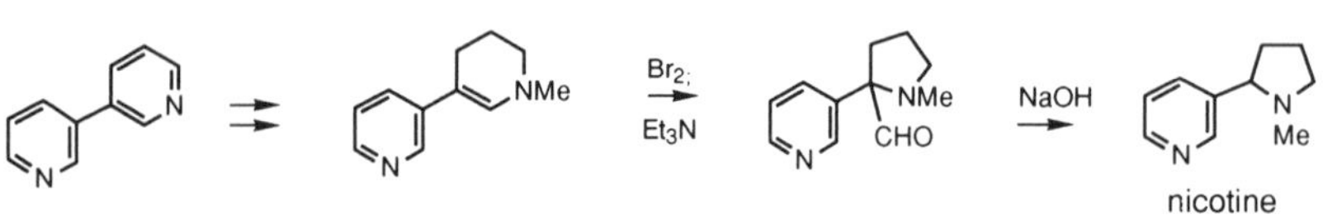

nicotine

Many pyridinium salts undergo ring opening on exposure to nucleophiles. The transformation of 4,4'-bipyridine to 6,6'-biazulene [Hanke, 1980] is a typical synthetic application.

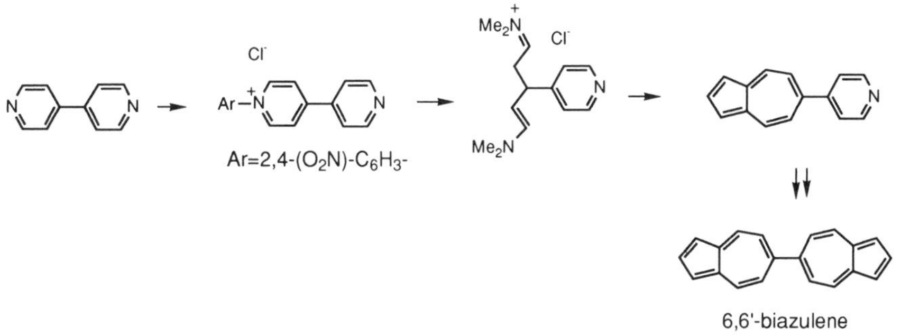

Ar=2,4-(O$_2$N)-C$_6$H$_3$-

6,6'-biazulene

Structural correlation of ascididemin with 1,10-phenanthroline-5,6-quinone has prompted a synthesis [Moody, 1990] which probably cannot be bettered in terms of brevity.

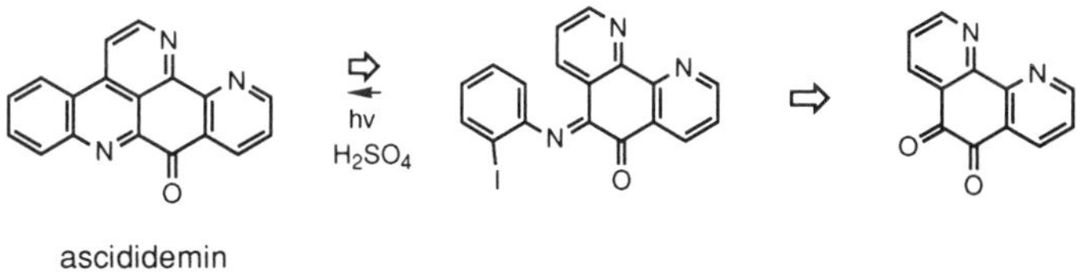

ascididemin

2,6-Diazaanthracene has been used as a building block for vineomycinone-B$_2$ methyl ester [Bolitt, 1991]. The two heterocycles were converted into substituted benzene moieties by the Bradsher reaction.

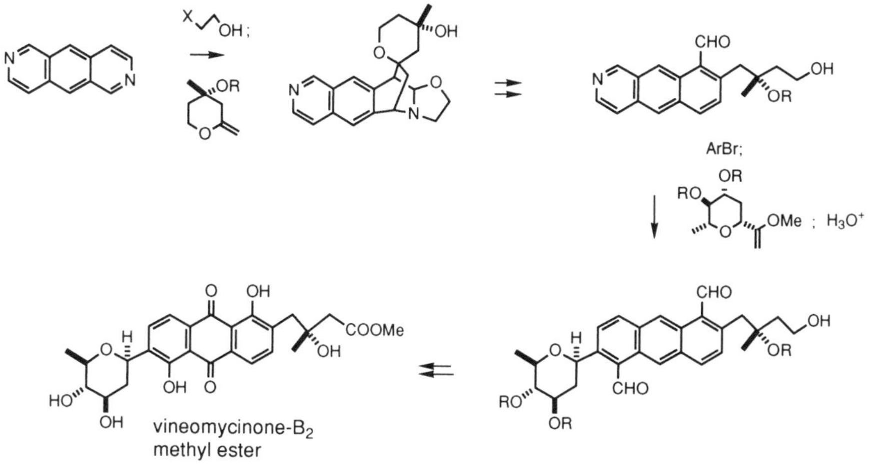

vineomycinone-B$_2$
methyl ester

Symmetrical pyridazines are not particularly useful in synthesis, but a number of unsymmetrical pyridazines derived from 1,2,4,5-tetrazines have been exploited further on the basis of the Diels–Alder/retro-Diels–Alder tandem reactions. Pyrimidines are also capable of undergoing Diels–Alder reactions to give pyridines upon in situ elimination of a nitrile (or HCN). Depending on the substituents of the pyrimidines, both cycloaddition modes of normal and inverse electron demand are achievable. An intramolecular cycloaddition of a symmetrical 4,6-dihydroxypyrimidine was applied to a synthesis of actinidine [Davies, 1981].

actinidine

1,3,5-Triazine-2,4,6-tricarboxylic esters undergo Diels–Alder reaction with amidines (in the 1,1,-diaminoalkene form) to give unsymmetrical pyrimidines. This was the basis of a synthesis of (−)-pyrimidoblamic acid (and deglycobleomycin-A2) [Boger, 1993].

(-)-pyrimidoblamic acid

Electron-deficient 1,2,4,5-tetrazines provide an excellent template for synthesis of aromatic and heteroaromatic substances based on their tandem Diels–Alder/retro-Diels–Alder reactions [Boger, 1986a]. They usually react with electron-rich dienophiles at room temperature, but with simple or electron-deficient dienophiles higher reaction temperatures are required. By adroit manipulation the two nuclear carbon atoms of dimethyl 1,2,4,5-tetrazine-3,6-dicarboxylate have been incorporated into the skeletons of streptonigrin [Boger, 1985], the dimethyl ester of the cAMP phosphodiesterase inhibitor PDE-II [Boger, 1986c], octamethyporphin [Boger, 1984], and prodigiosin [Boger, 1988]. Pyridazines (usually unsymmetrical) and 1,2,4-triazines were initially formed which were employed in further transformations.

streptonigrin

PDE-II methyl ester

prodigiosin

A synthesis of *cis*-trikentrin-A [Boger, 1991] is a more recent demonstration of the efficient assembly of an unsymmetrical molecule by stitching together three components from the symmetrical 3,6-bismethylthio-1,2,4,5-tetrazine, an enamine of *cis*-2,4-dimethylcyclopentanone, and an allenyl amine. All elements but the two nuclear carbon atoms of the tetrazine were shed in the process.

cis-trikentrin-A

The first condensation and dinitrogen extrusion still afforded a symmetrical compound. Desymmetrization occurred when the corresponding disulfone was treated with the allenyl amine. Subsequent heating in acetic anhydride effected, besides the trivial *N*-acetylation, an intramolecular Diels–Alder reaction, elimination of dinitrogen, and also elimination of methanesulfinic acid to bring about aromatization. Hydrolysis of the acetamide completed the synthesis.

A methyl group of the disulfonyl diazine intermediate could be epimerized by treatment by triethylamine. Subjecting the epimer to an analogous reaction sequence as above produced *trans*-trikentrin.

Methods for activation of piperidines are analogous to those of pyrrolidines, that is, by chemical oxidation or electrooxidation. Synthetic applications for such species include biomimetic cyclization of a quinolizidine system that resulted in allomatridine and α-isosparteine [Bohlmann, 1963], oxidative formation of the 8-oxosparteine [van Tamelen, 1960b] from 1,5-dipiperidinyl-3-pentanone, and a side chain attachment process toward synthesis of conhydrine [Shono, 1983].

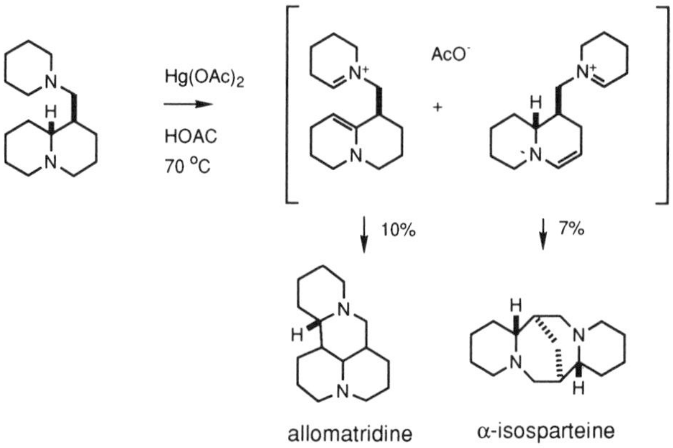

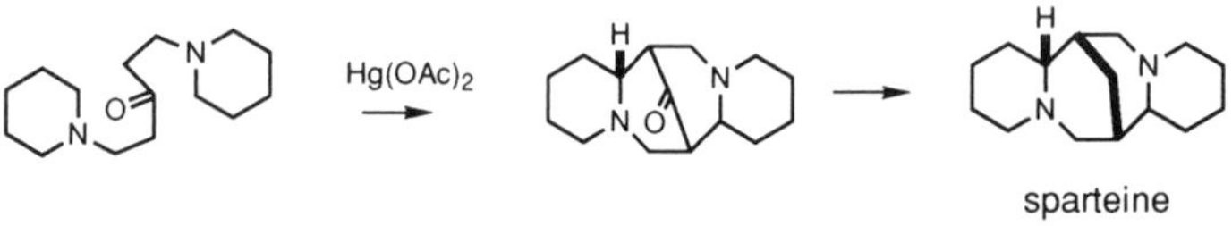

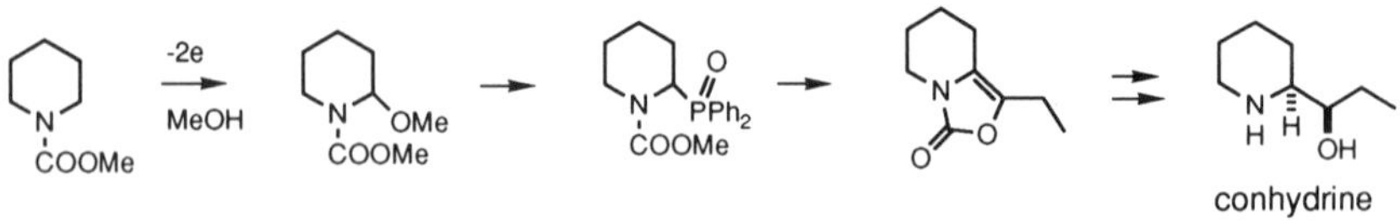

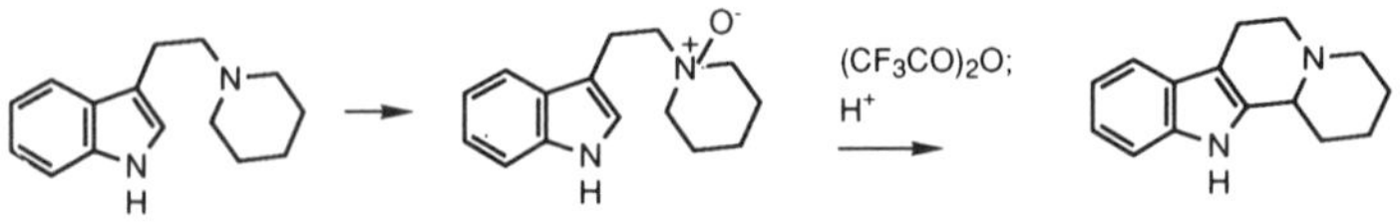

allomatridine   α-isosparteine

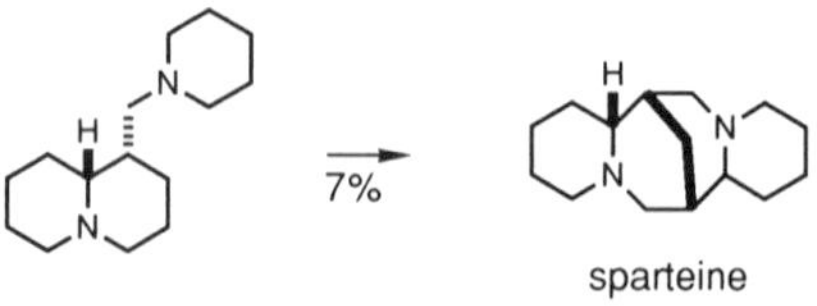

sparteine

sparteine

conhydrine

The classical method of desymmetrizing activation of piperidine is via *N*-chlorination and dehydrochlorination to give Δ¹-piperideine. Reaction of

Δ^1-piperideine with 3-pyridyllithium gave anabasine [Scully, 1980]. The fourfold Mannich reaction that served to build up 8-oxo-α-isosparteine [Schöpf, 1957] from Δ^1-piperideine is remarkable.

anabasine

α-isosparteine

Δ^1-Piperideine N-oxide reacts with dipolarophiles in the same manner as the lower homolog. Accordingly its potential for alkaloid synthesis, including sedrideine and allosedamine [Tufariello, 1978a], lupinine [Iwashita, 1982], epilupinine [Tufariello, 1976], myritine [Tufariello, 1978b], lasubine-I and -II [Iida, 1984a], vertaline [Shishido, 1985], porantheridine [Gössinger, 1980], and α-isoparteine [Oinuma, 1983], has been explored.

R = Me sedridine
R = Ph allosedamine

lupinine   epilupinine   myritine   porantheridine

lasubine-I   lasubine-II   vertaline   α-isosparteine

N-Benzyl-2,6-dicyanopiperidine can be alkylated at one or both α-positions, and the cyano groups are readily removed reductively, therefore it is a convenient starting material for many piperidine alkaloids, e.g., solenopsin-A [K. Takahashi, 1985].

The chemistry of substituted glutarimides is completely analogous to succinimides in terms of partial reduction and bond formation via trapping of N-acyliminium species. It has been possible to construct epilupinine [Chamberlin, 1984; Hiemstra, 1985], lasubine-I [Ent, 1988], and matrine [Chen, 1986].

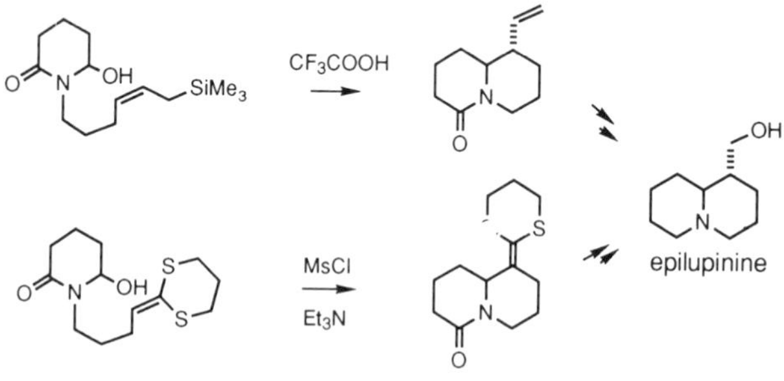

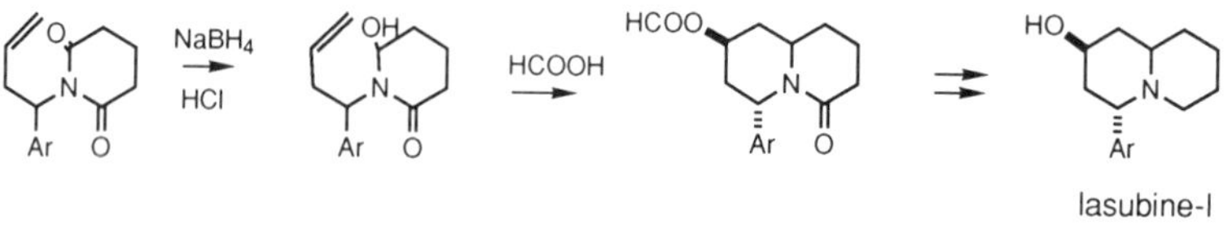

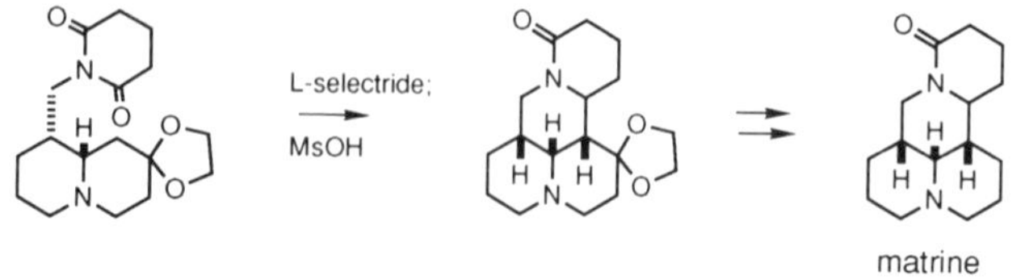

epilupinine

lasubine-I

matrine

It has been shown that a spirocyclic system of perhydrohistrionicotoxin is amenable to assembly by an intramolecular interception of the N-acyliminium ion in which the alkenyl chain is attached to C-5 [Evans, 1982b]. The required substrate was prepared from glutarimide by a Grignard reaction.

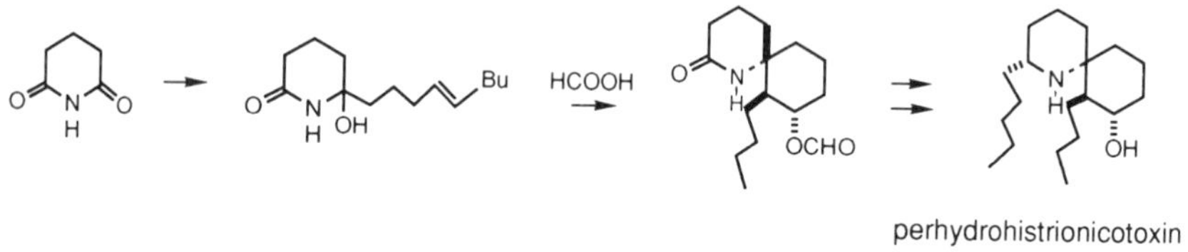

perhydrohistrionicotoxin

N-Alkyl and N-acyl-4-piperidones are very versatile compounds, often used in desymmetrizing operations via alkylation and the Robinson or Stork annulation to assemble many intermediates of alkaloids, e.g., ellipticine [Stilwell, 1964; LeGoffic, 1973] and yohimbine [Stork, 1972a].

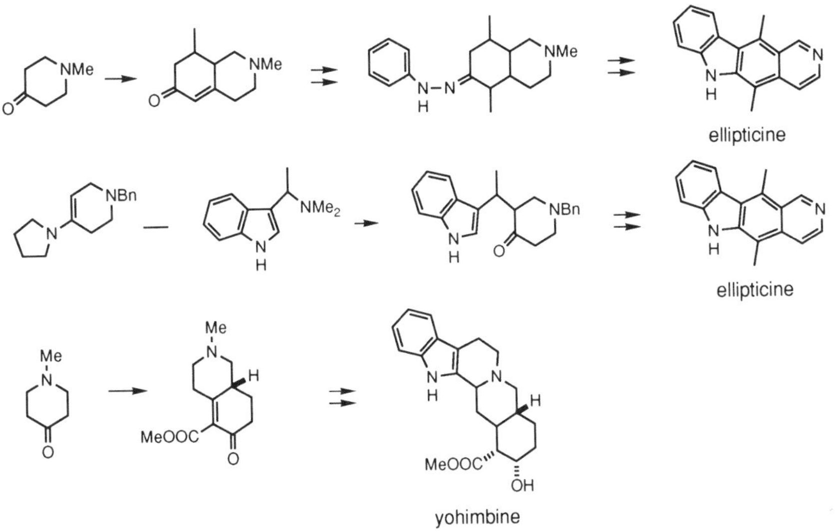

ellipticine

ellipticine

yohimbine

More elaborate applications and transformations of the *N*-methylpiperidone are illustrated in a preparation of an intermediate of condyfoline [A. Wu, 1975] and a formal synthesis of morphine [Evans, 1982a]. Because the piperidine ring in these target molecules is incorporated in bridged systems and far less conspicuous to casual examination, there is a surprise element associated with the usage of the building block.

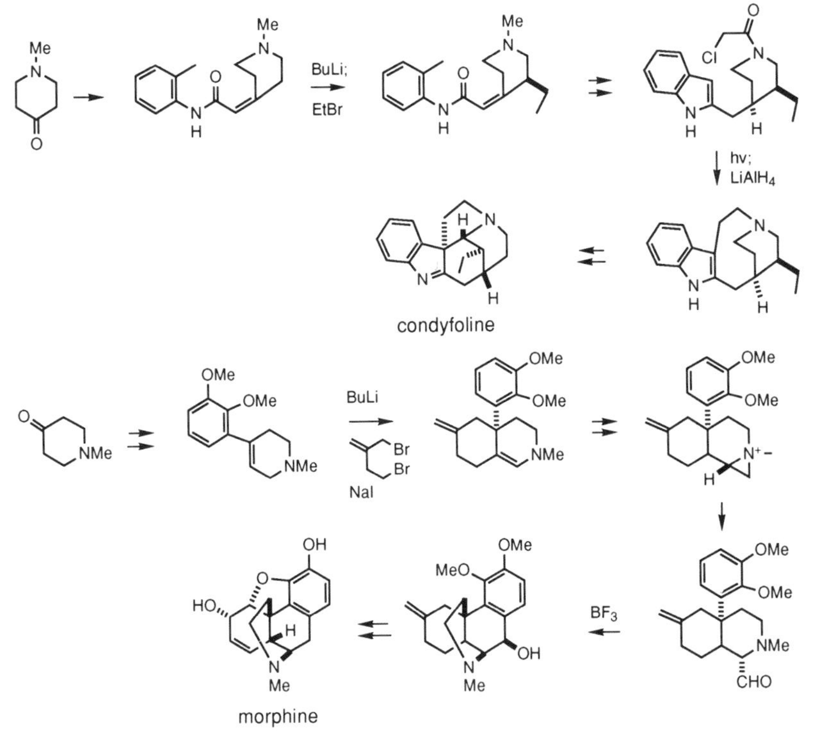

condyfoline

morphine

The ready availability of a bicyclic heterocycle provided an opportunity to synthesize *N*-benzoylmeroquinene and thence quinine and quinidine [Uskokovic, 1970]. Note that the *cis* stereochemistry of the β- and γ-pendants (to the nitrogen atom), contrary to that required for the synthesis of yohimbine, can be established by catalytic hydrogenation.

quinine

In a synthesis of catharanthine [Marazano, 1981] requiring *N*-benzyl-5-ethyl-1,2-dihydropyridine to condense with an α-(2-indolyl)acrylic ester in a Diels–Alder reaction, the source of the diene was a piperidone. The preparative method is superior to partial reduction of the corresponding pyridinium salt.

catharanthine

A very unusual approach to ellipticine [Differding, 1985] started from a Horner–Wadsworth–Emmons reaction of *N*-methyl-4-piperidone. The resulting ester was transformed into an anilide, and its subsequent dehydration to a ketene imine set up an intramolecular Diels–Alder reaction.

ellipticine

Introduction of carbon chains at other nuclear positions of the piperidine can be achieved at will, for example, at C-2 as needed for elaboration of streptazolin [Kozikowski, 1990], at C-5 with ketone group transposition to C-3 [Imanishi, 1978] for synthesis of tabersonine and catharanthine [Imanishi, 1980], ibogamine [Imanishi, 1981], and eburnamonine [Imanishi, 1983].

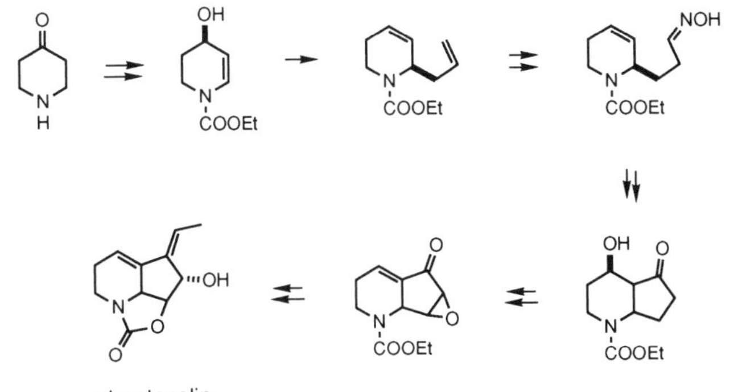

streptazolin

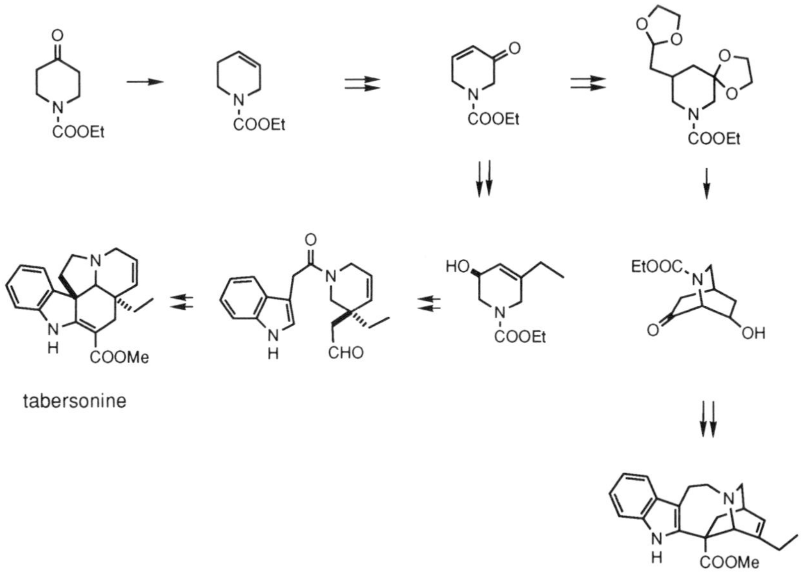

tabersonine

catharanthine

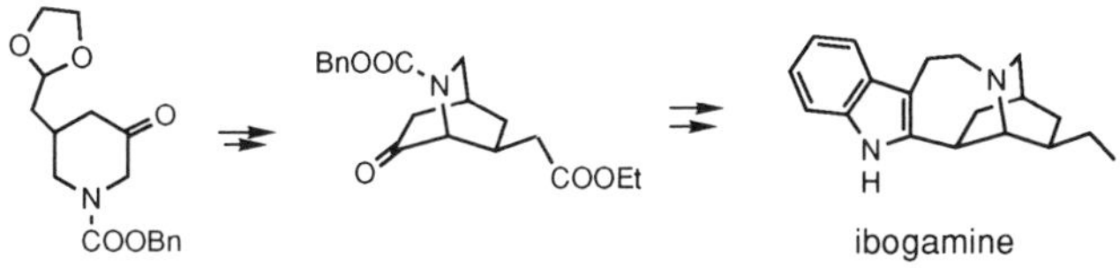

ibogamine

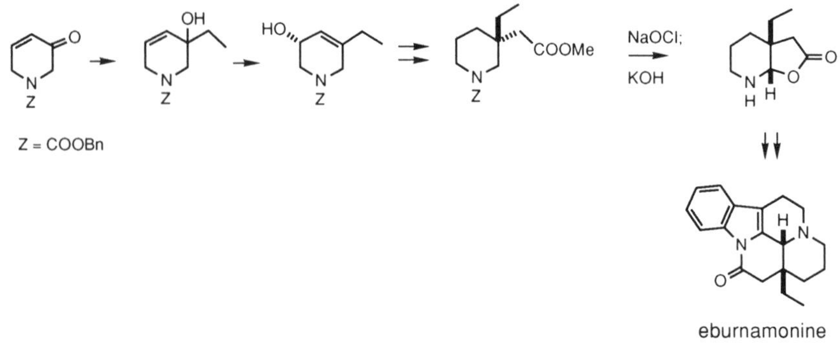

Z = COOBn

eburnamonine

Process variation led to useful precursors of corynantheidol and quinine [Imanishi, 1982].

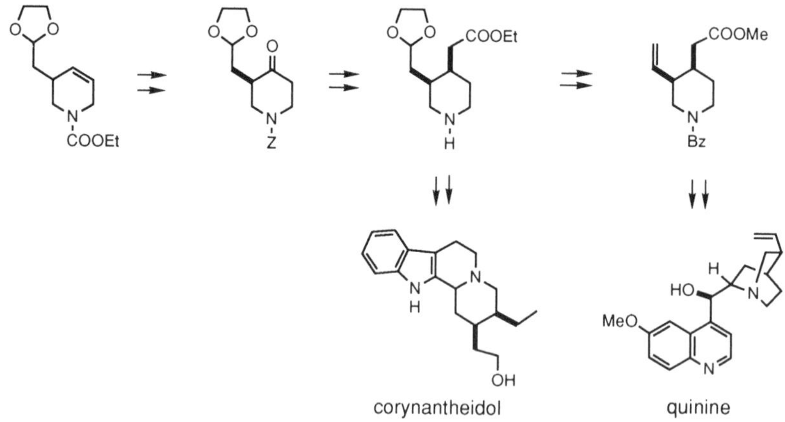

corynantheidol   quinine

N-Carbethoxy-4-piperidone has been converted into 2-piperazinecarboxylic acid [Merour, 1991] via a Schmidt reaction and Favorskii-type rearrangement.

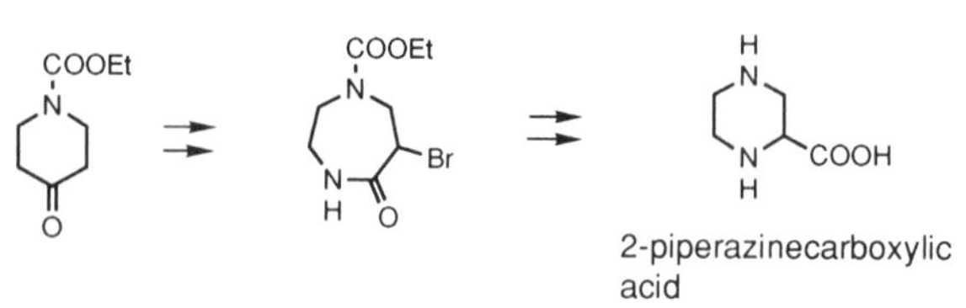

2-piperazinecarboxylic acid

Dimethyl *cis*-4-piperidone-2,6-dicarboxylate is readily available from the didehydro acid (chelidamic acid). Chain extension at the ketone site and dehydrogenation (one equivalent) provided an intermediate for the ester derivatives of indicaxanthin and betanidin [Hermann, 1977].

3,5-Piperidinediones with various substituents on the nitrogen atom have found uses in the elaboration of compounds as diverse as quebrachamine [Ziegler, 1970], I inokitiol and stipitatic acid [Tamura, 1977].

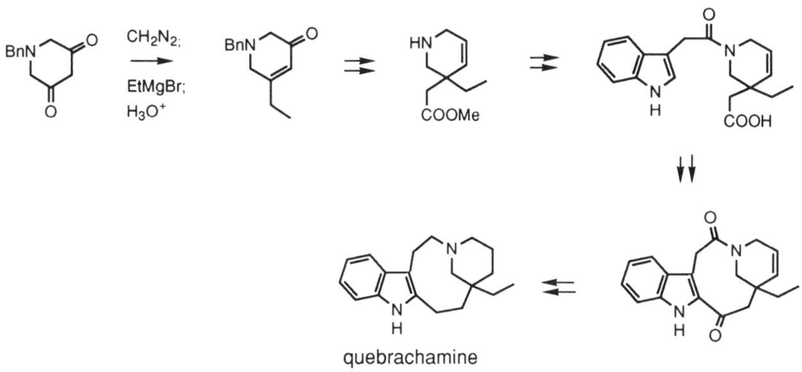
CH2N2;
EtMgBr;
H3O+
BnN
HN
COOMe
COOH
quebrachamine

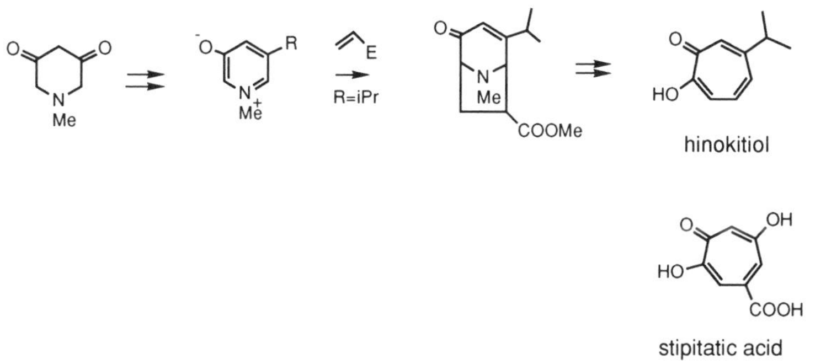
Me
R=iPr
COOMe
HO
hinokitiol
OH
COOH
stipitatic acid

These diketones existing mainly in the enol forms, are symmetrical in terms of reactivities. To the reagents all the atoms in each side of the plane lying over the nitrogen atom and C-4 are equivalent, although the intrinsic enolizability of these molecules renders their desymmetrization particularly easy. Interestingly a common intermediate or the synthesis of several lycopodium alkaloids was an all-*cis* 10b-substituted perhydro-6a-azaphenalene-1-3-dione [Wenkert, 1984], despite the noncorrespondence of stereochemical features of this ketone to the tricyclic core of the alkaloids. It is the possibility of stereochemical self-adjustment (further desymmetrization) prior to the tetracyclic stage of the synthetic operation that gave this approach its flexibility.

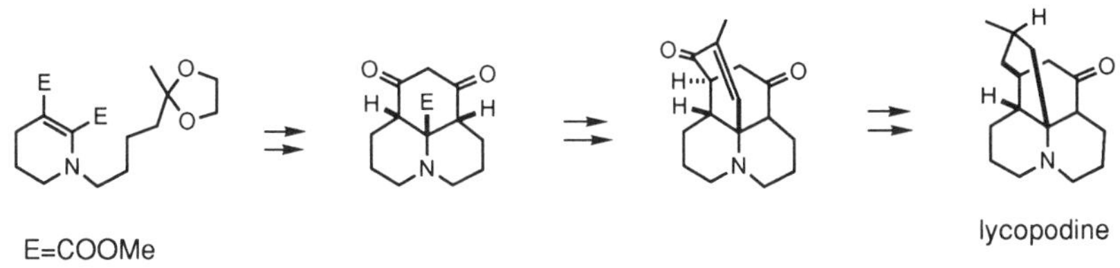
E=COOMe
lycopodine

Another synthesis of lycopodine [Ayer, 1968] also capitalized on the possibility of employing symmetrical precursors. Noteworthy is the regiochemistry of the intramolecular alkylation indicating Δ^1-enolization with respect to the *cis*-fused ring and Δ^2-enolization with respect to the *trans*-fused ring.

9-methoxyjulolidine

lycopodine

The electronic perturbation of an architecturally symmetric dibenzo-1-azabicyclo[3.3.1]nonane by a methoxy group in one of the aromatic rings was reflected in the selective Polonovski reaction of the *N*-oxide [Nomoto, 1984]. Consequently a cherylline synthesis was developed on this basis.

cherylline

An attractive approach to sarpargine-type alkaloids [Cloudsdale, 1982] focused on desymmetrizing manipulation of 2,6-dioxygenated 9-azabicyclo[3.3.1]nonanes, which are readily prepared from 1,5-cyclooctadiene diepoxide. The desymmetrization can be performed via monooxidation of the diol with poly(vinylpyridinium chlorochromate) or monoacetalization of the diketone. The latter procedure was nonselective, and chromatographic separation from the unreacted dione and the diacetal was required.

An enzymatic desymmetrization of a *meso*-1,3-glycol or its diacetate in which the functional groups are located in each of the two rings of 9-azabicyclo[3.3.1]nonane [Momose, 1992] gave enantiomeric all-*cis* 2,6-dialkyl-3-piperidinol synthons which are of apparent value in the synthesis of certain alkaloids, for example, (−)-cassine and (+)-spectaline [Momose, 1993]. The desymmetrization involved either a lipase-catalyzed transesterification of the diol or lipase-catalyzed hydrolysis of the diacetate.

The pseudosymmetrical *N*-(3-cyanopropyl)-6*exo*,7*exo*-dicyano-2-azabicyclo-[2.2.2]octane, obtained from 3-cyanopyridine in four stages, underwent a Ziegler cyclization. The derived tricyclic keto ester mixture was converted into two amino ketones each bearing an *exo*- and *endo*-oriented ethyl group (in a ratio of 2:3). Fischer indolization provided ibogamine and epiibogamine, respectively [Ikezaki, 1969]. A major defect of this synthesis is the epimerizability of the secondary cyano groups, the ester, and the ketone chain along the synthetic pathway.

ibogamine

The accessibility of the perboraphenalene system by hydroboration of 1,5,9-cyclododecatriene and the ready heteroatom exchange to give the aza-analog facilitated a synthetic approach to the ladybug defense agents such as hippodamine, hippocasine, and their *N*-oxides [Mueller, 1980]. The methyl group and/or the double bond were introduced via oxidative manipulation to generate a ketone.

A strategy, based on expansion of heterocyclic templates serving to construction of macrocyclic alkaloids, r..ay be illustrated by a synthesis of dihydroperiphylline [Wasserman, 1981], starting from piperidazine. The weak fusion bonds in two different intermediates were readily cleaved by reduction.

dihydroperiphylline

Retrosynthetic analysis of riboflavin (vitamin B$_2$) readily reveals that the terminal heterocycle may be derived from a symmetrical molecule. Both alloxan and barbituric acid have been successfully employed to condense with N-aribityl-3,4-xylidine and an azo derivative of the xylidine, respectively [Karrer, 1936; Tishler, 1947]. The symmetry of the heterocyclic building blocks and the pseudosymmetry of the target molecule are the key to a smooth transformation without generating other isomers.

R = H,O

riboflavin

Several natural products contain unsymmetrically substituted 2,5-dioxopiperazine subunits. Approaches to such complex molecules by stepwise alkylation or alkylidenation of the properly protected parent heterocycle are advantageous. Representative among them are bicyclomycin [S. Nakatsuka, 1983; R.M. Williams, 1985], saframycin-A [T. Fukuyama, 1990].

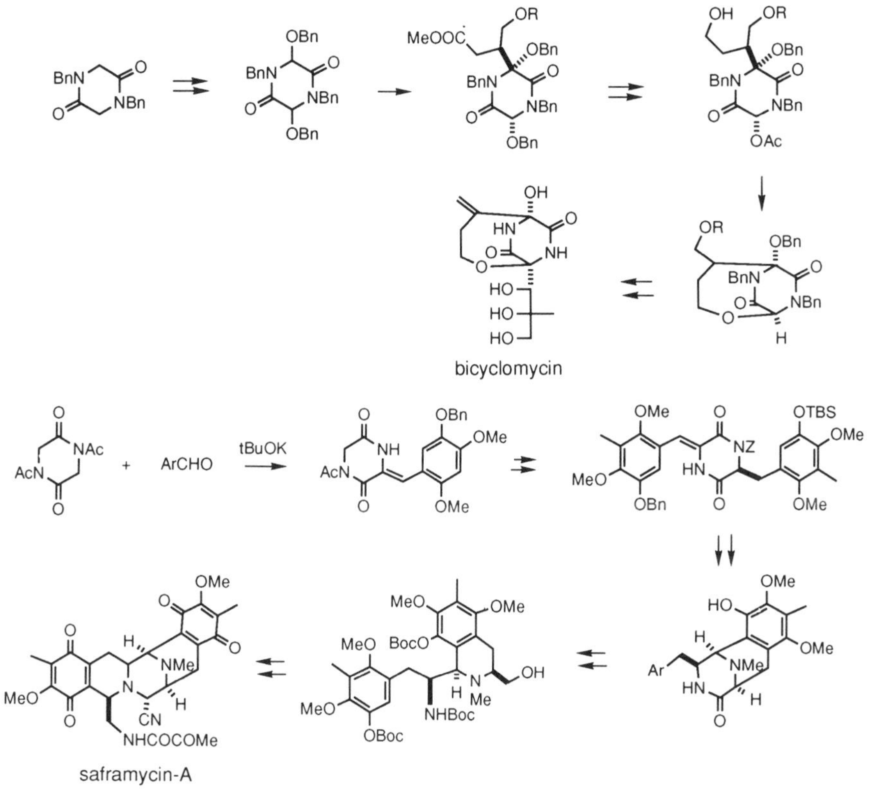

bicyclomycin

saframycin-A

Intramolecular oxidative phenolic coupling of a symmetrical 2,5-dioxopiperazine was the adopted synthetic method to form the macrocyclic ether of piperazinomycin [Nishiyama, 1986].

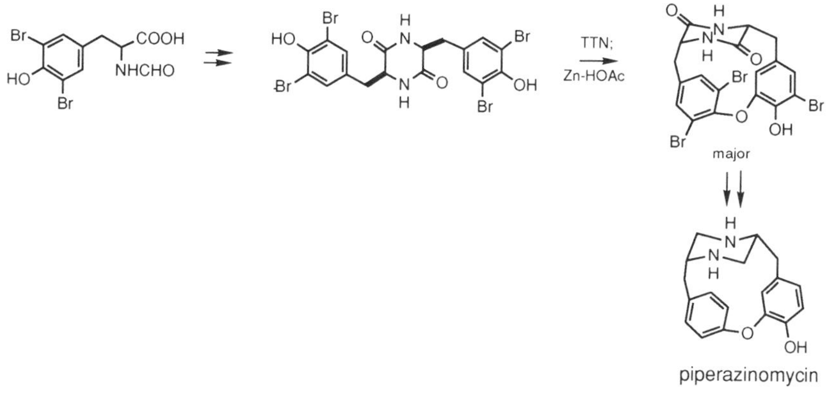

piperazinomycin

9.2. OXACYCLIC BUILDING BLOCKS

The furan ring is by far the most extensively used five-membered heteroaromatic system for skeletal construction of organic molecules because it is reactive toward electrophilic substitutions, α-metallation, oxidation, cycloaddition reactions including [2+2]-, [3+2]-, and [4+2]-versions, and the reaction products are readily transformed into hydroxyl and carbonyl compounds or other ring systems. Furan itself is very inexpensive, and it deserves consideration as a building block even more widely.

Introduction of a group to C-2 of furan by electrophilic reactions offers the choice of employing Friedel–Crafts-type conditions or lithiation. The choice is usually dictated by the nature of the substrate and the reagent. Both methods were involved in establishing the precursor of nonactic acid [Arco, 1976; Schmidt, 1976]. The synthetic strategy is quite obvious since the target molecule is a cis-2,5-disubstituted tetrahydrofuran.

nonactic acid

Although it is recognized that furans are masked 1,4-dicarbonyl compounds, the reactivity of the latter class of compounds is well concealed. Thus it is interesting to note that 2,5-dimethylfuran behaves as an enolate toward D-glyceraldehyde acetonide under high pressure [Jurczak, 1984]. The formal aldol adducts may be converted into 2-deoxypentitols.

Furan is transformable into (E)-butenedial monoacetals which are valuable building blocks for certain aliphatic compounds. Thus the sex pheromone of male dried bean beetle was readily prepared [Descoins, 1972].

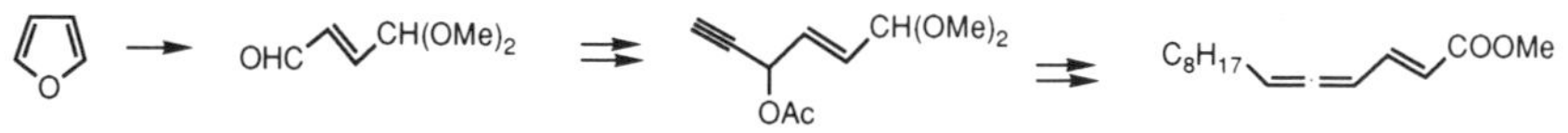

The Diels–Alder reactions of furan with various dienophiles have been exploited in the construction of intermediates containing the 7-oxabicyclo[2.2.1]heptane framework. A direct structural correlation of this type to cantharidin is inevitable [Stork, 1953; Dauben, 1985].

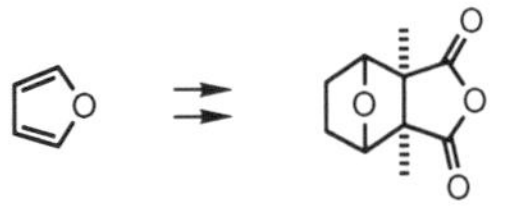

cantharidin

The adduct of furan and fumaryl chloride was converted into *trans*-maneonine-B [Holmes, 1984] via a tricyclic lactone.

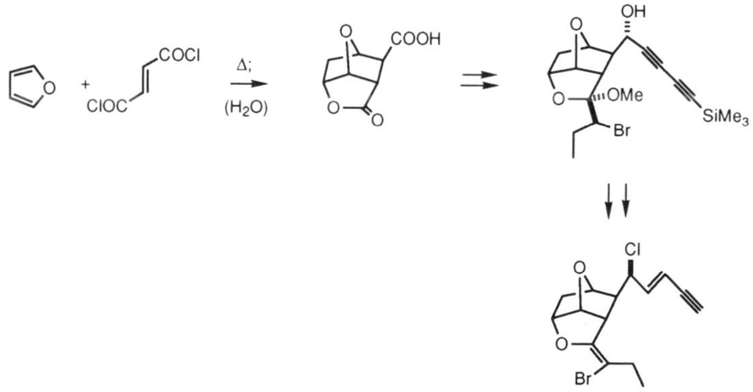

trans-maneonine-B

Another Diels–Alder reaction initiated a synthesis of (+)-compactin [Grieco, 1986]. In fact a second cycloaddition of the semihydrogenated adduct was employed to assemble all the skeletal elements of the target molecule. The oxabicyclic system was programmed to transmute into the required hydroxycyclohexene moiety by a Grob fragmentation.

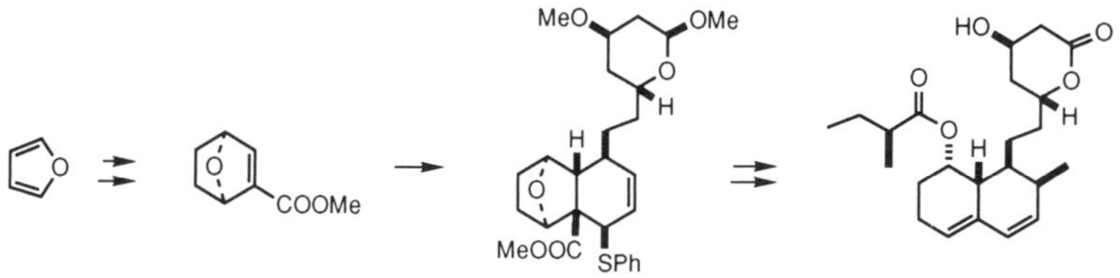

(+)-compactin

A convenient method for accessing 4,4-dialkylbutenolides [Grandguillot, 1979] involves methanolysis of the furan–maleic anhydride adduct, Grignard reaction, lactonization, and pyrolysis.

Furan also reacts as a diene with 1,3-dioxolen-2-one. Very simple reactions enabled transformation of these adducts into inositols [Kowarski, 1973].

exo + endo

neo-inositol

allo-inositol   myo-inositol

Electrodecarboxylation of 7-oxabicyclo[2.2.1]heptane-2-carboxylic acids is accompanied by skeletal rearrangement. The process, when applied to the saturated monoester derived from the furan–maleic ahydride adduct, gave a useful intermediate of hop ether [Imagawa, 1979]. Similarly several other iridoid monoterpenes such as matatabiether and neonepetalactone have been prepared [Imagawa, 1980].

R = H, Me

neonepetalactone

hop ether

Furan is an excellent interceptor of strain alkenes and alkynes. Thus its cycloaddition with cyclooctyne provided a unique opportunity of deriving muricatacin [Scholz, 1990, 1991]. Hydrogenation of the disubstituted double bond, oxidative cleavage of the remainder, and air oxidation led to a γ-lactone with useful functionality for further manipulation.

(-)-muricatacin

The ribose skeleton is readily prepared from Diels–Alder adducts of furan. Very important advances in this area are those concerning elaboration of chiral synthons by chemoenzymatic approaches. Using pig liver esterase *meso*-diesters are desymmetrized, and by enantiomer interchange maneuver both optical series are accessible, which on further elaboration have resulted in (+)-showdomycin [Ito, 1981] and (−)-cordycepin [Ohno, 1984].

(-)-cordycepin

(+)-showdomycin

Heterocyclic opening of such Diels–Alder adducts can lead to cyclohexadienols or benzenoid substances. The synthetic potential of such an adduct with acrylonitrile in the elaboration of shikimic acid and related metabolites has been recognized [Rajapaksa, 1984]. With 3,4-dibenzyloxyfuran as diene the oxygenation pattern was set in the adduct and the synthesis required hydrogenation instead of dihydroxylation [Koizumi, 1988; Koreeda, 1989]. A pleasing feature of these synthetic plans concerns atom economy—the preservation of the oxygen atom of the furan ring in the final products—with movement of a proton from carbon to oxygen.

(-)-methyl
triacetylshikimate

2,5-Bis(trimethylsiloxy)furan, readily obtained by silyation of succinic anhydride, reacts with various dienophiles to give substituted hydroquinones [Brownbridge, 1980]. 3,4-Dibenzyloxyfuran can act as a diene and dienophile in the Diels–Alder reaction. The latter role was manifested in the condensation with coumalic esters, therefore it has value in the construction of the bottom half of ivermectin [Jung, 1987].

Furan intercepts 2-metalloxyallyl cations which results in 8-oxabicyclo[3.2.1]oct-6-en-3-ones [H.M.R. Hoffmann, 1972]. These ketones are versatile intermediates, among the symmetrical representatives several have found applications such as in the synthesis of methyl nonactate [Arco, 1976], molecular segments of rifamycin-S [Rama Rao, 1985] and ionomycin [Lautens, 1993], damsin [Montana, 1990], and a synthon for related pseudoguaianolides [Cummins, 1988].

methyl nonactate

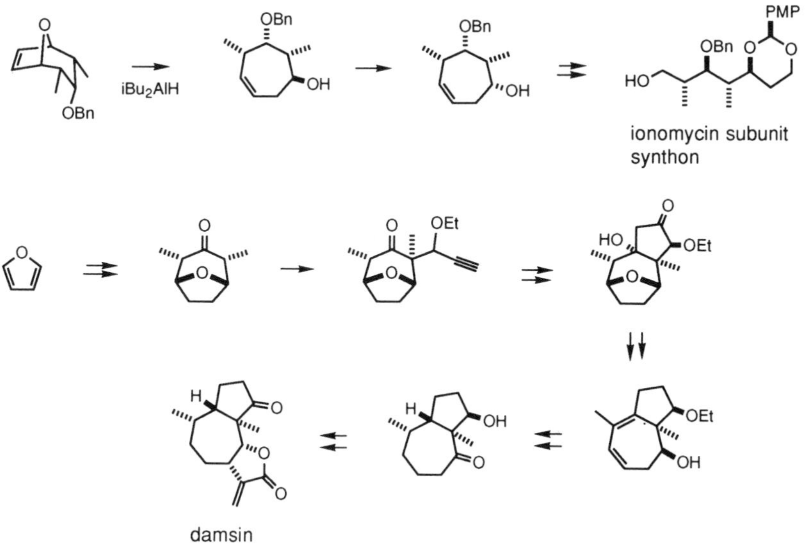

damsin

These examples show the retention of the heterocycle or its cleavage as a means of structural modification. The bicyclic system is also subject to S_N2' displacement on reaction with organocopper reagents [Lautens, 1990].

The dipolarophilicity of furan is further shown by reaction with nitrones. The cycloaddition can yield a precursor of nojirimycin [Vasella, 1982].

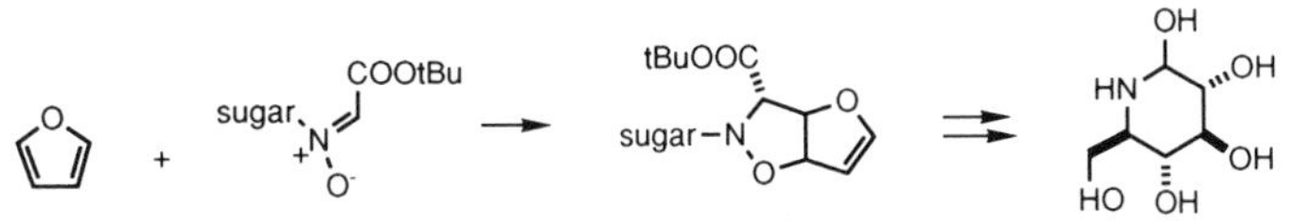

Furans also participate in the Paterno–Büchi reaction with carbonyl compounds. Such a reaction with 3,4-dimethylfuran proved to be the key to a synthesis of asteltoxin [Schreiber, 1984]. The photocycloadduct not only possessed all the framework carbon atoms of the fused heterocyclic system except an ethyl group, but also structurally imposed the stereochemical course in the establishment of two more asymmetric centers.

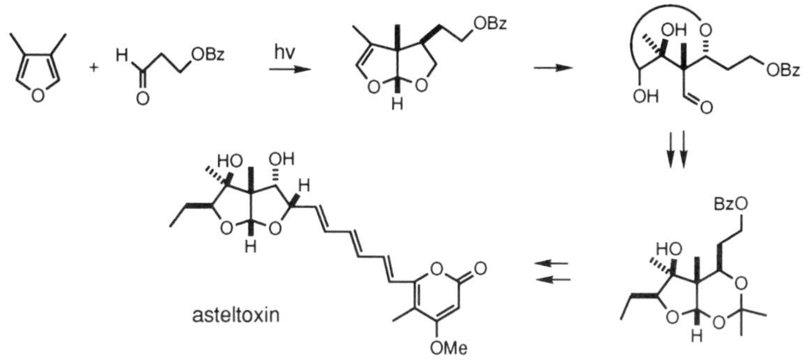

asteltoxin

Furan is also an excellent trap for carbenes/carbenoids. As the adducts obtained from diazoketones undergo facile ring opening. The reaction sequence is valuable to the preparation of hydroxyeicosatetraenoic acids (HETEs) [J. Adams, 1984].

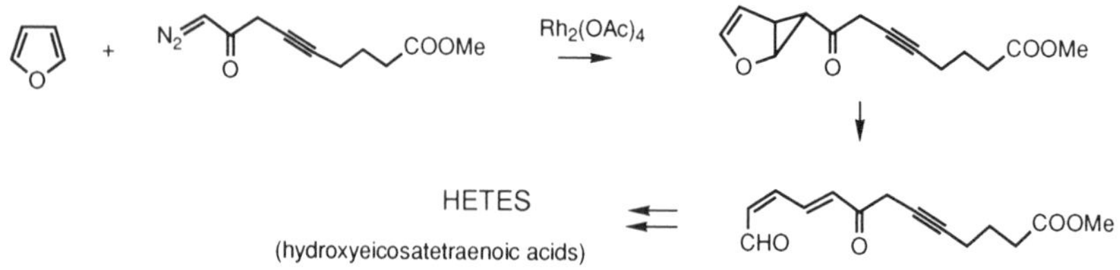

A synthetic route to cinnamodial [Burton, 1981] was devised to exploit the local symmetry of a furan which on oxidation afforded a 2,5-diacetoxy-2,5-dihydro derivative. Differentiation of the two α-positions was then achieved by a 1,4-elimination (only one possibility) which established the required oxygenation pattern.

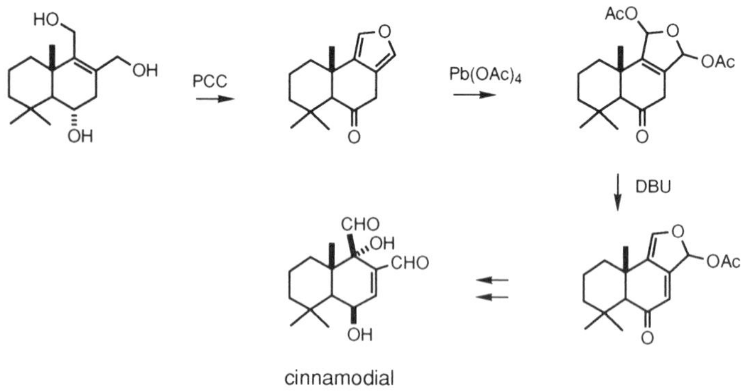

cinnamodial

The furan ring is a masked butenolide. Thus pyrovellerolactone has been synthesized from a furano precursor [Froborg, 1975]. This synthesis both profited and was penalized by symmetry. A pseudosymmetrical bridged ring intermediate was obtained by a twofold alkylation without complication, but the final oxidation also yielded an isomeric butenolide.

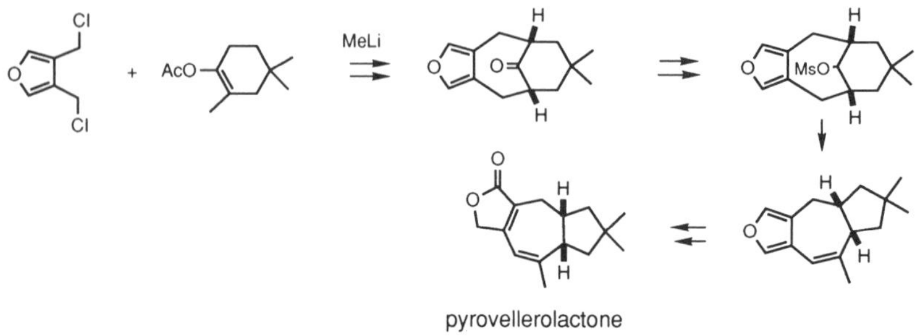

pyrovellerolactone

In one synthetic approach to (+)-camptothecin [Corey, 1975b] the two functional groups of diethyl 3,4-furandicarboxylate were differentiated and elaborated into a fused δ-lactone prior to converting the furan ring into a γ-chlorobutenolide, which was then used to unite with the quinolinopyrrolidine.

(+)-camptothecin

3,4-Furandicarboxylic esters are much more readily available, for example, from reaction of furan with acetylenedicarboxylic esters via the Diels–Alder/retro-Diels–Alder reaction tandem, and by proper functional group interchange operations other substituents may be installed. In fact a key component for a synthesis of confertin [Schultz, 1982] was derived from monomethyl 3,4-furandicarboxylate. Note that the oxidative maneuver on the furan ring at the final stages of the synthesis was not complicated by the generation of isomeric structures.

confertin

It is also evident that the diester is useful in the elaboration of lactaral [Froborg, 1974]. Desymmetrization was via monoprotection of the corresponding diol.

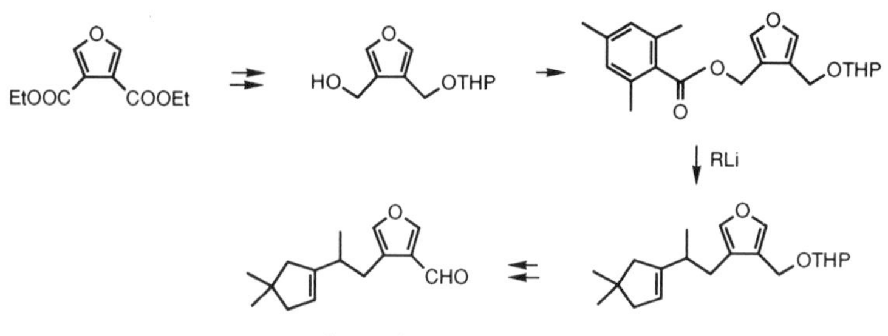

lactaral

The exploitability of local symmetry for the synthesis of pyridoxine [Elming, 1955; Firestone, 1967] is evident, too.

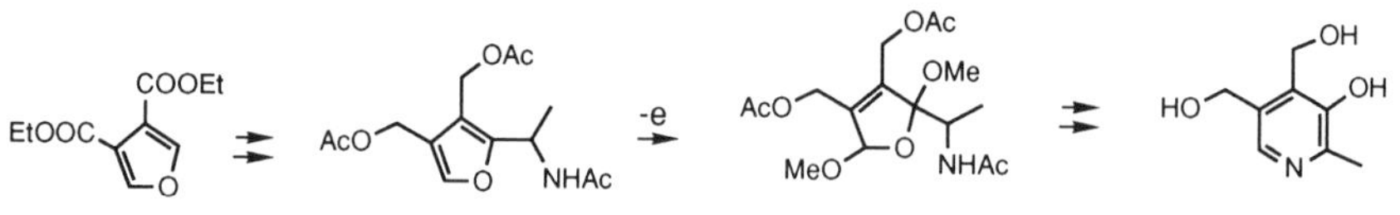

pyridoxine

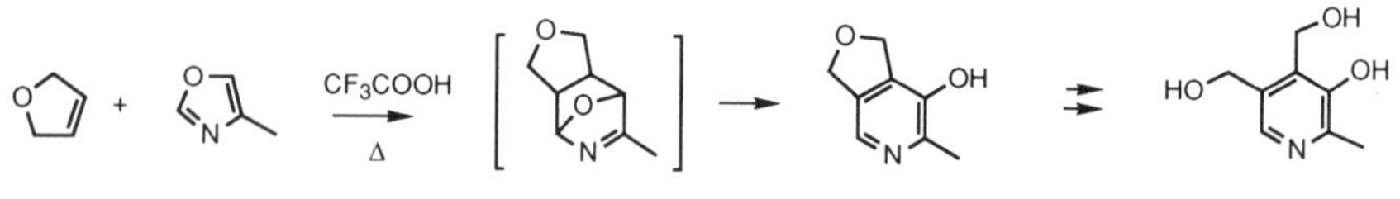

pyridoxine

A general method for synthesizing α-substituted γ-hydroxybutenolides from 2,5-dimethoxy-2,5-dihydrofuran is by ozonolysis, Wittig reaction, and acid treatment [Fell, 1990]. Actually the process involved chain elongation of a self-protected glyoxal.

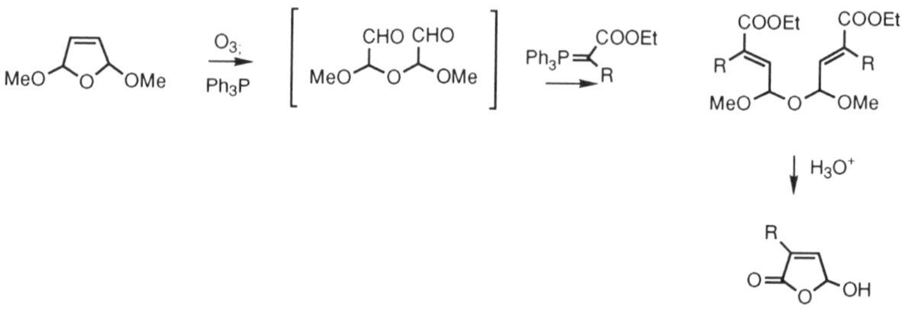

2,5-Dimethoxytetrahydrofuran is a synthetic equivalent of succinaldehyde. The reaction of this substance with primary amine hydrochloride led to N-substituted carbazoles [Kashima, 1986].

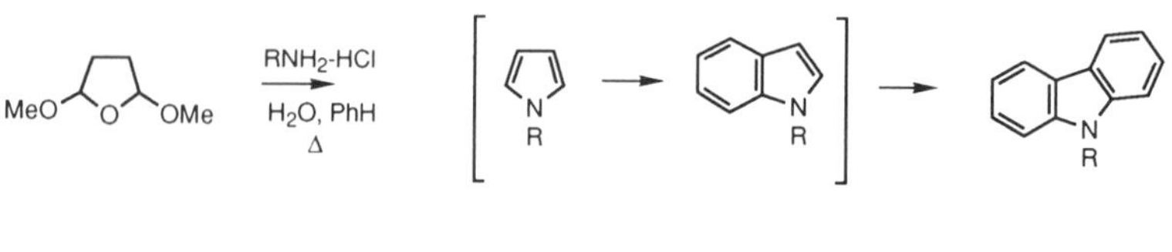

Under milder conditions pyrroles are isolated. Incorporation of such a process in routes to ipalbidine [Jefford, 1986], (−)-indolizidine-167B [Jefford, 1991], and (−)-monomorine [Jefford, 1994] are instructive.

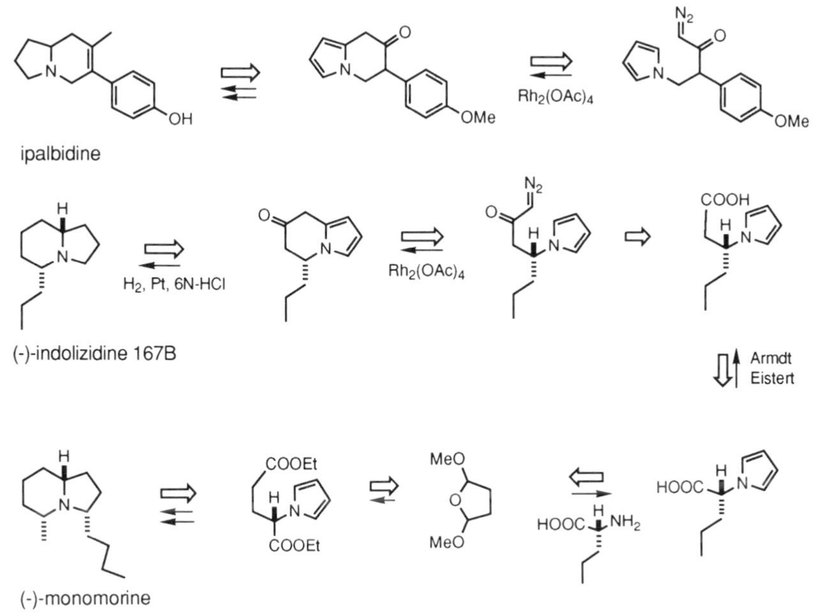

ipalbidine

(-)-indolizidine 167B

(-)-monomorine

A chiral pyrrolidine synthon is readily prepared from 2,5-dimethoxytetra-hydrofuran and phenylglycinol [Royer, 1987]. The versatility of this synthon is related to the fact that one of the α-carbon atoms acts as a nucleophile while the other acts as an electrophile [P.Q. Huang, 1987].

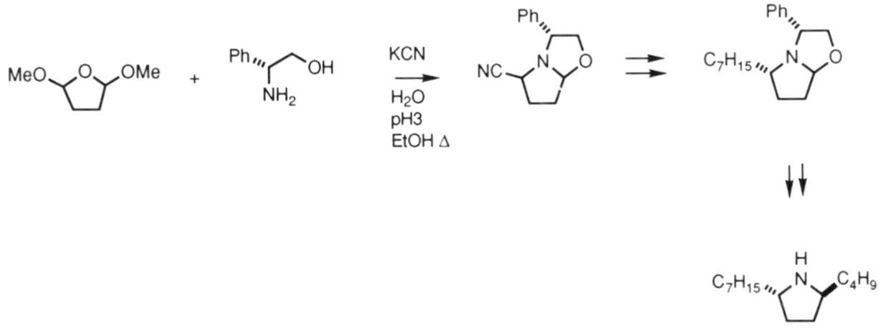

For a synthesis of 1,4-diketones 2,5-dimethoxytetrahydrofuran may be converted into the bisdithiane which is used in a stepwise alkylation [Ellison, 1972]. The differentially protected succinaldehyde is more versatile, for example in annulation [Lansbury, 1988b].

cis-jasmone

alliacol-A

An unusual but useful building block is tetrakis(methylene)-7-oxanorbornane. Two molecules of this compound and two molecules of 1,4:5,8-diepoxy-1,4,5,8-tetrahydroanthracene have been induced to form a unique [12]cyclacene derivative, kohnkene [Ashton, 1992]. The feasibility of assembling anthracycline antibiotics from this bisdiene by controlled Diels–Alder reactions with benzoquinone and methyl vinyl ketone in two separate stages has been demonstrated [Tamariz, 1983]. A less symmetrical bisdiene containing two phenylthio groups, derived from furan–maleic anhydride adduct, is also useful [Tornare, 1985].

Construction of complex structures containing a 4-oxygenated pyran ring should consider tetrahydro-4-pyrone as the building block. This building block concept was clearly demonstrated in the approaches to patulin [Woodward, 1950a, b]. A more recent utility of the heterocycle is its incorporation into (+)-phyllanthocin [A.B. Smith, 1991].

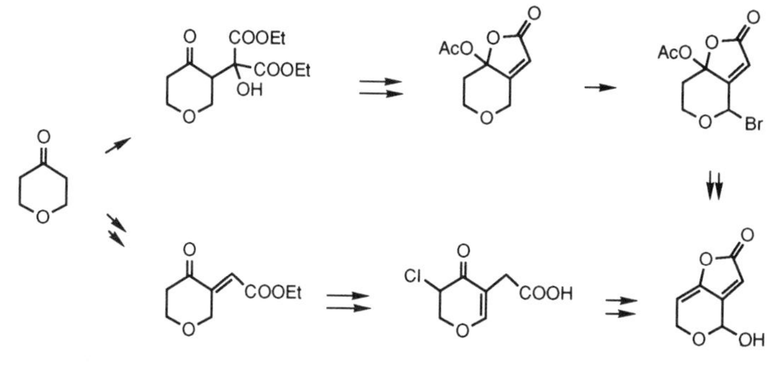

patulin

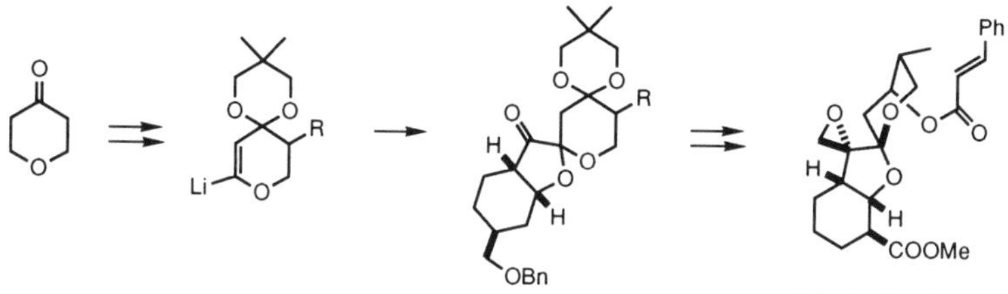

(+)-phyllanthocin

4-Methylenetetrahydropyran has found an unusual application in serving as a template for the elaboration of 1α,25-dihydroxyvitamin D₃ [Kabat, 1992].

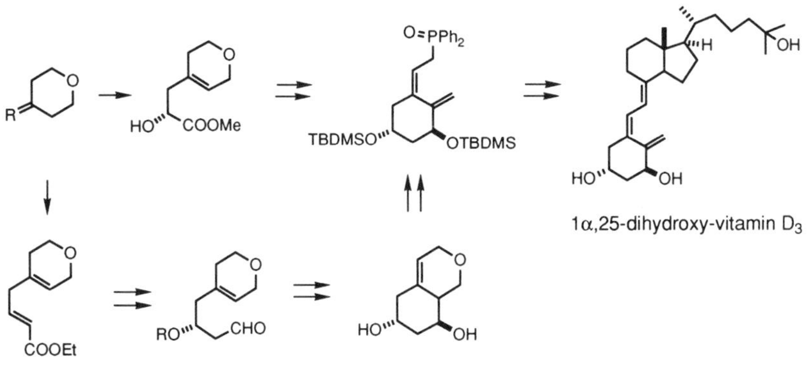

1α,25-dihydroxy-vitamin D₃

9.3. THIACYCLIC BUILDING BLOCKS

3-Sulfolene usually serves as a masked butadiene and a butadiene-1-anion synthon. Recently its role has been expanded to an unsymmetrical dienophile [S.F. Martin, 1993a], specifically addressing the synthetic need of breynolide.

breynolide

Although similarly to tetrahydro-4-pyrone, the thia analog has not been employed extensively in synthesis, the few applications that have been reported in the chemical literature are exceptionally elegant. A synthesis of the cecropia juvenile hormone JH-I, almost simultaneously developed by three research units [Kondo, 1972; Demoute, 1973; Stotter, 1973], took advantage of differential functionalization of the ketone to supply nucleophilic as well as electrophilic components for assembling a carbon chain masked in two heterocycles. The most important feature is the constraint of two trisubstituted double bonds in the desired geometry in these heterocyclic subunits which were to be cleaved by reductive desulfurization at a later stage.

JH-I

At the starting point of an erythromycin synthesis [Woodward, 1981] was tetrahydrothiapyran-4-one which was converted into two optically active dithiadecalin derivatives differing in one carbon atom, either ketonic or ethereal. The ethereal component was desulfurized and processed into an aldehyde to condense with the dithiadecalone. The product was essentially a protected and functionally modified seco acid of trisnorerythronolide-A, the array required redox transformations, desulfurization, and chain elongation to clear the first hurdle of the synthetic challenge.

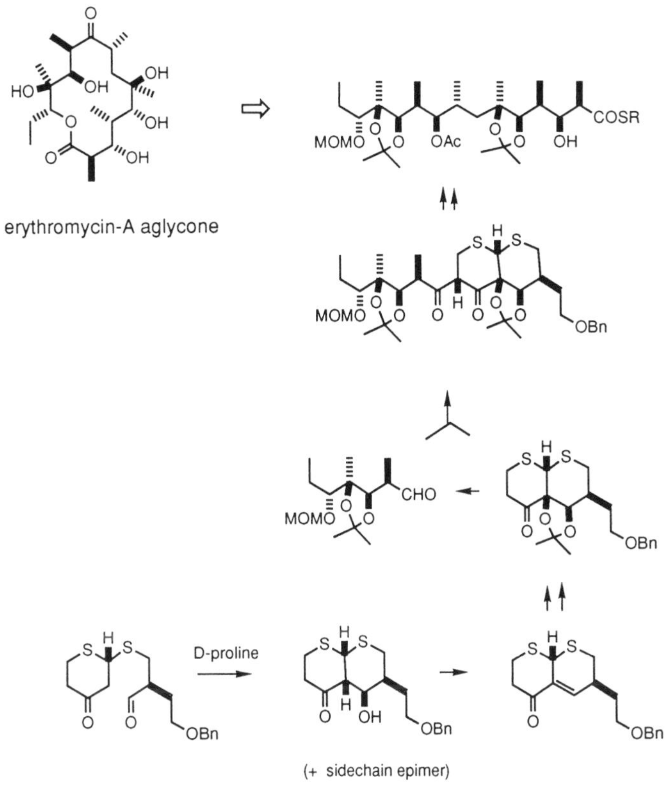

erythromycin-A aglycone

(+ sidechain epimer)

While 1,3-dithianes are generally employed in the umpolung purpose, a rare application involved oxidation and alkylation to elaborate α-lipoic acid [Menon, 1987]. Such derivatives of (+)- and (−)-menthones were transformed into (R)- and (S)-forms of α-lipoic acid, respectively.

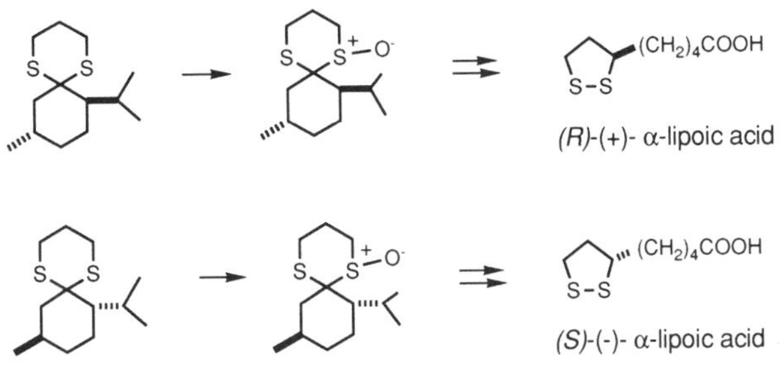

(R)-(+)- α-lipoic acid

(S)-(-)- α-lipoic acid

REFERENCES

Abelman, M.M., Overman, L.E., Tran, V.D. (1990) *J. Am. Chem. Soc.* **112:** 6959.

Achmatowicz, B., Wicha, J. (1988) *Liebigs Ann. Chem.* 1135.

Adams, C.E., Walker, F.J., Sharpless, K.B. (1985) *J. Org. Chem.* **50:** 420.

Adams, J., Rokach, J. (1984) *Tetrahedron Lett.* **25:** 35.

Adams, J.H., Brown, P.M., Gupta, P., Khan, M.S., Lewis, J.R. (1981) *Tetrahedron* **37:** 209.

Agami, C., Kadouri-Puchot, C., Le Guen, V. (1993) *Tetrahedron: Asymmetry* **4:** 641.

Aggarwal, V.K., Wang, M.F., Zaparucha, A. (1994) *Chem. Commun.* 87.

Ahmad, Z., Goswami, P., Venkateswaran, R.V. (1989) *Tetrahedron* **45:** 6833.

Akita, H., Yamada, H., Matsukura, H., Nakata, T., Oishi, T. (1988) *Tetrahedron Lett.* **29:** 6449.

Alberici, G.F., Andrieux, J., Adam, G., Plat, M.M. (1983) *Tetrahedron Lett.* **24:** 1937.

Alder, K., Rührmann, R. (1950) *Liebigs Ann. Chem.* **566:** 1.

Aleksejczyk, R.A., Berchtold, G.A. (1985) *J. Am. Chem. Soc.* **107:** 2554.

Alexakis, A., Chapdelaine, M.J., Posner, G.H., Runquist, A.W. (1978) *Tetrahedron Lett.* 4205, 4209.

Alexakis, A., Mangeney, P. (1990) *Tetrahedron: Asymmetry* **1:** 477.

Ali, A., Saxton, J.E. (1989) *Tetrahedron Lett.* **30:** 3197.

Ali, E., Chakraborty, P.K., Pakrashi, S.C. (1982) *Heterocycles* **19:** 1367, 1667.

Aljancic-Solaja, I., Rey, M., Dreiding, A.S. (1987) *Helv. Chim. Acta* **70:** 1302.

Altenbach, H.-J., Holzapfel, W. (1990) *Angew. Chem. Int. Ed. Engl.* **29:** 67.

Altman, J., Cohen, E., Maymon, T., Petersen, J.P., Reshef, N., Ginsburg, D. (1969) *Tetrahedron* **25:** 5115.

Alvarez, E., Cuvigny, T., du Penhoat, C.H., Julia, M. (1988) *Tetrahedron* **44:** 119.

Alvarez, E., Diaz, M.T., Rodriguez, M.L., Martin, J.D. (1990) *Tetrahedron Lett.* **31:** 1629.

Alvarez, E., Diaz, M.T., Perez, R., Martin, J.D. (1991a) *Tetrahedron Lett.* **32:** 2241.

Alvarez, E., Diaz, M.T., Zurita, D., Martin, J.D. (1991b) *Tetrahedron Lett.* **32:** 2245.

Alvarez, S., Serratosa, F. (1992) *J. Am. Chem. Soc.* **114:** 2623.

Amri, H., Rambaud, M., Villieras, J. (1990) *Tetrahedron* **46:** 3535.

Anderson, R.J., Henrick, C.A. (1975) *J. Am. Chem. Soc.* **97:** 4327.

Anet, F.A.L., Miura, S.S., Siegel, J., Mislow, K. (1983) *J. Am. Chem. Soc.* **105:** 1419.

Angle, S.R., Louie, M.S., Mattson, H.L., Yang, W. (1989a) *Tetrahedron Lett.* **30:** 1193.

Angle, S.R., Turnbull, K.D. (1989b) *J. Am. Chem. Soc.* **111:** 1136.

Angle, S.R., Arnaiz, D.O. (1990) *J. Org. Chem.* **55:** 3708.

Anjaneyulu, B., Govindarchari, T.R. (1969) *Tetrahedron Lett.* 2847.

Anthony, N.J., Clarke, T., Jones, A.B., Ley, S.V. (1987) *Tetrahedron Lett.* **28:** 5755.

Aoyagi, S., Fujimaki, S., Yamazaki, N., Kibayashi, C. (1991) *J. Org. Chem.* **56:** 815.

Arai, Y., Kamikawa, T., Kubota, T., Masuda, Y., Yamamoto, R. (1973) *Phytochem.* **12:** 2279.

Arai, Y., Kontani, T., Koizumi, T. (1991) *Chem. Lett.* 2279.

Araki, Y., Nagasawa, J., Ishido, Y. (1981) *J. Chem. Soc. Perkin Trans. I* 12.

Arata, Y., Tanaka, K., Yoshifugi, S., Kanamoto, S. (1979) *Chem. Pharm. Bull.* **27:** 981.

Aratani, T., Yoneyoshi, Y., Nagase, T. (1977) *Tetrahedron Lett.* 2599.

Arco, M.J., Trammell, M.H., White, J.D. (1976) *J. Org. Chem.* **41:** 2075.

Ardisson, J., Ferezou, J.P., Julia, M., Lenglet, L., Pancrazi, A. (1987) *Tetrahedron Lett.* **28:** 1997.

Arita, M., Adachi, K., Ito, Y., Sawai, H., Ohno, M. (1983) *J. Am. Chem. Soc.* **105:** 4049.

Arseniyadis, S., Huang, P.Q., Husson, H.P. (1988a) *Tetrahedron Lett.* **29:** 631.

Arseniyadis, S., Huang, P.Q., Husson, H.P. (1988b) *Tetrahedron Lett.* **29:** 1391.

Arseniyadis, S., Huang, P.Q., Piveteau, D., Husson, H.P. (1988c) *Tetrahedron* **44:** 2457.

Arseniyadis, S., Rodriguez, R., Cabrera, E., Thompson, A., Ourisson, G. (1991) *Tetrahedron* **47:** 7045.

Asami, M. (1985) *Tetrahedron Lett.* **26:** 5803.

Asaoka, M., Naito, S., Takei, H. (1985) *Tetrahedron Lett.* **26:** 2103.

Asaoka, M., Fujii, N., Takei, H. (1988a) *Chem. Lett.* 805.

Asaoka, M., Fujii, N., Takei, H. (1988b) *Chem. Lett.* 1665.

Asaoka, M., Shima, K., Fujii, N., Takei, H. (1988c) *Tetrahedron* **44:** 4757.

Asaoka, M., Takenouchi, K., Takei, H. (1988d) *Tetrahedron Lett.* **29:** 325.

Asaoka, M., Takenouchi, K., Takei, H. (1988e) *Chem. Lett.* 921.

Asaoka, M., Takenouchi, K., Takei, H. (1988f) *Chem. Lett.* 1225.

Asaoka, M., Sonoda, S., Takei, H. (1989a) *Chem. Lett.* 1847.

Asaoka, M., Takei, H. (1989b) *Heterocycles* **29:** 243.

Asaoka, M., Sakurai, M., Takei, H. (1990a) *Tetrahedron Lett.* **31:** 4759.

Ssaoka, M., Takei, H. (1990b) *J. Synth. Org. Chem. Jpn.* **48:** 216.

Asaoka, M., Sakurai, M., Takei, H. (1991) *Tetrahedron Lett.* **32:** 7567.

Ashton, P.R., Brown, G.R., Isaacs, N.S., Giuffrida, D., Kohnke, F.H., Mathias, J.P., Slawin, A.M.Z., Smith, D.R., Stoddard, J.F., Williams, D.J. (1992) *J. Am. Chem. Soc.* **114:** 6330.

Askani, R. (1970) *Tetrahedron Lett.* 3349.

Astles, P.C., Thomas, E.J. (1989) *Synlett* 42.

Attwood, S.V., Barrett, A.G.M. (1984). *J. Chem. Soc. Perkin Trans. I* 1315.

Aubé, J., Ghosh, S., Tanol, M. (1994) *J. Am. Chem. Soc.* **116:** 9009.

Auerbach, J., Weinreb, S.M. (1974) *Chem. Commun.* 298.

Axen, U., Lincoln, F.H., Thompson, J.L. (1969) *Chem. Commun.* 303.

Ayer, W.A., Bowman, W.R., Joseph, T.C., Smith, P. (1968) *J. Am. Chem. Soc.* **90:** 1648.

Ayer, W.A., Browne, L.M., Fung, S. (1976a) *Can. J. Chem.* **54:** 3276.

Ayer, W.A., Dawe, R., Eisner, R.A., Furuichi, K. (1976b) *Can. J. Chem.* **54:** 473.

Babine, R.E. (1986) *Tetrahedron Lett.* **27:** 5791.

Baer, E., Fischer, H.O.L. (1939) *J. Biol. Chem.* **128:** 463.

Baeyer, A., Landsberg, L. (1882) *Ber. Deut. Chem. Ges.* **15:** 57.

Baker, P.M., Bycroft, B.W., Roberts, J.C. (1967) *Chem. Commun.* 1913.

Baker R., Billington, D.C., Ekanayake, N. (1981) *Chem. Commun.* 1234.

Baker, R., Gibson, C.L., Swain, C.J., Tapolczay, D.J. (1984) *Chem. Commun.* 619.

Baldwin, J.E., Barden, T.C. (1981) *J. Org. Chem.* **46:** 2442.

Baldwin, J.E., Adlington, R.M., Mitchell, M.B. *Chem. Commun.* 1332.

Ban, Y., Sato, Y., Inoue, I., Nagai, M., Oishi, T., Terashima, M., Yonemitsu, O., Kanaoka, Y. (1965) *Tetrahedron Lett.* 2261.

Ban, Y., Yoshida, K., Goto, J., Oishi, T., Takeda, E. (1983) *Tetrahedron* **39:** 3657.

Banerjee, D.K., Angadi, V.B. (1973) *Indian J. Chem.* **11:** 511.

Banfi, L., Guanti, G. (1993a) *Synthesis* 1029.

Banfi, L., Guanti, G., Narisano, E. (1993b) *Tetrahedron* **49:** 7385.

Baraldi, P.G., Barco, A., Benetti, S., Pollini, G.P., Simoni, D. (1983) *Tetrahedron Lett.* **24:** 5669.

Barborak, J.C., Watts, L., Pettit, R. (1966) *J. Am. Chem. Soc.* **88:** 1328.

Barker, A.J., Pattenden, G. (1983) *J. Chem. Soc. Perkin Trans. I* 1901.

Barltrop, J.A., Littlehailes, J.D., Rushton, J.D., Rogers, N.A.J. (1962) *Tetrahedron Lett.* 429.

Barrett, A.G.M., Boys, M.L., Boehm, T.L. (1994) *Chem. Commun.* 1881.

Barriere, F., Barriere, J.-C., Barton, D.H.R., Cleophax, J., Gateau-Olesker, A., Gero, S.D., Tadj, F. (1985) *Tetrahedron Lett.* **26:** 3119, 3121..

Barrish, J.C., Lee, H.L., Baggiolini, E.G., Uskokovic, M.R. (1987) *J. Org. Chem.* **52:** 1372.

Bartlett, M.F., Taylor, W.I. (1960) *J. Am. Chem. Soc.* **82:** 5491.

Bartlett, P.A., Greene, F.R. III (1078) *J. Am. Chem. Soc.* **100:** 4858.

Bartlett, P.A., Myerson, J. (1979) *J. Org. Chem.* **44:** 1625.

Bartlett, P.A., Adams, J.L. (1980) *J. Am. Chem. Soc.* **102:** 337.

Bartlett, P.A., Meadow, J.D., Ottow, E. (1984) *J. Am. Chem. Soc.* **106:** 5304.

Bartlett, P.A., Holm, K.H., Morimoto, A. (1985) *J. Org. Chem.* **50:** 5179.

Barton, D.H.R., Deflorin, A.M., Edwards, O.E. (1956) *J. Chem. Soc.* 530.

Barton, D.H.R., James, R., Kirby, G.W., Widdowson, D.A. (1967) *Chem. Commun.* 266.

Barton, D.H.R., Donnelly, D.M.X., Finet, J.-P., Guiry, P.J. (1990) *Tetrahedron* **31:** 7449.

Bates, H.A., Rapoport, H. (1979) *J. Am. Chem. Soc.* **101:** 1259.

Batt, D.G., Ganem, B. (1978) *Tetrahedron Lett.* 3323.

Battersby, A.R., Binks, R. (1955) *J. Chem. Soc.* 2888.

Beck, J.R., Kwok, R., Booher, R.N., Brown, A.C., Patterson, L.E., Pranc, P., Rockey, B., Pohland, A. (1968) *J. Am. Chem. Soc.* **90:** 4706.

Becker, H.-D., Sandros, K., Arvidsson, A. (1979) *J. Org. Chem.* **44:** 1336.

Becking, L., Schäfer, H.J. (1988) *Tetrahedron Lett.* **29:** 2801.

Begley, M.J., Jackson, C.B., Pattenden, G. (1990) *Tetrahedron Lett.* **46:** 4907.

Belanger, A., Poupart, J., Deslongchamps, P. (1968) *Tetrahedron Lett.* 2127.

Belletire, J.L., Adams, K.G. (1983) *Tetrahedron Lett.* **24:** 5575.

Belletire, J.L., Fremont, S.L. (1986) *Tetrahedron Lett.* **27:** 127.

Bergel'son, L.D., Batrakov, S.G. (1983) *Izv. Akad. Nauk SSSR Ser. Khim.* 1259.

Bergman, R., Magnusson, G. (1986) *J. Org. Chem.* **51:** 212.

Bergman, R., Pelcman, B. (1989) *J. Org. Chem.* **54:** 824.

Bernasconi, S., Gariboldi, P., Jommi, G., Montanari, S., Sisti, M. (1981) *J. Chem. Soc. Perkin Trans. I* 2394.

Bertrand, M., Gras, J.L. (1974) *Tetrahedron* **30:** 793.

Bertz, S.H. (1983) *Tetrahedron Lett.* **24:** 5577.

Bertz, S.H. (1984) *Chem. Commun.* 218.

Bertz, S.H., Lannoye, G., Cook, J.M. (1985) *Tetrahedron Lett.* **26:** 4695.

Berubé, G., Deslongchamps, P. (1984) *Can. J. Chem.* **62:** 1558.

Berubé, G., Fallis, A.G. (1989) *Tetrahedron Lett.* **30:** 4045.

Besselievre, R., Thal, C., Husson, H.-P., Potier, P. (1975) *Chem. Commun.* 90.

Besselièvre, R., Husson, H.-P. (1976) *Tetrahedron Lett.* 1873.

Besselièvre, R., Husson, H.-P. (1981) *Tetrahedron (Suppl. 1)* **37:** 241.

Bestmann, H.-J., Moenius, T. (1986) *Angew. Chem. Int. Ed. Engl.* **25:** 994.

Bestmann, H.-J., Roth, D. (1990) *Angew. Chem. Int. Ed. Engl.* **29:** 99.

Bhalerao, U.T., Rapoport, H. (1971) *J. Am. Chem. Soc.* **93:** 5311.

Biellmann, J.F., Ducep, J.B. (1969) *Tetrahedron Lett.* 3707.

Biemann, K., Büchi, G., Walker, B.H. (1957) *J. Am. Chem. Soc.* **79:** 5558.

Billups, W.E., McCord, D.J., Maughon, B.R. (1994) *J. Am. Chem. Soc.* **116:** 8831.

Bindra, J.S., Grodski, A., Schaaf, T.K., Corey, E.J. (1973) *J. Am. Chem. Soc.* **95:** 7522.

Birch, A.J., Keeton, R. (1968) *J. Chem. Soc. (C)* 109.

Birch, A.J., Macdonald, P.L., Powell, V.H. (1970) *J. Chem. Soc. (C)* 1469.

Birch, A.M., Pattenden, G. (1980) *Chem. Commun.* 1195.

Björkling, F., Boutelje, J., Gatenbeck, S., Hult, K., Norin, T., Szmulik, P. (1985) *Tetrahedron* **41:** 1347.

Blackburn, G.M., Ollis, W.D., Smith, C., Sutherland, I.O. (1969) *Chem. Commun.* 99.

Blum, Z., Ekström, M., Wistrand, L.-G. (1984) *Acta Chem. Scand.* **B38:** 297.

Bocchi, V., Gardini, G.P. (1969) *Org. Prep. Proced. Int.* **1:** 271.

Boeckman, R.K., Ko, S.S. (1980) *J. Am. Chem. Soc.* **102:** 7146.

Boeckman, R.K., Sum, F.W. (1982) *J. Am. Chem. Soc.* **104:** 4604.

Boeckman, R.K., Cheon, S.H. (1983a) *J. Am. Chem. Soc.* **105:** 4112.

Boeckman, R.K., Napier, J.J., Thomas, E.W., Sato, R.I. (1983b) *J. Org. Chem.* **48:** 4153.

Boeckman, R.K., Starrett, J.E., Nickell, D.G., Sum, P.-E. (1986) *J. Am. Chem. Soc.* **108:** 5549.

Boeckman, R.K., Weidner, C.H., Perni, R.B., Napier, J.J. (1989) *J. Am. Chem. Soc.* **111:** 8036.

Boeckman, R.K., Charette, A.B., Asberom, T., Johnston, B.H. (1991) *J. Am. Chem. Soc.* **113:** 5337.

Boger, D.L., Mullican, M.D. (1982) *Tetrahedron Lett.* **23:** 4555.

Boger, D.L., Coleman, R.S., Panek, J.S., Yohannes, D. (1984) *J. Org. Chem.* **49:** 4405.

Boger, D.L., Panek, J.S. (1985) *J. Am. Chem. Soc.* **107:** 5745.

Boger, D.L. (1986a) *Chem. Rev.* 86: 891.

Boger, D.L., Brotherton, C.E. (1986b) *J. Am. Chem. Soc.* **108:** 6713.

Boger, D.L., Coleman, R.S. (1986c) *J. Org. Chem.* **51:** 3250.

Boger, D.L., Patel, M. (1988) *J. Org. Chem.* **53:** 1405.

Boger, D.L., Zhang, M. (1991) *J. Am. Chem. Soc.* **113:** 4230.

Boger, D.L., Menezes, R.M., Honda, T. (1993) *Angew. Chem. Int. Ed. Engl.* **32:** 273.

Bohlmann, F., Englisch, A., Ottawa, N., Sander, H., Weise, W. (1956) *Chem. Ber.* **89:** 792.

Bohlmann, F., Winterfeldt, E., Friese, U. (1963) *Chem. Ber.* **96:** 2251.

Bohlmann, F., Lonitz, M. (1980) *Chem. Ber.* **113:** 2410.

Bohlmann, F., Fiedler, L. (1981) *Chem. Ber.* **114:** 227.

Bohlmann, F., Rotard, W. (1982) *Liebigs Ann. Chem.* 1211.

Boland, W., Mertes, K., Jaenicke, L., Muller, D.G., Folster, E. (1983) *Helv. Chim. Acta* **66:** 1905.

Bold, G., Chao, S., Bgide, R., Wu, S.-H., Patel, D.V., Sih, C.J. (1987) *Tetrahedron Lett.* **28:** 1973.

Bolitt, V., Mioskowski, C., Kollah, R.O., Manna, S., Rajapaksa, D., Falck, J.R. (1991) *J. Am. Chem. Soc.* **113:** 6320.

Boons, G.-J., Entwistle, D.A., Ley, S.V., Woods, M. (1993) *Tetrahedron Lett.* **34:** 5649.

Borch, R.F., Ho, B.C. (1977) *J. Org. Chem.* **42:** 1225.

Borchardt, A., Fuchicello, A., Kilway, K.V., Baldridge, K.K., Siegel, J.S. (1992) *J. Am. Chem. Soc.* **114:** 1921.

Born, M., Tamm, C. (1989) *Tetrahedron Lett.* **30:** 2083.

Borzilleri, R.M., Weinreb, S.M., Parvez, M. (1994) *J. Am. Chem. Soc.* **116:** 9789.

Bosch, M.P., Camps, F., Coll, J., Guerrero, A., Tatsuoka, T., Meinwald, J. (1986) *J. Org. Chem.* **51:** 773.

Boschelli, D., Smith, A.B., III, Stringer, O.D., Jenkins, R.H., Davis, F.A. (1981) *Tetrahedron Lett.* **22:** 4385.

Bosse, D., deMeijere, A. (1974) *Angew. Chem. Int. Ed. Engl.* **13:** 663.

Böttger, D., Welzel, P. (1983) *Tetrahedron Lett.* **24:** 5201.

Boyer, F.-D., Lallemand, J.-Y. (1994) *Tetrahedron* **50:** 10443.

Branca, Q., Fischli, A. (1977) *Helv. Chim. Acta* **60:** 925.

Brandänge, S., Lundin, C. (1971) *Acta Chem. Scand.* **25:** 2447.

Braun, M., Büchi, G., Bushey, D.F. (1978) *J. Am. Chem. Soc.* **100:** 4208.

Braun, M., Bauer, C. (1991) *Liebigs Ann. Chem.* 1157.

Breitgoff, D., Laumen, K., Schneider, M.P. (1986) *Chem. Commun.* 1523.

Brenner, M., Rexhausen, H., Steffan, B., Steglich, W. (1988) *Tetrahedron* **44:** 2887.

Breuilles, P., Schmittenberger, T., Uguen, D. (1993) *Tetrahedron Lett.* **34:** 4205.

Briggs, M.A., Haines, A.H., Jones, H.F. (1985) *J. Chem. Soc. Perkin Trans. I* 795.

Bringmann, G., Jansen, J.R., Rink, H.-P. (1986) *Angew. Chem. Int. Ed. Engl.* **25:** 913.

Bringmann, G., Geuder, T. (1991) *Synlett* 829.

Broene, R.D., Diederich, F. (1991) *Tetrahedron Lett.* **32:** 5227.

Brooks, D.W., Grothaus, P.G., Palmer, J.T. (1982) *Tetrahedron Lett.* **23:** 4187.

Brooks, D.W., Mazdiyasni, H., Sallay, P. (1985) *J. Org. Chem.* **50:** 3411.

Brown, M.J., Harrison, T., Overman, L.E. (1991) *J. Am. Chem. Soc.* **113:** 5378.

Brown, S.M., Hudlicky, T. (1993) In Hudlicky, T., ed. *Organic Synthesis. Theory and Applications*, Vol. 2, JAI Press, Greenwich, CT., pp. 113–176.

Brownbridge, P., Chan, T.H. (1980) *Tetrahedron Lett.* **21:** 3423, 3427.

Brugidou, J., Chiche-Trinh, B.H., Christok, H., Poncet, J. (1979) *Tetrahedron Lett.* 1223.

Bryson, T.A., Reichel, C.J. (1980) *Tetrahedron Lett.* **21:** 2381.

Büchi, G., MacLeod, W.D., Padilla, J. (1964) *J. Am. Chem. Soc.* **86:** 4438.

Büchi, G., Foulkes, D.M., Kurono, M., Mitchell, G.F., Schneider R.S. (1967) *J. Am. Chem. Soc.* **89:** 6745.

Büchi, G., Carlson, J.A. (1969) *J. Am. Chem. Soc.* **91:** 6470.

Büchi, G., Carlson, J.A., Powell, J.E., Tietze, L.-F. (1970) *J. Am. Chem. Soc.* **92:** 2165.

Büchi, G., Egger, B. (1971a) *J. Org. Chem.* **36:** 2021.

Büchi, G., Minster, D., Young, J.C.F. (1971b) *J. Am. Chem. Soc.* **93:** 4319.

Büchi, G., Fliri, H., Shapiro, R. (1978) *J. Org. Chem.* **43:** 4765.

Büchi, G., Chu, P.S. (1979a) *J. Am. Chem. Soc.* **101:** 6767.

Büchi, G., Wüest, H. (1979b) *Helv. Chim. Acta* **62:** 2661.

Büchi, G., Rodriguez, A.D., Yakushijin, K. (1989) *J. Org. Chem.* **54:** 4494.

Buchschacher, P., Cassal, J.M., Fuerst, A., Meier, W. (1977) *Helv. Chim. Acta* **60:** 2747.

Bucourt, R., Nedelec, L., Gasc, J.-C., Weill-Raynal, J. (1967) *Bull. Soc. Chim. Fr.* 561.

Buisson, D., Azerad, R., Revial, G., d'Angelo, J. (1984) *Tetrahedron Lett.* **25:** 6005.

Bullimore, B.K., McOmie, J.F.W., Turner, A.B. (1966) *Chem. Commun.* 824.

Burgstahler, A.W., Bithos, Z.J. (1960) *J. Am. Chem. Soc.* **82:** 5466.

Burke, S.D., Murtiashaw, C.W., Dike, M.S., Strickland, S.M.S. (1981) *J. Org. Chem.* **46:** 2400.

Burke, S.D., Murtiashaw, C.W., Saunders, J.O., Dike, M.S. (1982) *J. Am. Chem. Soc.* **104:** 872.

Burke, S.D., Buchanan, J.L., Rovin, J.D. (1991) *Tetrahedron Lett.* **32:** 3961.

Burton, L.P.J., White, J.D. (1981) *J. Am. Chem. Soc.* **103:** 3226.

Buttery, C.D., Cameron, A.G., Dell, C.P., Knight, D.W. (1990) *J. Chem. Soc. Perkin Trans. I* 1601.

Bycroft, B.W., Hassaniali-Walji, A., Johnson, A.W., King, T.J. (1970) *J. Chem. Soc. (C)* 1686.

Caille, J.C., Bellamy, F., Guillard, R. (1984) *Tetrahedron Lett.* **25:** 2345.

Caine, D., Hasenhuettl, G. (1975) *Tetrahedron Lett.* 743.

Caine, D., Boucugnani, A.A., Chao, S.T., Dawson, J.B., Ingwalson, P.F. (1976) *J. Org. Chem.* **41:** 1539.

Callant, P., Storme, P., Van der Eycken, E., Vandewalle, M. (1983) *Tetrahedron Lett.* **24:** 5797.

Camps, P., Lozano, R., Miranda, M.A. (1986) *J. Chem. Res.* S 250.

Canet, J.-L., Fade, A., Salaün, J. (1992) *J. Org. Chem.* **57:** 3463.

Carcellar, E., Centellas, V., Moyano, A., Pericas, M.A., Serratosa, F. (1985) *Tetrahedron Lett.* **26:** 2475.

Carda, M., Van der Eycken, J., Vandewalle, M. (1990) *Tetrahedron: Asymmetry* **1:** 17.

Carda, M., Marco, J.A. (1991) *Tetrahedron Lett.* **32:** 5191.

Carless, H.A.J., Oak, O.Z. (1989) *Tetrahedron Lett.* **30:** 1719.

Carless, H.A.J., Busia, K. (1990) *Tetrahedron Lett.* **31:** 1617.

Carless, H.A.J., Busia, K., Oak, O.Z. (1993) *Synlett* 672.

Carlson, R.G., Zey, E.G. (1972) *J. Org. Chem.* **37:** 2468.

Carlström, A.-S., Frejd, T. (1991) *Chem. Commun.* 1216.

Carreño, M.C., Ruano, J.L.G., Garrido, M., Ruiz, P., Solladie, G. (1990) *Tetrahedron Lett.* **31:** 6653.

Cartier, D., Ouahrani, M., Levy, J. (1989) *Tetrahedron Lett.* **30:** 1951.

Cassar, L., Eaton, P.E., Halpern, J. (1970) *J. Am. Chem. Soc.* **92:** 6366.

Castanedo, R., Zetina, C.B., Maldonado, L.A. (1987) *Heterocycles* **25:** 175.

Cava, M.P., Noguchi, I. (1973) *J. Org. Chem.* **38:** 60.

Chakraborty, D.P., Das, K.C., Chowdhury, B.K. (1969) *Phytochem.* **8:** 773.

Challand, B.D., Hikino, H., Kornis, G., Lange, G., de Mayo, P. (1969) *J. Org. Chem.* **34:** 794.

Chamberlin, A.R., Nguyen, H.D., Chung, J.Y.L. (1984) *J. Org. Chem.* **49:** 1682.

Chao, C., Zhang, P. (1988) *Tetrahedron Lett.* **29:** 225.

Chapman, J.H., Holton, P.G., Ritchie, A.C., Walker, T., Webb, G.B., Whiting, K.D.E. (1962) *J. Chem. Soc.* 2471.

Chapman, O.L., Engel, M.R., Springer, J.P., Clardy, J.C. (1971) *J. Am. Chem. Soc.* **93:** 6696.

Chastanet, J., Roussi, G. (1985) *Heterocycles* **23:** 653.

Chen, J., Browne, L.J., Gonnela, N.C. (1986) *Chem. Commun.* 905.

Chenera, B., Chuang, C.-P., Hart, D.J., Lai, C.-S. (1992) *J. Org. Chem.* **57:** 2018.

Cheng, K.F., Kong, Y.C., Chan, T.Y. (1985) *Chem. Commun.* 48.

Chevolot, L., Husson, H.-P. (1975) *Tetrahedron* **31:** 2491.

Chida, N., Ohtsuka, M., Nakazawa, K., Ogawa, S. (1991) *J. Org. Chem.* **56:** 2976.

Chin, C.G., Cuts, H.W., Masamune, S. (1966) *Chem. Commun.* 880.

Chong, R., King, R.R., Whalley, W.B. (1969) *Chem. Commun.* 1512.

Chou, T.-C., Yang, M.S., Lin, C.-T. (1994) *J. Org. Chem.* **59:** 661.

Chou, T.-s., Lee, S.J., Yao, N.-K. (1989) *Tetrahedron* **45:** 4113.

Ciufolini, M.A., Byrne, N.E. (1991) *J. Am. Chem. Soc.* **113:** 8016.

Clark, D.E., Holton, P.G., Meredith, R.P.K., Walker, T., Whiting, K.D.E. (1962) *J. Chem. Soc.* 2479.

Clark, D.E., Meredith, R.P.K., Ritchie, A.C., Walker, T. (1962b) *J. Chem. Soc.* 2490.

Clayton, S.C., Regan, A.C. (1993) *Tetrahedron Lett.* **34:** 7493.

Cloudsdale, I.S., Kluge, A.F., McClure, N.L. (1982) *J. Org. Chem.* **47:** 919.

Coates, R.M., Shaw, J.E. (1970a) *J. Am. Chem. Soc.* **92:** 5657.

Coates, R.M., Shaw, J.E. (1970b). *J. Org. Chem.* **35:** 2597.

Coates, R.M., Shah, S.K., Mason, R.W. (1979) *J. Am. Chem. Soc.* **101:** 6765.

Coblens, K.E., Muralidharan, V.B., Ganem, B. (1982) *J. Org. Chem.* **47:** 5041.

Cocu, F.G., Posternak, T., Wolczunowicz, G. (1970) *Helv. Chim. Acta* **53:** 2275.

Cohen, N., Eichel, W.F., Lopresti, R.J., Neukom, C., Saucy, G. (1976) *J. Org. Chem.* **41:** 3505.

Cohen, N., Barrier, B.L., Lopresti, R.J. (1980) *Tetrahedron Lett.* **21:** 4163.

Colegate, S.M., Dorling, P.R., Huxtable, C.R. (1984) *Aust. J. Chem.* **37:** 1503.

Coleman, R.S., Grant, E.B. (1994) *J. Am. Chem. Soc.* **116:** 8795.

Collum, D.B., McDonald, J.H., III, Still, W.C. (1980) *J. Am. Chem. Soc.* **102:** 2117, 2118, 2120.

Colvin, E.W., Malchenko, S., Raphael, R.A., Roberts, J.H. (1973) *J. Chem. Soc. Perkin Trans. I* 1989.

Comins, D.L., Brown, J.D. (1986) *Tetrahedron Lett.* **27:** 2219.

Comins, D.L., Dehghani, A. (1991a) *Tetrahedron Lett.* **32:** 5697.

Comins, D.L., Weglarz, M.A. (1991b) *J. Org. Chem.* **56:** 2506.

Comins, D.L., Zeller, E. (1991c) *Tetrahedron Lett.* **32:** 5889.

Comins, D.L., Killpack, M.O. (1992) *J. Am. Chem. Soc.* **114:** 10972.

Confalone, P.N., Pizzolato, G., Confalone, D.L., Uskokovic, M.R. (1980) *J. Am. Chem. Soc.* **102:** 1954.

Corey, E.J., Hess, H.-J., Proskow, S. (1963) *J. Am. Chem. Soc.* **85:** 3979.

Corey, E.J., Nozoe, S. (1964) *J. Am. Chem. Soc.* **86:** 1652.

Corey, E.J., Andersen, N.H., Carlson, R.M., Paust, J., Vedejs, E., Vlattas, I., Winter, R.E.K. (1968a) *J. Am. Chem. Soc.* **90:** 3245.

Corey, E.J., Katzenellbogen, J.A., Gilman, N.W., Roman, S.A., Erickson, B.E. (1968b) *J. Am. Chem. Soc.* **90:** 5618.

Corey, E.J., Girotra, N.N., Mathew, C.T. (1969a) *J. Am. Chem. Soc.* **91:** 1557.

Corey, E.J., Vlattas, I., Harding, K. (1969b) *J. Am. Chem. Soc.* **91:** 535.

Corey, E.J., Weinshenker, N.M., Schaaf, T.K., Huber, W. (1969c) *J. Am. Chem. Soc.* **91:** 5675.

Corey, E.J., Caine, D.E., Libit, L. (1971) *J. Am. Chem. Soc.* **93:** 7016.

Corey, E.J., Fuchs, P.L. (1972a) *J. Am. Chem. Soc.* **94:** 4014.

Corey, E.J., Walinsky, S.W. (1972b) *J. Am. Chem. Soc.* **94:** 8932.

Corey, E.J., Balanson, R.D. (1973a) *Tetrahedron Lett.* 3153.

Corey, E.J., Watt, D.S. (1973b) *J. Am. Chem. Soc.* **95:** 2302.

Corey, E.J., Bock, M.G. (1975a) *Tetrahedron Lett.* 2643.

Corey, E.J., Crouse, D.E., Anderson, J.E. (1975b). *J. Org. Chem.* **40:** 2140.

Corey, E.J., Suggs, J.W. (1975c) *Tetrahedron Lett.* 3775.

Corey, E.J., Wollenberg, R.H. (1976) *Tetrahedron Lett.* 4701, 4705.

Corey, E.J., Wetter, H.F., Kozikowski, A.P., Rama Rao, A.V. (1977 *Tetrahedron Lett*. 777.

Corey, E.J., Trybulski, E.J., Melvin, L.S., Nicolaou, K.C., Secrist, J.A., Lett, R., Sheldrake, P.W., Falck, J.R., Brunelle, D.J., Haslanger, M.F., Kim, S., Yoo, S.-e. (1978) *J. Am. Chem. Soc.* **100:** 4618.

Corey, E.J., Arai, Y., Mioskowski, C. (1979a) *J. Am. Chem. Soc.* **101:** 6748.

Corey, E.J., Smith, J.G. (1979b) *J. Am. Chem. Soc.* **101:** 1038.

Corey, E.J., Kang, J. (1981) *J. Am. Chem. Soc.* **103:** 4618.

Corey, E.J., Niimura, K., Konishi, Y., Hashimoto, S., Hamada, Y. (1986) *Tetrahedron Lett*. **27:** 2199.

Corey, E.J., Su, W.-g. (1988) *Tetrahedron Lett*. **29:** 3423.

Cornforth, J.W., Cornforth, R.H., Mathew, K.K. (1959) *J. Chem. Soc*. 112, 2539.

Cortes, E., Salmon, M., Walls, F. 1965) *Bol. Inst. Quim. Univ. Nacl. Autom. Mex*. **17:** 19.

Cory, R.M., Chan, D.M.T., McLaren, F.R., Rasmussen, M.H., Renneboog, R.M. (1979) *Tetrahedron Lett*. 4133.

Cory, R.M., Anderson, P.C., Bailey, M.D., McLaren, F.R., Renneboog, R.M., Yamamoto, B.R. (1985) *Can. J. Chem*. **63:** 2618.

Cossy, J., Belotti, D. (1988) *Tetrahedron Lett*. **29:** 6113.

Covarrubias, A., Maldonado, L.A. (1991) *4th Chem. Congr. N. Am*. ORGN 176.

Cox, P.J., Simpkins, N.S. (1991) *Tetrahedron: Asymmetry* **2:** 1.

Crandall, T.G., Lawton, R.G. (1969) *J. Am. Chem. Soc*. **91:** 2127.

Cranwell, P.A., Saxton, J.E. (1962) *J. Chem. Soc*. 3842.

Criegee, R., Becher, P. (1957) *Chem. Ber*. **90:** 2516.

Crombie, L., Jones, R.C.F., Palmer, C.J. (1985) *Tetrahedron Lett*. **26:** 2929.

Cross, B.E., Hanson, J.R., Briggs, L.H., Cambie, R.C., Rutledge, P.S. (1963) *Proc. Chem. Soc*. 17.

Cummins, W.J., Drew, M.G.B., Markson, A.J. (1988) *Tetrahedron* **44:** 5151.

Cupas, C., Schleyer, P.v.R., Trecker, D.J. (1965) *J. Am. Chem. Soc*. **87:** 917.

Cupas, C.A., Hodakowski, L. (1974) *J. Am. Chem. Soc*. **96:** 4668.

Curran, D.P., Chen, M.H., (1987a) *J. Am. Chem. Soc*. **109:** 6558.

Curran, D.P., Kuo, S.-C. (1987b) *Tetrahedron Lett*. **43:** 5653.

Curran, D.P., Liu, H. (1992) *J. Am. Chem. Soc*. **114:** 5863.

Czarnocki, Z., MacLean, D.B., Szarek, W.A. (1987) *Chem. Commun*. 493.

Czeskis, B.A., Alexeev, I.G., Moiseenkov, A.M. (1993) *Mendeleev Commun*. 93.

d'Angelo, J., Revial, G. (1983) *Tetrahedron Lett*. **24:** 2103.

Danheiser, R.L., Morin, J.M., Salaski, E.J. (1985) *J. Am. Chem. Soc*. **107:** 8066.

Danieli, B., Lesma, G., Palmisano, G. (1980) *Chem. Commun*. 860.

Danieli, B., Lesma, G., Mauro, M., Palmisano, G., Passarella, D. (1991) *Tetrahedron: Asymmetry* **1:** 793.

Danieli, B., Lesma, G., Maurl, M., Palmisano, G., Passarella, D. (1994) *Tetrahedron* **50:** 8837.

Daniewski, A.R., Kocor, M. (1975) *J. Org. Chem*. **40:** 3136.

Daniewski, A.R., Valenta, Z. (1984) *Bull. Pol. Acad. Sci., Chem*. **32:** 115.

Daniewski, A.R., Uskokovic, M.R. (1990) *Tetrahedron Lett*. **31:** 5599.

Danishefsky, S., Dumas, D. (1968) *Chem. Commun.* 1287.

Danishefsky, S., Etheredge, S.J., Volkmann, R., Eggler, J., Quick, J. (1971) *J. Am. Chem. Soc.* **93:** 5575.

Danishefsky, S., Cain, P. (1974) *J. Org. Chem.* **39:** 2925.

Danishefsky, S., Cain, P. (1976) *J. Am. Chem. Soc.* **99:** 4975.

Danishefsky, S., Hirama, M., Gombatz, K., Harayama, T., Berman, E., Schuda, P. (1979) *J. Am. Chem. Soc.* **101:** 7020.

Danishefsky, S., Myles, D.C., Harvey, D.F. (1987a) *J. Am. Chem. Soc.* **109:** 862.

Danishefsky, S.J., Panek, J.S. (1987b) *J. Am. Chem. Soc.* **109:** 917.

Danishefsky, S.J., Mentio, N. (1988) *J. Am. Chem. Soc.* **110:** 8129.

Danishefsky, S.J., Cabal, M.P., Chow, K. (1989) *J. Am. Chem. Soc.* **111:** 3456.

Dankova, T.F., Siderova, E.A., Preobrazhenskii, N.A. (1941) *J. Gen. Chem.* (*USSR*) **11:** 934.

Dastur, K.P. (1974) *J. Am. Chem. Soc.* **96:** 2605.

Dauben, W.G., Whalen, D.L. (1966) *Tetrahedron Lett.* 3743.

Dauben, W.G., Hart, D.J., Ipaktschi, J., Kozikowski, A. (1973) *Tetrahedron Lett.* 4425.

Dauben, W.G., Kozikowski, A.P., Zimmerman, W.T. (1975) *Tetrahedron Lett.* 515.

Dauben, W.G., Hart, D.J. (1977) *J. Am. Chem. Soc.* **99:** 7307.

Dauben, W.G., Walker, D.M. (1981) *J. Org. Chem.* **46:** 1103.

Dauben, W.G., Cunningham, A.F. (1983) *J. Org. Chem.* **48:** 2842.

Dauben, W.G., Shapiro, G. (1984) *J. Org. Chem.* **49:** 4252.

Dauben, W.G., Gerdes, J.M., Smith, D.B. (1985) *J. Org. Chem.* **50:** 2576.

Dauben, W.G., Wang, T.Z., Stephens, R.W. (1990) *Tetrahedron Lett.* **31:** 2393.

Dauben, W.G., Farkas, I., Bridon, D.P., Chuang, C.-P., Henegar, K.E. (1991) *J. Am. Chem. Soc.* **113:** 5883.

Davey, A.E., Schaeffer, M.J., Taylor, R.J.K. (1991) *Chem. Commun.* 1137.

Davies, L.B., Greenberg, S.G., Sammes, P.G. (1091) *J. Chem. Soc. Perkin Trans. I* 1909.

De Bernardo, S., Weigele, M. (1976) *J. Org. Chem.* **41:** 287.

de Groot, A., Broekhuysen, M.P., Doddema, L.L., Vollering, M.C., Westerbeek, J.M.M. (1982) *Tetrahedron Lett.* **23:** 4831.

Dehmlow, E.V., Stiehm, T. (1990) *Tetrahedron Lett.* **31:** 1841.

de Meijere, A., Kaufmann, D., Schallner O. (1971) *Angew. Chem. Int. Ed. Engl.* **10:** 417.

Demoute J.-P., Hainaut, D., Toromanoff, E. (1973) *C.R. Acad. Sci. Paris* (*C*) **277:** 49.

Demuth, M., Hinsken, W. (1985) *Angew. Chem. Int. Ed. Engl.* **24:** 973.

Demuth, M., Ritterskamp, P., Weigt, E., Schaffner, K. (1986) *J. Am. Chem. Soc.* **108:** 4149.

Demuth, M., Hinsken, W. (1988) *Helv. Chim. Acta* **71:** 569.

Dener, J.M., Hart, D.J., Ramesh, S. (1988) *J. Org. Chem.* **53:** 6022.

Descoins, C., Henrick, C.A., Siddall, J.B. (1972) *Tetrahedron Lett.* 3777.

Deshpande, M.N., Jawdosiuk, M., Kubiak, G., Venkatachalam, M., Weiss, U., Cook, J.M. (1985) *J. Am. Chem. Soc.* **107:** 4786.

Devine, P.N., Oh, T. (1991) *Tetrahedron Lett.* **32:** 883.

Devos, M.J., Krief, A. (1978) *Tetrahedron Lett.* 1845.

Dickinson, R.A., Kubela, R., MacAlpine, G.A., Stojanac, Z., Valenta, Z. (1972) *Can. J. Chem.* **50:** 2377.

Differding, E., Ghosez, L. (1985) *Tetrahedron Lett.* **26:** 1647.

Disnayaka, B.W., Weedon, A.C. (1985) *Chem. Commun.* 1282.

Diwu, Z., Lown, J.W. (1992) *Tetrahedron* **48:** 45.

Dodd, J.H., Starratt, J.E., Weinreb, S.M. (1984) *J. Am. Chem. Soc.* **106:** 1811.

Doering, W.v.E., Roth, W.R. (1963) *Tetrahedron* **19:** 715.

Dolby, L.J., Biere, H. (1968) *J. Am. Chem. Soc.* **90:** 2699.

Dolby, L.J., Hanson, G. (1976) *J. Org. Chem.* **41:** 563.

Dowd, P., Choi, S.C. (1991) *Tetrahedron Lett.* **32:** 565.

Dower, W.V., Vollhardt, K.P.C. (1982) *J. Am. Chem. Soc.* **104:** 6878.

Dufour, M., Gramain, J.-C., Husson, H.-P., Sinibaldi, M.-E., Troin, Y. (1989) *Tetrahedron Lett.* **30:** 3429.

Duhamel, L., Ravard, A., Plaquevent, J.C., Davoust, D. (1987) *Tetrahedron Lett.* **28:** 5517.

Dumortier, L., Liu, P., Dobbelaere, S., Van der Eycken, J., Vandewalle, M. (1992) *Synlett* 243.

Duthaler, R.O., Maienfish, P. (1982) *Helv. Chim. Acta* **65:** 635; (1984) *Helv. Chim. Acta* **67:** 832.

Earl, E.A., Volhardt, K.P.C. (1983) *J. Am. Chem. Soc.* **105:** 6991.

Eaton, P.E., Cole, T.W. (1964) *J. Am. Chem. Soc.* **86:** 962, 3157.

Eaton, P.E., Hudson, R.A., Giordano, C. (1974) *Chem. Commun.* 978.

Eaton, P.E., Mueller, R.H., Carlson, G.R., Cullison, D.A., Cooper, G.F., Chou, T.-C., Krebs, E.-P. (1977) *J. Am. Chem. Soc.* **99:** 2751.

Eaton, P.E., Andrews, G.D., Krebs, E.-P., Kunai, A. (1979) *J. Org. Chem.* **44: 2824.**

Eaton, P.E., Or, Y.S., Branca, S.J. (1981) *J. Am. Chem. Soc.* **103:** 2134.

Eaton, P.E., Leipzig, B.D. (1983) *J. Am. Chem. Soc.* **105:** 1656.

Eder, U., Sauer, G., Wiechert, R. (1971) *Angew. Chem. Int. Ed. Engl.* **10:** 496.

Eilbracht, P., Balsz, E., Acker, M. (1985) *Chem. Ber.* **118:** 825.

Eilerman, R.G., Willis, B.J. (1981) *Chem. Commun.* 30.

Ellison, R.A., Woessner, W.D. (1972) *Chem. Commun.* 529.

Elming, N., Clauson-Kaas, N. (1955) *Acta Chem. Scand.* **9:** 23.

Emerson, G.F., Watts, L., Pettit, R. (1965) *J. Am. Chem. Soc.* **87:** 131.

Enhsen, A., Karabelas, K., Heerding, J.M., Moore, H.W. (1990) *J. Org. Chem.* **55:** 1177.

Ent, H., de Koning, H., Speckamp, W.N. (1986) *J. Org. Chem.* **51:** 1687.

Ent, H., de Koning, H., Speckamp, W.N. (1988) *Heterocycles* **27:** 237.

Erden, I. (1986) *Synth. Commun.* **16:** 117.

Esch, P.M., Hiemstra, H., Klaver, W.J., Speckamp, W.N. (1987) *Heterocycles* **26:** 75.

Eschler, B.M., Haynes, R.K., Ironside, M.D., Kremmydas, S., Ridley, D.D., Hambley, T.W. (1991) *J. Org. Chem.* **56:** 4760.

Evans, D.A., Cain, P.A., Wong, R.Y. (1977a) *J. Am. Chem. Soc.* **99:** 7983.

Evans, D.A., Wong, R.Y. (1977b) *J. Org. Chem.* **42:** 350.

Evans, D.A., Sacks, C.E., Kleschick, W.A., Taber, T.R. (1979) *J. Am. Chem. Soc.* **101:** 6789.

Evans, D.A., Tanis, S.P., Hart, D.J. (1981) *J. Am. Chem. Soc.* **103:** 5813.

Evans, D.A., Mitch, C.H. (1982a) *Tetrahedron Lett.* **23:** 285.

Evans, D.A., Thomas, E.W., Cherpeck, R.E. (1982b) *J. Am. Chem. Soc.* **104:** 3695.

Fang, J.-M. (1982) *J. Org. Chem.* **47:** 3464.

Fehr, C., Ohloff, G., Büchi, G. (1979) *Helv. Chim. Acta* **62:** 2655.

Felix, D., Schreiber, J., Ohloff G., Eschenmoser, A. (1971) *Helv. Chim. Acta* **54:** 2896.

Felix, G., Lapouyade, R., Castellan, A., Bouas-Laurent, H., Gautier, J., Hauw, C. (1975) *Tetrahedron Lett.* 409.

Fell, S.C.M., Harbridge, J.B. (1990) *Chem. Scand.* **31:** 4227.

Feng, F., Murai, A. (1992) *Chem. Lett.* 1587.

Ferezou, J.P., Gauchet-Prunet, J., Julia, M., Pancrazi, A. (1988) *Tetrahedron Lett.* **29:** 3667.

Fessner, W.-D., Prinzbach, H., Rihs, G. (1983) *Tetrahedron Lett.* **24:** 5857.

Fessner, W.-D., Prinzbach, H. (1986) *Tetrahedron* **42:** 1797.

Fessner, W.-D., Murty, B.A.R.C., Worth, J., Hunkler, D., Fritz, H., Prinzbach, H., Roth, W.D., Schleyer, P.v., McEwen, A.B., Maier, W.F. (1987) *Angew. Chem. Int. Ed. Engl.* **26:** 452.

Fetizon, M., De Lobelle, J. (1960) *Tetrahedron Lett.* #9, 16.

Fetizon, M., Golfier, M., Louis, J.-M. (1975) *Tetrahedron* **31:** 171.

Fex, T. (1981) *Tetrahedron Lett.* **22:** 2707.

Ficini, J., Touzin, A.M. (1977) *Tetrahedron Lett.* 1081.

Firestone, R.A., Harris, E.E., Reuter, W. (1967) *Tetrahedron* **23:** 943.

Fischer, H., Kirstahler, A. (1928) *Liebigs Ann. Chem.* **466:** 178.

Fischer, H., Zeile, K. (1929) *Liebigs Ann. Chem.* **468:** 98.

Fischli, A., Klaus, M., Mayer, H., Schönholzer, P., Ruegg, R. (1975) *Helv. Chim. Acta* **58:** 564.

Fisher, M.J., Myers, C.D., Joglar, J., Chen, S.-H., Danishefsky, S.J. (1991) *J. Org. Chem.* **56:** 5826.

Fitjer, L., Conia, J.M. (1973) *Angew. Chem. Int. Ed. Engl.* **12:** 334.

Fitjer, L. (1982) *Chem. Ber.* **115:** 1047.

Fitjer, L., Majewski, M., Kanschik, A. (1988) *Tetrahedron Lett.* **29:** 1263.

Fitjer, L., Quaback, U. (1989) *Angew. Chem. Int. Ed. Engl.* **28:** 94.

Flann, C.J., Overman, L.E. (1987) *J. Am. Chem. Soc.* **109:** 6115.

Fleming, I., Ghosh, S.K. (1992) *Chem. Commun.* 1775.

Fleming, I., Ghosh, S.K. (1994) *Chem. Commun.* 2285, 2287.

Flimm, W. (1890) *Ber. Deut. Chem. Ges.* **23:** 57.

Fliri, H.G., Scholz, D., Stütz, A. (1979) *Monatsh. Chem.* **110:** 245.

Flitsch, W., Russkamp, P. (1983) *Liebigs Ann. Chem.* 521.

Flitsch, W., Pandl, K. (1987) *Liebigs Ann. Chem.* 649.

Forsyth, C.J., Clardy, J. (1990) *J. Am. Chem. Soc.* **112:** 3497.

Foy, J.E., Ganem, B. (1977) *Tetrahedron Lett.* 775.

Franck-Neumann, M., Sedrati, M., Mokhi, M. (1986) *Tetrahedron Lett.* **27:** 3861.

Franck-Neumann, M., Miesch, M., Barth, F. (1989) *Tetrahedron Lett.* **30:** 3537.

Franke, L.R.R.A., Wolf, H., Wray, V. (1984) *Tetrahedron* **40**: 3491.

Franz, T., Hein, M., Veith, U., Jäger, V., Peters, E.-M., Peters, K., von Schnering, H.G. (1994) *Angew. Chem. Int. Ed. Engl.* **33**: 1298.

Frater, G. (1974) *Helv. Chim. Acta* **57**: 172.

Fried, J., Lin, C.H., Sih, J.C., Dalven, P., Cooper, G.F. (1972) *J. Am. Chem. Soc.* **94**: 4343.

Frincke, J.M., Faulkner, D.J. (1982) *J. Am. Chem. Soc.* **104**: 265.

Froborg, J., Magnusson, G., Thoren, S. (1974) *Angew. Chem. Scand.* **B28**: 265.

Froborg, J., Magnusson, G., Thoren, S. (1975) *J. Org. Chem.* **40**: 1595.

Froborg, J., Magnusson, G. (1978) *J. Am. Chem. Soc.* **100**: 6728.

Fuji, K., Ichikawa, K., Fujita, E. (1979) *Chem. Pharm. Bull.* **27**: 3183.

Fuji, K., Ichikawa, K., Fujita, E. (1980) *J. Chem. Soc. Perkin Trans. I* 1066.

Fuji, K., Node, M., Terada, S., Murata, M., Nagasawa, H., Taga, T., Machida, K. (1985) *J. Am. Chem. Soc.* **107**: 6404.

Fuji, K., Kawabata, T., Kiryu, Y., Sugiura, Y., Taga, T., Miwa, Y. (1990a) *Tetrahedron Lett.* **31**: 6663.

Fuji, K., Node, M., Naniwa, Y., Kawabata, T. (1990b) *Tetrahedron Lett.* **31**: 3175.

Fujimoto, R., Kishi, Y., Blount, J.F. (1980) *J. Am. Chem. Soc.* **102**: 7154.

Fujita, K., Nakai, H., Kobayashi, S., Inoue, K., Nojima, S., Ohno, M. (1982) *Tetrahedron Lett.* **23**: 3507.

Fujiwara, S., Smith, A.B., III. [1992) *Tetrahedron Lett.* **33**: 1185.

Fukamiya, N., Kato, M., Yoshikoshi, A. (1973) *J. Chem. Soc. Perkin Trans. I* 1843.

Fukuda, Y., Nakatani, K., Ito, Y., Terashima, S. (1990) *Tetrahedron Lett.* **31**: 6699.

Fukuyama, T., Dunkerton, L.V., Aratani, M., Kishi, Y. (1975) *J. Org. Chem.* **41**: 2011.

Fukuyama, T., Wang, C.-L.-J., Kishi, Y. (1979) *J. Am. Chem. Soc.* **101**: 260.

Fukuyama, T., Frank, R.K., Jewell, C.F. (1980) *J. Am. Chem. Soc.* **102**: 2122.

Fukuyama, T., Yang, L., Ajeck, K.L., Sachleben, R.A. (1990) *J. Am. Chem. Soc.* **112**: 3712.

Fukuyama, Y., Tokoroyama, T., Kubota, T. (1973) *Tetrahedron Lett.* 4869.

Fukuyama, Y., Kirkemo, C.L., White, J.D. (1977) *J. Am. Chem. Soc.* **99**: 646.

Fukuzawa, A., Sato, H., Miyamoto, M., Masamune, T. (1986) *Tetrahedron Lett.* **27**: 2901.

Fukuzawa, A., Sato, H., Masamune, T. (1987) *Tetrahedron Lett.* **28**: 4303.

Funk, R.L., Volhardt, K.P.C. (1979) *J. Am. Chem. Soc.* **101**: 215.

Funk, R.L., Abelman, M.M., Novak, P.M., Jellison, K.M. (1988) *Tetrahedron Lett.* **29**: 1493.

Furuichi, K., Miwa, T. (1974) *Tetrahedron Lett.* 3689.

Furuta, K., Ikeda, N., Yamamoto, H. (1984) *Tetrahedron Lett.* **25**: 675.

Gais, H.-J., Lied, T. (1984a) *Angew. Chem. Int. Ed. Engl.* **23**: 145.

Gais, H.-J., Lied, T., Lukas, K.L. (1984b) *Angew. Chem. Int. Ed. Engl.* **23**: 511.

Gais, H.-J., Lukas, K.L. (1984c) *Angew. Chem. Int. Ed. Engl.* **23**: 142.

Gambacorta, A., Botta, M., Turchetta, S. (1988) *Tetrahedron* **44**: 4837.

Gammill, R.B., Bryson, T.A. (1976) *Synth. Commun.* **6**: 209.

Gao, Y., Zepp, C.M. (1991) *Tetrahedron Lett.* **32**: 3155.

Garlaschelli, L., Vidari, G., Zanoni, G. (1992) *Tetrahedron* **48**: 9495.

Garner, P., Ho, W.B., Shin, H. (1993) *J. Am. Chem. Soc.* **115:** 10742.

Garratt, P.J., White, J.F. (1977) *J. Org. Chem.* **42:** 1733.

Garratt, P.J., Porter, J.R. (1986) *J. Org. Chem.* **51:** 5450.

Garver, L.C., van Tamelen, E.E. (1982) *J. Am. Chem. Soc.* **104:** 867.

Gateau-Olesker, A., Cleophax, J., Gero, S.D. (1986) *Tetrahedron Lett.* **27:** 41.

Gates, M. (1950) *J. Am. Chem. Soc.* **72:** 228.

Gates, M., Tschudi, M. (1956) *J. Am. Chem. Soc.* **78:** 1380.

Geiger, R.F., Lalonde, M., Stoller, H., Schleich, K. (1984) *Helv. Chim. Acta* **67:** 1274.

Geissman, T.A., Waiss, A.C. (1962) *J. Org. Chem.* **27:** 139.

Gellerman, G., Babad, M., Kashman, Y. (1993) *Tetrahedron Lett.* **34:** 1827.

Gênet, J.P., Piau, F., Ficini, J. (1980) *Tetrahedron Lett.* **21:** 3183.

Gensler, W.J., Solomon, P.H. (1973) *J. Org. Chem.* **38:** 1726.

Gerecke, M., Zimmermann, J.-P., Aschwanden, W. (1970) *Helv. Chim. Acta* **53:** 991.

Gerlach, H., Wetter, H. (1974) *Helv. Chim. Acta* **57:** 2306.

Gerlach, H., Oertle, K., Thalmann, A., Saervi, S. (1975) *Helv. Chim. Acta* **58:** 2036.

Gerlach, H., Oertle, K., Thalmann, A. (1976) *Helv. Chim. Acta* **59:** 755.

Germanas, J., Aubert, C., Volhardt, K.P.C. (1991) *J. Am. Chem. Soc.* **113:** 4006.

Gervay, J.E., McCapra, F., Money, T., Sharma, G.M. (1966) *Chem. Commun.* 142.

Ghera, E., Ben-David, Y. (1978) *Chem. Commun.* 480.

Ghera, E., Maurya, R., Ben-David, Y. (1986) *Tetrahedron Lett.* **27:** 3935.

Ghosez, L., Marko, I., Hesbain-Frisque, A.-M. (1986) *Tetrahedron Lett.* **27:** 5211.

Giese, B., Rupaner, R. (1988) *Synthesis* 219.

Gill, M., Rickards, R.W. (1979) *Chem. Commun.* 121.

Girotra, N.N., Wendler, N.L. (1982) *Tetrahedron Lett.* **23:** 5501.

Girotra, N.N., Reamer, R.A., Wender, N.L. (1984) *Tetrahedron Lett.* **25:** 5371.

Glanzmann, M., Karalai, C., Ostersehlt, B., Schön, U., Frese, C., Winterfeldt, E. (1982) *Tetrahedron Lett.* **38:** 2805.

Gleason, J.G., Bryan, D.B., Kinzig, C.M. (1980) *Tetrahedron Lett.* **21:** 1129.

Gleiter, R., Karcher, M. (1988) *Angew. Chem. Int. Ed. Engl.* **27:** 840.

Godleski, S.A., Meinhart, J.D., Miller, D.J., Wall, S.V. (1981) *Tetrahedron Lett.* **22:** 2247.

Goering, B.K., Ganem, B. (1994) *Tetrahedron Lett.* **35:** 6997.

Goldschmidt, G. (1886) *Monatsh. Chem.* **7:** 685.

Goldsmith, D.J., Sakano, I. (1976) *J. Org. Chem.* **41:** 2095.

Golinski, M., Vasudevan, S., Floresca, R., Brock, C.P., Watt, D.S. (1993) *Tetrahedron Lett.* **34:** 55.

Gopalan, A., Magnus, P. (1980) *J. Am. Chem. Soc.* **102:** 1756.

Gössinger, E. (1980) *Tetrahedron Lett.* 2229.

Gourcy, J.-G., Dauphin, G., Jeminet, G. (1991) *Tetrahedron: Asymmetry* **2:** 31.

Gramain, J.-C., Remuson, R., Vallee, D. (1985) *J. Org. Chem.* **50:** 710.

Grandguillot, J.C., Rouessac, F. (1979) *Synthesis* 607.

Grandjean, D., Pale, P., Cuche, J. (1991) *Tetrahedron Lett.* **32:** 3043.

Gray, R.W., Dreiding, A.S. (1977) *Helv. Chim. Acta* **60:** 1969.

Grayson, D.H., Wilson, J.R.H. (1984) *Chem. Commun.* 1695.

Greene, A.E. (1980) *Tetrahedron Lett.* **21:** 3059.

Greene, A.E., Luche, M.-J., Serra, A.A. (1985) *J. Org. Chem.* **50:** 3957.

Greene, A.E., Serra, A.A., Garreiro, E.J., Costa, P.R.R. (1987) *J. Org. Chem.* **52:** 1169.

Gribble, G.W. (1972) *J. Org. Chem.* **37:** 1833.

Grieco, P.A., Masaki, Y. (1975) *J. Org. Chem.* **40:** 150.

Grieco, P.A., Oguri, T., Wang, C.L.J., Williams, E. (1977a) *J. Org. Chem.* **42:** 4113.

Grieco, P.A., Ohfune, Y., Majetich, G. (1977b) *J. Am. Chem. Soc.* **99:** 7393.

Grieco, P.A., Oguri, T., Gilman, S., De Titta, G.T. (1978) *J. Am. Chem. Soc.* **100:** 1616.

Grieco, P.A., Ohfune, Y., Yokoyama, Y., Owens, W. (1979a) *J. Am. Chem. Soc.* **101:** 4749.

Grieco, P.A., Takigawa, T., Moore, D.R. (1979b) *J. Am. Chem. Soc.* **101:** 4380.

Grieco, P.A., Ferrino, S., Vidari, G. (1980a) *J. Am. Chem. Soc.* **102:** 7587.

Grieco, P.A., Takigawa, T., Schillinger, W.J. (1980b) *J. Org. Chem.* **45:** 2247.

Grieco, P.A., Srinivasn, C.V. (1981) *J. Org. Chem.* **46:** 2591.

Grieco, P.A., Inanaga, J., Lin, N.-L., Yanami, T. (1982a) *J. Am. Chem. Soc.* **104:** 5781.

Grieco, P.A., Majetich, G.F., Ohfune, Y. (1982b) *J. Am. Chem. Soc.* **104:** 4226.

Grieco, P.A., Ohfune, Y.Y., Majetich, G.F., Wang, C.-L.J. (1982c) *J. Am. Chem. Soc.* **104:** 4233.

Grieco, P.A., Flynn, D.L., Zelle, R.E. (1984) *J. Am. Chem. Soc.* **106:** 6414.

Grieco, P.A., Lis, R., Zelle, R.E., Finn, J. (1986) *J. Am. Chem. Soc.* **108:** 5908.

Grieco, P.A., Nunes, J.J., Gaul, M.D. (1990' *J. Am. Chem. Soc.* **112:** 4595.

Grieco, P.A., Collins, J.L., Moher, E.D., Fleck, T.J., Gross, R.S. (1993) *J. Am. Chem. Soc.* **115:** 6078.

Grierson, D.S., Royer, J., Guerrier, L., Husson, H.-P. (1986) *J. Org. Chem.* **51:** 4475.

Griffin, G.W., Peterson, L.I. (1962) *J. Am. Chem. Soc.* **84:** 3398.

Griffin, G.W., Peterson, L.I. (1963) *J. Am. Chem. Soc.* **85:** 2268.

Griffith, D.A., Danishefsky, S. (1991) *J. Am. Chem. Soc.* **113:** 5863.

Groth, U., Köhler, T., Taapken, T. (1991) *Tetrahedron Lett.* **47:** 7583.

Guanti, G., Banfi, L., Narisano, E. (1989) *Tetrahedron Lett.* **30:** 2697.

Guanti, G., Banfi, L., Ghiron, C., Narisano, E. (1991a) *Tetrahedron Lett.* **32:** 267.

Guanti, G., Banfi, L., Narisano, E. (1991b) *Tetrahedron Lett.* **32:** 6939.

Guanti, G., Banfi, L., Narisano, E., Thea, S. (1991c) *Tetrahedron Lett.* **32:** 6943.

Guerrier, L., Royer, J., Grierson, D.S., Husson, H.-P. (1983) *J. Am. Chem. Soc.* **105:** 7754.

Guile, S., Saxton, J.E., Thornton-Pett, M.T. (1991) *Tetrahedron Lett.* **32:** 1381.

Gunatilaka, A.A.L., Hirai, N., Kingston, D.G.I. (1983) *Tetrahedron Lett.* **24:** 5457.

Gupta, R.B., Kaloustian, M.K., Franck, R.W., Blount, J.F. (1991) *J. Am. Chem. Soc.* **113:** 359.

Gutowsky, H.S., Chen, J., Hajduk, P.J., Keen, J.D., Chuang, C., Emilsson, T. (1991) *J. Am. Chem. Soc.* **113:** 4747.

Gutzwiller, J., Buchschacher, P., Fürst, A. (1967) *Synthesis* 167.

Hagiwara, H., Okano, A., Uda, H. (1990) *J. Chem. Soc. Perkin Trans. I* 2109.

Hagiwara, H., Uda, H. (1988) *Chem. Commun.* 815.

Haiza, M., Lee, J., Snyder, J.K. (1990) *J. Org. Chem.* **55**: 5008.

Hajos, Z.G., Parrish, D.R. (1974) *J. Org. Chem.* **39**: 1615.

Hakam, K., Thielmann, M., Thielmann, T., Winterfeldt, E. (1987) *Tetrahedron* **43**: 2035.

Hall, E.S., McCapra, F., Scott, A.I. (1967) *Tetrahedron* **23**: 4131.

Han, Y.-K., Paquette, L.A. (1979) *J. Org. Chem.* **44**: 3731.

Hanke, M., Jutz, C. (1980) *Synthesis* 31.

Hanson, R.M., Sharpless, K.B. (1986) *J. Org. Chem.* **51**: 1922.

Hanson, R.M. (1991) *Chem. Rev.* **91**: 437.

Harada, N., Sugioka, T., Ando, Y., Uda, H., Kuriki, T. (1988) *J. Am. Chem. Soc.* **110**: 8483.

Harada, N., Sugioka, T., Uda, H., Kuriki, T. (1990) *J. Org. Chem.* **55**: 3158.

Harada, T., Mukaiyama, T. (1981) *Chem. Lett.* 1109.

Harada, T., Hayashiya, T., Wada, I., Iwa-ake, N., Oku, A. (1987a) *J. Am. Chem. Soc.* **109**: 527.

Harada, T., Wada, I., Oku, A. (1987b) *Tetrahedron Lett.* **28**: 4181.

Harada, T. (1989) *J. Synth. Org. Chem. Jpn.* **47**: 113.

Harada, T., Matsuda, Y., Wada, I., Uchimura, J., Oku, A. (1990) *Chem. Commun.* 21.

Harada, T., Wada, I., Uchimura, J., Inoue, A., Tanaka, S., Oku, A. (1991) *Tetrahedron Lett.* **32**: 1219.

Harada, T., Kagamihara, Y., Tanaka, S., Sakamoto, K., Oku, A. (1992a) *J. Org. Chem.* **57**: 1637.

Harada, T., Takahashi, T., Takahashi, S. (1992b) *Tetrahedron Lett.* **33**: 369.

Harayama, T., Takatani, M., Inubushi, Y. (1980) *Chem. Pharm. Bull.* **28**: 2394.

Harding, K.E., Clement, B.A., Moreno, L., Peter-Kalinic, J. (1981) *J. Org. Chem.* **46**: 940.

Harley-Mason, J., Jackson, A.H. (1954) *J. Chem. Soc.* 3651.

Harley-Mason, J., Taylor, C.G. (1969) *Chem. Commun.* 281.

Harris, T.M., Wittek, P.J. (1975) *J. Am. Chem. Soc.* **97**: 3270.

Harris, T.M., Harris, C.M. (1977) *Tetrahedron* **33**: 2159.

Harris, T.M., Webb, A.D., Harris, C.M., Wittek, P.J., Murray, T.P. (1976) *J. Am. Chem. Soc.* **98**: 6065.

Harris, T.M., Harris, C.M. (1986) *Pure Appl. Chem.* **58**: 283.

Harruff, L.G., Brown, M., Boekelheide, V. (1978) *J. Am. Chem. Soc.* **100**: 2893.

Hart, D.J., Cain, P.A., Evans, D.A. (1978) *J. Am. Chem. Soc.* **100**: 1548.

Hart, D.J., Kanai, K. (1983) *J. Am. Chem. Soc.* **105**: 1255.

Hart, D.J., Tsai, Y.-M. (1984) *J. Am. Chem. Soc.* **106**: 8209.

Hart, D.J., Yang, T.-K. (1985) *J. Org. Chem.* **50**: 235.

Hashimoto, H., Tsuzuki, K., Sakan, F., Shirahama, H., Matsumoto, T. (1974) *Tetrahedron Lett.* 3745.

Hashimoto, M., Harigaya, H., Yanagiya, M., Shirahama, H. (1991) *J. Org. Chem.* **56**: 2299.

Hassall, C.H., Morgan, B.A. (1970) *Chem. Commun.* 1345.

Hatakeyama, S., Kuniya, S., Numata, H., Ochi, N., Takano, S. (1988) *J. Am. Chem. Soc.* **110**: 5201.

Hatakeyama, S., Sakurai, K., Saijo, K., Takano, S. (1985a) *Tetrahedron Lett.* **26**: 1333.

Hatakeyama, S., Sakurai, K., Takano, S. (1985b) *Chem. Commun.* 1759.

Hatakeyama, S., Sakurai, K., Takano, S. (1986) *Tetrahedron Lett.* **27:** 4485.

Hatakeyama, S., Sato, K., Takano, S. (1993) *Tetrahedron Lett.* **34:** 7425.

Hatch, R.P., Shringarpure, J., Weinreb, S.M. (1978) *J. Org. Chem.* **43:** 4172.

Hauser, F.M., Mal, D. (1984) *J. Am. Chem. Soc.* **106:** 1862.

Hayakawa, K., Ohsuki, S., Kanematsu, K. (1986) *Tetrahedron Lett.* **27:** 4205.

Hayashi, T., Yamamoto, A., Ito, Y. (1988a) *Tetrahedron Lett.* **29:** 99.

Hayashi, T., Yamamoto, A., Ito, Y. (1988b) *Tetrahedron Lett.* **29:** 669.

Hayashi, Y., Nishizawa, M., Harita, S., Sakan, T. (1972) *Chem. Lett.* 375.

Hayashi, Y., Nishizawa, M., Sakan, T. (1975) *Chem. Lett.* 387.

Heathcock, C.H. (1966) *J. Am. Chem. Soc.* **88:** 4110.

Heathcock, C.H., Ratcliffe, R. (1971) *J. Am. Chem. Soc.* **93:** 1746.

Heathcock, C.H., Delmar, E.G., Graham, S.L. (1982a) *J. Am. Chem. Soc.* **104:** 1907.

Heathcock, C.H., Kleinman, E.F., Binkley, E.S. (1982b) *J. Am. Chem. Soc.* **104:** 1054.

Heathcock, C.H., von Geldern, T.W. (1987) *Heterocycles* **25:** 75.

Heathcock, C.H., Blumenkopf, T.A., Smith, K.M. (1989) *J. Org. Chem.* **54:** 1548.

Heathcock, C.H., Clark, R.D. (Unpublished) quoted in ApSimon, J., ed. (1983) *The Total Synthesis of Natural Products*, Vol. 5, p. 184, Wiley, New York.

Heerding, J.M., Moore, H.W. (1991) *J. Org. Chem.* **56:** 4048.

Heitz, M.P., Overman, L.E. (1989) *J. Org. Chem.* **54:** 2591.

Hendrickson, J.B., Göschke, R., Rees, R. (1964) *Tetrahedron* **20:** 565.

Hendrickson, J.B. (1977) *J. Am. Chem. Soc.* **99:** 5439.

Herdeis, C., Hartke-Karger, C. (1979) *Heterocycles* **29:** 287.

Herdeis, C., Hartke-Karger, C. (1991) *Liebigs Ann. Chem.* 99.

Hermann, K., Dreiding, A.S. (1977) *Helv. Chim. Acta* **60:** 673.

Herz, W., Iyer, V.S., Nair, M.G. (1975) *J. Org. Chem.* **40:** 3519.

Hewgill, F.R., Pass, M.C. (1985) *Aust. J. Chem.* **38:** 537.

Hidalgo-Del Vecchio, G., Oehleschlager, A.C. (1994) *J. Org. Chem.* **59:** 4853.

Hiemstra, H., Sno, M.H.A.M., Vijn, R.J., Speckamp, W.N. (1985) *J. Org. Chem.* **50:** 4014.

Hirama, M., Gomibuchi, T., Fujiwara, K., Sugiura, Y., Uesugi, M. (1991) *J. Am. Chem. Soc.* **113:** 9851.

Hirao, K., Yonemitsu, O. (1980) *Chem. Commun.* 423.

Hird, N.W., Lee, T.V., Leigh, A.J., Maxwell, J.R., Peakman, T.M. (1989) *Tetrahedron Lett.* **30:** 4867.

Hirota, H., Yokoyama, A., Miyagi, K., Nakamura, T., Takahashi, T. (1987) *Tetrahedron Lett.* **28:** 435.

Hirota, H., Yokoyama, A., Miyagi, K., Nakamura, T., Igarashi, M., Takahashi, T. (1991) *J. Org. Chem.* **56:** 1119.

Hirsenkorn, R. (1990) *Tetrahedron Lett.* **31:** 7591.

Hizuka, M., Fang, C., Suemune, H., Sakai, K. (1989) *Chem. Pharm. Bull.* **37:** 1185.

Hlubucek, J.R., Ritchie, E., Taylor, W.C. (1970) *Aust. J. Chem.* **23:** 1881.

Ho, P.-T. (1982) *Can. J. Chem.* **60:** 90.

Ho, T.-L. (1971) *J. Chin. Chem. Soc. Ser. II* **18:** 153.

Ho, T.-L. (1972) *Can. J. Chem.* **50:** 1098.

Ho, T.-L. (1973) *J. Chem. Soc. Perkin Trans. I* 2579.

Ho, T.-L., Liu, S.-H. (1982a) *Chem. Ind.* 721.

Ho, T.-L., Liu, S.-H. (1982b) *Synth. Commun.* **12:** 501.

Ho, T.-L., Liu, S.-H. (1983) *Synth. Commun.* **13:** 1125.

Ho, T.-L., Yeh, W.-L., Yule, J., Liu, H.-J. (1992) *Can. J. Chem.* **70:** 1375.

Ho, T.-L., Chang, M.H. (1994) *Tetrahedron Lett.* **35:** 4819.

Hodgson, D.M., Whitherington, J., Moloney, B.A. (1994) *Tetrahedron: Asymmetry* **5:** 337.

Hoffmann, H.M.R., Clemens, K.E., Smithers, R.H. (1972) *J. Am. Chem. Soc.* **94:** 3940.

Hoffmann, R.W., Zeiss, H.-J., Ladner, W., Tabche, S. (1982) *Chem. Ber.* **115:** 2357.

Hoffmann, R.W., Ladner, W., Ditrich, K. (1989) *Liebigs Ann. Chem.* 883.

Hofmann, P., Beck, E., Hoffmann, M.D., Sieber, A. (1986) *Liebigs Ann. Chem.* 1779.

Holland, D., Stoddart, J.F. (1982) *Tetrahedron Lett.* **23:** 5367.

Holmes, A.B., Jennings-White, C.L.B., Kendrick, D.A. (1984) *Chem. Commun.* 1594.

Holmes, A.B., Smith, A.L., Williams, S.F., Hughes, L.R., Lidert, Z., Swithenbank, C. (1991) *J. Org. Chem.* **56:** 1393.

Hölscher, P., Knölker, H.-J., Winterfeldt, E. (1990) *Tetrahedron Lett.* **31:** 2705.

Honda, M., Hirata, K., Sueoka, H., Katsuki, T., Yamaguchi, M. (1981) *Tetrahedron Lett.* **22:** 2679.

Honda, M., Katsuki, T., Yamaguchi, M. (1984) *Tetrahedron Lett.* **25:** 3857.

Honda, T., Toya, T. (1992) *Heterocycles* **33:** 291.

Honda, T., Kimura, N., Tsubuki, M. (1993) *Tetrahedron: Asymmetry* **4:** 21.

Honda, T., Kimura, N. (1994) *Chem. Commun.* 77.

Horiguchi, Y., Nakamura, E., Kuwajima, I. (1989) *J. Am. Chem. Soc.* **111:** 6257.

Hoye, T.R., Peck, D.R., Trumper, P.K. (1981) *J. Am. Chem. Soc.* **103:** 5618.

Hoye, T.R., Peck, D.R., Swanson, T.A. (1984) *J. Am. Chem. Soc.* **106:** 2738.

Hoye, T.R., Suhadolnik, J.C. (1985) *J. Am. Chem. Soc.* **107:** 5312.

Hoye, T.R., Jenkins, S.A. (1987) *J. Am. Chem. Soc.* **109:** 6196.

Hoye, T.R., Hanson, P.R., Kovelesky, A.C., Ocain, T.D., Zhuang, Z. (1991) *J. Am. Chem. Soc.* **113:** 9369.

Hoye, T.R., Witowski, N.E. (1992) *J. Am. Chem. Soc.* **114:** 7291.

Hoye, T.R., North, J.T., Yao, L.J. (1994) *J. Am. Chem. Soc.* **116:** 2617.

Hsu, C.-T., Wang, N.-Y., Latimer, L.H., Sih, C.J. (1983) *J. Am. Chem. Soc.* **105:** 593.

Huang, F.-C., Lee, L.F.H., Mittal, R.S.D., Ravikumar, R.R., Chan, J.A., Sih, C.J., Caspi, E., Eck, C.R. (1975) *J. Am. Chem. Soc.* **97:** 4144.

Huang, P.Q., Arseniyadis, S., Husson, H.P. (1987) *Tetrahedron Lett.* **28:** 547.

Hubert, J.C., Wijnberg, J.B.P.A., Speckamp, W.N. (1975) *Tetrahedron* **31:** 1437.

Hudlicky, T., Koszyk, F.J., Kutchan, T.M., Sheth, J.P. (1980) *J. Am. Chem. Soc.* **45:** 5020.

Hudlicky, T., Sinai-Zingde, G., Natchus, M.G., Ranu, B.C., Papadopolous, P. (1987) *Tetrahedron* **43:** 5685.

Hudlicky, T., Luna, H., Barbieri, G., Kwart, L.D. (1988) *J. Am. Chem. Soc.* **110:** 4735.

Hudlicky, T., Price, J.D., Fan, R., Tsunoda, T. (1990) *J. Am. Chem. Soc.* **112:** 9439.

Hughes, I., Raphael, R.A. (1983) *Tetrahedron Lett.* **24:** 1441.

Hungerbühler, E., Seebach, D., Wasmuth, D. (1979) *Angew. Chem. Int. Ed. Engl.* **18:** 958.

Hutmacher, H.-M., Fritz, H.-G., Musso, H. (1975) *Angew. Chem. Int. Ed. Engl.* **14:** 180.

Ichihara, A., Kimura, R., Oda, K., Sakamura, S. (1976) *Tetrahedron Lett.* 4741.

Ichikawa, Y., Isobe, M., Goto, T. (1987) *Tetrahedron* **43:** 4749.

Ihara, M., Kawaguchi, A., Chihiro, M., Fukumoto, K., Kametani, T. (1986a) *Chem. Commun.* 671.

Ihara, M., Toyoto, M., Fukumoto, K., Kametani, T. (1986b) *J. Chem. Soc. Perkin Trans. I* 2151.

Ihara, M., Taniguchi, N., Fukumoto, K., Kametani, T. (1987) *Chem. Commun.* 1438.

Ihara, M., Yasui, K., Fukumoto, K., Kametani, T. (1988) *Tetrahedron Lett.* **29:** 4963.

Ihara, M., Suzuki, M., Fukumoto, K., Kabuto, C. (1990a) *J. Am. Chem. Soc.* **112:** 1164.

Ihara, M., Tokunaga, Y., Fukumoto, K. (1990b) *J. Org. Chem.* **55:** 4497.

Ihara, M., Tokunaga, Y., Taniguchi, N., Fukumoto, K., Kabuto, C. (1991) *J. Org. Chem.* **56:** 5281.

Iida, H., Tanaka, M., Kibayashi, C. (1984a) *J. Org. Chem.* **49:** 1909.

Iida, H., Watanabe, Y., Kibayashi, C. (1984b) *Tetrahedron Lett.* **25:** 5094.

Iida, H., Watanabe, Y., Kibayashi, C. (1985a) *J. Chem. Soc. Perkin Trans. I* 261.

Iida, H., Yamazaki, N., Kibayashi, C. (1985b) *Tetrahedron Lett.* **26:** 3255.

Iida, H., Yamazaki, N., Kibayashi, C. (1986a) *J. Org. Chem.* **51:** 1069.

Iida, H., Yamazaki, N., Kibayashi, C. (1986b) *J. Org. Chem.* **51:** 4245.

Iida, H., Yamazaki, N., Kibayashi, C. (1987) *J. Org. Chem.* **52:** 1956.

Iimori, T., Takahashi, Y., Izawa, T., Kobayashi, S., Ohno, M. (1983) *J. Am. Chem. Soc.* **105:** 1659.

Ikeda, S., Weinhouse, M.I., Janda, K.D., Lemer, R., Danishefsky, S.J. (1991) *J. Am. Chem. Soc.* **113:** 7763.

Ikeda, T., Hutchinson, C.R. (1984) *J. Org. Chem.* **49:** 2837.

Ikemoto, N., Schreiber, S.L. (1990) *J. Am. Chem. Soc.* **112:** 9657.

Ikezaki, M., Wakamatsu, T., Ban, Y. (1969) *Chem. Commun.* 88.

Imagawa, T., Murai, N., Akiyama, T., Kawanisi, M. (1979) *Chem. Commun.* 734.

Imagawa, T., Sonobe, T., Ishiwari, H., Akiyama, T., Kawanisi, M. (1980) *J. Org. Chem.* **45:** 2005.

Imanishi, T., Imanishi, I., Momose, T. (1978) *Synth. Commun.* **8:** 99.

Imanishi, T., Shin, H., Yagi, N., Hanaoka, M. (1980) *Tetrahedron Lett.* **21:** 3285.

Imanishi, T., Yagi, N., Shin, H., Hanaoka, M. (1981) *Tetrahedron Lett.* **22:** 4001.

Imanishi, T., Inoue, M., Wada, Y., Hanaoka, M. (1982) *Chem. Pharm. Bull.* **30:** 1925.

Imanishi, T., Miyashita, K., Nakai, A., Inoue, M., Hanaoka, M. (1983) *Chem. Pharm. Bull.* **31:** 1191.

Inhoffen, H.H., Irmscher, K., Hirschfeld, H., Stache, U., Kreutzer, A. (1958) *Chem. Ber.* **91:** 2309.

Inoue, T., Kuwajima, I. (1980) *Chem. Commun.* 251.

Inubushi, Y., Kikuchi, T., Tanaka, K., Saji, I., Tokane, K. (1974) *Chem. Pharm. Bull.* **22:** 349.

Ireland, R.E., Baldwin, S.W., Dawson, D.J., Dawson, M.I., Dolfini, J.E., Newbould, J., Johnson, W.S., Brown, M., Crawford, R.J., Hudrlik, P.F., Rasmussen, G.H., Schmiegel, K.K. (1970) *J. Am. Chem. Soc.* **92:** 5743.

Ireland, R.E., Bey, P., Cheng, K.F., Czarney, R.J., Moser, J.-F., Trust, R.I. (1975a) *J. Org. Chem.* **40:** 1000.

Ireland, R.E., Dawson, M.I., Kowalski, C.J., Lipinski, C.A., Marshall, D.R., Tilley, J.W., Bordner, J., Trus, B.L. (1975b) *J. Org. Chem.* **40:** 973.

Ireland, R.E., Thaisrivongs, S., Dussault, P.H. (1988) *J. Am. Chem. Soc.* **110:** 5768.

Ishibashi, H., Sato, K., Ikeda, M., Maeda, H., Akai, S., Tamura, Y. (1985) *J. Chem. Soc. Perkin Trans. I* 605.

Ishibashi, H., Uehara, C., Komatsu, H., Ikeda, M. (1987) *Chem. Pharm. Bull.* **35:** 2750.

Ishibashi, H., So, T.S., Nakatani, H., Minami, K., Ikeda, M. (1988) *Chem. Commun.* 827.

Ishibashi, H., Sato, T., Takahashi, M., Hayashi, M., Ikeda, M. (1991a) *Heterocycles* **27:** 2787.

Ishibashi, H., So, T.S., Okochi, K., Sato, T., Nakamura, N., Nakatani, H., Ikeda, M. (1991b) *J. Org. Chem.* **56:** 95.

Ishii, Y., Ikariya, T., Saburi, M., Yoshikawa, S. (1986) *Tetrahedron Lett.* **27:** 365.

Ishizaki, M., Hoshino, O., Iitaka, Y. (1991) *Tetrahedron Lett.* **32:** 7079.

Isler, O., Lindlar, H., Montavon, M., Ruegg, R., Zeller, P. (1956) *Helv. Chim. Acta* **39:** 249.

Isler, O., Schudel, P. (1963) *Adv. Org. Chem.* **4:** 115.

Isobe, M., Iio, H., Kawai, T., Goto, T. (1977) *Tetrahedron Lett.* 703.

Ito, Y., Saegusa, T. (1977) *J. Org. Chem. J. Org. Chem.* **42:** 2326.

Ito, Y., Shibata, T., Arita, M., Sawai, H., Ohno, M. (1981) *J. Am. Chem. Soc.* **103:** 6739.

Iwadare, S., Shizuri, Y., Yamada, K., Hirata, Y. (1978) *Tetrahedron Lett.* **34:** 1457.

Iwashima, M., Nagaoka, H., Kobayashi, K., Yamada, Y. (1992) *Tetrahedron Lett.* **33:** 81

Iwashita, T., Suzuki, M., Kusumi, T., Kakisawa, H. (1980) *Chem. Lett.* 383.

Iwashita, T., Kusumi, T., Kakisawa, H. (1982) *J. Org. Chem.* **47:** 230.

Iwata, C., Fusaka, T., Fujiwara, T., Tomita, K., Yamada, M. (1981) *Chem. Commun.* 463.

Iwata, C., Fujita, M., Hattori, K., Uchida, S., Imanishi, T. (1985) *Tetrahedron Lett.* **26:** 2221.

Iyoda, M., Kushida, T., Kitami, S., Oda, M. (1986) *Chem. Commun.* 1049.

Izawa, H., Shirai, R., Kawasaki, H., Kim, H., Koga, K. (1989) *Tetrahedron Lett.* **30:** 7221.

Jacobi, P.A., Kaczmarek, C.S.R., Udodong, U.E. (1987) *Tetrahedron* **43:** 5475.

Jäger, V., Wehner, V. (1989). *Angew. Chem. Int. Ed. Engl.* **28:** 469.

Jakovic, I.J., Goodbrand, H.B., Kok, K.P., Jones, J.B. (1982) *J. Am. Chem. Soc.* **104:** 4659.

Jarglis, P., Lichtenthaler, F.W. (1982) *Angew. Chem. Int. Ed. Engl.* **21:** 141.

Jefford, C.W., Kubota, T., Zaslona, A. (1986) *Helv. Chim. Acta* **69:** 2048.

Jefford, C.W., Tang, Q., Zaslona, A. (1991) *J. Am. Chem. Soc.* **113:** 3513.

Jefford, C.W., Sienkiewicz, K., Thornton, S.R. (1994) *Tetrahedron Lett.* **35:** 4759.

Jeffs, P.W., Cortese, N.A., Wolfram, J. (1982) *J. Org. Chem.* **47:** 3881.

Johnson, C.R., Penning, T.D. (1986) *J. Am. Chem. Soc.* **108:** 5655.

Johnson, C.R., Golebiowski, A., McGill, T.K., Steensma, D.H. (1991a) *Tetrahedron Lett.* **32:** 2597.

Johnson, C.R., Golebiowski, A., Steensma, D.H. (1991b) *Tetrahedron Lett.* **32:** 3931.

Johnson, C.R., Plé, P.A., Adams, J.P. (1991) *Chem. Commun.* 1006.

Johnson, F.J., Paul, K.G., Favara, D. (1982) *J. Org. Chem.* **47:** 4254.

Johnson, W.S., Jensen, N.P., Hooz, J. (1966) *J. Am. Chem. Soc.* **88:** 3859.

Johnson, W.S., Werthemann, L., Bartlett, W.R., Brocksom, T.J., Li, T.-t., Faulkner, D.J., Peterson, M.R. (1970) *J. Am. Chem. Soc.* **92:** 741.

Johnson, W.S., Gravestock, M.B., McCarry, B.E. (1971) *J. Am. Chem. Soc.* **93:** 4332.

Johnson, W.S., Shenvi, A.B., Boots, S.G. (1982) *Tetrahedron* **38:** 1397.

Jones, G., Raphael, R.A., Wright, S. (1974) *J. Chem. Soc. Perkin Trans. I* 1676.

Jones, J.B., Finchi, M.A.W., Jakovac, I.J. (1982) *Can. J. Chem.* **60:** 2007.

Jones, M.I., Froussidus, C., Evans, D.A. (1976] *Chem. Commun.* 472.

Jones, T.K., Reamer, R.A., Desmond, R., Mills, S.G. (1990). *J. Am. Chem. Soc.* **112:** 2998.

Joshi, B.S., Jiang, Q., Rho, T., Pelletier, S.W. (1994) *J. Org. Chem.* **59:** 8220.

Jung, M.E., Hudspeth, J.P. (1977) *J. Am. Chem. Soc.* **99:** 5508.

Jung, M.E., McCombs, C.A. (1978) *J. Am. Chem. Soc.* **100:** 5207.

Jung, M.E., Shaw, T.J. (1980) *J. Am. Chem. Soc.* **102:** 6304.

Jung, M.E., Miller, S.J. (1981) *J. Am. Chem. Soc.* **103:** 1984.

Jung, M.E., Usui, Y., Vu, C.T. (1987) *Tetrahedron Lett.* **28:** 5977.

Jung, M.E., Choi, Y.M. (1991) *J. Org. Chem.* **56:** 6729.

Jurczak, J., Pikul, S. (1984) *Tetrahedron Lett.* **25:** 3107.

Jurczak, J., Pikul, S., Bauer, T. (1986) *Tetrahedron* **42:** 447.

Just, G., Simonovitch, C., Lincoln, F.H., Schneider, W.P., Axen, U., Spero, G.B., Pike, J.E. (1969) *J. Am. Chem. Soc.* **91:** 5364.

Kabta, M.M., Lange, M., Wovkulich, P.M., Uskokovic, M.R. (1992) *Tetrahedron Lett.* **33:** 7701.

Kaga, H., Kobayashi, S., Ohno, M. (1988) *Tetrahedron Lett.* **29:** 1057.

Kagechika, K., Shibasaki, M. (1991) *J. Org. Chem.* **56:** 4093.

Kagechika, K., Ohshima, T., Shibasaki, M. (1993) *Tetrahedron* **49:** 1773.

Kaino, M., Naruse, Y., Ishihara, K., Yamamoto, H. (1990) *J. Org. Chem.* **55:** 5814.

Kaiser, G., Musso, H. (1985) *Chem. Ber.* **118:** 2266.

Kametani, T., Shibuya, S., Seino, S., Fukumoto, K. (1974) *J. Chem. Soc.* 4146.

Kametani, T., Fukumoto, K. (1972) *Synthesis* 657.

Kametani, T., Suzuki, K., Nemoto, H. (1980) *J. Org. Chem.* **45:** 2204.

Kametani, T., Ohsawa, T., Ihara, M. (1981) *J. Chem. Soc. Perkin Trans. I* 1563.

Kametani, T., Suzuki, K., Nemoto, H. (1982a) *J. Org. Chem.* **47:** 2331.

Kametani, T., Takagi, N., Kanaya, N., Honda, T., Fukumoto, K. (1982b) *J. Heterocycl. Chem.* **19:** 1217.

Kametani, T., Higashiyama, K., Otomasu, H., Honda, T. (1984) *Heterocycles* **22:** 729.

Kametani, T., Nagahara, T., Honda, T. (1985) *J. Org. Chem.* **50:** 2327.

Kametani, T., Yukawa, H., Honda, T. (1986) *Chem. Commun.* 651.

Kane, V.V., Doyle, D.L. (1981) *Tetrahedron Lett.* **22:** 3027, 3031.

Kaneko, T., Wong, H., Okamoto, K.T., Clardy, J. (1985) *Tetrahedron Lett.* **26:** 4015.

Kann, N., Rein, T. (1993) *J. Org. Chem.* **58:** 3802.

Kano, S., Sugino, E., Shibuya, S., Hibino, S. (1981) *J. Org. Chem.* **46:** 2979.

Karanewsky, D.S., Malley, M.F., Gougoutas, J.Z. (1991) *J. Org. Chem.* **56:** 3744.

Karrer, P., Meerwein, H.F. (1936) *Helv. Chim. Acta* **19:** 264.

Kashihara, H., Suemune, H., Kawahara, T., Sakai, K. (1987) *Tetrahedron Lett.* **28:** 6489.

Kashima, C., Mukai, N., Yamamoto, Y., Tsuda, Y., Omote, Y. (1977) *Heterocycles* **7:** 241.

Kashima, C., Hibi, S., Maruyama, T., Omote, Y. (1986) *Tetrahedron Lett.* **27:** 2131.

Kasturi, T.R., Chandra, R. (1988) *J. Org. Chem.* **53:** 3178.

Kato, M., Kosugi, H., Yoshikoshi, A. (1970a) *Chem. Commun.* 185.

Kato, M., Kosugi, H., Yoshikoshi, A. (1970b) *Chem. Commun.* 934.

Kato, M., Kido, F., Wu, M.-D., Yoshikoshi, A. (1974) *Bull. Chem. Soc. Jpn.* **47:** 1516.

Kato, M., Kurihara, H., Yoshikoshi, A. (1979) *J. Chem. Soc. Perkin Trans. I* 2740.

Kato, M., Heima, K., Matsumura, Y., Yoshikoshi, A. (1981) *J. Am. Chem. Soc.* **103:** 2434.

Katoh, T., Kirihara, M., Nagata, Y., Kobayashi, Y., Kobayashi, Y., Ari, K., Minami, J., Terashima, S. (1993) *Tetrahedron Lett.* **34:** 5747.

Katsube, J., Shimomura, H., Murayama, E., Toki, K., Matsui, M. (1971) *Agric. Biol. Chem.* **35:** 1768.

Katsuki, T., Lee, A.W.M., Ma, P., Martin, V.S., Masamune, S., Sharpless, K.B., Tuddenham, D., Walker, F.J. (1982) *J. Org. Chem.* **47:** 1373.

Katz, T.J., Acton, N. (1973) *J. Am. Chem. Soc.* **95:** 2738.

Kawada, K., Kim, M., Watt, D.S. (1989) *Tetrahedron Lett.* **30:** 5989.

Kawamata, T., Harimaya, K., Inayama, S. (1988') *Bull. Chem. Soc. Jpn.* **61:** 3770.

Kawasaki, M., Matsuda, F., Terashima, T. (1988) *Tetrahedron Lett.* **29:** 791.

Kay, I.T., Bartholomew, D. (1984) *Tetrahedron Lett.* **25:** 2035.

Keck, G.E., Webb, R.R. (1982) *J. Org. Chem.* **47:** 1302.

Keck, G.E., Enholm, E.J. (1985) *Tetrahedron Lett.* **26:** 3311.

Keck, G.E., Park, M., Krishnamurthy, D. (1993) *J. Org. Chem.* **58:** 3787.

Kelly, R.C., Van Rheenen, V., Schletter, I., Pillai, M.D. (1973) *J. Am. Chem. Soc.* **95:** 2746.

Kelly, T.R., Vaya, J., Ananthasubramanian, L. (1980) *J. Am. Chem. Soc.* **101:** 5983.

Kelly, T.R., Echavarren, A., Chandrakumar, N.S., Köksal, Y. (1984) *Tetrahedron Lett.* **25:** 2127.

Kelly, T.R., Bell, S.H., Ohashi, N., Armstrong-Chong, R.J. (1988) *J. Am. Chem. Soc.* **110:** 6471.

Kende, A.S., Liebeskind, L.S. (1976) *J. Am. Chem. Soc.* **98:** 267.

Kende, A.S., Liebeskind, L.S., Mills, J.E., Rutledge, P.S., Curran, D.P. (1977) *J. Am. Chem. Soc.* **99:** 7082.

Kende, A.S., Roth, B., Kubo, I. (1982) *Tetrahedron Lett.* **23:** 1751.

Kende, A.S., DeVita, R.J. (1988) *Tetrahedron Lett.* **29:** 2521.

Kende, A.S., Fujii, Y., Mendoza, J.S. (1990) *J. Am. Chem. Soc.* **112:** 9645.

Kenny, M.J., Mander, L.N., Sethi, S.P. (1986) *Tetrahedron Lett.* **27:** 3923, 3927.

Kent, G.J., Godleski, S.A., Osawa, E., Schleyer, P.v.R. (1977) *J. Org. Chem.* **42:** 3852.

Kerscher, V., Kreiser, W. (1987) *Tetrahedron Lett.* **28:** 531.

Kido, F., Noda, Y., Maruyama, T., Kabuto, C., Yoshikoshi, A. (1981) *J. Org. Chem.* **46:** 4264.

Kieczykowski, G.R., Schlessinger, R.H. (1978) *J. Am. Chem. Soc.* **100:** 1928.

Kienzle, F., Holland, G.W., Lernow, J.L., Kwoh, S., Rosen, P. (1973) *J. Org. Chem.* **38:** 3440.

Kienzle, F., Stadlwieser, J., Rank, W., Mergelsberg, I. (1988) *Tetrahedron Lett.* **29:** 6479.

Kigoshi, H., Imamura, Y., Niwa, H., Yamada, K. (1989a) *J. Am. Chem. Soc.* **111:** 2302.

Kigoshi, H., Sawada, A., Nakayama, Y., Niwa, H., Yamada, K. (1989b) *Tetrahedron Lett.* **30:** 1983.

Kim, G., Chu-Moyer, M.Y., Danieshefsky, S.J. (1990) *J. Am. Chem. Soc.* **112:** 2003.

Kim, M., Gross, R.S., Sevestre, H., Dunlap, N.K., Watt, D.S. (1988) *J. Org. Chem.* **53:** 93.

Kim, S., Bando, Y., Takahashi, N., Horii, Z. (1978) *Chem. Oharm. Bull.* **26:** 3150.

Kimura, H., Miyamoto, S., Shinkai, H., Kato, T. (1982) *Chem. Pharm. Bull.* **30:** 723.

Kinast, G., Tietze, L.-F. (1976) *Chem. Ber.* **109:** 3626.

King, F.E., King, T.J., Topliss, J.G. (1957) *J. Chem. Soc.* 573.

King, S.B., Ganem, B. (1991) *J. Am. Chem. Soc.* **113:** 5089.

Kinoshita, M., Nakata, M. (1986) *J. Synth. Org. Chem. Jpn.* **46:** 206.

Kinsella, M.A., Kalish, V.J., Weinreb, S.M. (1990) *J. Org. Chem.* **55:** 105.

Kishi, Y. (1981) *Pure Appl. Chem.* **53:** 1163.

Kitagawa, O., Hanano, T., Tanabe, K., Shiro, M., Taguchi, T. (1992) *Chem. Commun.* 1005.

Kitahara, T., Mori, K., Matsui, M. (1979) *Tetrahedron Lett.* 3021.

Kitahara, T., Mori, K. (1985) *Tetrahedron Lett.* **26:** 451.

Kitahara, Y., Yoshikoshi, A., Oida, S. (1964) *Tetrahedron* **26:** 1763.

Klein, L.L., McWhorter, W.W., Ko, S.S., Pfaff, K.-P., Kishi, Y., Uemura, D., Hirata, Y. (1982) *J. Am. Chem. Soc.* **104:** 7362.

Klunder, A.J.H., Bos, W., Zwanenburg, B. (1981' *Tetrahedron Lett.* **22:** 4557.

Klünenberg, H., Schäfer, H.J. (1978) *Angew. Chem. Int. Ed. Engl.* **17:** 47.

Knapp, S., Ornof, R.M., Rodriguez, K.E. (1983) *J. Am. Chem. Soc.* **105:** 5494.

Ko, S.Y., Lee, A.W.M., Masamune, S., Reed, L.A., III, Sharpless, K.B., Walker, F.J. (1990) *Tetrahedron* **46:** 245.

Kobayashi, S., Kamiyama, K., Iimori, T., Ohno, M. (1984) *Tetrahedron Lett.* **25:** 2557.

Kobayashi, S., Kamiyama, K., Ohno, M. (1990a) *J. Org. Chem.* **55:** 1169.

Kobayashi, S., Shibata, J., Shimada, M., Ohno, M. (1990b) *Tetrahedron Lett.* **31:** 1577.

Kobayashi, S., Nakada, M., Ohno, M. (1992) *Pure Appl. Chem.* **64:** 1121.

Kobayashi, Y., Kato, N., Shimazaki, T., Sato, F. (1988) *Tetrahedron Lett.* **29:** 6297.

Kobayashi, Y., Kitano, Y., Matsumoto, T., Sato, F. (1986) *Tetrahedron Lett.* **27:** 4775.

Köbrich, G., Heinemann, H. (1965) *Angew. Chem. Int. Ed. Engl.* **4:** 594.

Kocienski, P., Ostrow, R.W. (1976) *J. Org. Chem.* **41:** 398.

Kocienski, P., Yates, C. (1984) *Chem. Commun.* 151.

Kocienski, P.J., Street, S.D.A., Yeates, C., Campbell, S.F. (1987) *J. Chem. Soc. Perkin Trans. I* 2189.

Kocienski, P., Stocks, M., Donald, D., Perry, M. (1990) *Synlett* 38.

Kodama, M., Kurihara, T., Sasaki, J., Ito, S. (1979) *Can. J. Chem.* **57:** 3343.

Koft, E.R., Smith, A.B., III (1982) *J. Am. Chem. Soc.* **104:** 5568.

Kohnke, F.H., Stoddart, J.F. (1989) *Pure Appl. Chem.* **61:** 1581.

Koizumi, T., Namika, T., Takahashi, T., Takeuchi, Y. (1988) *Chem. Pharm. Bull.* **36:** 3213.

Köksal, Y., Raddatz, P., Winterfeldt, E. (1980) *Angew. Chem. Int. Ed. Engl.* **19:** 472.

Kolaczkowski, L., Reusch, W. (1985) *J. Org. Chem.* **50:** 4766.

Kometani, T., Takeuchi, Y., Yoshii, E. (1983) *J. Org. Chem.* **48:** 2630.

Komppa, G. (1903) *Ber.* **36:** 4332.

Kondo, K., Negishi, A., Matsui, K., Tunemoto, D., Masamune, S. (1972) *Chem. Commun.* 1311.

Kondo, K., Sodeoka, M., Mori, M., Shibasaki, M. (1993) *Synthesis* 1311.

Kondo, K., Sodeoka, M., Mori, M., Shibasaki, M. (1993) *Synthesis* 920.

Kongkathip, B., Kongkathip, N. (1984) *Tetrahedron Lett.* **25:** 2175.

Koreeda, M., Chen, Y.P.L. (1981) *Tetrahedron Lett.* **22:** 15.

Koreeda, M., Mislankar, S.G. (1983) *J. Am. Chem. Soc.* **105:** 7203.

Koreeda, M., Ricca, D.J., Luengo, J.I. (1988) *J. Org. Chem.* **53:** 5586.

Koreeda, M., Jung, K.Y., Ichita, J. (1989) *J. Chem. Soc. Perkin Trans. I* 2129.

Koser, S., Hoffmann, H.M.R., Williams, D.J. (1993) *J. Org. Chem.* **58:** 6163.

Köster, F.H., Wolf, H. (1981) *Tetrahedron Lett.* **22:** 3937.

Kotsuki, H., Kadota, I., Ochi, M. (1989) *Tetrahedron Lett.* **30:** 3999.

Kotsuki, H., Kadota, I., Ochi, M. (1990) *J. Org. Chem.* **55:** 4417.

Kotsuki, H., Nishikawa, H., Mori, Y., Ochi, M. (1992) *J. Org. Chem.* **57:** 5036.

Kowarski, C.R., Sarel, S. (1973) *J. Org. Chem.* **38:** 117.

Kozhushkov, S.I., Haumann, T., Boese, R., de Meijere, A. (1993) *Angew. Chem. Int. Ed. Engl.* **32:** 401.

Kozikowski, A.P., Greco, M.N., Springer, J.P. (1982) *J. Am. Chem. Soc.* **104:** 7622.

Kozikowski, A.P., Chen, Y.-Y., Wang, B.C., Xu, Z-B. (1984a) *Tetrahedron* **40:** 2345.

Kozikowski, A.P., Scripko, J.G. (1984b) *J. Am. Chem. Soc.* **106:** 353.

Kozikowski, A.P., Li, C.-S. (1985) *J. Org. Chem.* **50:** 778.

Kozikowski, A.P., Park, P. (1990) *J. Org. Chem.* **55:** 4668.

Kozikowski, A.P., Xia, Y., Reddy, E.R., Tückmantel, W., Hanin, I., Tang, X.C. (1991) *J. Org. Chem.* **56:** 4636.

Krapcho, A., Vivelo, J.A. (1985) *Chem. Commun.* 233.

Kraus, G.A., Neuenschwander, K. (1980) *Heterocycles* 841.

Kraus, G.A., Hon, Y.-S. (1987) *Heterocycles* **25:** 377.

Kraus, G.A., Wu, Y. (1991) *Tetrahedron Lett.* **32:** 3803.

Krief, A., Dumont, W., Pasau, P. (1988) *Tetrahedron Lett.* **29:** 1079.

Krief, A., Surleraux, D., Ropson, N. (1993) *Tetrahedron: Asymmetry* **4:** 289.

Kubiak, G., Fu, X., Gupta, A.K., Cook, J.M. (1990) *Tetrahedron Lett.* **31:** 4285.

Kuck, D., Bögge, H. (1986) *J. Am. Chem. Soc.* **108:** 8107.

Kuck, D., Schuster, A. (1988) *Angew. Chem. Int. Ed. Engl.* **27:** 1192.

Kuck, D., Paisdor, B., Gestmann, D. (1994) *Angew. Chem. Int. Ed. Engl.* **33:** 1251.

Kuehne, M.E. (1964) *J. Am. Chem. Soc.* **86:** 2946.

Kuehne, M.E., Bayha, C. (1966) *Tetrahedron Lett.* 1311.

Kuehne, M.E., Podhorez, D.E. (1985) Takano, S., Inomata, K., Kurotaki, A., Ohkawa, T., Ogasawara, K. (1987b) *Chem. Commun.* 124.

Kuhn, T., Tamm, C., Riesen, A., Zehnder, M. (1989) *Tetrahedron Lett.* **30:** 693.

Kumar, A., Devaprabhakara, D. (1976) *Synthesis* 461.

Kupchan, S.M., Dhingra, O.P., Kim, C. (1978) *J. Org. Chem.* **43:** 4076.

Kurek-Tyrlik, A., Wicha, J., Snatzke, G. (1988) *Tetrahedron Lett.* **29:** 4001.

Kurihara, M., Kamiyama, K., Kobayashi, S., Ohno, M. (1985) *Tetrahedron Lett.* **26:** 5831.

Kuroda, C., Theramongkol, P., Engebrecht, J.R., White, J.D. (1986) *J. Org. Chem.* **51:** 956.

Kurth, M.J., Brown, E.G. (1987) *J. Am. Chem. Soc.* **109:** 6844.

Kuwahara, Y., Fukami, H., Howard, R., Ishii, S., Matsumura, F., Burkholder, W.E. (1978) *Tetrahedron* **34:** 1769.

Lamberton, J.A., Geewananda, Y.A., Gunawandana, P., Bick, I.R.C. (1983) *J. Nat. Prod.* **46:** 235.

Landheer, I.J., deWolf, W.H., Bickelhaupt, F. (1975) *Tetrahedron Lett.* 349.

Lansbury, P.T., Serelis, A.K. (1978) *Tetrahedron Lett.* 1909.

Lansbury, P.T., Galbo, J.P., Springer, J.P. (1988a) *Tetrahedron Lett.* **29:** 147.

Lansbury, P.T., Zhi, B. (1988b) *Tetrahedron Lett.* **29:** 179, 5735.

Lansbury, P.T., Spagnuolo, C.J., Zhi, B., Grimm, E.L. (1990) *Tetrahedron Lett.* **31:** 3965.

Larock, R.C., Lee, N.H. (1991) *J. Am. Chem. Soc.* **113:** 7815.

Laronze, J.Y., Laronze-Fontaine, J., Levy, J., Le Men, J. (1974) *Tetrahedron Lett.* 491.

Laronze, J.Y., El Boukili, R., Cartier, D., Laronze, J., Levy, J. (1989) *Tetrahedron Lett.* **30:** 2229.

Laumen, K., Schneider, M. (1984) *Tetrahedron Lett.* **25:** 5875.

Lautens, M., Abd-El-Aziz, A.S., Lough, A. (1990) *J. Org. Chem.* **55:** 5305.

Lautens, M., Chiu, P., Colucci, J.T. (1993) *Angew. Chem. Int. Ed. Engl.* **32:** 281.

Lavielle, S., Bory, S., Moreau, B., Luche, M.J., Marquet, A. (1978) *J. Am. Chem. Soc.* **100:** 1558.

Lee, C.-H., Liang, S., Haumann, T., Boese, R., de Meijere, A. (1993) *Angew. Chem. Int. Ed. Engl.* **32:** 559.

Leete, E. (1972) *Chem. Commun.* 1091.

LeGoffic, F., Gouyette, A., Ahond, A. (1973) *Tetrahedron* **29:** 3357.

Lehmkuhl, H., Mehler, K. (1982) *Liebigs Ann. Chem.* 2244.

le-Hocine, M.B., Khac, D.D., Fetizon, M., Guir, F., Guo, Y., Prange, T. (1992) *Tetrahedron Lett.* **33:** 1443.

Lemal, D.M., McGregor, S.D. (1966) *J. Am. Chem. Soc.* **88:** 1335.

Leonard, J., Ouali, D., Rahman, S.K. (1990) *Tetrahedron Lett.* **31:** 739.

Le Perchec, P., Conia, J.M. (1970) *Tetrahedron Lett.* 1587.

Levine, S.G., Heard, N.E. (1988) *3rd Chem. Congr. N. Am.* ORGN 167.

Ley, S.V., Sternfeld, F., Taylor, S. (1987) *Tetrahedron Lett.* **28:** 225.

Ley, S.V., Anthony, N.J., Armstrong, A., Brasca, M.G., Clarke, T., Culshaw, T., Greck, C., Grice, P., Jones, A.B., Lygo, B., Madin, A., Sheppard, R.N., Slawin, A.M.Z., Williams, D.J. (1989) *Tetrahedron* **45:** 7161.

Li, T., Wu, Y.L. (1981) *J. Am. Chem. Soc.* **103:** 7007.

Li, Y., Mak, T.C.W., Wong, H.N.C., Chan, T.-L. (1991) *4th Chem. Congr. N. Am.* ORGN 181.

Lipshutz, B.H., Kozlowski, J.A. (1984) *J. Org. Chem.* **49:** 1147.

Little, R.D., Muller, G.W. (1979) *J. Am. Chem. Soc.* **101:** 7129.

Liu, H.-J., Llinas-Brunet, M. (1984) *Can. J. Chem.* **62:** 1747.

Liu, H.-J., Wynn, H. (1985) *Tetrahedron Lett.* **26:** 4843.

Liu, L., Katz, T.J. (1990) *Tetrahedron Lett.* **31:** 3983.

Liu, Z.-Y., He, L., Zheng, H. (1993) *Tetrahedron: Asymmetry* **4:** 2277.

Livinghouse, T., Stevens, R.V. (1978) *Chem. Commun.* 754.

Lohrisch, H.J., Schmidt, H., Steglich, W. (1986) *Liebigs Ann. Chem.* 195.

Lukes, R.M., Poos, G.I., Sarett, L.H. (1952) *J. Am. Chem. Soc.* **74:** 1401.

Luyten, M., Müller, S., Herzog, B., Keese, R. (1987) *Helv. Chim. Acta* **70:** 1250.

Lythgoe, B., Nambudiry, M.E.N., Tideswell, J. (1977) *Tetrahedron Lett.* 3685.

MacConnell, J.G., Blum, M.S., Fales, H.M. (1971) *Tetrahedron* **27:** 1129.

Macdonald, T.L. (1980) *J. Org. Chem.* **45:** 193.

Machinaga, N., Kibayashi, C. (1990) *Tetrahedron Lett.* **31:** 3637.

Machinaga, N., Kibayashi, C. (1991a) *J. Org. Chem.* **56:** 1386.

Machinaga, N., Kibayashi, C. (1991b) *Chem. Commun.* 405.

Machinaga, N., Kibayashi, C. (1993) *Tetrahedron Lett.* **34:** 5739.

MacLeod, J.K., Monahan, L.C. (1988) *Tetrahedron Lett.* **29:** 391.

Maggio, J.E., Simmons, H.E., Kouba, J.K. (1981) *J. Am. Chem. Soc.* **103:** 1578.

Magnusson, G., Lindqvist, F. (1990) *Chem. Commun.* 1080.

Maier, G., Matusch, R., Pfriem, S., Schäfer, I. (1978) *Angew. Chem. Int. Ed. Engl.* **17:** 520.

Maier, G., Lage, H.W., Reisenauer, H.P. (1981) *Angew. Chem. Int. Ed. Engl.* **20:** 976.

Maier, G. (1988) *Angew. Chem. Int. Ed. Engl.* **27:** 309.

Maiti, S.C., Thomsen, R.H., Mahendran, M. (1978) *J. Chem. Res.* 1682.

Majetich, G., Grieco, P.A., Nishizawa, M. (1977) *J. Org. Chem.* **42:** 2327.

Majetich, G., Behnke, M., Hull, K. (1985) *J. Org. Chem.* **50:** 3615.

Majetich, G., Reingold, C. (1987) *Heterocycles* **25:** 271.

Majetich, G., Lowery, D., Khetani, V., Song, J.-S., Hull, K., Ringold, C. (1991a) *J. Org. Chem.* **56:** 3988.

Majetich, G., Song, J.-S., Ringold, C., Nemeth, G.A., Newton, M.G. (1991b) *J. Org. Chem.* **56:** 3973.

Mandell, L., Singh, K.P., Gresham, J.T., Freeman, W. (1963) *J. Am. Chem. Soc.* **85:** 2682.

Mandel'shtam, T.V., Kharicheva, E.M. (1973) *Zh. Org. Khim.* **9:** 1648.

Mander, L.N., Turner, J.V. (1973) *J. Org. Chem.* **38:** 2915.

Mander, L.N., Robinson, R.P. (1991) *J. Org. Chem.* **56:** 3595.

Mangoni, L., Adinoff, M., Laonigro, G., Caputo, R. (1972) *Tetrahedron Lett.* **28:** 611.

Marazano, C., Le Goff, M.-T., Fourrey, J.-L., Das, B.C. (1981) *Chem. Commun.* 389.

Marchand, A.P., Chou, T.-c., Ekstrand, J.D., van der Helm, D. (1976) *J. Org. Chem.* **41:** 1438.

Marotta, E., Rastelli, E., Righi, P., Rosini, G. (1993) *Tetrahedron: Asymmetry* **4:** 735.

Marshall, J.A., Faubl, H., Warne, T.M. (1967) *Chem. Commun.* 753.

Marshall, J.A., Hochestetler, A.R. (1968) *J. Org. Chem.* **33:** 2593.

Marshall, J.A., Partridge, J.J. (1969) *Tetrahedron Lett.* **25:** 2159.

Marshall, J.A., Johnson, P.C. (1970) *J. Org. Chem.* **35:** 102.

Marshall, J.A., Ruden, R.A. (1971) *J. Org. Chem.* **36:** 594.

Marshall, J.A., Ruth, J.A. (1974) *J. Org. Chem.* **39:** 1971.

Marshall, J.A., Ellison, R.H. (1976) *J. Am. Chem. Soc.* **98:** 4312.

Marshall, J.A., Flynn, G.A. (1979) *J. Org. Chem.* **44:** 1391.

Marshall, J.A., Shearer, B.G. Crooks, S.L. (1987) *J. Org. Chem.* **52:** 1236.

Marshall, J.A., Salovich, J.M., Shearer, B.G. (1990) *J. Org. Chem.* **55:** 2398.

Marshall, J.A., Luke, G.P. (1991) *J. Org. Chem.* **56:** 483.

Marshall, J.A., Crute, T.D., III, Hsi, J.D. (1992) *J. Org. Chem.* **57:** 115.

Martin, S.F., Campbell, C.L. (1987) *Tetrahedron Lett.* **28:** 503.

Martin, S.F., Daniel, D. (1993a) *Tetrahedron Lett.* **34:** 4281.

Martin, S.F., Tso, H.-H. (1993b) *Heterocycles* **35:** 85.

Martinez, G.R., Grieco, P.A., Williams, E., Kanai, K., Srinivasan, C.V. (1982) *J. Am. Chem. Soc.* **104:** 1436.

Maruyama, K., Ishihara, Y., Yamamoto, Y. (1981) *Tetrahedron Lett.* **22:** 4235.

Marx, J.N., Bih, Q.-R. (1987) *J. Org. Chem.* **52:** 336.

Marx, J.N., Dobrowolski, P.J. (1982) *Tetrahedron Lett.* **23:** 4457.

Marx, M.H., Wiley, R.A. (1985) *Tetrahedron Lett.* **26:** 1379.

Masaki, Y., Nagata, K., Serizawa, Y., Kaji, K. (1982) *Tetrahedron Lett.* **23:** 5553.

Masaki, Y., Nagata, K., Serizawa, Y., Kaji, K. (1983a) *Chem. Lett.* 1835.

Masaki, Y., Serizawa, Y., Nagata, K., Kaji, K. (1983b) *Chem. Lett.* 1601.

Masaki, Y., Nagata, K., Serizawa, Y., Kaji, K. (1984) *Tetrahedron Lett.* **25:** 95.

Masaki, Y., Serizawa, Y., Nagata, K., Oda, H., Nagashima, H., Kaji, K. (1986) *Tetrahedron Lett.* **27:** 231.

Masaki, Y., Imaeda, T., Nagata, K., Oda, H., Itoh, A. (1989) *Tetrahedron Lett.* **30:** 6395.

Masaki, Y., Imaeda, T., Oda, H., Itoh, A., Shiro, M. (1992) *Chem. Lett.* 1209.

Masamune, S., Cuts, H., Hogben, M.G. (1966) *Tetrahedron Lett.* 1017.

Masamune, S., Ang, S.K., Egli, C., Nakatsuka, N., Sarkar, S.K., Yasunari, Y. (1967) *J. Am. Chem. Soc.* **89:** 2506.

Masamune, S., Kim, C.U., Wilson, K.E., Spessard, G.O., Georghiou, P.E., Bates, G.S. (1975) *J. Am. Chem. Soc.* **97:** 3512.

Masamune, S., Ali, S.A., Snitman, D.L., Garvey, D.S. (1980) *Angew. Chem. Int. Ed. Engl.* **19:** 557.

Masamune, S., Hirama, M., Mori, S., Ali, S.A., Garvey, D.S. (1981) *J. Am. Chem. Soc.* **103:** 1568.

Matos, J.R., Wong, C.-H. (1986) *J. Org. Chem.* **51:** 2388.

Matsuda, K., Nomura, K., Yoshii, E. (1989) *Chem. Commun.* 221.

Matsukuma, A., Nagumo, S., Suemune, H., Sakai, K. (1991) *Tetrahedron Lett.* **32:** 7559.

Matsumoto, M., Kuroda, K. (1981) *Tetrahedron Lett.* **22:** 4437.

Matsumoto, T., Katsuki, M., Jona, H., Suzuki, K. (1991) *J. Am. Chem. Soc.* **113:** 6982.

Matthews, R.S., Whitesell, J.K. (1975) *J. Org. Chem.* **40:** 3312.

Matus, I., Fischer, J. (1985) *Tetrahedron Lett.* **26:** 385.

McCurry, P.M., Singh, R.K., Link, S. (1973) *Tetrahedron Lett.* 1155.

McDonald, I.A., Dreiding, A.S., Hutmacher, H.-M., Musso, H. (1973) *Helv. Chim. Acta* **56:** 1385.

McGuire, H.M., Odom, H.C., Pinder, A.R. (1974) *J. Chem. Soc. Perkin Trans. I* 1879.

McKennis, J.S., Brenner, L., Ward, J.S., Pettit, R. (1971) *J. Am. Chem. Soc.* **93:** 4957.

McKillop, A., Perry, D.H., Edwards, M., Antus, S., Farkas, L., Nogradi, M., Taylor, E.C. (1976) *J. Org. Chem.* **41:** 282.

McMurry, J.E. (1968) *J. Am. Chem. Soc.* **90:** 6821.

McMurry, J.E., Isser, S.J. (1972) *J. Am. Chem. Soc.* **94:** 7132.

McMurry, J.E., Fleming, M.P. (1974) *J. Am. Chem. Soc.* **96:** 4708.

McMurry, J.E., Andrus, A., Ksander, G.M., Musser, J.H., Johnson, M.A. (1979) *J. Am. Chem. Soc.* **101:** 1330.

McMurry, J.E., Farina, V., Scott, W.J., Davidson, A.H., Summers, D.R., Shenvi, A. (1984) *J. Org. Chem.* **49:** 3803.

McMurry, J.E., Haley, G.J., Matz, J.R., Clardy, J.C., Van Duyne, G., Gleiter, R., Schäfer, W., White, D.H. (1986) *J. Am. Chem. Soc.* **108:** 2932.

McMurry, J.E., Dushin, R.G. (1989) *J. Am. Chem. Soc.* **111:** 8928.

Mechoulam, R., Braun, P., Gaoni, Y. (1967) *J. Am. Chem. Soc.* **89:** 4552.

Medich, J.R., Kunnen, K.B., Johnson, C.R. (1987) *Tetrahedron Lett.* **28:** 4131.

Meerwein, H., Emster, K.V. (1920) *Ber.* **53:** 1815.

Mehmandoust, M., Marazano, C., Das, B.C. (1989) *Chem. Commun.* 1185.

Mehta, G., Murthy, A.N. (1984) *Chem. Commun.* 1058.

Mehta, G., Nair, M.S. (1985) *Chem. Commun.* 629.

Mehta, G., Rao, K.S. (1986a) *J. Am. Chem. Soc.* **108:** 8015.

Mehta, G., Murthy, A.N., Reddy, A.V. (1986b) *J. Am. Chem. Soc.* **108:** 3443.

Mehta, G., Padma, S. (1987) *J. Am. Chem. Soc.* **109:** 7230.

Mehta, G., Rao, K.S., Reddy, K.R. (1988a) *Tetrahedron Lett.* **29:** 5025.

Mehta, G., Reddy, K.R. (1988b) *Tetrahedron Lett.* **29:** 3607.

Mehta, G., Reddy, M.S., Radhakrishnan, R., Manjula, M.V., Viswamitra M.A. (1991) *Tetrahedron Lett.* **32:** 6219.

Meijer, E.W., Wynberg, H. (1988) *Angew. Chem. Int. Ed. Engl.* **27:** 975.

Menon, R.B., Kumar, M.A., Ravindranathan, T. (1987) *Tetrahedron Lett.* **28:** 5313.

Merlini, L., Nasini, G. (1967) *Gazz. Chim. Ital.* **97:** 1915.

Merour, J.Y., Coadou, J.Y. (1991) *Tetrahedron Lett.* **32:** 2469.

Meyer, H., Seebach, D. (1975) *Liebigs Ann. Chem.* 2261.

Meyer, H.H. (1977) *Liebigs Ann. Chem.* 732.

Meyers, A.I., Dupre, B. (1987a) *Heterocycles* **25:** 113.

Meyers, A.I., Lefker, B.A. (1987b) *Tetrahedron* **43:** 5663.

Mikami, K., Narisawa, S., Shimizu, M., Terada, M. (1992) *J. Am. Chem. Soc.* **114:** 6566.

Miller, L.S., Grohmann, K., Dannenberg, J.J., Todaro, L. (1981) *J. Am. Chem. Soc.* **103:** 6249.

Miller, R.B., Behare, E.S. (1974) *J. Am. Chem. Soc.* **96:** 8102.

Minato, H., Nagasaki, T. (1968) *J. Chem. Soc. C* 621.

Misumi, A., Furuta, K., Yamamoto, H. (1984) *Tetrahedron Lett.* **25:** 671.

Misumi, A., Iwanaga, K., Furuta, K., Yamamoto, H. (1985) *J. Am. Chem. Soc.* **107:** 3343.

Mitschka, R., Cook, J.M., Weiss, U. (1978) *J. Am. Chem. Soc.* **100:** 3973.

Miyagi, K., Nakamura, T., Hirota, H., Igarashi, M., Takahashi, T. (1984) *Tetrahedron Lett.* **25:** 5299.

Miyashita, M., Yoshikoshi, A. (1974) *J. Am. Chem. Soc.* **96:** 1917.

Miyashita, M., Kumazawa, T., Yoshikoshi, A. (1981) *Chem. Lett.* 593.

Miyata, O., Hirata, Y., Naito, T., Ninomiya, I. (1983) *Chem. Commun.* 1231.

Miyaura, N., Satoh, Y., Hara, S., Suzuki, A. (1986) *Bull. Chem. Soc. Jpn.* **59:** 2029.

Möens, L., Baizer, M.M., Little, R.D. (1986) *J. Org. Chem.* **51:** 4497.

Mohr, P., Tori, M., Grossen, P., Herold, P., Tamm, C. (1982) *Helv. Chim. Acta* **65:** 1412.

Moiseenkov, A.M., Czeskis, B.A., Semenovsky, A.V. (1982) *Chem. Commun.* 109.

Molander, G.A., Hoberg, J.O. (1992) *J. Am. Chem. Soc.* **114:** 3123.

Momose, T., Atarashi, S., Eugster, C.H. (1979) *Heterocycles* **12:** 41.

Momose, T., Toyooka, N., Seki, S., Hirai, Y. (1990) *Chem. Pharm. Bull.* **38:** 2072.

Momose, T., Toyooka, N., Jin, M. (1992) *Tetrahedron Lett.* **33:** 5389.

Momose, T., Toyooka, N. (1993) *Tetrahedron Lett.* **34:** 5785.

Mondon, A., Ehrhardt, M. (1966) *Tetrahedron Lett.* 2557.

Mondon, A., Hansen, K.E., Bohlme, K., Faro, H.P., Nestler, H.J., Vilhuber, H.G., Böttcher, K. (1970) *Chem. Ber.* **103:** 615.

Montana, A.M., Nicholas, K.M. (1990) *J. Org. Chem.* **55:** 1569.

Monti, S.A., Larsen, S.D. (1978) *J. Org. Chem.* **43:** 2282.

Moody, C.J., Rees, C.W., Thomas, R. (1990) *Tetrahedron Lett.* **31:** 4375.

Mori, A., Ishihara, K., Arai, I., Yamamoto, H. (1987) *Tetrahedron* **43:** 755.

Mori, K. (1974) *Tetrahedron* **30:** 4223.

Mori, K., Oda, M., Matsui, M. (1976) *Tetrahedron Lett.* 3173.

Mori, K., Takigawa, T., Matsui, M. (1979a) *Tetrahedron* **35:** 833.

Mori, K., Takigawa, T., Matsui, M. (1979b) *Tetrahedron* **35:** 933.

Mori, K., Iwasawa, H. (1980) *Tetrahedron* **36:** 87.

Mori, K., Ueda, H. (1982) *Tetrahedron* **38:** 1227.

Mori, K., Uematsu, T., Minobe, M., Yanagi, K. (1983) *Tetrahedron* **39:** 1735.

Mori, K., Uematsu, T., Watanabe, H., Yanagi, K., Minobe, M. (1984) *Tetrahedron Lett.* **25:** 3875.

Mori, K., Mori, H. (1985a) *Tetrahedron* **41:** 5487.

Mori, K., Senda, S. (1985b) *Tetrahedron* **41:** 541.

Mori, K., Takechi, S. (1985c) *Tetrahedron* **41:** 3049.

Mori, K., Mori, H. (1986a) *Tetrahedron* **42:** 5531.

Mori, K., Nakazono, Y. (1986b) *Tetrahedron* **42:** 283.

Mori, K., Seu, Y.-B. (1986c) *Liebigs Ann. Chem.* 205.

Mori, K., Tamura, H. (1986d) *Tetrahedron* **42:** 2643.

Mori, K., Watanabe, H. (1986e) *Tetrahedron* **42:** 273.

Mori, K., Ikunaka, M. (1987a) *Tetrahedron* **43:** 45.

Mori, K., Komatsu, M. (1987b) *Tetrahedron* **43:** 3409.

Mori, K., Fujiwhara, M. (1988a) *Tetrahedron* **44**: 343.

Mori, K., Komatsu, M. (1988b) *Liebigs Ann. Chem.* 107.

Mori, K., Takeuchi, T. (1988c) *Tetrahedron* **44**: 333.

Mori, K., Tamura, H. (1988d) *Liebigs Ann. Chem.* 97.

Mori, K., Tsuji, M. (1988e) *Tetrahedron* **44**: 2835.

Mori, K., Chiba, N. (1989a) *Liebigs Ann. Chem.* 957.

Mori, K., Takaishi, H. (1989b) *Liebigs Ann. Chem.* 695.

Mori, K., Takaishi, H. (1989c) *Liebigs Ann. Chem.* 939.

Mori, K., Takeuchi, T. (1989d) *Liebigs Ann. Chem.* 453.

Mori, K., Uno, T. (1989e) *Tetrahedron* **45**: 1945.

Mori, K., Watanabe, H. (1989f) *Pure Appl. Chem.* **61**: 543.

Mori, K., Harada, H., Zagatti, P., Cork, A., Hall, D.R. (1991a) *Liebigs Ann. Chem.* 259.

Mori, K., Wu, J. (1991b) *Liebigs Ann. Chem.* 782.

Mori, K., Takayama, S., Yoshimura, S. (1993a) *Liebigs Ann. Chem.* 91.

Mori, K., Takikawa, H., Kido, M. (1993b) *J. Chem. Soc. Perkin Trans. I* 169.

Mori, K. (1994a) *Pure Appl. Chem.* **66**: 1991.

Mori, K., Matsushima, Y. (1994b) *Synthesis* 417.

Morizawa, Y., Oda, H., Oshima, K., Nozaki, H. (1984) *Tetrahedron Lett.* **25**: 1163.

Muchowski, J.M., Nelson, P.H. (1980) *Tetrahedron Lett.* **21**: 4585.

Mueller, R.H., Thompson, M.E. (1980) *Tetrahedron Lett.* **21**: 1093.

Mueller, R.H., Thompson, M.E., DiPardo, R.M. (1984) *J. Org. Chem.* **49**: 2217.

Mukaiyama, T., Sakito, Y., Asami, M. (1979) *Chem. Lett.* 705.

Mukaiyama, T., Suzuki, K., Yamada, T., Tabusa, F. (1990) *Tetrahedron* **46**: 265.

Müller, H., Melder, J.-P., Fessner, W.-D., Hunkler, D., Fritz, H., Prinzbach, H. (1988) *Angew. Chem. Int. Ed. Engl.* **27**: 1103.

Mulzer, J., Kappert, M. (1983) *Angew. Chem. Int. Ed. Engl.* **22**: 63.

Mulzer, J., Angermann, A., Münch, W. (1986a) *Liebigs Ann. Chem.* 825.

Mulzer, J., Angermann, A., Seilz, C., Schubert, B. (1986b) *J. Org. Chem.* **51**: 5294.

Mulzer, J., de Lasalle, P., Freissler, A. (1986c) *Liebigs Ann. Chem.* 1152.

Mulzer, J., Angermann, A., Münch, W., Schlichthörl, S., Hentzschel, A. (1987a) *Liebigs Ann. Chem.* 7.

Mulzer, J., Autenrieth-Ansorge, L., Kirstein, H., Matsuoka, T., Münch, W. (1987b) *J. Org. Chem.* **52**: 3784.

Mulzer, J., Büttelmann, A., Münch, W. (1988) *Liebigs Ann. Chem.* 445.

Murai, A., Sato, S., Masamune, T. (1981a) *Chem. Lett.* 429.

Murai, A., Sato, S., Masamune, T. (1981b) *Chem. Commun.* 904.

Murai, A., Sato, S., Masamune, T. (1981c) *Tetrahedron Lett.* 1033.

Murai, A. (1989) *Pure Appl. Chem.* **61**: 393.

Muratake, H., Takahashi, T., Natsume, M. (1983) *Heterocycles* **20**: 1963.

Muratake, H., Natsume, M. (1985) *Heterocycles* **23**: 1111.

Muratake, H., Natsume, M. (1987) *Tetrahedron Lett.* **28**: 2265.

Muratake, H., Natsume, M. (1989) *Tetrahedron Lett.* **30**: 5771.

Muratake, H., Natsume, M. (1990) *Tetrahedron* **46:** 6331.

Murphy, R.A., Cava, M.P. (1984) *Tetrahedron* **25:** 803.

Musso, H., Biethan, U. (1967) *Chem. Ber.* **100:** 119.

Musso, H., Klusacek, H. (1970) *Chem. Ber.* **103:** 3076.

Muxfeldt, H., Hardtmann, G., Kathawala, F., Vedejs, E., Mopberry, J.B. (1968) *J. Am. Chem. Soc.* **90:** 6534.

Myers, A.G., Fraley, M.E., Tom, N.J., Cohen, S.B., Madar, D.J. (1995) *Chem. and Biol.* **2:** 33.

Nagao, Y., Inoue, T., Hashimoto, K., Hagiwara, Y., Ochiai, M., Fujita, E. (1985) *Chem. Commun.* 1419.

Nagao, Y., Dai, W.-M., Ochiai, M., Tsukagoshi, S., Fujita, E. (1988) *J. Am. Chem. Soc.* **110:** 289.

Nagao, Y., Tohjo, T., Ochiai, M., Shiro, M. (1992) *Chem. Lett.* 335.

Nagaoka, H., Kishi, Y. (1981) *Tetrahedron* **37:** 3873.

Nagaoka, H., Miyakoshi, T., Yamada, Y. (1984) *Tetrahedron Lett.* **25:** 3621.

Nagaoka, H., Kobayashi, K., Matsui, T., Yamada, Y. (1987) *Tetrahedron Lett.* **28:** 2021.

Nagata, W., Hirai, S., Okumura, T., Kawata, K. (1968) *J. Am. Chem. Soc.* **90:** 1650.

Nair, M.S.R., Mathur, H.H., Bhattacharyya, S.C. (1964) *J. Chem. Soc.* 4154.

Naito, T., Miyata, O., Ninomiya, I. (1979) *Chem. Commun.* 517.

Naito, T., Iida, N., Ninomiya, I. (1981) *Chem. Commun.* 44.

Naito, T., Doi, E., Miyata, O., Ninomiya, I. (1986) *Heterocycles* **24:** 903.

Nakajima, M., Tomida, I., Takei, S. (1957) *Chem. Ber.* **90:** 246.

Nakajima, M., Tomida, I., Takei, S. (1959) *Chem. Ber.* **92:** 163.

Nakamura, E., Isaka, M., Matsuzawa, S. (1988) *J. Am. Chem. Soc.* **110:** 1297.

Nakano, A., Takimoto, S., Inanaga, J., Katsuki, T., Ouchida, S., Inoue, K., Aiga, M., Okudo, N., Yamaguchi, M. (1979) *Chem. Lett.* 1019.

Nakatsuka, M., Ragan, J.A., Sammakia, T., Smith, D.B. Uehling, D.E., Schreiber, S.L. (1990) *J. Am. Chem. Soc.* **112:** 5583.

Nakatsuka, S., Yamada, K., Yoshida, K., Asano, O., Murakami, Y., Goto, T. (1983) *Tetrahedron Lett.* **24:** 5627.

Nambudiry, M.E.N., Krishna Rao, G.S. (1974) *J. Chem. Soc. Perkin Trans. I* 317.

Narasaka, K., Sakakura, T., Uchimaru, T., Guedin-Vuong, D. (1984) *J. Am. Chem. Soc.* **106:** 2954.

Naruse, Y., Yamamoto, H. (1988) *Tetrahedron* **44:** 6021.

Naruta, Y., Nagai, N., Yokota, T., Maruyama, K. (1986) *Chem. Lett.* 1185.

Natsume, M., Kitagawa, Y. (1980a) *Tetrahedron Lett.* **21:** 839.

Natsume, M., Ogawa, M. (1980b) *Heterocycles* **14:** 169.

Natsume, M., Ogawa, M. (1980c) *Heterocycles* **14:** 615.

Natsume, M., Muratake, H. (1981a) *Heterocycles* **16:** 375.

Natsume, M., Muratake, H., Kanda, Y. (1981b) *Heterocycles* **16:** 959.

Natsume, M., Ogawa, M. (1981c) *Heterocycles* **16:** 973.

Natsume, M., Ogawa, M. (1983) *Heterocycles* **20:** 601.

Natsume, M. (1986) *J. Synth. Org. Chem. Jpn.* **44:** 326.

Nemoto, H., Kurobe, H., Kametani, T. (1984) *Tetrahedron Lett.* **25:** 4669.

Nemoto, H., Kurobe, H., Fukumoto, K., Kametani, T. (1985) *Chem. Lett.* 259.

Nemoto, H., Shitara, E., Fukumoto, K., Kametani, T. (1987) *Heterocycles* **25:** 51.

Nesmeyanova, O.A., Rudashevskaya, T.Y., Dyachenko, A.I., Savilova, S.F., Nefedov, O.M. (1982) *Synthesis* 296.

Newkome, G.R., Roach, L.C., Montelaro, R.C., Hill, R.K. (1972) *J. Org. Chem.* **37:** 2098.

Newkome, G.R., Lee, H.W. (1983) *J. Am. Chem. Soc.* **105:** 5956.

Nickon, A., Kwasnik, H., Swartz, T., Williams, R.O., DiGiorgio, J.B. (1965) *J. Am. Chem. Soc.* **87:** 1615.

Nickon, A., Pandit, G.D. (1968) *Tetrahedron Lett.* 3663.

Nicolaou, K.C., Claremon, D.A., Papahatjis, D.P., Magolda, R.L. (1981a) *J. Am. Chem. Soc.* **103:** 6969.

Nicolaou, K.C., Papahatjis, D.P., Claremon, D.A., Dolle, III, R.E. (1981b) *J. Am. Chem. Soc.* **103:** 6967.

Nicolaou, K.C., Petasis, N.A., Zipkin, R.E., Uenishi, J. (1982) *J. Am. Chem. Soc.* **104:** 5555.

Nicolaou, K.C., Zipkin, R.E., Dolle, III, R.E., Harris, B.D. (1984) *J. Am. Chem. Soc.* **106:** 3548.

Nicolaou, K.C., Papahatjis, D.P., Claremon, D.A., Magolda, R.L., Dolle, III, R.E. (1985) *J. Org. Chem.* **50:** 1440.

Nicolaou, K.C., Webber, S.E., Ramphal, J., Abe, Y. (1987) *Angew. Chem. Int. Ed. Engl.* **26:** 1019.

Nicolaou, K.C., Ogilvie, W.W. (1990) *Chemtracts-Org. Chem.* **3:** 327.

Nicolaou, K.C., Chakarborty, T.K., Piscopio, A.D., Minow, N., Bertinato, P. (1993) *J. Am. Chem. Soc.* **115:** 4419.

Nicolosi, G., Patti, A., Piattelli, M., Sanfilippo, C. (1994) *Tetrahedron: Asymmetry* **5:** 283.

Nishiyama, S., Nakamura, K., Suzuki, Y., Yamamura, S. (1986) *Tetrahedron Lett.* **27:** 4481.

Niwa, H., Nishiwaki, M., Tsukada, I., Ishigaki, T., Ito, S., Wakamatsu, K., Mori, T., Igawa, M., Yamada, K. (1990) *J. Am. Chem. Soc.* **112:** 9001.

Nokami, J., Ohkura, M., Dan-Oh, Y., Sakamoto, Y. (1991) *Tetrahedron Lett.* **32:** 2409.

Nomoto, T., Nasui, N., Takayama, H. (1984) *Chem. Commun.* 1646.

Nossin, P.M.M.M., Speckamp, W.N. (1979) *Tetrahedron Lett.* 4411.

Noyori, R., Baba, Y., Makino, S., Takaya, T. (1978) *J. Am. Chem. Soc.* **100:** 1786.

Noyori, R., Tomino, I., Yamada, M., Nishizawa, M. (1984) *J. Am. Chem. Soc.* **106:** 6717.

Noyori, R., Suzuki, M. (1990) *Chemtracts-Org. Chem.* **3:** 173.

Noyori, R. (1994) *Asymmetric Catalyses in Organic Synthesis*, Wiley, New York.

Nozaki, H., Yamamoto, H., Mori, T. (1969) *Can. J. Chem.* **47:** 1107.

O'Brien, M.K., Pearson, A.J., Pinkerton, A.A., Schmidt, W., Willman, K. (1989) *J. Am. Chem. Soc.* **111:** 1499.

O'Connor, S.J., Williard, P.G. (1989) *Tetrahedron Lett.* **30:** 4637.

Oda, O., Sakai, K. (1975) *Tetrahedron Lett.* 3705.

Odinokov, V.N., Kukovinets, O.S., Sakharova, N.I., Tolstikov, G.A. (1985) *Zh. Org. Khim.* **21:** 1180.

Ogawa, M., Kuriya, N., Natsume, M. (1984a) *Tetrahedron Lett.* **25:** 969.

Ogawa, M., Natsume, M. (1984b) *Heterocycles* **21:** 769.

Ogawa, M., Natsume, M. (1985) *Heterocycles* **23:** 831.

Ogura, K., Yamashita, M., Tsuchihashi, G. (1976) *Tetrahedron Lett.* 759.

Ohashi, M., Muruishi, T., Kakisawa, H. (1968) *Tetrahedron Lett.* 719.

Ohfune, Y., Grieco, P.A., Wang, C.-L., Majetich, G. (1978) *J. Am. Chem. Soc.* **100:** 5946.

Ohkata, K., Isako, T., Hanafusa, T. (1978) *Chem. Ind.* 274.

Ohno, M., Okamoto, M., Kawabe, N., Umezawa, H., Takeuchi, T., Iinuma, H., Takahashi, S. (1971) *J. Am. Chem. Soc.* **93:** 1285.

Ohno, M., Ito, Y., Arita, M., Shibata, T., Adachi, K., Sawai, H. (1984) *Tetrahedron* **40:** 145.

Ohno, M., Otsuka, M. (1989) *Org. React.* **37:** 1.

Ohnuma, T., Tabe, M., Shiita, K., Ban, Y. (1983) *Tetrahedron Lett.* **24:** 4249.

Ohsawa, T., Ihara, M., Fukumoto, K., Kametani, T. (1982) *Heterocycles* **19:** 1605.

Ohwa, M., Kogure, T., Eliel, E.L. (1986) *J. Org. Chem.* **51:** 2599.

Oinuma, H., Dan, S., Kakisawa, H. (1983) *Chem. Commun.* 654.

Okamoto, S., Kobayashi, Y., Kato, H., Hori, K., Takahashi, T., Tsuji, J., Sato, F. (1988) *J. Org. Chem.* **53:** 5590.

Olson, G.L., Cheung, H.-C., Morgan, K., Saucy, G. (1980) *J. Org. Chem.* **45:** 803.

O'Malley, G.J., Murphy, R.A., Cava, M.P. (1985) *J. Org. Chem.* **50:** 5533.

Openshaw, H.T., Whittaker, N. (1963) *J. Chem. Soc.* 1461.

Oppolzer, W., Hauth, H., Pfäffli, P., Wenger, R. (1977) *Helv. Chim. Acta* **60:** 1801.

Oppolzer, W., Briner, P.H., Snowden, R.L. (1980a) *Helv. Chim. Acta* **63:** 967.

Oppolzer, W., Wylie, R.D. (1980b) *Helv. Chim. Acta* **63:** 1198.

Oppolzer, W., Bättig, K. (1981) *Helv. Chim. Acta* **64:** 2489.

Oppolzer, W., Strauss, H.F., Simmons, D.P. (1982) *Tetrahedron Lett.* **23:** 4673.

Oppolzer, W., Zutterman, F., Bättig, K. (1983) *Helv. Chim. Acta* **66:** 522.

Orito, K., Yorita, K., Suginome, H. (1991) *Tetrahedron Lett.* **32:** 5999.

Osakada, K., Obana, M., Ikariya, T., Saburi, M., Yoshikawa, S. (1981) *Tetrahedron Lett.* **22:** 4297.

Osawa, E., Schleyer, P.v.R., Chang, L.W.K., Kane, V.V. (1974) *Tetrahedron Lett.* 4189.

O'Sullivan, P.J., Moreno, R., Murphy, W.S. (1992) *Tetrahedron Lett.* **33:** 535.

Otomasu, H., Higashiyama, K., Honda, T., Kametani, T. (1982a) *J. Chem. Soc. Perkin Trans. I* 2399.

Otomasu, H., Takatsu, N., Honda, T., Kametani, T. (1982b) *Tetrahedron* **38:** 2627.

Overman, L.E. (1975) *Tetrahedron Lett.* 1149.

Overman, L.E., Jacobsen, E.J. (1982) *Tetrahedron Lett.* **23:** 2741.

Overman, L.E., Malone, T.C., Meier, G.P. (1983) *J. Am. Chem. Soc.* **105:** 6993.

Padwa, A., Murphree, S.S., Yeske, P.E. (1990) *J. Org. Chem.* **55:** 4241.

Paisdor, B., Kuck, D. (1991) *J. Org. Chem.* **56:** 4753.

Paquette, L.A. (1970) *J. Am. Chem. Soc.* **92:** 5765.

Paquette, L.A., Stowell, J.C. (1971) *J. Am. Chem. Soc.* **93:** 2459.

Paquette, L.A., Davis, R.F., James, D.R. (1974) *Tetrahedron Lett.* 1615.

Paquette, L.A., Wyvratt, M.J., Berk, H.C., Moerck, R.E. (1978) *J. Am. Chem. Soc.* **100:** 5845.

Paquette, L.A., Snow, R.A., Muthard, J.L., Cynkowski, T. (1979) *J. Am. Chem. Soc.* **101:** 6991.

Paquette, L.A., Crouse, G.D., Sharma, A.K. (1980) *J. Am. Chem. Soc.* **102:** 3972.

Paquette, L.A., Han, Y.K. (1981) *J. Am. Chem. Soc.* **103:** 1831.

Paquette, L.A., Browne, A.R., Doecke, C.W., Williams, R.V. (1983) *J. Am. Chem. Soc.* **105:** 4113.

Paquette, L.A. (1984) In *Strategies and tactics in organic synthesis*, (Lindberg, T., ed.), Academic Press, Orlando, FL.

Paquette, L.A., Ham, W.H., Dime, D.S. (1985) *Tetrahedron Lett.* **26:** 4983.

Paquette, L.A., Sugimura, T. (1986) *J. Am. Chem. Soc.* **108:** 3841.

Paquette, L.A., Poupart, M.A. (1988) *Tetrahedron Lett.* **29:** 273.

Paquette, L.A., Sauer, D.R., Cleary, D.G., Kinsella, M.A., Blackwell, C.M., Anderson, L.G. (1992) *J. Am. Chem. Soc.* **114:** 7375.

Paquette, L.A., L.A., Wang, T.-Z., Vo, N.H. (1993) *J. Am. Chem. Soc.* **115:** 1676.

Park, H., King, P.F., Paquette, L.A. (1979) *J. Am. Chem. Soc.* **101:** 4773.

Park, P.-U., Broka, C.A., Johnson, B.F., Kishi, Y. (1987) *J. Am. Chem. Soc.* **109:** 6205.

Parker, K.A., Kallmerten, J. (1980) *J. Am. Chem. Soc.* **102:** 5881.

Parker, K.A., Cohen, I.D., Padwa, A., Dent, W. (1984) *Tetrahedron Lett.* **25:** 4917.

Parker, K.A., Kim, H.-J. (1992) *J. Org. Chem.* **57:** 752.

Parker, W., Raphael, R.A., Wilkinson, D.I. (1959) *J. Chem. Soc.* 2433.

Parker, W.L., Johnson, F. (1973) *J. Org. Chem.* **38:** 2489.

Parsons, W.H., Schlessinger, R.H., Quesada, M.L. (1980) *J. Am. Chem. Soc.* **102:** 889.

Partridge, J.J., Chadha, N.K., Uskokovic, M.R. (1973a) *J. Am. Chem. Soc.* **95:** 532; (1973b) *J. Am. Chem. Soc.* **95:** 7171.

Patel, D.V., Van Middlesworth, F., Donaubauer, J., Gannett, P., Sih, C.J. (1986) *J. Am. Chem. Soc.* **108:** 4603.

Pattenden, G., Teague, S.J. (1984) *Tetrahedron Lett.* **25:** 3021.

Pattenden, G., Teague, S.J. (1987) *Tetrahedron* **43:** 5637.

Pattenden, G., Smith, G.F. (1990) *Tetrahedron Lett.* **31:** 6557.

Patterson, J.W. (1993) *Tetrahedron* **49:** 4789.

Pearlman, B.A. (1979) *J. Am. Chem. Soc.* **101:** 6404.

Pearson, A.J., Bansal, H.S. (1986) *Tetrahedron Lett.* **27:** 283.

Pearson, A.J., Bansal, H.S., Lai, Y.-S. (1987) *Chem. Commun.* 519.

Pearson, A.J., Lai, Y.-S. (1988) *Chem. Commun.* 442.

Pearson, W.H., Walavalkar, R. (1994) *Tetrahedron* **50:** 12293.

Pedrocchi-Fantoni, G., Servi, S. (1991) *J. Chem. Soc. Perkin Trans. I* 1764.

Peel, R., Sutherland, J.K. (1974) *Chem. Commun.* 151.

Pelletier, S.W., Chappell, R.W., Prabhakar, S. (1966) *Tetrahedron Lett.* 3489.

Pelletier, S.W., Chappell, R.W., Prabhakar, S. (1968a) *J. Am. Chem. Soc.* **90:** 2889.

Pelletier, S.W., Prabhakar, S. (1968b) *J. Am. Chem. Soc.* **90:** 5318.

Penco, S.F., Angelucci, M., Ballabio, M., Barchelli, G., Suarato, A., Vanotti, E., Vigevani, A., Arcamone, F. (1984) *Tetrahedron* **40:** 4677.

Pesaro, M., Bozzato, G., Schudel, P. (1968) *Chem. Commun.* 1152.

Petrzilka, T., Haefliger, W., Sikemeier, C., Ohloff, G., Eschenmoser, A. (1967) *Helv. Chim. Acta* **50:** 719.

Pfenninger, A., Roesle, A., Keese, R. (1985) *Helv. Chim. Acta* **68:** 493.

Piers, E., de Waal, W., Britton, R.W. (1971) *J. Am. Chem. Soc.* **93:** 5113.

Piers, E., Phillips-Johnson, W.M. (1975) *Can. J. Chem.* **53:** 1281.

Piers, E., Ruedinger, E.H. (1979) *Chem. Commun.* 166.

Piers, E., Winter, M. (1982) *Liebigs Ann. Chem.* 973.

Piers, E., Abeysekera, B.F., Herbert, D.J., Suckling, I.D. (1985a) *Can. J. Chem.* **63:** 3418.

Piers, E., Moss, N. (1985b) *Tetrahedron Lett.* **26:** 2735.

Piers, E., Friesen, R.W. (1986) *J. Org. Chem.* **51:** 3405.

Piers, E., Jean, M., Marrs, P.S. (1987) *Tetrahedron Lett.* **28:** 5075.

Piers, E., Karunaratne, V. (1989) *Can. J. Chem.* **67:** 160.

Piettre, S., Heathcock, C.H. (1990) *Science* **248:** 1532.

Pikul, S., Kozlowska, M., Jurczak, J. (1987) *Tetrahedron Lett.* **28:** 2627.

Pilet, O., Birbaum, J.-L., Vogel, P. (1983) *Helv. Chim. Acta* **66:** 19.

Pirrung, M.C. (1981) *J. Am. Chem. Soc.* **103:** 82.

Pirrung, M.C., Thomson, S.A. (1987) *J. Org. Chem.* **52:** 227.

Pirrung, M.C., Brown, W.L., Rege, S., Laughton, P. (1991) *J. Am. Chem. Soc.* **113:** 8561.

Pirrung, M.C., Lee, Y.R. (1994a) *Tetrahedron Lett.* **35:** 6231.

Pirrung, M.C., Zhang, J., Morehead, A.T. (1994b) *Tetrahedron Lett.* **35:** 6229.

Plaquevent, J., Chichaoui, I. (1993) *New J. Chem.* **17:** 383.

Plavac, F., Heathcock, C.H. (1979) *Tetrahedron Lett.* 2115.

Plieninger, H., Schmalz, D. (1976a) *Chem. Ber.* **109:** 2140.

Plieninger, H., Schmalz, D., Westphal, J., Völkl, A. (1976b) *Chem. Ber.* **109:** 2126.

Plugge, M.F.C., Mol, J.C. (1991) *Synlett* 507.

Polniaszek, R.P., Belmont, S.E. (1990) *J. Org. Chem.* **55:** 4688.

Ponaras, A.A., Meah, M.Y. (1990) *200th ACS Nat Meet.* ORGN 296.

Pond, D.M., Thweatt, J.G. (1975) *U.S. Patent* 3917710 (assigned to Eastman Kodak).

Poppe, L., Novak, L., Kolonits, P., Bata, A., Szantay, C. (1986) *Tetrahedron Lett.* **27:** 5769.

Poss, C.S., Schreiber, S.L. (1994) *Acc. Chem. Res.* **27:** 9.

Potts, K.T., Mattingly, G.S. (1968) *J. Org. Chem.* **33:** 3985.

Prasad, K., Repic, O. (1984a) *Tetrahedron Lett.* **25:** 2435.

Prasad, K., Repic, O. (1984b) *Tetrahedron Lett.* **25:** 3391.

Prelog, V., Acklin, W. (1956) *Helv. Chim. Acta* **39:** 748.

Proksch, E., de Meijere, A. (1976) *Tetrahedron Lett.* 4851.

Quallich, G.J., Schlessinger, R.H. (1979) *J. Am. Chem. Soc.* **101:** 7627.

Quast, H., von der Saal, W. (1982) *Tetrahedron Lett.* **23:** 3653.

Quinkert, G., Schmieder, K.R., Dürner, G., Hache, K., Stegk, A., Barton, D.H.R. (1977) *Chem. Ber.* **110:** 3582.

Quiron, J.-C., Grierson, D.S., Royer, J., Husson, H.-P. (1988) *Tetrahedron Lett.* **29:** 3311.

Rajapaksa, D., Keay, B.A., Rodrigo, R. (1984) *Can. J. Chem.* **62:** 826.

Rama Rao, A.V., Yadav, J.S., Reddy, K.B., Mehendale, A.R. (1984) *Tetrahedron* **40:** 4643.

Rama Rao, A.V., Yadav, J.S., Vidyasagar, V. (1985) *Chem. Commun.* 55.

Rama Rao, A.V., Mysorekar, S.V., Gurjar, M.K., Yadav, J.S. (1987a) *Tetrahedron Lett.* **28:** 2183.

Rama Rao, A.V., Reddy, E.R., Joshi, B.V., Yadav, J.S. (1987b) *Tetrahedron Lett.* **28:** 6497.

Ratcliffe, A.H., Smith, G.F., Smith, G.N. (1973) *Tetrahedron Lett.* 5179.

Ratovelomanana, V., Royer, J., Husson, H.-P. (1985) *Tetrahedron Lett.* **26:** 3803.

Rawal, V.H., Cava, M.P. (1983) *Tetrahedron Lett.* **24:** 5581.

Razdan, R.K., Handrick, G.R., Dalzell, H.C. (1975) *Experientia* **31:** 16.

Reetz, M.T., Heimbach, H., Schwellnus, K. (1984) *Tetrahedron Lett.* **25:** 511.

Reginato, G., Ricci, A., Roelens, S., Scapecchi, S. (1990) *J. Org. Chem.* **55:** 5132.

Reingold, I.D., Drake, J. (1989) *Tetrahedron Lett.* **30:** 1921.

Reusch, W., Grimm, K., Karoglan, J.E., Martin, J., Subrahamanian, K.P., Venkataramani, P.S., Yordy, J.D. (1977) *J. Am. Chem. Soc.* **99:** 1958.

Reuvers, J.T.A., de Groot, A. (1986) *J. Org. Chem.* **51:** 4594.

Rigby, J.H., Senanayake, C. (1987) *J. Am. Chem. Soc.* **109:** 3147.

Rigby, J.H., Qabar, M. (1991) *J. Am. Chem. Soc.* **113:** 8975.

Ripoll, J.L., Limasset, J.C., Conia, J.M. (1971) *Tetrahedron* **27:** 2431.

Riss, B.P., Muckensturm, B. (1986) *Tetrahedron Lett.* **27:** 4979.

Riva, R., Banfi, L., Danieli, B., Guanti, G., Lesma, G., Palmisano, G. (1987) *Chem. Commun.* 299.

Roberts, M.R., Schlessinger, R.H. (1981) *J. Am. Chem. Soc.* **103:** 724.

Roberts, W.P., Shoham, G. (1981) *Tetrahedron Lett.* 4895.

Robinson, R. (1917) *J. Chem. Soc.* 762.

Rokach, J., Young, R.N., Kakushima, M., Lau, C.-K., Seguin, R., Frenette, R., Guidon, Y. (1981) *Tetrahedron Lett.* **22:** 979.

Rosen, T., Heathcock, C.H. (1985) *J. Am. Chem. Soc.* **107:** 3731.

Roush, W.R., Palkowitz, A.D., Ando, K. (1990) *J. Am. Chem. Soc.* **112:** 6348.

Roush, W.R., Straub, J.A., VanNieuwenhze, M.S. (1991) *J. Org. Chem.* **56:** 1636.

Royer, J., Husson, H.-P. (1985a) *J. Org. Chem.* **50:** 670.

Royer, J., Husson, H.-P. (1985b) *Tetrahedron Lett.* **26:** 1515.

Royer, J., Husson, H.-P. (1987) *Tetrahedron Lett.* **28:** 6175.

Rubin, Y., Kahr, M., Knobler, C.B., Diederich, F., Wilkins, C.L. (1991) *J. Am. Chem. Soc.* **113:** 495.

Ruppert, J., Eder, U., Wiechert, R. (1973) *Chem. Ber.* **106:** 3636.

Rychnovsky, S.D., Griesgraber, G., Zeller, S., Skalitzky, D.J. (1991) *J. Org. Chem.* **56:** 5161.

Rychnovsky, S.D., Griesgraber, G., Kim, J. (1994) *J. Am. Chem. Soc.* **116:** 2621.

Ryckman, D.M., Stevens, R.V. (1987) *J. Am. Chem. Soc.* **109:** 4940.

Safaryn, J.E., Chiarello, J., Chen, K.-M., Joullie, M.M. (1986) *Tetrahedron* **42:** 2635.

Saito, S., Morikawa, Y., Moriwake, T. (1988) *J. Org. Chem.* **55:** 5424.

Saito, S., Hamano, S., Moriyama, H., Okada, K., Morikawa, T. (1990) *Tetrahedron Lett.* **29:** 1157.

Sakan, F., Hashimoto, H., Ichihara, A., Shirahama, H., Matsumoto, T. (1971) *Tetrahedron Lett.* 3703.

Sakito, Y., Mukaiyama, T. (1979) *Chem. Lett.* 1207.

Sallay, S.J. (1967) *J. Am. Chem. Soc.* **89:** 6762.

Salmony, A., Simonis, H. (1905) *Ber. Deut. Chem. Ges.* **38:** 2580.

Sampath, V., Lund, E.C., Knudsen, M.J., Olmstead, M.M., Schore, N.E. (1987) *J. Org. Chem.* **52:** 3595.

Samson, M., De Clercq, P., De Wilde, H., Vandewalle, M. (1977) *Tetrahedron Lett.* 3195.

Sanchez, I.H., Mendoza, M.T. (1980) *Tetrahedron Lett.* **21:** 3651.

Santaniello, E., Ferraboschi, P., Grisenti, P. (1990) *Tetrahedron Lett.* **31:** 5657.

Santelli-Rouvier, C. (1984) *Tetrahedron Lett.* **25:** 4371.

Sarett, L.H., Lukes, R.L., Poos, G.I., Robinson, J.M., Beyler, R.E., Vandegrift, J.M., Arth, G.E. (1952) *J. Am. Chem. Soc.* **74:** 1393.

Sarkar, T.K.S., Ghosh, S.K., Subba Rao, P.S.V., Mamdapur, V.R. (1990) *Tetrahedron Lett.* **31:** 3461, 3465.

Sarma, A.S., Chattopadhyay, P. (1982) *J. Org. Chem.* **47:** 1727.

Sasai, H., Shibasaki, M. (1987) *Tetrahedron Lett.* **28:** 333.

Sato, T., Kaneko, H., Yamaguchi, S. (1980) *J. Org. Chem.* **45:** 3778.

Saucy, G., Borer, R. (1971) *Helv. Chim. Acta* **54:** 2121.

Savoia, D., Concialini, V., Roffia, S., Tarsi, L. (1991) *J. Org. Chem.* **56:** 1822.

Scheffold, R., Reves, L., Aebersold, J., Schaltegger, A. (1976) *Chimia* **30:** 57.

Schell, F.M., Ganguly, R.N. (1980) *J. Org. Chem.* **45:** 4069.

Schenck, G.O., Wirtz, R. (1953) *Naturwiss.* **40:** 581.

Scheuing, G., Winterhalder, L. (1929) *Liebigs Ann. Chem.* **473:** 126.

Schiehser, G.A., White, J.D. (1980) *J. Org. Chem.* **45:** 1864.

Schiess, P., Heitzmann, M. (1978) *Helv. Chim. Acta* **61:** 844.

Schink, H.E., Pettersson, H., Bäckvall, J.-E. (1991) *J. Org. Chem.* **56:** 2769.

Schink, H.E., Bäckvall, J.-E. (1992) *J. Org. Chem.* **57:** 1588.

Schlessinger, R.H., Nugent, R.A. (1982) *J. Am. Chem. Soc.* **104:** 1116.

Schleyer, P.v.R., Donaldson, M.M. (1960) *J. Am. Chem. Soc.* **82:** 4645.

Schmidt, U., Gambos, J., Haslinger, E., Zak, H. (1976) *Chem. Ber.* **109:** 2628.

Schmidt, U., Meyer, R., Leitenberger, V., Lieberknecht, A., Griesser, H. (1991) *Chem. Commun.* 275.

Schneider, M., Engel, N., Boensmann, ?. (1984) *Angew. Chem. Int. Ed. Engl.* **23:** 64.

Schneider, W.P., Axen, U., Lincoln, F.H., Pike, J.E., Thompson, J.L. (1968) *J. Am. Chem. Soc.* **90:** 5895.

Schneiders, G.E., Stevenson, R. (1981) *J. Org. Chem.* **46:** 2969.

Schnurrenberger, P., Hungerbühler, E., Seebach, D. (1984) *Tetrahedron Lett.* **25:** 2209.

Schöpf, C., Lehmann, G. (1935) *Liebigs Ann. Chem.* **581:** 1.

Schöpf, C., Lehmann, G., Arnold, W. (1937) *Angew. Chem.* **50:** 203.

Schöpf, C., Benz, G., Braun, F., Hinkel, H., Rokohl, R. (1957) *Angew. Chem.* **69:** 69.

Scholz, G., Konusch, J., Tochtermann, W. (1990) *Liebigs Ann. Chem.* 593.

Scholz, G., Tochtermann, W. (1991) *Tetrahedron Lett.* **32:** 5535.

Schore, N.E. (1979) *Synth. Commun.* **9:** 41.

Schore, N.E., Najdi, S.D. (1987) *J. Org. Chem.* **52:** 5296.

Schreiber, S.L., Satake, K. (1980) *J. Am. Chem. Soc.* **102:** 6163.

Schreiber, S.L., Claus, R.E., Reagan, J. (1982) *Tetrahedron Lett.* **23:** 3867.

Schreiber, S.L., Sommer, T.J. (1982) *Tetrahedron Lett.* **24:** 4781.

Schreiber, S.L., Satake, K. (1984) *J. Am. Chem. Soc.* **106:** 4186.

Schreiber, S.L., Wang, Z. (1985) *J. Am. Chem. Soc.* **107:** 5303.

Schreiber, S.L., Goulet, M.T., Schulte, G. (1987) *J. Am. Chem. Soc.* **109:** 4718.

Schreiber, S.L., Meyers, H.V. (1988) *J. Am. Chem. Soc.* **110:** 5198.

Schröder, G. (1963) *Angew. Chem. Int. Ed. Engl.* **2:** 481.

Schuda, P.F., Phillips, J.L., Morgan, T.M. (1986) *J. Org. Chem.* **51:** 2742.

Schulte, L.D., Rieke, R.D. (1988) *Tetrahedron Lett.* **29:** 5483.

Schulte-Elte, K.H., Hauser, A., Ohloff, G. (1979) *Helv. Chim. Acta* **62:** 2673.

Schultz, A.G., Motyka, L.A. (1982) *J. Am. Chem. Soc.* **104:** 5800.

Schwartz, E., Shanzer, A. (1982) *Tetrahedron Lett.* **23:** 979.

Schwesinger, R., Prinzbach, H. (1975) *Angew. Chem. Int. Ed. Engl.* **14:** 630.

Scott, A.I., McCapra, F., Buchanan, R.L., Day, A.C., Young, D.W. (1965) *Tetrahedron* **21:** 3605.

Scott, L.T., Hashemi, M.M., Meyer, D.T., Warren, H.B. (1991) *J. Am. Chem. Soc.* **113:** 7082.

Scott, L.T., Hashemi, M.M., Bratcher, M.S. (1992) *J. Am. Chem. Soc.* **114:** 1920.

Scully, F.E. (1980) *J. Org. Chem.* **45:** 1515.

Secen, H., Sütbeyaz, Y., Balci, M. (1990) *Tetrahedron Lett.* **31:** 1323.

Seebach, D., Kalinowski, H.-O., Bastani, B., Crass, G., Daum, H., Dörr, H., DuPreez, N.P., Ehrig, V., Langer, W., Nüssler, C., Oei, H.-A., Schmitt, M. (1977) *Helv. Chim. Acta* **60:** 301.

Seebach, D., Hungerbühler, H. (1980) in Scheffold, R. (ed.) *Modern Synth. Methods,* **2:** 91.

Semmelhack, M.F., Brickner, S.J. (1981a) *J. Am. Chem. Soc.* **103:** 3945.

Semmelhack, M.F., Helquist, P., Jones, L.D., Keller, L., Mendelson, L., Ryono, L.S., Smith, J.G., Stauffer, R.D. (1981b) *J. Am. Chem. Soc.* **103:** 6460.

Semmelhack, M.F., Tomoda, S., Nagaoka, H., Boettger, S.D., Hurst, K.M. (1982) *J. Am. Chem. Soc.* **104:** 747.

Serratosa, F., Lopez, F., Font, J. (1972) *Tetrahedron Lett.* 2589.

Seu, Y.-B., Mori, K. (1986) *Agric. Biol. Chem.* **50:** 2923.

Seu, Y.-B., Kho, Y.-H. (1994) *Nat. Prod. Lett.* **4:** 61.

Shair, M.D., Yoon, T., Danishefsky, S.J. (1994) *J. Org. Chem.* **59:** 3755.

Shanzer, A. (1980) *Tetrahedron Lett.* **21:** 221.

Shealy, Y.F., Clayton, J.D. (19866) *J. Am. Chem. Soc.* **88:** 3885.

Shemyakin, M.M., Ovchinnikov, Y.A., Antonov, V.K., Kiryushkin, A.A., Ivanov, V.T., Shchelokov, V.I., Shkrob, A.M. (1964) *Tetrahedron Lett.* 47.

Shibasaki, M., Ogawa, Y. (1985) *Tetrahedron Lett.* **26:** 3841.

Shimizu, I., Nakagawa, H. (1992) *Tetrahedron Lett.* **33:** 4957.

Shimshock, S.J., Waltermire, R.E., DeShong, P. (1991) *J. Am. Chem. Soc.* **113:** 8791.

Shishido, K., Tanaka, F., Fukumoto, K., Kametani, T. (1985) *Chem. Pharm. Bull.* **33:** 532.

Shono, T., Nishiguchi, I., Nitta, M. (1976) *Chem. Lett.* 1319.

Shono, T., Matsumura, Y., Tsubata, K. (1981) *J. Am. Chem. Soc.* **103:** 1172.

Shono, T., Matsumura, Y., Kanazawa, T. (1983) *Tetrahedron Lett.* **24:** 4577.

Shono, T., Matsumura, Y., Uchida, K., Tsubata, K., Makino, A. (1984) *J. Org. Chem.* **49:** 300.

Shono, T., Matsumura, Y., Uchida K., Tagami, K. (1987) *Chem. Lett.* 919.

Sih, C.J., Heather, H.B., Sood, R., Price, P., Peruzzotti, G., Lee, L.F.H., Lee, S.S. (1975) *J. Am. Chem. Soc.* **97:** 865.

Sih, C.J., Massuda, D., Corey, P., Gleim, R.D., Suzuki, F. (1979) *Tetrahedron Lett.* 1285.

Simonet, B., Rousseau, G. (1993) *Tetrahedron Lett.* **34:** 5723.

Simpkins, N.S. (1986) *Chem. Commun.* 88.

Singh, P., Weinreb, S.M. (1976) *Tetrahedron* **32:** 2379.

Smith, A.B., III, Richmond, R.E. (1983) *J. Am. Chem. Soc.* **105:** 575.

Smith, A.B., III, Mewshaw, R. (1984) *J. Org. Chem.* **49:** 3685.

Smith, A.B., III, Mewshaw, R. (1985) *J. Am. Chem. Soc.* **107:** 1769.

Smith, A.B., III, Dorsey, B.D., Visnick, M., Maeda, T., Malamas, M.S. (1986) *J. Am. Chem. Soc.* **108:** 3110.

Smith, A.B., III, Hale, K.J. (1989) *Tetrahedron Lett.* **30:** 1037.

Smith, A.B., III, Fukui, M., Vaccaro, H.A., Empfield, J.R. (1991) *J. Am. Chem. Soc.* **113:** 2071.

Smith, A.B., III, Rano, T.A., Chida, N., Sulikowski, G.A., Wood, J.L. (1992) *J. Am. Chem. Soc.* **114:** 8008.

Smith, E.C., Barborak, J.C. (1976) *J. Org. Chem.* **41:** 1433.

Snider, B.B., Phillipps, G.B. (1982) *J. Am. Chem. Soc.* **104:** 1113.

Snowden, R.L., Sonnay, P., Ohloff, G. (1981) *Helv. Chim. Acta* **64:** 25.

Snowden, R.L. (1986) *Tetrahedron* **42:** 3277.

Snowden, R.L., Linder, S.M., Muller, B.L., Schulte-Elte, K.H. (1987) *Helv. Chim. Acta* **70:** 1858.

Solas, D., Wolinsky, J. (1983) *J. Org. Chem.* **48:** 670.

Solé, D., Bonjoch, J. (1991) *Tetrahedron Lett.* **32:** 5183.

Solladié, G., Hutt, J. (1987) *Tetrahedron Lett.* **28:** 797.

Solladié, G., Huser, N. (1994) *Tetrahedron: Asymmetry* **5:** 255.

Somei, M., Kawasaki, T. (1989) *Chem. Pharm. Bull.* **37:** 3426.

Sowell, C.G., Wolin, R.L., Little, R.D. (1990) *Tetrahedron Lett.* **31:** 485.

Speckamp, W.N., DeBoer, J.J.J. (1983) *Recl. Trav. Chim. Pays-Bas* **102:** 405.

Spencer, T.A., Smith, R.A.J., Storm, D.L., Villarica, R.M. (1971) *J. Am. Chem. Soc.* **93:** 4856.

Spurr, P.R., Murty, B.A.R.C., Fessner, W.-D., Fritz, H., Prinzbach, H. (1987) *Angew. Chem. Int. Ed. Engl.* **26:** 455.

Srikrishna, A., Krishnan, K. (1988) *Tetrahedron Lett.* **29:** 4995.

Stafford, J.A., Heathcock, C.H. (1990) *J. Org. Chem.* **55:** 5433.

Steglich, W., Reininger, W. (1970) *Chem. Commun.* 178.

Stetter, H., Thomas, H.G. (1968) *Chem. Ber.* **101:** 1115.

Stevens, R.V., Fitzpatrick, J.M., Kaplan, M., Zimmerman, R.L. (1971) *Chem. Commun.* 857.

Stevens, R.V., Lee, A.W.M. (1979) *J. Am. Chem. Soc.* **101:** 7032.

Still, W.C., Schneider, M.J. (1977) *J. Am. Chem. Soc.* **99:** 948.

Still, W.C. (1979) *J. Am. Chem. Soc.* **101:** 2493.

Still, W.C., Darst, K.P. (1980a) *J. Am. Chem. Soc.* **102:** 7385.

Still, W.C., Schneider, J.A. (1980b) *J. Org. Chem.* **45:** 3375.

Still, W.C., Tsai, M.-Y. (1980c) *J. Am. Chem. Soc.* **102:** 3654.

Still, W.C., Barrish, J.C. (1983a) *J. Am. Chem. Soc.* **105:** 2487.

Still, W.C., Murata, S., Revial, G., Yoshihara, K. (1983b) *J. Am. Chem. Soc.* **105:** 625.

Stillwell, R.N. (1964) *Ph.D. Thesis, Harvard Univ.*

Stojanac, N., Stojanac, Z., White, P.S., Valenta, Z. (1979) *Can. J. Chem.* **57:** 3346.

Stojanac, N., Valenta, Z. (1991) *Can. J. Chem.* **69:** 853.

Stone, M.J., Maplestone, R.A., Rahman, S.K., Williams, D.H. (1991) *Tetrahedron Lett.* **32:** 2663.

Stork, G., van Tamelen, E.E., Friedman, L.J., Burgstahler, A.W. (1953) *J. Am. Chem. Soc.* **75:** 384.

Stork, G., Clarke, F.H. (1961) *J. Am. Chem. Soc.* **83:** 3114.

Stork, G., Meisels, A., Davies, J.E. (1963) *J. Am. Chem. Soc.* **85:** 3419.

Stork, G., Tomasz, M. (1964) *J. Am. Chem. Soc.* **86:** 471.

Stork, G., Danishefsky, S., Ohashi, M. (1967a) *J. Am. Chem. Soc.* **89:** 5459.

Stork, G., McMurry, J.E. (1967b) *J. Am. Chem. Soc.* **89:** 5464.

Stork, G., Kretchmer, R.A., Schlessinger, R.H. (1968) *J. Am. Chem. Soc.* **90:** 1647.

Stork, G., Guthikonda, R.N. (1972a) *J. Am. Chem. Soc.* **94:** 5109.

Stork, G., Tabak, J.M., Blount, J.F. (1972b) *J. Am. Chem. Soc.* **94:** 4735.

Stork, G., Danheiser, R.L., Ganem, B. (1973) *J. Am. Chem. Soc.* **95:** 3414.

Stork, G., Singh, J. (1974) *J. Am. Chem. Soc.* **96:** 6181.

Stork, G., Isobe, M. (1975) *J. Am. Chem. Soc.* **97:** 6260.

Stork, G., Kraus, G. (1976) *J. Am. Chem. Soc.* **98:** 6747.

Stork, G., Takahashi, T. (1977) *J. Am. Chem. Soc.* **99:** 1275.

Stork, G., Boeckman, R.K., Taber, D.F., Still, W.C., Singh, J. (1979a) *J. Am. Chem. Soc.* **101:** 7107.

Stork, G., Nair, V. (1979b) *J. Am. Chem. Soc.* **101:** 1315.

Stork, G., Paterson, I., Lee, F.K.C. (1982) *J. Am. Chem. Soc.* **104:** 4686.

Stork, G., Sher, P.M., Chen, H.-L. (1986) *J. Am. Chem. Soc.* **108:** 6384.

Stork, G., Gardner, J.O. (Unpublished) quoted in [Stork, 1979a].

Stotter, P.L., Hornish, R.E. (1973) *J. Am. Chem. Soc.* **95:** 4444.

Strauss, H.F., Wiechers, A. (1978) *Tetrahedron* **34:** 127.

Street, S.D.A., Yeates, C., Kocienski, P., Campbell, S.F. (1985) *Chem. Commun.* 1386.

Strunz, G.M., Lal, G.S. (1982) *Can. J. Chem.* **60:** 2528.

Strunz, G.M., Giguere, P., Ebacher, M. (1983) *Synth. Commun.* **13:** 823.

Stumpp, M.C., Schmidt, R.R. (1986) *Tetrahedron* **42:** 5941.

Subrahamanian, K.P., Reusch, W. (1978) *Tetrahedron Lett.* 3789.

Suemune, H., Hizuka, M., Kamashita, T., Sakai, K. (1989) *Chem. Pharm. Bull.* **37**: 1379.

Suemune, H., Takahashi, M., Maeda, S., Xie, Z.-F., Sakai, K. (1990) *Tetrahedron: Asymmetry* **1**: 425.

Suginome, H., Yamada, S. (1987) *Tetrahedron Lett.* **28**: 3963.

Sütbeyaz, Y., Secen, H., Balci, M. (1988) *Chem. Commun.* 1330.

Sutherland, J.K. (1968) *Chem. Commun.* 1192.

Suzuki, F., Trenbeath, S., Gleim, R.D., Sih, C.J. (1978) *J. Am. Chem. Soc.* **100**: 2272.

Suzuki, K., Miyazawa, M., Shimazaki, M., Tsuchihashi, G. (1986) *Tetrahedron Lett.* **27**: 6237.

Suzuki, K., Matsumoto, T., Tomooka, K., Matsumoto, K., Tsuchihashi, G. (1987) *Chem. Lett.* 113.

Suzuki, M., Koyano, H., Morita, Y., Noyori, R. (1989) *Synlett.* 22.

Suzuki, T., Sato, E., Unno, K., Kametani, T. (1985a) *Heterocycles* **23**: 835.

Suzuki, T., Sato, E., Unno, K., Kametani, T. (1985b) *Heterocycles* **23**: 839.

Suzuki, T., Sato, E., Unno, K., Kametani, T. (1986) *J. Chem. Soc. Perkin Trans. 1* 2263.

Suzuki, T., Sato, E., Matsuda, Y., Yada, T., Koizumi, S., Unno, K., Kametani, T. (1988) *Chem. Commun.* 1531.

Suzuki, T., Uozumi, Y., Shibasaki, M. (1991) *Chem. Commun.* 1593.

Szychowski, J., MacLean, D.B. (1979) *Can. J. Chem.* **57**: 1631.

Taber, D.F., Schuchardt, J.L. (1985) *J. Am. Chem. Soc.* **107**: 5289.

Taber, D.F., Hoerner, R.S., Hagen, M.D. (1991) *J. Org. Chem.* **56**: 1287.

Tada, M., Sugimoto, Y., Takahashi, T. (1980) *Bull. Chem. Soc. Jpn.* **53**: 2966.

Tada, M., Ohtsu, K., Chiba, K. (1994) *Chem. Pharm. Bull.* **42**: 2167.

Takahashi, I., Nomura, A., Kitajima, H, (1990) *Synth. Commun.* **20**: 1569.

Takahashi, K., Kurita, H., Ogura, K., Iida, H. (1985) *J. Org. Chem.* **50**: 4368.

Takahashi, K., Aihara, T., Ogura, K. (1987) *Chem. Lett.* 2359.

Takahashi, K., Brossi, A. (1989) *Synth. Commun.* **19**: 835.

Takahashi, S., Kusumi, T., Kakisawa, H. (1979) *Chem. Lett.* 515.

Takahashi, S., Oritani, T., Yamashita, K. (1988) *Tetrahedron Lett.* **44**: 7081.

Takahashi, T., Ikeda, H., Tsuji, J. (1980a) *Tetrahedron Lett.* 3885.

Takahashi, T., Nagashima, T., Tsuji, J. (1980b) *Chem. Lett.* 369.

Takano, S., Hatakeyama, S., Ogasawara, K. (1976a) *J. Am. Chem. Soc.* **98**: 3022.

Takano, S., Tanigawa, K., Ogasawara, K. (1976b) *Chem. Commun.* 189.

Takano, S., Kubodera, N., Ogasawara, K. (1977) *J. Org. Chem.* **42**: 786.

Takano, S., Takahashi, Y., Hatakeyama, S., Ogasawara, K. (1979) *Heterocycles* **12**: 765.

Takano, S., Chiba, K., Yonaga, M., Ogasawara, K. (1980a) *Chem. Commun.* 616.

Takano, S., Hatakeyama, S., Ogasawara, K. (1980b) *J. Chem. Soc. Perkin Trans. 1* 457.

Takano, S., Murakata, C., Ogasawara, K. (1980c) *Chem. Commun.* **14**: 1301.

Takano, S., Yonaga, M., Chiba, K., Ogasawara, K. (1980d) *Tetrahedron Lett.* **21**: 3697.

Takano, S., Imamura, Y., Ogasawara, K. (1981a) *Tetrahedron Lett.* **22**: 4479.

Takano, S., Murakata, C., Ogasawara, K. (1981b) *Heterocycles* **16**: 247.

Takano, S., Ogawa, N., Ogasawara, K. (1981c) *Heterocycles* **16**: 915.

Takano, S., Yonaga, M., Ogasawara, K. (1981d) *Chem. Commun.* 1153.

Takano, S., Goto, E., Hirama, M., Ogasawara, K. (1982a) *Chem. Pharm. Bull.* **30**: 2641.

Takano, S., Goto, E., Ogasawara, K. (1982b) *Tetrahedron Lett.* **23**: 5567.

Takano, S., Goto, E., Ogasawara, K. (1982c) *Chem. Lett.* 1913.

Takano, S., Hatakeyama, S., Takahashi, Y., Ogasawara, K. (1982d) *Heterocycles* **17**: 263.

Takano, S., Ogasawara, K. (1982e) *J. Synth. Org. Chem. Jpn.* **40**: 1037.

Takano, S., Uchida, W., Hatakeyama, S., Ogasawara, K. (1982f) *Chem. Lett.* 733.

Takano, S., Hirama, M., Ogasawara, K. (1983a) *Heterocycles* **20**: 1363.

Takano, S., Kasahara, C., Ogasawara, K. (1983b) *Chem. Lett.* 175.

Takano, S., Yamada, S., Numata, H., Ogasawara, K. (1983c) *Chem. Commun.* 760.

Takano, S., Morimoto, M., Ogasawara, K. (1984) *Chem. Commun.* 82.

Takano, S., Akiyama, M., Ogasawara, K. (1985a) *J. Chem. Soc. Perkin Trans.* **I** 2447.

Takano, S., Tanaka, M., Seo, K., Hirama, M., Ogasawara, K. (1985b) *J. Org. Chem.* **50**: 931.

Takano, S., Sato, S., Goto, E., Ogasawara, K. (1986) *Chem. Commun.* 156.

Takano, S., Chiba, H., Kudo, J., Takahashi, M., Ogasawara, K. (1987a) *Heterocycles* **26**: 2461.

Takano, S., Inomata, K., Kurotaki, A., Ohkawa, T., Ogasawara, K. (1987b) *Chem. Commun.* 1720.

Takano, S., Iwabuchi, Y., Ogasawara, K. (1987c) *J. Am. Chem. Soc.* **109**: 5523.

Takano, S., Kurotaki, A., Ogasawara K. (1987d) *Tetrahedron Lett.* **28**: 3991.

Takano, S., Iwabuchi, Y., Ogasawara, K. (1988a) *Chem. Commun.* 1204.

Takano, S., Iwabuchi, Y., Ogasawara, K. (1988b) *Chem. Commun.* 1527.

Takano, S., Ohkawa, T., Tamori, S., Satoh, S., Ogasawara, K. (1988c) *Chem. Commun.* 189.

Takano, S., Satoh, S., Ogasawara, K. (1988d) *Chem. Commun.* 59.

Takano, S., Shimazaki, Y., Sekiguchi, Y., Ogasawara, K. (1988e) *Chem. Lett.* 2041.

Takano, S., Shimazaki, Y., Takahashi, M., Ogasawara, K. (1988f) *Chem. Commun.* 1004.

Takano, S., Sugihara, T., Satoh, S., Ogasawara, K. (1988g) *J. Am. Chem. Soc.* **110**: 6467.

Takano, S., Yanase, M., Sugihara, T., Ogasawara, K. (1988h) *Chem. Commun.* 1538.

Takano, S., Inomata, K., Ogasawara, K. (1989a) *Chem. Commun.* 271.

Takano, S., Iwabuchi, Y., Ogasawara, K. (1989b) *Heterocycles* **29**: 1861.

Takano, S., Iwabuchi, Y., Ogasawara, K. (1989c) *Chem. Commun.* 1371.

Takano, S., Sekiguchi, Y., Ogasawara, K. (1989d) *Heterocycles* **29**: 445.

Takano, S., Shimazaki, Y., Ogasawara, K. (1989e) *Heterocycles* **29**: 2101.

Takano, S., Shimazaki, Y., Sekiguchi, Y., Ogasawara, K. (1989f) *Synthesis* 539.

Takano, S., Sugihara, T., Samizu, K., Akiyama, M., Ogasawara, K. (1989g) *Chem. Lett.* 1781.

Takano, S., Tomita, S., Iwabuchi, Y., Ogasawara, K. (1989h) *Heterocycles* **29**: 1473.

Takano, S., Moriya, M., Iwabuchi, Y., Ogasawara, K. (1990a) *Chem. Lett.* 109.

Takano, S., Sekiguchi, Y., Setoh, M., Yoshimitsu, T., Inomata, K., Takahashi, M., Ogasawara, K. (1990b) *Heterocycles* **31**: 1715.

Takano, S., Shimazaki, Y., Iwabuchi, Y., Ogasawara, K. (1990c) *Tetrahedron Lett.* **31:** 3619.

Takano, S., Sugihara, T., Ogasawara, K. (1990d) *Heterocycles* **31:** 1721.

Takano, S., Iwabuchi, Y., Ogasawara, K. (1991a) *Chem. Commun.* 820.

Takano, S., Sugihara, T., Kamikubo, T., Ogasawara, K. (1991b) *Heterocycles* **32:** 1587.

Takano, S., Sugihara, T., Ogasawara, K. (1991c) *Synlett.* 279.

Takano, S., Sugihara, T., Ogasawara, K. (1991d) *Tetrahedron Lett.* **32:** 2797.

Takano, S., Sekiguchi, Y., Ogasawara, K. (1992a) *Heterocycles* **33:** 59.

Takano, S., Sekiguchi, Y., Ogasawara, K. (1992b) *Heterocycles* **33:** 743.

Takano, S., Moriya, M., Ogasawara, K. (1993a) *Synlett.* 601.

Takano, S., Moriya, M., Ogasawara, K. (1993b) *Chem. Commun.* 614.

Takano, S., Samizu, K., Ogasawara, K. (1993c) *Synlett.* 393.

Takase, S., Itoh, Y., Uchida, I., Tanaka, H., Aoki, H. (1986) *Tetrahedron* **42:** 5887.

Takeshita, H., Iwabuchi, H., Kouno, I., Iino, M., Nomura, D. (1979) *Chem. Lett.* 649.

Takeya, T., Okubo, T., Nishida, S., Tobinaga, S. (1985) *Chem. Pharm. Bull.* **33:** 3599.

Takle, A., Kocienski, P.J. (1990) *Tetrahedron* **46:** 4503.

Tamao, K., Kodama, S., Nakatsuka, T., Kiso, Y., Kumada, M. (1975) *J. Am. Chem. Soc.* **97:** 4405.

Tamao, K., Tohma, T., Inui, N., Nakayama, O., Ito, Y. (1990) *Tetrahedron Lett.* **31:** 7333.

Tamariz, J., Schwager, L., Stilbard, J.H.A., Vogel, P. (1983) *Tetrahedron Lett.* **24:** 1497.

Tamariz, J., Vogel, P. (1984) *Tetrahedron* **40:** 4549.

Tamura, Y., Saito, T., Kiyokawa, H., Chen, L.C., Ishibashi, H. (1977) *Tetrahedron Lett.* **4075:** Tanaka, A., Yamashita, K. (1980) *Agric. Biol. Chem.* **44:** 199.

Tanaka, H., Yoshioka, T., Shimauchi, Y., Yoshimoto, A., Ishikura, T., Naganawa, H., Takeuchi, T., Umezawa, H. (1984) *Tetrahedron Lett.* **25:** 3351.

Tanaka, K., Uchiyama, F., Sakamoto, K., Inubushi, Y. (1982) *J. Am. Chem. Soc.* **104:** 4965.

Tanaka, K., uMatsuura, H., Funaki, I., Suzuki, H. (1991) *Chem. Commun.* 1145.

Tanaka, K., Ohta, Y., Fuki, K., Taga, T. (1993) *Tetrahedron Lett.* **34:** 4071.

Tanaka, M., Murakami, T., Suemune, H., Sakai, K. (1992a) *Heterocycles* **33:** 697.

Tanaka, M., Yoshioka, M., Suemune, H., Sakai, K. (1992b) *Chem. Commun.* 1454.

Tanaka, T., Iijima, I. (1970) *Tetrahedron Lett.* 3963.

Tanner, D., Birgersson, C., Dhaliwal, H.K. (1990) *Tetrahedron Lett.* **31:** 1903.

Tanoue, Y., Terada, A., Sugyo, Y. (1987) *J. Org. Chem.* **52:** 1437.

Taschner, M.J., Black, D.J. (1988) *J. Am. Chem. Soc.* **110:** 6892.

Tatsuta, K., Akimoto, K., Kinoshita, M. (1979) *J. Am. Chem. Soc.* **101:** 6116.

Taylor, M.D., Minaskanian, G., Winzenberg, K.N., Santone, P., Smith, A.B., III (1982) *J. Org. Chem.* **47:** 3960.

Teisseire, P., Pesnelle, P., Corbier, B., Plattier, M., Monpetit, P. (1974) *Recherches* **19:** 69.

Teisseire, P., Giraudi, E. (1980) *VIIIth Int. Congr. Essent. Oils* paper 166.

Terada, A., Tanoue, Y., Hatada, H., Sakamoto, H. (1987) *Bull. Chem. Soc. Jpn.* **60:** 205.

Terao, Y., Imai, N., Achiwa, K., Sekiya, M. (1982) *Chem. Pharm. Bull.* **30:** 3167.

Terashima, S., Yamada, S., Nara, M. (1977) *Tetrahedron Lett.* 1001.

Ternansky, R.J., Balogh, D.W., Paquette, L.A. (1982) *J. Am. Chem. Soc.* **104:** 4503.

Thanning, M., Wistrand, L.-G. (1990) *J. Org. Chem.* **55:** 1406.

Tiecco, M., Tingoli, M., Testaferri, L., Chianelli, D., Wenkert, E. (1986) *Tetrahedron* **42:** 1475.

Tishler, M., Pfister, K., III, Babson, R.D., Ladenburg, K., Fleming, A.J. (1947) *J. Am. Chem. Soc.* **69:** 1487.

Tius, M.A., Gomez-Galeno, J., Gu, X., Zaidi, J.H. (1991) *J. Am. Chem. Soc.* **113:** 5775.

Tokuda, M., Mimura, N., Yoshioka, K., Karasawa, T., Fujita, H., Suginome, H. (1993) *Synthesis* 1086.

Tokunaga, Y., Nagano, H., Shiota, M. (1986) *J. Chem. Soc. Perkin Trans. I* 581.

Tolbert, L.M., Gregory, J.C., Brock, C.P. (1985) *J. Org. Chem.* **50:** 548.

Tomioka, K., Ishiguro, T., Koga, K. (1980) *Tetrahedron Lett.* **21:** 2973.

Tomioka, K., Masumi, F., Yamashita, T., Koga, K. (1989) *Tetrahedron* **45:** 643.

Tömösközi, I., Gruber, C., Gulacsi, E. (1985) *Tetrahedron Lett.* **26:** 3141.

Torii, S., Tanaka, H., Mandai, T. (1975) *J. Org. Chem.* **40:** 2221.

Torii, S., Okamoto, T., Kadono, S. (1977) *Chem. Lett.* 459.

Torii, S., Inokuchi, T., Ogawa, H. (1979) *J. Org. Chem.* **44:** 3412.

Torii, S., Inokuchi, T., Oi, R. (1983) *J. Org. Chem.* **48:** 1944.

Torisawa, Y., Hashimoto, A., Nakayama, M., Seki, H., Hara, R., Hino, T. (1991) *Tetrahedron* **47:** 8067.

Tornare, J.M., Vogel, P. (1985) *Helv. Chim. Acta* **68:** 1069.

Toth, J.E., Fuchs, P.L. (1987) *J. Org. Chem.* **52:** 473.

Toyooka, N., Nishino, A., Momose, T. (1993) *Tetrahedron Lett.* **34:** 4539.

Toyota, M., Seishi, T., Fukumoto, K. (1993) *Tetrahedron Lett.* **34:** 5947.

Trabert, L., Hopf, H. (1980) *Liebigs Ann. Chem.* 1786.

Troin, Y., Sinibaldi, M.E., Gramain, J.-C., Rubiralta, M., Diez, A. (1991) *Tetrahedron Lett.* **32:** 6129.

Trost, B.M., Hiroi, K. (1976) *J. Am. Chem. Soc.* **98:** 4313.

Trost, B.M., Timko, J.M., Stanton, J.L. (1978) *Chem. Commun.* 436.

Trost, B.M., Bernstein, P.R., Funfschilling, P.C. (1979a) *J. Am. Chem. Soc.* **101:** 4378.

Trost, B.M., Nishimura, Y., Yamamoto, K. (1979b) *J. Am. Chem. Soc.* **101:** 1328.

Trost, B.M., Shuey, C.D., DiNinno, F. (1979c) *J. Am. Chem. Soc.* **101:** 1284.

Trost, B.M. Curran, D.P. (1980) *J. Am. Chem. Soc.* **102:** 5699.

Trost, B.M., McDougal, P.G. (1982a) *J. Am. Chem. Soc.* **104:** 6110.

Trost, B.M., Renaut, P. (1982b) *J. Am. Chem. Soc.* **104:** 6668.

Trost, B.M., Fray, M.J. (1984) *Tetrahedron Lett.* **25:** 4605.

Trost, B.M. (1986a) *Angew. Chem. Int. Ed. Engl.* **25:** 1.

Trost, B.M., Lynch, J., Renault, P., Dteinman, D.H. (1986b) *J. Am. Chem. Soc.* **108:** 284.

Trost, B.M. (1988) *Pure Appl. Chem.* **60:** 1615.

Trost, B.M., Van Vranken, D.L. (1990) *J. Am. Chem. Soc.* **112:** 1261.

Trost, B.M., Kondo, Y. (1991) *Tetrahedron Lett.* **32:** 1613.

Tse, I., Snieckus, V. (1976) *Chem. Commun.* 505.

Tsuda, Y., Sano, T., Kawaguchi, K., Inubushi, Y. (1964) *Tetrahedron Lett.* 1279.

Tsuji, J., Mizutani, K., Shimizu, I., Yamamoto, K. (1976) *Tetrahedron Lett.* 773.

Tsuji, J., Mandai, T. (1977) *Tetrahedron Lett.* 3285.

Tsuji, J. (1979a) *Pure Appl. Chem.* **51:** 1235.

Tsuji, J., Yamada, T., Kaito, M., Mandai, T. (1979b) *Tetrahedron Lett.* 2257.

Tsuji, J., Hashiguchi, S. (1980a) *Tetrahedron Lett.* **21:** 2955.

Tsuji, J., Yamada, T., Shimizu, I. (1980b) *J. Org. Chem.* **45:** 5209.

Tufariello, J.J., Tette, J.P. (1975) *J. Org. Chem.* **40:** 3866.

Tufariello, J.J., Tegeler, J.J. (1976) *Tetrahedron Lett.* 4037.

Tufariello, J.J., Ali, S.K. (1978a) *Tetrahedron Lett.* 4647.

Tufariello, J.J., Gastrone, R.G. (1978b) *Tetrahedron Lett.* 2753.

Tufariello, J.J. (1979a) *Acc. Chem. Res.* **12:** 396.

Tufariello, J.J., Ali, S.K. (1979b) *J. Am. Chem. Soc.* **101:** 7116.

Tufariello, J.J., Mullen, G.B., Tegeler, J.J., Trybulski, E.J., Wong, S.C., Ali, S.K. (1979c) *J. Am. Chem. Soc.* **101:** 2435.

Tufariello, J.J., Meckler, H., Pushpananda, K., Senaratne, A. (1985) *Tetrahedron* **41:** 3447.

Tufariello, J.J., Dyszlewski, A.D. (1987a) *Chem. Commun.* 1138.

Tufariello, J.J., Milowsky, A., Al-Nuri, M., Goldstein, S. (1987b) *Tetrahedron Lett.* **28:** 263, 267.

Uhler, R.O. (1960) *Dissert. Abstr.* **21:** 765.

Underwood, G.R., Ramamoorthy, B. (1970) *Tetrahedron Lett.* 4125.

Uskokovic, M.R., Gutzwiller, J., Henderson, T. (1970) *J. Am. Chem. Soc.* **92:** 203.

Utaka, M., Makino, H., Oota, Y., Tsuboi, S., Takeda, A. (1983) *Tetrahedron Lett.* **24:** 2567.

Uyehara, T., Kabasawa, Y., Kao, T., Furuta, T. (1985) *Tetrahedron Lett.* **26:** 2343.

Valverde, S., Herradon, B., Martin-Lomas, M. (1985) *Tetrahedron Lett.* **26:** 3731.

Van Bac, N., Langlois, Y. (1988) *Tetrahedron Lett.* **29:** 2819.

Van der Eycken, E., Van der Eycken, J., Vandewalle, M. (1985) *Chem. Commun.* 1719.

Van Helden, R., De Jong, A.J. (1977) Ger. Offen. 2652202 (assigned to Shell Int. Res. M.B.V.).

Van Hijfte, L., Vandewalle, M. (1982) *Tetrahedron Lett.* **23:** 2229.

van Tamelen, E.E., Hidahl, G.T. (1956) *J. Am. Chem. Soc.* **78:** 4405.

van Tamelen, E.E., Shamma, M., Burgstahler, A.W., Wolinsky, J., Tamm, R., Aldrich, P.E. (1958) *J. Am. Chem. Soc.* **80:** 5006.

van Tamelen, E.E., Dolby, L.J., Lawton, L.G. (1960a) *Tetrahedron Lett.* **19:** 30.

van Tamelen, E.E., Foltz, R.L. (1960b) *J. Am. Chem. Soc.* **82:** 2400.

van Tamelen, E.E., Foltz, R.L. (1969a) *J. Am. Chem. Soc.* **91:** 7372.

van Tamelen, E.E., Hester, J.B. (1969b) *J. Am. Chem. Soc.* **91:** 7342.

van Tamelen, E.E., Placeway, C., Schiemenz, G.P., Wright, I.G. (1969c) *J. Am. Chem. Soc.* **91:** 7359.

van Tamelen, E.E., Wright, I.G. (1969d) *J. Am. Chem. Soc.* **91:** 7349.

van Tamelen, E.E., Yardley, J.P., Miyano, M., Hinshaw, W.B. (1969e) *J. Am. Chem. Soc.* **91:** 7333.

van Tamelen, E.E., Oliver, L.K. (1970) *J. Am. Chem. Soc.* **92:** 2136.

van Tamelen, E.E., Seiler, M.P., Wierenga, W. (1972) *J. Am. Chem. Soc.* **94:** 8229.

van Tamelen, E.E., (1981) *Pure Appl. Chem.* **53:** 1259.

Vasella, A., Voeffray, R. (1982) *Helv. Chim. Acta* **65**: 1134.

Vedejs, E., Larsen, S.D. (1984) *J. Am. Chem. Soc.* **106**: 3030.

Velluz, L., Valls, J., Nomine, G. (1965) *Angew. Chem. Int. Ed. Engl.* **4**: 181.

Velluz, L., Valls, J., Mathieu, J. (1967) *Angew. Chem. Int. Ed. Engl.* **6**: 778.

Venkatachalam, M., Deshpande, M.N., Jawdosiuk, M., Kubiak, G., Wehrli, S., Cook, J.M., Weiss, U. (1986) *Tetrahedron* **42**: 1597.

Vidari, G., Ferrino, S., Grieco, P.A. (1984) *J. Am. Chem. Soc.* **106**: 3539.

Vigneron, J.P., Kagan, H., Horeau, A. (1968) *Tetrahedron Lett.* 5681.

Villagomez-Ibarra, R., Joseph-Nathan, P. (1994) *Tetrahedron Lett.* **35**: 4771.

Vite, G.D., Spencer, T.A. (1988a) *J. Org. Chem.* **53**: 2555; (1988b) *J. Org. Chem.* **53**: 2560.

Vogel, E., Lex, J., Schieb, T., Schmickler, H., Schmidt, K., Schulz, W.H. (1986) *Angew. Chem. Int. Ed. Engl.* **25**: 723.

Volkmann, R., Danishefsky, S., Eggler, J., Solomon, D.M. (1971) *J. Am. Chem. Soc.* **93**: 5576.

Vyas, D.M., Chaing, Y., Doyle, T.W. (1984) *Tetrahedron Lett.* **25**: 487.

Wada, M., Nishihara, Y., Akiba, K. (1985) *Tetrahedron Lett.* **26**: 3267.

Waitkus, P.A., Peterson, L.I., Griffin, G.W. (1966) *J. Am. Chem. Soc.* **88**: 181.

Wakamatsu, T., Ban, Y. (1980) *Kagaku no Ryoiki, Zokan* **128**: 149.

Wakamatsu, T., Akasaka, K. Ban, Y. (1974) *Tetrahedron Lett.* 3883.

Wakamatsu, T., Hara, H. Ban, Y. (1985a) *J. Org. Chem.* **50**: 108.

Wakamatsu, T., Yamada, S., Ozaki, F., Ban, Y. (1985b) *Tetrahedron Lett.* **26**: 1989.

Wakamatsu, T., Hara, H., Taira, K., Ban, Y. (1987) *Heterocycles* **26**: 1203.

Wakamatsu, T., Miyachi, N., Ozaki, F., Shibasaki, M., Ban, Y. (1988) *Tetrahedron Lett.* **29**: 3829.

Walba, D.M., Richards, R.M., Haltiwanger, R.C. (1982a) *J. Am. Chem. Soc.* **104**: 3219.

Walba, D.M., Wand, M.D. (1982b) *Tetrahedron Lett.* **23**: 4995.

Walba, D.M. (1985) *Tetrahedron* **41**: 3161.

Wang, T.-F., Yang, C.-F. (1989) *Chem. Commun.* 1876.

Wang, Y.-F., Sih, C.J. (1984a) *Tetrahedron Lett.* **25**: 4999.

Wang, Y.-F., Chen, C.-S., Girdaukas, G., Sih, C.J. (1984b) *J. Am. Chem. Soc.* **106**: 3695.

Wang, Z., Schreiber, S.L. (1990) *Tetrahedron Lett.* **31**: 31.

Wang, Z., Deschenes, D. (1992) *J. Am. Chem. Soc.* **114**: 1090.

Ward, J.S., Pettit, R. (1971) *J. Am. Chem. Soc.* **93**: 262.

Wasserman, H.H., Matsuyama, H. (1961) *J. Am. Chem. Soc.* **103**: 461.

Wasserman, H.H., Keehn, P.M. (1967) *J. Am. Chem. Soc.* **89**: 2770.

Wasserman, H.H., Gambale, R.J., Pulwer, M.J. (1981) *Tetrahedron* **37**: 4059.

Wasserman, H.H., Berger, G.D., Cho, K.R. (1982) *Tetrahedron Lett.* **23**: 465.

Watabe, T., Hosoda, Y., Okada, K., Oda, M. (1987) *Bull. Chem. Soc. Jpn.* **60**: 3801.

Watanabe, M., Snieckus, V. (1980) *J. Am. Chem. Soc.* **102**: 1457.

Watt, D.S., Corey, E.J. (1972) *Tetrahedron Lett.* 4651.

Watts, L., Fitzpatrick, J.D., Pettit, R. (1965) *J. Am. Chem. Soc.* **87**: 3253.

Watts, L., Fitzpatrick, J.D., Pettit, R. (1966) *J. Am. Chem. Soc.* **88**: 623.

Webb, T.H., Wilcox, C.S. (1990) *J. Org. Chem.* **55**: 363.

Weber-Schilling, C.A., Wanzlick, H.-W. (1969) *Tetrahedron Lett.* 2345.

Wehle, D., Fitjer, L. (1987) *Angew. Chem. Int. Ed. Engl.* **26**: 130.

Weinreb, S.M., Khatri, N.A., Shringapure, J. (1979) *J. Am. Chem. Soc.* **101**: 5073.

Weiss, U., Edwardd, J.M. (1968) *Tetrahedron Lett.* 4885.

Welch, S.C., Rao, A.S.C.P. (1978) *J. Org. Chem.* **43**: 1957.

Welch, S.C., Hagan, C.P., White, D.H., Fleming, W.P., Trotter, J.W. (1977) *J. Am. Chem. Soc.* **99**: 549.

Welch, S.C., Assercq, J.-M., Loh, J.-P. (1986) *Tetrahedron Lett.* **27**: 1115.

Wender, P.A., Eissenstat, M.A., Filosa, M.P. (1979) *J. Am. Chem. Soc.* **101**: 2196.

Wender, P.A., Schaus, J.M., White, A.W. (1987) *Heterocycles* **25**: 263.

Wender, P.A., Singh, S.K. (1990) *Tetrahedron Lett.* **31**: 2517.

Wenger, R.M. (1985) *Helv. Chim. Acta* **24**: 77.

Wenkert, E., Strike, D.P. (1964) *J. Am. Chem. Soc.* **86**: 2044.

Wenkert, E., Broka, C.A. (1984) *Chem. Commun.* 714.

Wenkert, E., Kim, H.-S. (1989) in *Studies in Natural Product Chemistry*, Atta-ur-Rahman, ed. Col 3, p. 287, Elsevier, Amsterdam.

Whitesell, J.K., Minton, M.A. (1986) *J. Am. Chem. Soc.* **108**: 6802.

Whitesell, J.K., Minton, M.A. (1987) *J. Am. Chem. Soc.* **109**: 6403.

Whitesell, J.K., Allen, D.E. (1988) *J. Am. Chem. Soc.* **110**: 3585.

Whitesell, J.K. (1989) *Chem.Rev.* **89**: 1581.

Whitlock, H.W., Smith, G.L. (1967) *J. Am. Chem. Soc.* **89**: 3600.

Whitney, R.A. (1981) *Can. J. Chem.* **59**: 2650.

Wiberg, K.B., Walker, F.H. (1982) *J. Am. Chem. Soc.* **104**: 5239.

Wiberg, K.B., Matturro, M.G., Okarma,P.J., Jason, M.E. (1984) *J. Am. Chem. Soc.* **106**: 2194.

Wiesner, K., Musil, V., Wiesner, K.J. (1968) *Tetrahedron Lett.* 5643.

Wijnberg, B.P., Speckamp, W.N. (1981) *Tetrahedron Lett.* **22**: 5079.

Wilds, A.L., Ralls, J.W., Wildman, W.C., McCaleb, K.E. (1950) *J. Am. Chem. Soc.* **72**: 5794.

Wilds, A.L., Ralls, J.W., Tyner, D.A., Daniels, R., Kraychy, S., Harnik, M. (1953) *J. Am. Chem. Soc.* **75**: 4878.

Williams, D.R., Barner, B.A., Nishitani, K., Phillips, J.G. (1982) *J. Am. Chem. Soc.* **104**: 4708.

Williams, D.R., Benbow, J.W. (1988) *J. Org. Chem.* **53**: 4643.

Williams, D.R., Brown, D.L., Benbow, J.W. (1989) *J. Am. Chem. Soc.* **111**: 1923.

Williams, D.R., Armstrong, R.W., Dung, J.-S. (1985) *J. Am. Chem. Soc.* **107**: 3253.

Williams, V.Z., Schleyer, P.v.R., Gleicher, G.J., Rodewald, C.B. (1966) *J. Am. Chem. Soc.* **88**: 3862.

Willstätter, R., Bommer, M. (1921) *Liebigs Ann. Chem.* **422**: 15.

Wilson, S.R., Mao, D.T. (1978) *J. Am. Chem. Soc.* **100**: 6289.

Wilson, S.R., Sawicki, R.A. (1979) *J. Org. Chem.* **44**: 330.

Wilson, S.R., Misra, R.N. (1980) *J. Org. Chem.* **45**: 5079.

Wilson, S.R., Sawicki, R.A., Human, J.C. (1981) *J. Org. Chem.* **46**: 3887.

Winkler, J.D., Hershberger, P.M. (1989) *J. Am. Chem. Soc.* **111**: 4852.

Winkler, J.D., Kim, S., Condroski,K.R., Asensio, A., Houk, K.N. (1994) *J. Org. Chem.* **59**: 6879.

Wirz, B., Schmid, R., Walter, W. (1990) *Biocatalysis* **3**: 159.

Wiseman, J.R., Lee, S.Y. (1986) *J. Org. Chem.* **51**: 2485.

Wohl, A., Johnson, A. (1907) *Ber.* **40**: 4712.

Wolinsky, J., Lau, R., Hamsher, J.J., Cimarusti, C.M. (1982) *Synth. Commun.* **2**: 327.

Wong, C.M., Buccini, J., Chang, I., TeRaa, J., Schwenk, R. (1969') *Can. J. Chem.* **47**: 2421.

Wong, C.M., Ho, T.-L., Niemczura, W.P. (1975) *Can. J. Chem.* **53**: 3144.

Wong, C.M., Singh, R., Singh, K., Lam, H.Y.P. (1979) *Can. J. Chem.* **57**: 3304.

Wood, J.L., Jeong, S., Salcedo, A., Jenkins, J. (1995) *J. Org. Chem.* **60**: 286.

Woodward, R.B., Singh, G.B. (1950a) *J. Am. Chem. Soc.* **72**: 1428; (1950b) *J. Am. Chem. Soc.* **72**: 5351.

Woodward, R.B., Sondheimer, F., Taub, D., Heusler, K., McLamore, W.M. (1952) *J. Am. Chem. Soc.* **74**: 4223.

Woodward, R.B., Bader, F.E., Bickel, H., Frey, A.J., Kierstead, R.W. (1958) *Tetrahedron* **2**: 1.

Woodward, R.B. (1961) *Pure Appl. Chem.* **2**: 383.

Woodward, R.B., Fukunaga, T., Kelly, R.C. (1964) *J. Am. Chem. Soc.* **86**: 3162.

Woodward, R.B. (1966' *Science* **153**: 487.

Woodward, R.B., Gosteli, J., Ernst,I., Friary, R.J., Nestler, G., Raman, H., Sitrin, R., Suter, C., Whitesell, J.K. (1973) *J. Am. Chem. Soc.* **95**: 6853.

Woodward, R.B., Hoye, T.R. (1977) *J. Am. Chem. Soc.* **99**: 8007.

Woodward, R.B., Logusch, E., Nambiar, K.P., Sakan,K., Ward, D.E., Au-Yeung, B.-W., Balaram, P.,Browne, L.J., Card, P.J., Chen, C.H., Chenevert, R.B., Fliri, A., Frobel, K., Gais, H.-J., Garratt, D.G., Hayakawa, K., Heggie, W., Hesson, D.P., Hoppe, D., Hoppe, I., Hyatt, J.A., Ikeda, D., Jacobi, P.A.,Kim, K.S., Kobuke, Y., Kojima, K., Krowicki, K., Lee, V.J., Leutert, T., Malchenko, S., Martens, J., Matthews, R.S., Ong, B.S., Press, J.B., Rajan Babu, T.V., Rousseau, G., Sauter, H.M., Suzuki, M., Tatsuta, K., Tolbert, L.M., Truesdale, E.A., Uchida, I., Ueda, Y., Uyehara, T., Vasella, A.T., Vladuchick, W.C., Wade, P.A., Williams, R.M., Wong, H.N.-C. (1981) *J. Am. Chem. Soc.* **103**: 3210, 3213, 3215.

Wright, J.J.K., Cooper, A.B., McPhail, A.T., Merrill, Y., Nagabhushan, T.L., Puar, M.S. (1982) *Chem. Commun.* 1188.

Wrobel, J., Takahashi, K., Honkan, V., Lannoye, G., Bertz, S.H., Cook, J.M. (1983) *J. Org. Chem.* **48**: 139.

Wu, A., Snieckus, V. (1975) *Tetrahedron Lett.* 2057.

Wu, X.-M., Funakoshi, K., Sakai, K. (1993) *Tetrahedron Lett.* **34**: 5927.

Wu, Y.-J., Burnell, D.J. (1988) *Tetrahedron Lett.* **29**: 4369.

Wu, Y.-J., Burnell, D.J. (1991) *Chem. Commun.* 764.

Wulff, W.D., Tang, P.-C. (1984) *J. Am. Chem. Soc.* **106**: 434.

Xie, J., Zhou, J., Zhang, C. (1983) *Kexue Tongbao* **28**: 430.

Xie, Z.-F., Suemune, H., Nakamura, I., Sakai, K. (1987) *Chem. Pharm. Bull.* **35**: 4454.

Xie, Z.-F., Suemune, H., Sakai, K. (1988) *Chem. Commun.* 1638.

Yadav, J.S., Corey, P.,Hsu, C.T., Perlman, K., Sih, C.J. (1981) *Tetrahedron Lett.* **22**: 811.

Yadav, J.S., Cahnder, M.C.,Joshi, B.V. (1988a) *Tetrahedron Lett.* **29:** 2737.

Yadav, J.S., Reddy, P.S., Joshi, B.V. (1988b) *Tetrahedron* **44:** 7243.

Yadav, J.S., Rao, E.S., Rao, V.S., Choudary, B.M. (1990) *Tetrahedron Lett.* **31:** 2491.

Yamada, K., Aoki, K., Kato, T., Uemura, D., van Tamelen, E.E.(1974) *Chem. Commun.* 908.

Yamada, S., Ono, F., Katagiri, T., Tanaka, J. (1978) *Synth. Commun.* **8:** 241.

Yamaguchi, M., Hasebe, K., Minami, T. (1986) *Tetrahedron Lett.* **27:** 2401.

Yamaguchi, R., Hamasaki, T., Sasaki, T., Kozima, S., Takaya, H. (1991) *Synlett.* 719.

Yamakawa, K., Izuta, I., Oka, H., Sakaguchi, R. (1974) *Tetrahedron Lett.* 2187.

Yamamoto, K., Nishioka, T., Oda, J., Yamamoto, Y. (1988) *Tetrahedron Lett.* **29:** 1717.

Yamamoto, Y., Furuta, T. (1990) *J. Org. Chem.* **55:** 3971.

Yamamoto, Y., Ohno, M., Eguchi, S. (1994a) *Tetrahedron* **50:** 7783.

Yamamoto, Y., Ohno, M., Eguchi, S. (1994b) *J. Org. Chem.* **59:** 4707.

Yamazaki, N., Kibayashi, C., (1989) *J. Am. Chem. Soc.* **111:** 1396.

Yang, S.S., Dawson, B.T., Rieke, R.D. (1991) *Tetrahedron Lett.* **32:** 3341.

Yao, Z.-J., Wu, Y.-L. (1995) *J. Org. Chem.* **60:** 1170.

Yee, Y.K., Schultz, A.G. (1979) *J. Org. Chem.* **44:** 719.

Yerxa, B.R., Yang,K., Moore, H.W. (1994) *Tetrahedron* **50:** 6173.

Yoda, H., Kitayama, H., Katagiri, T., Takabe,K. (1993) *Tetrahedron: Asymmetry* **4:** 1455.

Yokokawa, F., Hamada, Y., Shiori, T. (1993) *Tetrahedron Lett.* **34:** 6559.

Yokomatsu, T., Iwasawa, H., Shibuya, S. (1992) *Chem. Commun.* 728.

Yokoyama, Y.,Kawashima, H., Kohno, M., Ogawa, Y., Uchida, S. (1991) *Tetrahedron Lett.* **32:** 1479.

Yue, C.W., Royer, J., Husson, H.-P. (1990) *J. Org. Chem.* **55:** 1140.

Yue, C., Nicolay, J.-F., Royer, J., Husson, H.-P. (1994) *Tetrahedron* **50:** 3139.

Yurev, Y.K., Zefirov, N.S. (1961) *Zh. Obshch. Khim.* **31:** 685.

Zarraga, M., Rodriguez, M.L., Ruiz-Perez, C., Martin, J.D. (1989) *Tetrahedron Lett.* **30:** 3725.

Zarraga, M., Martin, J.D. (1991) *Tetrahedron Lett.* **32:** 2249.

Zefirov, N.S., Kozhushkov, S.I., Ugrak, B.I., Lukin,K.A., Kokoreva, O.V., Yufit, D.S., Struchkov, Y.T., Zoellner, S., Boese, R., deMeijere, A. (1992) *J. Org. Chem.* **57:** 701.

Zegarski, J., Howard, B.M. (1985) *Tetrahedron Lett.* **26:** 1363.

Zhao, S., Mehta, G., Helquist, P. (1991) *Tetrahedron Lett.* **32:** 5753.

Zhu, J., Wang, Q., Li, Y. (1988) *Chem. Commun.* 1549.

Zhu, J., Royer, J., Quiron, J.-C., Husson, H.-P. (1991) *Tetrahedron Lett.* **32:** 2485.

Ziegler, F.E., Bennett, G.B. (1970) *Tetrahedron Lett.* 2545.

Ziegler, F.E., Zhang, Z. (1987) *Tetrahedron Lett.* **28:** 5973.

Ziegler, F.E., Sobolov, S.B. (1990) *J. Am. Chem. Soc.* **112:** 2749.

Zimmerman, H.E., Gruneweld, G.L., Paufler, R.M., Sherwin, M.A. (1969) *J. Am. Chem. Soc.* **91:** 2330.

Zoretic, P.A., Bhakta, C., Khan, R.H. (1983) *Tetrahedron Lett.* **24:** 1125.

Zurflüh, R., Wall, E.N., Siddall, J.B., Edwards, J.A. (1968) *J. Am. Chem. Soc.* **92:** 425.

Zutterman, F., deWilde,H., Mijngheer, R., deClercq, P., Vandewalle, M. (1979) *Tetrahedron* **35:** 2389.

INDEX